Geophysical Monograph Series

Including

IUGG Volumes

Maurice Ewing Volumes

Mineral Physics Volumes

GEOPHYSICAL MONOGRAPH SERIES

Geophysical Monograph Volumes

1. **Antarctica in the International Geophysical Year** *A. P. Crary, L. M. Gould, E. O. Hulburt, Hugh Odishaw, and Waldo E. Smith (Eds.)*
2. **Geophysics and the IGY** *Hugh Odishaw and Stanley Ruttenberg (Eds.)*
3. **Atmospheric Chemistry of Chlorine and Sulfur Compounds** *James P. Lodge, Jr. (Ed.)*
4. **Contemporary Geodesy** *Charles A. Whitten and Kenneth H. Drummond (Eds.)*
5. **Physics of Precipitation** *Helmut Weickmann (Ed.)*
6. **The Crust of the Pacific Basin** *Gordon A. Macdonald and Hisashi Kuno (Eds.)*
7. **Antarctic Research: The Matthew Fontaine Maury Memorial Symposium** *H. Wexler, M. J. Rubin, and J. E. Caskey, Jr. (Eds.)*
8. **Terrestrial Heat Flow** *William H. K. Lee (Ed.)*
9. **Gravity Anomalies: Unsurveyed Areas** *Hyman Orlin (Ed.)*
10. **The Earth Beneath the Continents: A Volume of Geophysical Studies in Honor of Merle A. Tuve** *John S. Steinhart and T. Jefferson Smith (Eds.)*
11. **Isotope Techniques in the Hydrologic Cycle** *Glenn E. Stout (Ed.)*
12. **The Crust and Upper Mantle of the Pacific Area** *Leon Knopoff, Charles L. Drake, and Pembroke J. Hart (Eds.)*
13. **The Earth's Crust and Upper Mantle** *Pembroke J. Hart (Ed.)*
14. **The Structure and Physical Properties of the Earth's Crust** *John G. Heacock (Ed.)*
15. **The Use of Artificial Satellites for Geodesy** *Soren W. Henricksen, Armando Mancini, and Bernard H. Chovitz (Eds.)*
16. **Flow and Fracture of Rocks** *H. C. Heard, I. Y. Borg, N. L. Carter, and C. B. Raleigh (Eds.)*
17. **Man-Made Lakes: Their Problems and Environmental Effects** *William C. Ackermann, Gilbert F. White, and E. B. Worthington (Eds.)*
18. **The Upper Atmosphere in Motion: A Selection of Papers With Annotation** *C. O. Hines and Colleagues*
19. **The Geophysics of the Pacific Ocean Basin and Its Margin: A Volume in Honor of George P. Woollard** *George H. Sutton, Murli H. Manghnani, and Ralph Moberly (Eds.)*
20. **The Earth's Crust: Its Nature and Physical Properties** *John C. Heacock (Ed.)*
21. **Quantitative Modeling of Magnetospheric Processes** *W. P. Olson (Ed.)*
22. **Derivation, Meaning, and Use of Geomagnetic Indices** *P. N. Mayaud*
23. **The Tectonic and Geologic Evolution of Southeast Asian Seas and Islands** *Dennis E. Hayes (Ed.)*
24. **Mechanical Behavior of Crustal Rocks: The Handin Volume** *N. L. Carter, M. Friedman, J. M. Logan, and D. W. Stearns (Eds.)*
25. **Physics of Auroral Arc Formation** *S.-I. Akasofu and J. R. Kan (Eds.)*
26. **Heterogeneous Atmospheric Chemistry** *David R. Schryer (Ed.)*
27. **The Tectonic and Geologic Evolution of Southeast Asian Seas and Islands: Part 2** *Dennis E. Hayes (Ed.)*
28. **Magnetospheric Currents** *Thomas A. Potemra (Ed.)*
29. **Climate Processes and Climate Sensitivity (Maurice Ewing Volume 5)** *James E. Hansen and Taro Takahashi (Eds.)*
30. **Magnetic Reconnection in Space and Laboratory Plasmas** *Edward W. Hones, Jr. (Ed.)*
31. **Point Defects in Minerals (Mineral Physics Volume 1)** *Robert N. Schock (Ed.)*
32. **The Carbon Cycle and Atmospheric CO_2: Natural Variations Archean to Present** *E. T. Sundquist and W. S. Broecker (Eds.)*
33. **Greenland Ice Core: Geophysics, Geochemistry, and the Environment** *C. C. Langway, Jr., H. Oeschger, and W. Dansgaard (Eds.)*
34. **Collisionless Shocks in the Heliosphere: A Tutorial Review** *Robert G. Stone and Bruce T. Tsurutani (Eds.)*
35. **Collisionless Shocks in the Heliosphere: Reviews of Current Research** *Bruce T. Tsurutani and Robert G. Stone (Eds.)*
36. **Mineral and Rock Deformation: Laboratory Studies —The Paterson Volume** *B. E. Hobbs and H. C. Heard (Eds.)*
37. **Earthquake Source Mechanics (Maurice Ewing Volume 6)** *Shamita Das, John Boatwright, and Christopher H. Scholz (Eds.)*
38. **Ion Acceleration in the Magnetosphere and Ionosphere** *Tom Chang (Ed.)*
39. **High Pressure Research in Mineral Physics (Mineral Physics Volume 2)** *Murli H. Manghnani and Yasuhiko Syono (Eds.)*
40. **Gondwana Six: Structure, Tectonics, and Geophysics** *Gary D. McKenzie (Ed.)*
41. **Gondwana Six: Stratigraphy, Sedimentology, and Paleontology** *Garry D. McKenzie (Ed.)*
42. **Flow and Transport Through Unsaturated Fractured Rock** *Daniel D. Evans and Thomas J. Nicholson (Eds.)*
43. **Seamounts, Islands, and Atolls** *Barbara H. Keating, Patricia Fryer, Rodey Batiza, and George W. Boehlert (Eds.)*

44 **Modeling Magnetospheric Plasma** *T. E. Moore and J. H. Waite, Jr. (Eds.)*

45 **Perovskite: A Structure of Great Interest to Geophysics and Materials Science** *Alexandra Navrotsky and Donald J. Weidner (Eds.)*

46 **Structure and Dynamics of Earth's Deep Interior (IUGG Volume 1)** *D. E. Smylie and Raymond Hide (Eds.)*

47 **Hydrological Regimes and Their Subsurface Thermal Effects (IUGG Volume 2)** *Alan E. Beck, Grant Garven, and Lajos Stegena (Eds.)*

48 **Origin and Evolution of Sedimentary Basins and Their Energy and Mineral Resources (IUGG Volume 3)** *Raymond A. Price (Ed.)*

49 **Slow Deformation and Transmission of Stress in the Earth (IUGG Volume 4)** *Steven C. Cohen and Petr Vaníček (Eds.)*

50 **Deep Structure and Past Kinematics of Accreted Terranes (IUGG Volume 5)** *John W. Hillhouse (Ed.)*

51 **Properties and Processes of Earth's Lower Crust (IUGG Volume 6)** *Robert F. Mereu, Stephan Mueller, and David M. Fountain (Eds.)*

52 **Understanding Climate Change (IUGG Volume 7)** *Andre L. Berger, Robert E. Dickinson, and J. Kidson (Eds.)*

53 **Plasma Waves and Instabilities at Comets and in Magnetospheres** *Bruce T. Tsurutani and Hiroshi Oya (Eds.)*

54 **Solar System Plasma Physics** *J. H. Waite, Jr., J. L. Burch, and R. L. Moore (Eds.)*

55 **Aspects of Climate Variability in the Pacific and Western Americas** *David H. Peterson (Ed.)*

56 **The Brittle-Ductile Transition in Rocks** *A. G. Duba, W. B. Durham, J. W. Handin, and H. F. Wang (Eds.)*

57 **Evolution of Mid Ocean Ridges (IUGG Volume 8)** *John M. Sinton (Ed.)*

58 **Physics of Magnetic Flux Ropes** *C. T. Russell, E. R. Priest, and L. C. Lee (Eds.)*

59 **Variations in Earth Rotation (IUGG Volume 9)** *Dennis D. McCarthy and Williams E. Carter (Eds.)*

60 **Quo Vadimus Geophysics for the Next Generation (IUGG Volume 10)** *George D. Garland and John R. Apel (Eds.)*

61 **Cometary Plasma Processes** *Alan D. Johnstone (Ed.)*

62 **Modeling Magnetospheric Plasma Processes** *Gordon R. Wilson (Ed.)*

63 **Marine Particles: Analysis and Characterization** *David C. Hurd and Derek W. Spencer (Eds.)*

64 **Magnetospheric Substorms** *Joseph R. Kan, Thomas A. Potemra, Susumu Kokubun, and Takesi Iijima (Eds.)*

65 **Explosion Source Phenomenology** *Steven R. Taylor, Howard J. Patton, and Paul G. Richards (Eds.)*

66 **Venus and Mars: Atmospheres, Ionospheres, and Solar Wind Interactions** *Janet G. Luhmann, Mariella Tatrallyay, and Robert O. Pepin (Eds.)*

67 **High-Pressure Research: Application to Earth and Planetary Sciences (Mineral Physics Volume 3)** *Yasuhiko Syono and Murli H. Manghnani (Eds.)*

68 **Microwave Remote Sensing of Sea Ice** *Frank Carsey, Roger Barry, Josefino Comiso, D. Andrew Rothrock, Robert Shuchman, W. Terry Tucker, Wilford Weeks, and Dale Winebrenner*

69 **Sea Level Changes: Determination and Effects (IUGG Volume 11)** *P. L. Woodworth, D. T. Pugh, J. G. DeRonde, R. G. Warrick, and J. Hannah*

70 **Synthesis of Results from Scientific Drilling in the Indian Ocean** *Robert A. Duncan, David K. Rea, Robert B. Kidd, Ulrich von Rad, and Jeffrey K. Weissel (Eds.)*

71 **Mantle Flow and Melt Generation at Mid-Ocean Ridges** *Jason Phipps Morgan, Donna K. Blackman, and John M. Sinton (Eds.)*

72 **Dynamics of Earth's Deep Interior and Earth Rotation (IUGG Volume 12)** *Jean-Louis Le Mouël, D.E. Smylie, and Thomas Herring (Eds.)*

73 **Environmental Effects on Spacecraft Positioning and Trajectories (IUGG Volume 13)** *A. Vallance Jones (Ed.)*

74 **Evolution of the Earth and Planets (IUGG Volume 14)** *E. Takahashi, Raymond Jeanloz, and David Rubie (Eds.)*

75 **Interactions Between Global Climate Subsystems: The Legacy of Hann (IUGG Volume 15)** *G. A. McBean and M. Hantel (Eds.)*

76 **Relating Geophysical Structures and Processes: The Jeffreys Volume (IUGG Volume 16)** *K. Aki and R. Dmowska (Eds.)*

77 **The Mesozoic Pacific: Geology, Tectonics, and Volcanism—A Volume in Memory of Sy Schlanger** *Malcolm S. Pringle, William W. Sager, William V. Sliter, and Seth Stein (Eds.)*

78 **Climate Change in Continental Isotopic Records** *P. K. Swart, K. C. Lohmann, J. McKenzie, and S. Savin (Eds.)*

79 **The Tornado: Its Structure, Dynamics, Prediction, and Hazards** *C. Church, D. Burgess, C. Doswell, R. Davies-Jones (Eds.)*

80 **Auroral Plasma Dynamics** *R. L. Lysak (Ed.)*

81 **Solar Wind Sources of Magnetospheric Ultra-Low Frequency Waves** *M. J. Engebretson, K. Takahashi, and M. Scholer (Eds.)*

82 Gravimetry and Space Techniques Applied to Geodynamics and Ocean Dynamics (IUGG Volume 17) *Bob E. Schutz, Allen Anderson, Claude Froidevaux, and Michael Parke (Eds.)*

83 Nonlinear Dynamics and Predictability of Geophysical Phenomena (IUGG Volume 18) *William I. Newman, Andrei Gabrielov, and Donald L. Turcotte (Eds.)*

84 Solar System Plasmas in Space and Time *J. Burch, J. H. Waite, Jr. (Eds.)*

85 The Polar Oceans and Their Role in Shaping the Global Environment *O. M. Johannessen, R. D. Muench, and J. E. Overland (Eds.)*

86 Space Plasmas: Coupling Between Small and Medium Scale Processes *Maha Ashour-Abdalla, Tom Chang, and Paul Dusenbery (Eds.)*

87 The Upper Mesosphere and Lower Thermosphere: A Review of Experiment and Theory *R. M. Johnson and T. L. Killeen (Eds.)*

88 Active Margins and Marginal Basins of the Western Pacific *Brian Taylor and James Natland (Eds.)*

89 Natural and Anthropogenic Influences in Fluvial Geomorphology *John E. Costa, Andrew J. Miller, Kenneth W. Potter, and Peter R. Wilcock (Eds.)*

90 Physics of the Magnetopause *Paul Song, B.U.Ö. Sonnerup, and M.F. Thomsen (Eds.)*

91 Seafloor Hydrothermal Systems: Physical, Chemical, Biological, and Geological Interactions *Susan E. Humphris, Robert A. Zierenberg, Lauren S. Mullineaux, and Richard E. Thomson (Eds.)*

92 Mauna Loa Revealed: Structure, Composition, History, and Hazards *J. M. Rhodes and John P. Lockwood (Eds.)*

Maurice Ewing Volumes

1 Island Arcs, Deep Sea Trenches, and Back-Arc Basins *Manik Talwani and Walter C. Pitman III (Eds.)*

2 Deep Drilling Results in the Atlantic Ocean: Ocean Crust *Manik Talwani, Christopher G. Harrison, and Dennis E. Hayes (Eds.)*

3 Deep Drilling Results in the Atlantic Ocean: Continental Margins and Paleoenvironment *Manik Talwani, William Hay, and William B. F. Ryan (Eds.)*

4 Earthquake Prediction—An International Review *David W. Simpson and Paul G. Richards (Eds.)*

5 Climate Processes and Climate Sensitivity *James E. Hansen and Taro Takahashi (Eds.)*

6 Earthquake Source Mechanics *Shamita Das, John Boatwright, and Christopher H. Scholz (Eds.)*

IUGG Volumes

1 Structure and Dynamics of Earth's Deep Interior *D. E. Smylie and Raymond Hide (Eds.)*

2 Hydrological Regimes and Their Subsurface Thermal Effects *Alan E. Beck, Grant Garven, and Lajos Stegena (Eds.)*

3 Origin and Evolution of Sedimentary Basins and Their Energy and Mineral Resources *Raymond A. Price (Ed.)*

4 Slow Deformation and Transmission of Stress in the Earth *Steven C. Cohen and Petr Vaníček (Eds.)*

5 Deep Structure and Past Kinematics of Accreted Terrances *John W. Hillhouse (Ed.)*

6 Properties and Processes of Earth's Lower Crust *Robert F. Mereu, Stephan Mueller, and David M. Fountain (Eds.)*

7 Understanding Climate Change *Andre L. Berger, Robert E. Dickinson, and J. Kidson (Eds.)*

8 Evolution of Mid Ocean Ridges *John M. Sinton (Ed.)*

9 Variations in Earth Rotation *Dennis D. McCarthy and William E. Carter (Eds.)*

10 Quo Vadimus Geophysics for the Next Generation *George D. Garland and John R. Apel (Eds.)*

11 Sea Level Changes: Determinations and Effects *Philip L. Woodworth, David T. Pugh, John G. DeRonde, Richard G. Warrick, and John Hannah (Eds.)*

12 Dynamics of Earth's Deep Interior and Earth Rotation *Jean-Louis Le Mouël, D.E. Smylie, and Thomas Herring (Eds.)*

13 Environmental Effects on Spacecraft Positioning and Trajectories *A. Vallance Jones (Ed.)*

14 Evolution of the Earth and Planets *E. Takahashi, Raymond Jeanloz, and David Rubie (Eds.)*

15 Interactions Between Global Climate Subsystems: The Legacy of Hann *G. A. McBean and M. Hantel (Eds.)*

16 Relating Geophysical Structures and Processes: The Jeffreys Volume *K. Aki and R. Dmowska (Eds.)*

17 Gravimetry and Space Techniques Applied to Geodynamics and Ocean Dynamics *Bob E. Schutz, Allen Anderson, Claude Froidevaux, and Michael Parke (Eds.)*

18 Nonlinear Dynamics and Predictability of Geophysical Phenomena *William I. Newman, Andrei Gabrielov, and Donald L. Turcotte (Eds.)*

Mineral Physics Volumes

1 Point Defects in Minerals *Robert N. Schock (Ed.)*

2 High Pressure Research in Mineral Physics *Murli H. Manghnani and Yasuhiko Syona (Eds.)*

3 High Pressure Research: Application to Earth and Planetary Sciences *Yasuhiko Syono and Murli H. Manghnani (Eds.)*

Geophysical Monograph 93

Cross-Scale Coupling in Space Plasmas

James L. Horwitz
Nagendra Singh
James L. Burch
Editors

American Geophysical Union

Published under the aegis of the AGU Books Board.

Library of Congress Cataloging-in-Publication Data

Cross-scale coupling in space plasmas / James L. Horwitz, Nagendra Singh, James L. Burch, editors.
  p. cm. — (Geophysical monograph ; 93)
  Includes bibliographical references.
  ISBN 0-87590-075-5 (alk. paper)
  1. Space plasmas. 2. Transport theory.
I. Horwitz, James L. II. Singh, Nagendra, 1944– . III. Burch, James L. IV. Series.
QC809.P5C76 1996
523.01—dc20

95-46316
CIP

ISBN 0-87590-075-5
ISSN 0065-8448

Copyright 1995 by the American Geophysical Union
2000 Florida Avenue, N.W.
Washington, DC 20009

Figures, tables, and short excerpts may be reprinted in scientific books and journals if the source is properly cited.

Authorization to photocopy items for internal or personal use, or the internal or personal use of specific clients, is granted by the American Geophysical Union for libraries and other users registered with the Copyright Clearance Center (CCC) Transactional Reporting Service, provided that the base fee of $1.00 per copy plus $0.20 per page is paid directly to CCC, 222 Rosewood Dr., Danvers, MA 01923. 0065-8448/95/$01.00+0.20
This consent does not extend to other kinds of copying, such as copying for creating new collective works or for resale. The reproduction of multiple copies and the use of full articles or the use of extracts, including figures and tables, for commercial purposes requires permission from AGU.

Printed in the United States of America.

CONTENTS

Preface
James L. Horwitz, Nagendra Singh, and James L. Burch xi

General Issues and Methods in Micro/Mesoscale Coupling

Can We Find Useful Algorithms for Anomalous Transport?
C. T. Dum 1

An Upper Bound for the Proton Temperature Anisotropy
S. Peter Gary 13

Interrelationship of Local and Global Physics in the Low Altitude Ionosphere
G. Ganguli 23

Microscale Effects From Global Hot Plasma Imagery
T. E. Moore, M.-C. Fok, J. D. Perez, and J. P. Keady 37

Development of NonMaxwellian Velocity Distributions as a Consequence of Nonlocal Coulomb Collisions
G. R. Wilson 47

Modeling of Spatial and Temporal Scales in Turbulent Flows and Their Relevance to Space Plasma Transport
Ram K. Avva and Ashok K. Singhal 61

Regional Particle Simulations and Global Two-Fluid Modeling of the Magnetospheric Current System
R. M. Winglee 71

Frequency Range and Spectral Width of Waves Associated With Transverse-Velocity Shear
V. Gavrishchaka, M. E. Koepke, J. J. Carroll III, W. E. Amatucci, and G. Ganguli 81

Micro/Mesoscale Coupling in Auroral and Polar Wind Plasma Phenomena

Micro/Mesoscale Coupling in the Auroral Region: Observations
J. L. Burch 87

Semikinetic Simulation of Effects of Ionization by Precipitating Auroral Electrons on Ionospheric Plasma Transport
D. G. Brown, P. G. Richards, J. L. Horwitz, and G. R. Wilson 97

High Latitude Outflow of Centrifugally Accelerated Ions Through the Collisional/Collisionless Transition Region
C. W. Ho, J. L. Horwitz, G. R. Wilson, and D. G. Brown 105

Observations of Lower-Hybrid Spikelet Phenomena: Topaz3 Particle Data
Kristina A. Lynch, Roger L. Arnoldy, and Paul M. Kintner 111

Effects of Auroral Electron Precipitation on Topside Ion Outflows
Phil G. Richards 121

Fine Scale Auroral Beams and Conics
J. D. Perez, Chao Liu, Lynne Lawson, and T. E. Moore 127

CONTENTS

Anisotropic Kinetic Effects of Photoelectrons on Polar Wind Transport
Sunny W. Y. Tam, Fareed Yasseen, Tom Chang, Supriya B. Ganguli, and John M. Retterer   133

Coupling of Micro- and Mesoscale Processes in the Polar Wind Plasma Transport: A Generalized Fluid Model With Microprocesses
Supriya B. Ganguli   141

Single Ion Dynamics and Multiscale Phenomena
P. L. Rothwell, M. B. Silevitch, Lars P. Block, and Carl-Gunne Fälthammar   151

Micro/Mesoscale Coupling in Plasmaspheric Phenomena

Problems in Simulating Ion Temperatures in Low Density Flux Tubes
R. H. Comfort, P. G. Richards, P. D. Craven, and M. O. Chandler   155

Ring Current-Plasmasphere Coupling Through Coulomb Collisions
Mei-Ching Fok, Paul D. Craven, Thomas E. Moore, and Philip G. Richards   161

Plasmasphere Modeling With Ring Current Heating
S. M. Guiter, M.-C. Fok, and T. E. Moore   173

Equatorial Warm Ion Thermalization by Coulomb Collisions With Cool Outer Plasmaspheric Ions
Jinsoo Lee, J. L. Horwitz, G. R. Wilson, J. Lin, and D. G. Brown   177

Nonsteady State Coupling Processes in Superthermal Electron Transport
M. W. Liemohn and G. V. Khazanov   181

Micro/Mesoscale Coupling in Middle Magnetospheric Phenomena

Proton Cyclotron Wave-Ion Interactions Observed by AMPTE/CCE
Brian J. Anderson   193

Aspects of Mesoscale Phenomena in the Middle Magnetosphere and Speculations on the Role of Microscale Processes
Barry H. Mauk   201

Relative Contribution of the Solar Wind and the Auroral Zone to Near-Earth Plasmas
Vahé Peroomian and Maha Ashour-Abdalla   213

Micro/Mesoscale Coupling in Dayside Magnetopause Phenomena

Coupling Between Microscale and Mesoscale Processes in the Dayside Magnetosheath, Magnetopause, and Boundary Layer Regions
L. C. Lee and J. G. Hawkins   219

Micro/Mesoscale Phenomena in the Dayside Magnetopause: A Tutorial
Paul Song   235

Anomalous Plasma Diffusion Due to Kinetic Alfvén Wave Fluctuations at the Dayside Magnetopause
Manju Prakash   249

Structure of Reconnection Layers at the Magnetopause and in the Magnetotail
Y. Lin   255

CONTENTS

Micro/Mesoscale Coupling in Magnetotail Phenomena

Micro/Mesoscale Coupling in Magnetotail Current Sheet: Observations
A. T. Y. Lui 261

The Role of Microprocesses in Macroscale Magnetotail Dynamics
Joachim Birn, Michael Hesse, and S. Peter Gary 275

Irreducible Cross-Scale Coupling in the Magnetotail Current Sheet: A Tutorial
J. B. Harold and J. Chen 287

Ion Energization and Cross-Scale Coupling During Magnetotail Reconnection
G. R. Burkhart 299

PREFACE

A principal goal of space plasma researchers is to understand the influence of various transport processes on each other, even when such processes operate at widely varying spatial and temporal scales. We know that large-scale plasma flows in space lead to unstable conditions with small spatial (centimeters to meters) and temporal (microseconds to seconds) scales. The large-scale flows, for example in the magnetosphere-ionosphere system, involve scale lengths of kilometers to several Earth radii and temporal scales of minutes to hours. We must know specific contextual answers to the questions: Do the small-scale waves (microprocesses) modify the large-scale flows? Do these modifications significantly affect the transport of mass, momentum, and energy? How can such coupling processes and their influences be revealed observationally? And, perhaps most challenging of all, how do we incorporate the microprocesses into theoretical models of larger-scale space plasma transport?

In this monograph we have challenged a number of today's leading space plasma researchers to provide the very latest thinking and information—observational, theoretical and modeling—on these questions of cross-scale transport in space plasmas. The monograph is focused on the phenomena arising in the magnetosphere, and to some extent in its coupling with the ionosphere. We have encouraged the authors to describe their ideas and work such that the volume will provide a coherent overview of the techniques and insights needed by even a beginning magnetospheric physicist to understand the current status of the field and its future directions. Toward this goal, the collection of 33 papers is anchored by 15 tutorial-style review articles.

The monograph treats issues in magnetospheric cross-scale coupling within six sections. We first examine general issues and methods. Here Christian Dum outlines his views of how a transport theory with anomalous effects might be constructed, and notes that the structure of the requisite transport equations will be often quite different from those dominated by classical collision processes. An interesting problem in which theoretical treatment of microprocesses in transport appears to be verified by space plasma data, described by Peter Gary, is that of limiting temperature anisotropies in the magnetosheath, and possibly the magnetosphere, by driven electromagnetic ion cyclotron waves. Guru Ganguli discusses in his tutorial the nature of shear effects in space plasmas, and some of his recent predictions have been verified in the laboratory by Gavrishchaka et al. The extensive and successful experience of fluid dynamics researchers in incorporating turbulence into their transport descriptions is discussed by Singhal and Avva as a means of guiding space plasma researchers toward methods that may be applicable to our own field. Gordon Wilson demonstrates how complex collision effects can be in space plasmas, showing that they can in fact drive plasmas into non-Maxwellian configurations under some circumstances. Moore et al. present their fascinating simulations of energetic neutral atom images, which hold out the prospect of experimentally extracting detailed microphysics from proposed magnetospheric imaging spacecraft missions. Robert Winglee demonstrates, through both particle simulations and two-fluid modeling, techniques to elucidate the physics of the overall magnetospheric current system as well as mesoscale systems such as the magnetosheath-cleft interface.

In the section on auroral and polar wind plasma phenomena, Jim Burch describes the detection of mesoscale and microscale phenomena in the auroral plasma from current spacecraft, and notes that the upcoming FAST mission promises to allow significantly clearer identification of the microscale processes which drive many mesoscale phenomena. Phil Richards and Brown et al. describe the effects of precipitating auroral electrons on the ionosphere and resulting outflows, through fluid and semikinetic models, respectively. Ho et al. and Tam et al. employ differing simulation techniques to explore the effects of frictional ion heating and centrifugal acceleration on the dynamic ionospheric transition region, and the effects of photoelectrons on the polar wind, respectively. Supriya Ganguli discusses fluid simulations of field-aligned plasma outflow as influenced by microprocesses, including recent work on cross-field effects of the D'Angelo instability on the field-aligned flow. Rothwell et al. use single-particle trajectory techniques to illustrate how electric field spatial gradients interact with ion gyromotions to cause auroral density striations and chaotic O^+ motions. Lynch et al. and Perez et al. describe observations of auroral ion conic creation processes from rocket and spacecraft measurements, respectively. Lynch et al. describe some of the very latest results on ionospheric ion acceleration in lower hybrid spikelet wave packets, while Perez et al. present their Dynamics Explorer-1 findings regarding currents which may lead to perpendicular ion heating and dipolarization events producing significant ionospheric outflows.

The papers on plasmaspheric micro/mesoscale coupling deal chiefly with effects of thermal energy transfer between hot or warm ion populations and the cold plasmaspheric plasma. Guiter et al. and Fok et al. examine the heating of the plasmasphere specifically through Coulomb collisions with ring current ions, with Fok et al. demonstrating good agreement between modeled and observed plasmaspheric ion temperatures. However, Comfort et al. note that the classical description of thermal conduction in the plasmasphere may be inadequate, and suggest that non-Maxwellian features of the ion distributions be considered regarding the thermal conduction properties. Fok et al. also find important effects of the plasmasphere on the ring current evolution, including a collisional "smoothing" of the drift holes found in the ring current ion energy spectra. Lee et al., on the other hand, consider the warm equatorially trapped ions often observed in the outer plasmasphere and simulate their thermalization by collisions with the background cool, dense, plasmaspheric ion population. Though the details of the electron characteristics are often neglected in plasmaspheric research, Khazanov and Liemohn develop a kinetic treatment for the evolution of suprathermal electrons (such as photoelectrons) in the plasmasphere and calculate new heating rates for their heat deposition into the plasmasphere.

Moving out from the plasmasphere to the middle magnetosphere, the next section contains timely articles by Brian Anderson, Barry Mauk, and Peroomian and Ashour-Abdalla. Anderson's paper describes some of the exciting observations from AMPTE/CCE on ion interactions with electromagnetic ion cyclotron (EMIC) waves. His discussions in several respects provide the observational basis for the theoretical analysis of driven anisotropy-EMIC wave relationships covered earlier by Peter Gary. Particularly interesting for micro/mesoscale coupling in the daytime middle magnetosphere ($L=5-9$) are his discussions of AMPTE/CCE near-equatorial results showing the perpendicular heating of protons up to a few eV in 90°-peaked distributions, while He^+ ions appear in the form of "X" pitch angle distributions, with flux peaks lying between parallel and perpendicular directions to the magnetic field. He concludes that these He^+ distributions are perpendicularly/resonantly heated at off-equatorial latitudes where the helium cyclotron frequency matches that of the same equatorially generated proton cyclotron waves that have propagated to such latitudes. Mauk examines various mesoscale phenomena observed in the middle magnetosphere at geosynchronous orbit, including the substorm injection boundary, propagating injection fronts, and dipolarization effects, which appear to lead to electrical discharges with large parallel electric fields which may accelerate auroral electrons, and discusses possible microscale processes which may play important roles in these phenomena. Peroomian and Ashour-Abdalla perform three-dimensional particle trajectory tracings to understand how the ring current population might contain contributions from the solar wind and the auroral ionospheric injection, and conclude that the dominant source must vary with local time and radial distance of the ring current.

In the next section, Lee and Hawkins lead off with a tutorial on micro/mesoscale coupling in the dayside magnetopause region, emphasizing theory and simulations. Paul Song's tutorial then stresses the observational side of such coupling there. The Lee and Hawkins work examines several scenarios in which mesoscale phenomena, such as driven temperature anisotropies, reconnection, and MHD surface waves, generate mirror waves, ion cyclotron waves, and kinetic Alfvén waves (KAWs), which in turn affect the particle distributions and modify the plasma transport processes. Song's treatment covers ISEE and AMPTE observations of the magnetopause, focussing on flux transfer events and the waves observed in connection with them. Yu Lin discusses the theory and simulation of magnetic field reconnection layers in the dayside magnetopause, magnetopause flanks, and the magnetotail plasma sheet, with particular attention to how the physics of these layers depends on the flow Mach numbers. To understand plasma diffusion across the dayside magnetopause, Manju Prakash calculates the anomalous diffusion coefficients associated with KAWs.

The final section covers magnetotail phenomena, with particular emphasis on the magnetotail current sheet. The Harold and Chen tutorial describes recent research on the collisionless but often chaotic motions of charged particles in the magnetotail current sheet, and suggests that highly structured ion distribution functions in this region can be understood through analysis of the classes of trajectories of ions orbiting through the neutral sheet. Grant Burkhart also examines single particle trajectories for this region, including electron effects, and describes a scenario for the substorm expansion phase involving the prior formation of a thin current sheet during the growth phase. Birn et al., on the other hand, emphasize MHD simulations of large-scale magnetotail dynamics, but seek to examine how different assumptions about the actual effects of microphysics on the transport lead to different scenarios in the macroscale dynamics. Tony Lui's tutorial discusses observations of micro/mesoscale coupling within the current sheet, the major focus being on current disruptions as they may originate from a cross-field current instability.

To maintain high standards for this monograph, the submitted papers were each rigorously reviewed by two referees. To ensure a fair and objective review, all Huntsville papers were handled independently by J. L. Burch, while Horwitz and Singh handled the review process for all other papers submitted for consideration. We would like to thank the reviewers who contributed their time to this vital refereeing process.

We would also to like to express our appreciation to several individuals who helped in the preparation of this volume. Marilyn Hargrave, Andrea Haller and Linda Kirkham of UAH/CSPAR and Bill Lewis of SwRI assisted with author and reviewer correspondence and tracking of papers for this volume. Finally, we would like to thank Odile de la Beaujardiere of NSF, Bob Carovillano of NASA, and the Alabama Space Grant Consortium for funding assistance which helped support the October 1994 meeting in Guntersville, Alabama, which inspired this monograph.

James L. Horwitz
Nagendra Singh
The University of Alabama in Huntsville
Huntsville, Alabama

James L. Burch
Southwest Research Institute
San Antonio, Texas

Editors

Can We Find Useful Algorithms for Anomalous Transport?

C. T. Dum[1]

Max Planck Institut für Extraterrestrische Physik, Garching, Germany

Classical transport theory of a collision dominated plasma is based upon the relaxation of particle distribution functions to Maxwellians. There is no such universal relaxation process in a turbulent plasma. In general, anomalous transport would have to be found from the simultaneous solution of the coupled kinetic equations for the wave spectrum and the particle distribution functions. The specific case of instability and turbulence must be examined for possible simplifications of this very difficult task. A number of examples will be discussed. Simplifications may arise, for example, if one can show that distribution functions approach a self-similar shape which can be characterized by a few parameters, or if nearly isotropic distributions are maintained by strong pitch angle scattering. Both conditions are satisfied for the interaction of electrons with ion acoustic turbulence. Construction of a transport theory for this case is considered in some detail. Simplifications arise also from the drift approximation for transport transport across a strong magnetic field. Approach to marginal stability by relaxation of distributions due to intense turbulence can also greatly simplify the problem. As a result a transport model will be obtained which usually differs not only in effective collision frequencies, but also in structure from the classical transport relations.

1. INTRODUCTION

One of the earliest indications of anomalous transport came from *Bohm et al.* [1949] when investigating magnetized arcs for uranium isotope separation. Their empirical cross field diffusion coefficient

$$D\perp = \frac{1}{16}\frac{cT_e}{eB} \quad (1)$$

was confirmed by experiments on the Princeton C stellarator and other plasma devices and is still being used as a measure for anomalous transport. It typically exceeds classical diffusion by several orders of magnitude. A possible 'derivation' comes from elementary kinetic theory if we assume that particles will suffer a displacement by the electron Larmor radius in collisions with an *anomalous* collision frequency which is some fraction ($f = 1/16$) of the electron gyro frequency. Rather than assuming a scattering process by plasma turbulence we can also obtain the cross field loss by $\mathbf{E} \times \mathbf{B}$ convection in an electric field whose potential is of the order $e\Phi/T_c = f = 1/16$ and has scale lengths determined by the density gradient. This explanation has the advantage that in contrast to the elementary kinetic model its guarantees *ambipolarity* of the cross field flux. Still another explanation of Bohm diffusion invokes magnetic field errors which partially destroy the magnetic surfaces confining the plasma [*Chen*, 1984]. Even if magnetic surfaces were perfect and scattering were only due to Coulomb collisions, toroidal devices would still show enhanced diffusion. This effect is related to the magnetic field inhomogeneity and is described by *neoclassical* transport theory. An elementary kinetic model would replace the Larmor radius by the much larger characteristic size of the magnetically trapped particle orbits (bananas). If Coulomb collisions are frequent enough to prevent trapping (Pfirsch-Schlüter regime), there is still enhanced cross field convection due to electric fields which necessarily arise in this geometry. Transport becomes essentially independent of collision frequency in an intermediate collisionality regime, the so-called plateau regime. Even in current devices plasma losses far exceed the rates predicted by classical or neoclassical transport theory. There is ample evidence that these losses are related to various

[1] Also at the Center for Space Research, Massachusetts Institute of Technology, Cambridge, MA.

micro-instabilities which are driven by gradients in density, temperatures, etc. [*Liewer*, 1985]. Many theoretical efforts have been made to derive expressions and scaling laws for anomalous diffusivities, ranging from elementary kinetic models, to quasilinear theory and various nonlinear theories [*Connor and Hastie*, 1994]. More recent theories try to also include the effects of magnetic field geometry which became evident from neoclassical transport theory. The aim of these theories usually is to derive a so-called *transport matrix* which relates fluxes of mass, momentum and energy to thermodynamic forces (gradients). This generally nondiagonal matrix could then be included in transport codes for predictions and comparison with experiment. The transport matrix is usually found from quasilinear theory. There still remains, however, the much more difficult task of also finding the wave spectrum to be used in these calculations. The particle distributions may also differ significantly from Maxwellians and should be determined self-consistently with the turbulent fluctuation spectrum.

Much of the excitement in space physics comes from the fact that we generally deal not with a magnetically confined plasma but with various regions of space which exchange particles, momentum, energy and are coupled by electromagnetic fields. The effects of transport are then perhaps less obvious than for confined laboratory plasmas. In fact, the term is often used for purely convective phenomena. We restrict here the term transport to microscopic processes such as heat conduction, viscosity and resistivity which are controlled by particle collisions, or in the case of anomalous transport, by scattering of particles by turbulent electric and magnetic fields. A combination of convection, classical and anomalous transport processes may, of course, be operative in a given case. Classical Coulomb collisions and collisions with neutrals certainly play an important role in the ionosphere, but we would expect anomalous processes to completely dominate transport across the magnetopause, for example. Both diffusion and reconnection have been invoked for the entry of solar wind plasma into the magnetosphere. Measured fluctuations seem to be sufficiently intense to produce the required diffusion coefficients [*Tsurutani and Thorne*, 1982; *LaBelle and Treumann*, 1988; *Treumann et al.*, 1991; *Thorne and Tsurutani*, 1991; *Treumann et al.*, 1992]. Reconnection also requires some kind of dissipative mechanism but ideas on the required rate seem to be less developed.

Interplanetary solar wind flow is driven by heat conduction along the magnetic field. The classical formulas for collision dominated heat conductivity which are of the form

$$\mathbf{q} = -\kappa \nabla T \qquad (2)$$

appear to fail for the most part, by predicting a heat flow which is much too large. This does not necessarily mean that heat flux is controlled by anomalous scattering processes. It means first of all that, contrary to the assumptions of classical transport theory, Coulomb collisions are not frequent enough to maintain a distribution function which is very close to a Maxwellian, especially for the high energy particles which predominately carry the heat flux. For very low collisionality a substantial fraction of the energetic particles may be nearly free streaming. Heat flux thus may in effect become *nonlocal*, i.e. no longer controlled by the local temperature gradient, as assumed in (2). A strongly anisotropic distribution with an extended tail will result. Depending on plasma parameters, such distributions may be subject to various micro-instabilities, e.g. the ion acoustic instability if the electron to ion temperature ratio is sufficiently large. Ion acoustic turbulence scatters energetic electrons in pitch angle, just like electron-ion collisions. A detailed analysis shows that observed levels of ion acoustic turbulence are high enough to effectively limit anisotropies and heat flux [*Dum*, 1983]. Turbulence levels, however, do not correspond to a steady state, but to short bursts of ion acoustic wave activity, as is also characteristic of many other observations of wave activity in space.

Many other situations can be found where anomalous transport may be important, usually in conjunction with macroscopic convection and electromagnetic field configurations. A transport theory analogous to classical transport theory thus would be desirable. The essential aim of such a theory would be to provide a simplified description, relating macroscopic variables, rather than having to rely on the detailed microscopic description which is provided by kinetic theories. In classical transport the possibility of such a description arises from the existence of a universal relaxation process, causing any distribution function to relax to a Maxwellian which is specified by the local fluid variables, density, mean velocity and mean energy (temperature). If collisions are sufficiently frequent then deviations from a Maxwellian will remain small and can be found by (first order) perturbation theory. By taking moments of the perturbed distribution functions one obtains transport relations which will close the set of fluid equations.

No universal relaxation process exists for a turbulent plasma. There is usually no good reason other than simplicity for the customary assumption of Maxwellian distributions. Moreover, the wave spectrum which determines effective collision frequencies usually evolves separately in space and time, unlike the collisional spectrum which is a known functional of the distribution functions. In a generalized sense macroscopic anomalous transport equations are obtained if it is possible to specify distribution functions and the wave spectrum by a set of parameters which will include the fluid variables. This set will also have to include the fluctuation level, unless one assumes, as is frequently done, that some nonlinear effect determines a quasi-steady level from the local parameters. It is far from obvious, however, that the kinetic equations can be reduced in this manner, especially for resonant micro-instabilities for which evolution usually depends on details of the distribution functions and wave spectrum. Certainly, each case of instability must be examined individually for possible simplifications of the kinetic equations. It is not surprising that a lot of investigations have been confined to rather elementary estimates of effective collision frequencies which are supposed to replace the Coulomb collision frequencies. More detailed investigations show, however, that the structure of the transport relations is likely to also change [Dum, 1978a].

Fluid equations which have the same formal structure as in the classical case may be obtained by taking moments of the kinetic equation for the distribution function. They provide a general framework and allow to deduce some useful properties. Closure of this set of equations, however, requires information about the particle distribution functions. These distributions will usually no longer be close to a Maxwellian, thus the need for a self-consistent determination should even be more obvious than in classical transport theory. The methods evidently depend on the nature of the turbulent spectra. The case of isotropization by pitch angle scattering will be discussed in some detail. Some developments concerning the difficult problem of determining the fluctuation spectra will then be reviewed. The discussion includes the marginal stability approach, which largely circumvents this problem, by assuming that turbulence is very intense and by making use of the generally large disparity of scales. Finally, we summarize our conclusions.

2. MOMENT EQUATIONS AND THE TURBULENT COLLISION TERM

Elementary kinetic models such as mentioned in Section 1 are often used for estimating anomalous transport, although such arguments apply at best to diffusion of a test particle, and, for example, do not guarantee ambipolarity, i.e. that net electron and ion fluxes are the same. In order to describe the net fluxes of mass, momentum, and energy one should use fluid equations. Just as in classical transport theory, they may be obtained by taking moments of the kinetic equation for the distribution function. Transforming velocities by $\mathbf{w} = \mathbf{v} - \mathbf{u}$, where $\mathbf{u} = \mathbf{u}(\mathbf{x}, t)$ is space and time dependent, the kinetic equation for particle distribution functions takes the form

$$\frac{df}{dt} + \mathbf{w} \cdot \frac{\partial f}{\partial \mathbf{x}} + \mathbf{w} \times \mathbf{\Omega} \cdot \frac{\partial f}{\partial \mathbf{w}} - \mathbf{w} \cdot \frac{\partial \mathbf{u}}{\partial \mathbf{x}} \cdot \frac{\partial f}{\partial \mathbf{w}} + \mathbf{a} \cdot \frac{\partial f}{\partial \mathbf{w}} = Cf \quad (3)$$

where $\frac{d}{dt} = \frac{\partial}{\partial t} + \mathbf{u} \cdot \frac{\partial}{\partial \mathbf{x}}$, $\mathbf{\Omega} = e_j \mathbf{B}/m_j c$, $\mathbf{a} = \frac{e_j}{m_j}[\mathbf{E} + \frac{\mathbf{u}}{c} \times \mathbf{B}] - \frac{d\mathbf{u}}{dt}$. The collision term Cf may describe collisions between particles and scattering by turbulence. Moments of the distribution function are defined by

$$n_j \langle \Phi \rangle = \int d\mathbf{w} \Phi(\mathbf{w}) f(\mathbf{w}) \quad (4)$$

where n_j is the density ($\Phi = 1$). Taking moments of (3) gives

$$\frac{dn_j \langle \Phi \rangle}{dt} + n_j \langle \Phi \rangle \frac{\partial}{\partial \mathbf{x}} \cdot \mathbf{u} + \frac{\partial}{\partial \mathbf{x}} \cdot n_j \langle \mathbf{w} \Phi \rangle$$
$$+ n_j \langle \mathbf{w} \frac{\partial}{\partial \mathbf{w}} \Phi \rangle : \frac{\partial \mathbf{u}}{\partial \mathbf{x}} - n_j \mathbf{a} \cdot \langle \frac{\partial}{\partial \mathbf{w}} \Phi \rangle$$
$$- n_j \langle (\mathbf{w} \times \mathbf{\Omega}) \cdot \frac{\partial}{\partial \mathbf{w}} \Phi \rangle = n_j \frac{\delta \langle \Phi \rangle}{\delta t} \quad (5)$$

where the collisional transfer rate is given by

$$n_j \frac{\delta \langle \Phi \rangle}{\delta t} = \int d\mathbf{w} \Phi Cf \quad (6)$$

An especially convenient choice for $\mathbf{u}$ is the particle mean velocity $\mathbf{u}_j$. In this frame $\langle \mathbf{w} \rangle = 0$ and $\Phi = 1$ gives the continuity equation

$$\frac{dn_j}{dt} + n_j \frac{\partial}{\partial \mathbf{x}} \cdot \mathbf{u}_j = 0 \quad (7)$$

The next moment equation determines acceleration

$$n_j m_j \mathbf{a} = \frac{\partial}{\partial \mathbf{x}} \cdot \mathcal{P}_j - \mathbf{R}_j \quad (8)$$

which is equivalent to the equation of motion

$$n_j m_j \frac{d\mathbf{u}_j}{dt} + \frac{\partial}{\partial \mathbf{x}} \cdot \mathcal{P}_j = n_j e_j [\mathbf{E} + (\frac{\mathbf{u}}{c}) \times \mathbf{B}] + \mathbf{R}_j \quad (9)$$

The equation for the temperature is obtained by taking a moment with $\Phi = w^2/2$, noting that the pressure tensor $\mathcal{P}_j = n_j m_j \langle \mathbf{ww} \rangle$ can be decomposed as $\mathcal{P}_j = p_j \mathcal{I} + \mathbf{\Pi}_j$, $p_j = n_j T_j$.

$$\frac{3}{2} n_j \frac{dT_j}{dt} + p_j \frac{\partial}{\partial \mathbf{x}} \cdot \mathbf{u}_j + \mathbf{\Pi}_j : \frac{\partial \mathbf{u}_j}{\partial \mathbf{x}} + \frac{\partial}{\partial \mathbf{x}} \cdot \mathbf{q}_j = Q_j \quad (10)$$

where the heat flux vector is determined by the next order moment

$$\mathbf{q}_j = n_j m_j \langle \mathbf{w} w^2/2 \rangle \tag{11}$$

The continuity equation was used to rewrite to corresponding first two terms of (5) as a convective derivative

$$\frac{dn_j \langle \Phi \rangle}{dt} + n_j \langle \Phi \rangle \frac{\partial}{\partial \mathbf{x}} \cdot \mathbf{u}_j = n_j \frac{d\langle \Phi \rangle}{dt} \tag{12}$$

The turbulent collision term is obtained by ensemble averaging the kinetic equation, i.e. the Vlasov equation in case of a collisionless plasma, over fluctuations.

$$Cf = -\frac{\partial}{\partial \mathbf{v}} \cdot \frac{e_j}{m_j} \langle (\delta \mathbf{E} + \frac{\mathbf{v}}{\mathbf{c}} \times \delta \mathbf{B}) \delta f \rangle \tag{13}$$

By also considering the short wave length stable fluctuations Coulomb collisions can be included in this formulation, although they are usually described by a separate simplified collision term, such as the Landau collision integral which is an explicit functional of the distribution function. The collision term conserves particles and the rate of momentum and energy transfer are given by

$$\mathbf{R}_j = \langle \delta \mathbf{E} \delta \sigma_j + \frac{\delta \mathbf{J}_j}{c} \times \delta \mathbf{B} \rangle \tag{14}$$

$$K_j = \langle \delta \mathbf{E} \cdot \delta \mathbf{J}_j \rangle \tag{15}$$

where $\delta \sigma_j$, $\delta \mathbf{J}_j$ are the charge and current fluctuations. The heating rate in (10) is given by

$$Q_j = K_j - \mathbf{R}_j \cdot \mathbf{u}_j \tag{16}$$

Maxwell's equations give

$$\mathbf{R} = \sum_j \mathbf{R}_j = -\frac{\partial \mathbf{G}^{em}}{\partial t} - \frac{\partial}{\partial \mathbf{x}} \cdot \mathcal{T}^{em} \tag{17}$$

$$K = \sum K_j = -\frac{\partial U^{em}}{\partial t} - \frac{\partial}{\partial \mathbf{x}} \cdot \mathbf{S}^{em} \tag{18}$$

which are the usual momentum conservation law and Poynting's theorem applied to the fluctuations, with $\mathbf{G}^{em} = (1/c^2)\mathbf{S}^{em} = (1/4\pi c)\langle \delta \mathbf{E} \times \delta \mathbf{B} \rangle$, $U^{em} = \langle \delta E^2 + \delta B^2 \rangle/8\pi$, $\mathcal{T}^{em} = U^{em}\mathcal{I} - \langle \delta \mathbf{E}\delta \mathbf{E} + \delta \mathbf{B}\delta \mathbf{B} \rangle/4\pi$. For homogeneous electrostatic turbulence we obtain $\mathbf{R}_e = -\mathbf{R}_i$, i.e. anomalous momentum transfer between electrons and ions has the character of a friction force, just like classical transfer. It can only arise from low frequency fluctuations in which the ions also participate. High frequency fluctuations, like electron-electron collisions in the classical case, can only have an indirect effect by shaping the electron distribution. However, unlike electron-electron collisions micro-instabilities usually affect only limited regions of velocity space. Poynting's theorem for electrostatic fluctuations, $-K_e = K_i + \frac{\partial}{\partial t}\langle \frac{\delta E^2}{8\pi} \rangle$, also implies that net heating by a stationary spectrum is possible only if both electrons and ions participate. If the electrons drift with respect to the ions, as in current driven instabilities, the electron heating rate

$$Q_e = -Q_i - \mathbf{R}_e \cdot (\mathbf{u}_e - \mathbf{u}_i) - \frac{\partial}{\partial t}\langle \frac{\delta E^2}{8\pi} \rangle \tag{19}$$

can be substantially larger then the ion heating rate Q_i. The rate of momentum transfer $\mathbf{R}_e$ determines *anomalous resistivity* by counteracting the acceleration provided by the electric field in (9). It also determines diffusion across the magnetic field. By balancing the Lorentz force we obtain

$$n_j e_j \mathbf{u}_{j\perp}^a = \frac{c}{B^2} \mathbf{R}_j \times \mathbf{B} \tag{20}$$

for the flux across the magnetic field. The conservation law (17) assures (approximate) ambipolarity. Using (14) and neglecting magnetic fluctuations we see that cross field transport arises from the correlation between fluctuations of the $\mathbf{E} \times \mathbf{B}$ drift and density fluctuations. The usual quasineutrality of low frequency fluctuations thus also assures ambipolarity. If we assume a slab plasma with a density gradient in the x direction and a magnetic field in the z direction, then particle flux in the x direction requires momentum transfer in the y direction. It can be produced by the diamagnetic drifts which flow in the y direction as a result of the pressure gradients in the x direction,

$$\mathbf{u}_{j\perp}^d = \frac{c}{n_j e_j B^2} \mathbf{B} \times \nabla \cdot \mathcal{P} \tag{21}$$

Finally, if we assume that momentum transfer is given by a friction law

$$\mathbf{R}_e = -\nu n_e m (\mathbf{u}_e - \mathbf{u}_i) \tag{22}$$

and scalar pressures, the continuity equation (7) becomes a diffusion equation with the diffusion coefficient

$$D = \nu \frac{T_e + T_i}{m\Omega_e^2} \tag{23}$$

where Ω_e is the electron cyclotron frequency. This derivation, which explicitly implies ambipolarity should be compared with elementary kinetic models mentioned in connection with (1). It is especially important to distinguish between this diffusion coefficient for the entire plasma and models which may apply to the Brownian motion of a selected group of test particles. If ν is the electron-ion collision frequency then (23) represents the classical diffusion coefficient. For anomalous transport we would have to prove (22) and find an effective collision frequency. Drift waves which are driven unstable by the diamagnetic drifts are a plausible candidate for anomalous diffusion, much as current driven instabilities are a plausible cause for anomalous resistivity.

We can express the transfer rates in terms of the wave spectrum if we use a conductivity relation $\delta \mathbf{J}_j = \kappa \cdot \delta \mathbf{E}$, or $\delta \sigma_j = -k^2 \epsilon_j \Phi_\mathbf{k}$ in case of electrostatic fluctuations with $\delta \mathbf{E}_\mathbf{k} = -i\mathbf{k}\Phi_\mathbf{k}$. In the latter case we obtain

$$\mathbf{R}_j = 2Im \int d\mathbf{k} W(k) \epsilon_j(\omega_\mathbf{k}, \mathbf{k}) \mathbf{k} \quad (24)$$

and

$$K_j = 2Im \int d\mathbf{k} W(k) \epsilon_j(\omega_\mathbf{k}, \mathbf{k}) \omega_\mathbf{k} \quad (25)$$

If we use the linear susceptibilities than these relations must be the same as obtained by taking moments of the quasilinear diffusion term. While useful, relations (24-25), however, are no short cut to solving kinetic equations such as the quasilinear equations, as both the wave spectrum and the actual distribution functions are needed. For micro-instabilities the imaginary part of the susceptibility depends on details (the slope) of the distribution function at resonant velocities. For fully developed turbulence in a collisionless plasma there is no particular reason to assume that these are described by a Maxwellian, as is usually done for investigations of linear instability. We may nevertheless see how a relation of the form (22) may arise if we assume a drifting isotropic distribution $F(w)$ for which (neglecting the static magnetic field)

$$Im\epsilon_e(\omega, \mathbf{k}) = \frac{8\pi^3 e^2}{mk^2} v'_{ph} F(v'_{ph}) \quad (26)$$

where $v'_{ph} = (\omega - \mathbf{k} \cdot \mathbf{u})/k$. For ion sound waves and not too large drifts we can assume $v'_{ph} \ll v_e$ where v_e is the electron thermal velocity and thus $F(v'_{ph}) \approx F(0)$. It is convenient to define a form factor by comparing with a Maxwellian of the same density and temperature, $F(0) = a_{-3} F_M(0)$. The rate of momentum transfer to the electrons becomes then

$$\mathbf{R}_e = -\omega_e \frac{W}{nT_e} a_{-3} (2\pi)^{1/2} nm \langle \frac{\omega_e}{kv_e} \hat{\mathbf{k}}\hat{\mathbf{k}} \cdot (\mathbf{u} - \frac{\omega \mathbf{k}}{k}\hat{\mathbf{k}}) \rangle \quad (27)$$

where the angular brackets denote an average over the (shape) of the spectrum, $\hat{\mathbf{k}} = \mathbf{k}/k$, ω_e is the electron plasma frequency, and W is the total electric energy. We see that the effective collision frequency is in general a tensor and that the ion velocity is replaced by a mean phase velocity. The heating rate may be obtained in a similar manner. Ion sound turbulence not only produces anomalous resistivity (by scattering of electrons in pitch angle) but also flattens the electron energy distribution. It can be shown that F(w) rapidly approaches a self similar distribution of the form

$$F(w) = \frac{n}{v_o^3} C_s exp[-(w/v_0)^s] \quad (28)$$

with s=5 [*Dum*, 1978a]. The form factor in this case is reduced to $a_{-3} = 0.44563(0.2815$ for 2D diffusion), compared to unity for a Maxwellian. These relations provide a reasonable description if current flow is across the magnetic field. For flow along the magnetic field, however, we have the phenomenon of electron runaway, just as for Coulomb collisions. The effective collision frequency for pitch angle scattering by ion sound turbulence in effect has the same v^{-3} speed dependence. As a result the electron distribution becomes strongly distorted, with an elongated tail, even if actual electron runaway is not significant. Anomalous resistivity along the magnetic field thus may be strongly reduced [*Dum*, 1978b]. Expressions of the form (27) also lose much of their significance if the wave-electron interaction is magnetized, as is the case for electrostatic ion cyclotron waves, for example. The slope of the velocity distribution along the field determines $Im\epsilon_e$ in this case and it will be significantly reduced compared to any initial Maxwellian, as a result of quasilinear diffusion. The asymptotic shape of this distribution and the residual slope can be obtained from quasilinear theory by balancing the ion damping rate. It agrees well with numerical solutions of the quasilinear equations [*Muschietti and Dum*, 1990], implying that the system indeed approaches marginal stability. Still another expression is obtained if we use linear instability theory to evaluate the electron susceptibility for drift waves [*Horton*, 1984; *Itoh*, 1992]. The imaginary part becomes

$$Im\epsilon(\omega, \mathbf{k}) = -\frac{4\pi^2 e_j^2}{m_j k^2} \int d\mathbf{v} J_0^2(\frac{k_\perp v_\perp}{\Omega_j}) \delta(\omega - k_z v_z)$$
$$\cdot [k_z \frac{\partial}{\partial v_z} + \frac{1}{\Omega_j}(-k_x \frac{\partial}{\partial y} + k_y \frac{\partial}{\partial x})]F \quad (29)$$

making use of the fact that the average distribution function must be of the form

$$f = F(v_z, v_\perp, x + \frac{v_y}{\Omega_j} - \frac{v_x}{\Omega_j}) \approx F(v_z, v_\perp, x, y)$$
$$+ \frac{1}{\Omega_j}(v_y \frac{\partial}{\partial x} - v_x \frac{\partial}{\partial y})F \quad (30)$$

and that frequencies are much smaller than gyro frequencies. Using (29) in (24) we see that momentum transfer required for diffusion across the magnetic field is proportional to the gradient of F. (Cross terms can arise if the wave spectrum is anisotropic.) If we set

$$F_j(\mathbf{v}, \mathbf{x}) = \frac{n_j}{v_j^3} \hat{F}(\frac{\mathbf{v}}{\mathbf{v_j}}, \mathbf{x}) \qquad (31)$$

then this gradient becomes

$$\nabla F = F \nabla ln n_j - \frac{1}{2}(3 + \mathbf{v} \cdot \frac{\partial F}{\partial \mathbf{v}}) \nabla ln T_j + \frac{n_j}{v_j^3} \nabla \hat{F} \qquad (32)$$

In addition to the gradients of density and temperature, changes in the shape of the distribution function thus also enter transport. A term proportional to the diamagnetic drift as in our simple model (22) may be extracted, of course, but even in the classical case there remains an additional term, the thermal force, which is proportional to the temperature gradient. Again, the question of local reduction in the slope of the distribution by quasilinear effects should be examined critically. Quasilinear diffusion will tend to reduce the effective slope [square bracket in (29)] of the resonant particle distribution to zero. Coulomb collisions can be highly effective in this small region of velocity space in maintaining a residual slope, even if the plasma is nearly collisionless [*Galeev and Rudakov*, 1964].

Just as for particle flux, an expression for anomalous heat flux across the magnetic field can be found if we consider the moment equation for $\Phi = mw^2\mathbf{w}/2$

$$\frac{d\mathbf{q}_j}{dt} + \mathbf{q}_j \nabla \cdot \mathbf{u}_j + \mathbf{q}_j \cdot \nabla \mathbf{u}_j + \mathcal{Q}_j : \nabla \mathbf{u}_j$$
$$+ \nabla \cdot n_j m_j \langle \mathbf{ww}\frac{w^2}{2}\rangle = \frac{3}{2}p_j \mathbf{a}_j + \mathbf{a}_j \cdot \mathcal{P}_j + \mathbf{q}_j \times \mathbf{\Omega_j}$$
$$+ \frac{\delta \mathbf{q}_j}{\delta t} \qquad (33)$$

and balance the Lorentz term against the collisional terms

$$\mathbf{q}_{j\perp} = -\frac{\mathbf{e}_0}{\Omega_j} \times [\frac{\delta \mathbf{q}_j}{\delta t} - \frac{1}{n_j m_j}\mathbf{R}_j \cdot (\frac{5}{2}p_j \mathcal{I} + \mathbf{\Pi}_j)] \qquad (34)$$

The term proportional to $\mathbf{R}_j$ arises from the collisional part of $\mathbf{a}_j$, eqn. (8) and represents collisional enthalpy flux, cf. (20). According to the definition (11) it is to be subtracted from energy flow in order to get the purely conductive heat flux. The collision term may be written as

$$\frac{\delta \mathbf{q}_j}{\delta t} = \frac{e_j}{m_j}\langle (\delta \mathbf{E} + \frac{\mathbf{u}_j}{c} \times \delta \mathbf{B}) \cdot (\frac{5}{2}\delta p_j \mathcal{I} + \delta \mathbf{\Pi}_j)$$
$$+ \frac{1}{c}\delta \mathbf{q}_j \times \delta \mathbf{B}\rangle \qquad (35)$$

where the velocity moments are to be evaluated for the fluctuating part δf_j of the distribution function. Again, cross field transport depends on the correlation between the fluctuations of the $\mathbf{E} \times \mathbf{B}$ drift and appropriate fluid variables (plus terms arising from magnetic fluctuations).

It is not difficult to see that cross field viscosity (cross field momentum transport) is obtained by solving

$$\Omega_j \times \mathbf{\Pi} - \mathbf{\Pi} \times \Omega_j = \frac{\delta \mathbf{\Pi}_j}{\delta t} \qquad (36)$$

which are the dominant terms in the moment equation for $\mathbf{\Pi}_j$, besides collisionless contributions to viscosity. (There are also collisionless cross field heat flux terms.) The collision integral is

$$\frac{\delta \mathbf{\Pi}_j}{\delta t} = m_j \langle \delta f_j (\delta \mathbf{aw} + \mathbf{w}\delta \mathbf{a} - \frac{2}{3}\mathcal{I}\delta \mathbf{a} \cdot \mathbf{w})\rangle \qquad (37)$$

where the brackets now indicate velocity integration, in addition to ensemble averaging and $\delta \mathbf{a}$ is the fluctuating Lorentz force. If we neglect the fluctuating magnetic field $\delta \mathbf{B}$ and use quasilinear theory, then (37) may be expressed in terms of the linear conductivity tensor and the wave spectrum. We wish to emphasize again, however, that the ensemble averaged distribution function must be known with sufficient accuracy in order to calculate δf_j or the conductivity tensor. This fact has often been overlooked, apparently because the usual calculations with a Maxwellian or bi-Maxwellian distribution may give a nonzero result (if the spectrum is anisotropic, as it usually is). A comparison with classical transport makes it obvious, however, that in the case of viscosity, for example, the ensemble averaged distribution should include the effects of velocity shear. This is most easily done by working with the kinetic equation in the frame of the mean velocity, (3), else drift related transport will appear only in a second order perturbation theory. In calculating transport across the magnetic field we have assumed that these processes are slow compared to the cyclotron frequency. Our method of using moment equations is equivalent to an iterative solution of the kinetic equation which starts with the collisionless solution and uses an expansion in $1/\Omega_j$.

Before concluding this section, we would like to mention that the drift kinetic equation is a convenient starting point for transport in a magnetized plasma for which the characteristic scales are long compared to gyro frequencies and gyro radii. We indirectly already made use of the corresponding expansions, in (29) for example, and in the expressions for transport across the magnetic field. The theory can be quite complicated if the magnetic field is nonuniform, but allows for example, to combine the neoclassical effects mentioned in Section 1 with anomalous transport [*Shaing*, 1988; *Balescu*, 1990]. The simple case of a uniform magnetic field may suffice here to illustrate a few points. The particle distribution function is written in the form

$$f(\mathbf{v}, \mathbf{x}) = F(x, v_\parallel, v_\perp) + \tilde{f}(\mathbf{x}, v_\parallel, v_\perp, \phi) \quad (38)$$

The equation for F is obtained by averaging the kinetic equation over gyration angle, after inserting the solution for the angle dependent part in terms of F, which to lowest order follows from

$$\Omega_j \frac{\partial \tilde{f}}{\partial \phi} = \mathbf{v} \cdot \frac{\partial}{\partial \mathbf{x}_\perp} F + \frac{e_j}{m_j} \mathbf{E} \cdot \frac{\partial}{\partial \mathbf{v}_\perp} F$$

and yields

$$\frac{\partial F}{\partial t} + v_\parallel \frac{\partial}{\partial x_\parallel} F + \mathbf{v}_E \cdot \frac{\partial}{\partial \mathbf{x}_\perp} F + \frac{e_j}{m_j} E_\parallel \frac{\partial}{\partial v_\parallel} F = CF \quad (39)$$

where $\mathbf{v}_E = c(\mathbf{E} \times \mathbf{B})/B^2$. If the fluctuations also satisfy this equation, as will be the case for long wavelength low frequency drift waves, then the turbulent collision term can be obtained by ensemble averaging this equation. (More generally, gyrokinetic equations which retain finite Larmor radius effects can be used.) Taking velocity moments of (39) produces a continuity equation for n_j which already contains the cross field diffusion term. The equation for the parallel momentum contains a perpendicular viscosity term, in addition to parallel momentum transfer. Although one may also derive an equation for the total energy, it makes sense to consider parallel and perpendicular energy separately. Some confusion seems to have arisen in the literature regarding cross field heat flux [*Ross*, 1989; *Balescu*, 1990]. One problem arises because some authors include portions of the convective energy and convective energy flux in what they call pressure and heat flux. This problem can easily be avoided by developing the drift kinetic equations from (3). The other problem is that some authors simply state that the perpendicular heat fluxes are given by the correlation between the fluctuating $\delta \mathbf{E} \times \mathbf{B}$ drift and the fluctuations in parallel or perpendicular thermal energy. Comparing with (35) we see that a term proportional to $\delta p_{j\perp}$ is missing. This problem can be resolved by noting that in order to calculate perpendicular heat flux and the perpendicular part of the heating term, $\langle \delta \mathbf{E} \cdot \delta \mathbf{J}_j \rangle$, one must also consider the higher order angle dependent fluctuation

$$\delta \tilde{f} = \frac{1}{\Omega_j}(\mathbf{v} \times \mathbf{e}_0) \cdot \frac{\partial}{\partial \mathbf{x}} \delta F \quad (40)$$

c.f. (30). It yields a perpendicular heating rate which exactly cancels the corresponding term in the heat flux.

We have made extensive use of expansions in $1/\Omega_j$ in calculating transport across the magnetic field. Distribution functions were in effect determined iteratively, starting with the collisionless solution. No such simplifications are available for transport along the magnetic field. Other than by ignoring this transport, a solution of the kinetic equation for the ensemble averaged distribution cannot be avoided. The situation is no different from classical transport in this respect, in particular with regard to the notorious problem with longitudinal heat flux [*Dum*, 1990].

3. ANOMALOUS TRANSPORT BY ION SOUND TURBULENCE AND NEARLY ISOTROPIC DISTRIBUTION FUNCTIONS

We have reiterated the need for a self-consistent determination of particle distribution functions. In a nearly collisionless plasma with fully developed plasma turbulence distribution functions will as a rule deviate significantly from a Maxwellian. The effect of wave-particle interactions evidently depends on the nature of turbulence. We consider here ion sound turbulence and related spectra which are characterized by small phase velocities $\omega/kv_e \ll 1$ and short wavelength $kv_e/\Omega_e \gg 1$. It has been demonstrated in simulation experiments of current driven ion sound instability that the interaction is in essence quasilinear [*Dum et al.*, 1974]. For current flow across the magnetic field, as in perpendicular shocks, the electron distribution flattens by assuming a selfsimilar shape given by (28), with $s = 5$. The ions develop a high energy tail. Landau damping from this tail compensates the residual electron contribution to wave growth, leading to quenching of the instability.

The (resonant) quasilinear diffusion tensor for the wave-electron interaction is given by

$$\mathcal{D}_{ew} = \frac{8\pi^2 e^2}{m^2} \int d\mathbf{k} W(\mathbf{k}) \delta(\omega_\mathbf{k} - \mathbf{k} \cdot \mathbf{v}) \hat{\mathbf{k}} \hat{\mathbf{k}} \quad (41)$$

where $\hat{k} = \mathbf{k}/k$ is the wave vector direction and $W(\mathbf{k})$ is the spectrum of the electric field energy. The collision term in (3) due to waves thus is the diffusion term

$$Cf = \frac{\partial}{\partial \mathbf{v}} \cdot \mathcal{D}_{ew} \cdot \frac{\partial}{\partial \mathbf{v}} f \quad (42)$$

Because $\omega/kv_e \ll 1$, wave-particle resonance occurs for most electrons with $\hat{k} \cdot \mathbf{v} \approx 0$, that is the interaction is essentially elastic, just like the scattering of electrons by electron-ion collisions, except that rates will usually be much higher. The interaction therefore tends to isotropize the electron distribution and it makes sense to write it in the form

$$f(\mathbf{w}) = F(w) + \hat{f}(\mathbf{w}) \qquad (43)$$

where if perturbing, drifts, gradients, and electric fields are weak, anisotropies should be small, $\hat{f} << F$, and can be determined from perturbation theory. Taking a spherical average of (3) we obtain [*Dum*, 1978a]

$$\frac{dF}{dt} - \frac{w}{3}\nabla \cdot \mathbf{u} \cdot \frac{\partial F}{\partial w} + \frac{\partial}{\partial \mathbf{x}} \cdot \langle \mathbf{w}\hat{f}\rangle + \frac{1}{w^2}\frac{\partial}{\partial w}w\langle(\mathbf{a}\cdot\mathbf{w} - \mathcal{U}:\mathcal{W})\hat{f}\rangle - <\hat{C}\hat{f}> = <C>F \quad (44)$$

where

$$U_{ik} = \frac{1}{2}\left(\frac{\partial u_i}{\partial x_k} + \frac{\partial u_k}{\partial x_i}\right) - \frac{1}{3}\frac{\partial}{\partial \mathbf{x}}\cdot \mathbf{u}\delta_{ik} \qquad (45)$$

is the shear tensor and

$$\mathcal{W} = \mathbf{ww} - \frac{w^2}{3}\mathcal{I}$$

The anisotropic part satisfies

$$\mathbf{w}\cdot\frac{\partial F}{\partial x} + (\mathbf{a}\cdot\mathbf{w} - \mathcal{U}:\mathcal{W})\frac{1}{w}\frac{\partial F}{\partial w} - \hat{C}F$$
$$= (\mathbf{w}\times\mathbf{\Omega})\cdot\frac{\partial \hat{f}}{\partial \mathbf{w}} + C\hat{f} - <\hat{C}\hat{f}> \qquad (46)$$

where the time dependence is neglected to lowest order, by virtue of the assumed dominance of isotropization. In a strong magnetic field it is useful to make the further decomposition

$$\hat{f}(\mathbf{w}) = \bar{f}(w_\parallel, w_\perp) + \tilde{f}(w_\parallel, w_\perp, \phi) \qquad (47)$$

and split (46) by taking a ϕ average

$$w_\parallel \frac{\partial F}{\partial w_\parallel} + (a_\parallel w_\parallel - \mathcal{U}_0:\mathcal{W}_0)\frac{1}{w}\frac{\partial F}{\partial w}$$
$$-\bar{C}F = \bar{C}\bar{f} - <\bar{C}\bar{f}> \qquad (48)$$

$$\mathbf{w}_\perp \cdot \frac{\partial F}{\partial \mathbf{x}}(\mathbf{a}_\perp \cdot \mathbf{w}_\perp - \mathcal{U}_\perp:\mathcal{W}_\perp)\frac{1}{w}\frac{\partial F}{\partial w} - \tilde{C}F$$
$$= -\Omega\frac{\partial \hat{f}}{\partial \phi} + \tilde{C}\bar{f} + C\tilde{f} - \overline{C\tilde{f}} \qquad (49)$$

A finite Larmor radius expansion may then be applied to (49), as discussed in the previous section.

If the wave spectrum were (nearly) isotropic in some frame, elastic scattering could be described by a Lorentz collision term, which is obtained by using spherical coordinates in (42). The effective collision frequency

$$\nu(w) = \pi\frac{W}{nT_e}\langle\frac{\omega_e}{kv_e}\rangle(\frac{v_e}{v})^3 \qquad (50)$$

has the same w^{-3} speed dependence, but usually is much larger than the electron-ion collision frequency. The solution of (46) in this case is straightforward, especially if electron-electron collisions which introduce an integral dependence can be neglected. The solution of (46) may be found by expanding $\hat{f}$ in spherical harmonics [*Dum*, 1978b].

$$\hat{f}(\mathbf{w}) = \mathbf{w}\cdot\mathbf{f}_1 + \mathcal{W}:\mathbf{f}_2 + \cdots \qquad (51)$$

Only the l=1 and l=2 components are needed in the transport equations, but in the case of anisotropic spectra the various l components are coupled by the diffusion term and collision frequencies become tensors, c.f. (27) for ν_{11}. A strong magnetic field effectively decouples the various components. By truncating the expansion at l=2 an approximate solution can be obtained which is correct in the limits of an isotropic spectrum or a strong magnetic field. The longitudinal perturbation $\bar{f}(w_\parallel, w_\perp)$ can be obtained by direct integration of (48). Once the distributions are known, transport relations are obtained by carrying out the velocity integrations required in the moment equations. The shape of the energy distribution has a profound effect on the magnitude of transport coefficients. It also affects the structure of the transport relations. Only for a Maxwellian energy distribution is momentum transfer and heat flux which is directly related to density gradients absent and one has the Onsager relation between the thermal force and drift related heat flow. For a flat topped distribution (28) with $s = 5$, on the other hand, there is a symmetry relation between density gradient and drift related transport. (Note that by drift related transport we do not mean convection but transport connected with the skewing of the distribution due to speed dependent friction.)

Inserting the solution for $\hat{f}$ in terms of F and the perturbing forces (gradients etc.) into (44) completes the equation for the self-consistent determination of the energy distribution. (Note the similarity with the quasilinear approach for the ensemble average distribution function.) The resulting equation which contains transport related diffusion terms, in addition to local heating is quite complicated and will not be given here, see *Dum* [1978b]. If local heating, the term $<C>F$, dominates then it can be shown [*Dum*, 1978a] that for self-similar solutions the diffusion coefficient must be of the form

$$D_o(w,t) = D(t)w^{(s-2)} \qquad (52)$$

and that the solutions of (44) are of the form (28). The Green's functions which show the approach to this asymptotic form and the heating rates can also be found. Using spherical coordinates in (42) we find that for ion sound turbulence the energy diffusion coefficient in a frame with drift **u** is

$$\langle D_{ew}^{ww} \rangle = \frac{1}{w^3} \int d\mathbf{k} W(\mathbf{k}) \Theta(\hat{\omega}) \frac{4\pi^2 e^2}{m^3 k^2} (\omega_\mathbf{k} - \mathbf{k} \cdot \mathbf{u})^2 \qquad (53)$$

where the resonance condition $\Theta(\hat{\omega}) = 1$ for $\hat{\omega} = |\omega - \mathbf{k} \cdot \mathbf{u}/kw| < 1$ and zero otherwise, may be dropped for most electrons. The w^{-3} speed dependence thus leads to a self-similar distribution (28) with $s = 5$. Energy diffusion from electron-ion collisions has the same speed dependence, but in the absence of turbulence and unless charge numbers are very high, electron-electron collisions will easily succeed in maintaining a distribution close to a Maxwellian. The Chapman-Enskog method for calculating classical transport is less general than our approach in that it assumes that the dominant process is relaxation to a Maxwellian. In our case the lowest order distribution function is determined by a kinetic equation which itself contains transport processes. The method of solution in this Section was based upon the assumed dominance of isotropization by elastic scattering. If gradients are very large then this approach too may break down for a significant portion of the high speed particles because of the w^{-3} speed dependence of scattering. A nonperturbative solution of the kinetic equations is then required to determine heat flux, etc. An expansion of the distribution function in spherical harmonics is still useful in this case, even though harmonics are coupled by nonlinearities and anisotropies [*Dum, 1990*].

4. DETERMINATION OF THE WAVE SPECTRUM AND MARGINAL STABILITY

In obtaining transport relations, we have so far assumed that the wave spectrum is given. Unless one has a measured wave spectrum for which transport is to be determined, the analysis will have to start from linear instability theory. It will give information about growth rates, frequencies and wave phase velocities which is important in determining the effect on transport. The prevalent idea in the past, at least, seems to have been that some nonlinear effect will limit wave growth and a (quasi-)steady wave spectrum will result. For example, a well known resistivity formula for ion sound turbulence has been derived by balancing the linear growth rate (for drifting Maxwellians) against nonlinear scattering of waves by ions [*Galeev and Sagdeev, 1979*]. Computer simulation [*Dum et al., 1974*] shows, however, that quasilinear evolution of the distribution functions, and along with it of wave growth usually plays the dominant role. And there are by now countless other examples in which one finds that wave growth rates evolve with the distribution functions, in the sense of quasilinear theory. This is not to say that there are no situations in which nonlinear effects such as mode coupling have an important effect in shaping the wave spectrum, especially if distribution functions can be more or less maintained by particle influx, binary collisions, etc.

For the various drift waves, where one might expect a steady state regime of turbulence if Coulomb collisions are sufficiently frequent, extensive calculations based on nonlinear mode coupling have been carried out [*Horton, 1984*]. Many calculations, however, content themselves with a *mixing length* estimate of the fluctuation level. The fluctuation level is estimated by balancing the linear growth rate against a turbulent damping term [*Kadomtsev, 1965*]

$$\gamma = k_\perp^2 D_\perp \qquad (54)$$

where $D_\perp$ is the cross field diffusion coefficient. More general expressions of this kind may be derived from resonance broadening theory [*Dupree, 1967; Dum and Dupree, 1970*].

Although a detailed evaluation of anomalous transport requires a knowledge of the wave spectrum, a simplified approach may be possible due to the usually large difference between microscopic and macroscopic scales. Only some average properties of the microscopic processes are required for the solution of macroscopic equations. This can be seen by considering fluid equations in the form of a conservation law

$$\frac{\partial U}{\partial t} + \nabla \cdot \mathbf{F} = Q \qquad (55)$$

A finite difference scheme is obtained by integrating over macroscopically small time and space intervals, using some interpolation scheme between grid points.

$$\frac{U_m^{n+1} - U_M^n}{\Delta t} = \frac{1}{\Delta t \Delta V} \int dt [-\int d\mathbf{S} \cdot F + \int d\mathbf{V} Q] \qquad (56)$$

The flux is thus time averaged and integrated over the surface of the volume element and the source is both time and space averaged. Typical grid sizes are often much larger than the size of the entire system used in microscopic simulations [*Dum, 1985*].

A frequently occurring case is that the plasma, even though initially strongly unstable, approaches a state of marginal stability, possibly with (relaxation-) oscillations about this state [*Dum*, 1985]. To macroscopically model such behavior it may be sufficient to switch on large anomalous transport terms when instability thresholds are exceeded. As long as transport rates are within reasonable bounds and the interrelationship between transport processes is correctly given, results may not be too sensitive to a particular choice on how the fluctuation level increases. Anyway, this self-regulatory effect can be tested in the macroscopic simulation. This *marginal stability* approach to transport has been used for a number of instabilities [*Manheimer and Boris*, 1972; *Manheimer and Antonsen*, 1979; *Lysak and Dum*, 1983; *Gary et al.*, 1994].

Although the marginal stability approach largely reduces transport calculations to a calculation of instability thresholds, some understanding of the underlying microscopic physics may still be essential. The instability threshold for kinetic instabilities depends not only on macroscopic parameters, but also on the detailed shape of the distribution functions. This shape often changes as the plasma approaches marginal stability. Quasilinear theory which describes such relaxation processes says that wave growth is to be evaluated for the evolving ensemble averaged distribution function. This is the exactly the same distribution which also determines anomalous transport.

In the case of ion sound turbulence the relationship between effective drift velocity for instability and electron transport is remarkably simple and leads to interesting effects [*Dum*, 1978b]. Using the expansion (51) of the distribution function in spherical harmonics it can be shown that the effect of even l harmonics on $Im\epsilon_e$ and of odd harmonics on $Re\epsilon_e$ is reduced by powers of the small factor $\hat{\omega}$ introduced in (53). The effect of the $l = 1$ component is to introduce an effective drift velocity $\mathbf{u}^*$ in

$$Im\epsilon_e(\omega, \mathbf{k}) = \left(\frac{\omega_e}{kv_e}\right)^2 \left(\frac{\pi}{2}\right)^{1/2} a_{-3} \frac{\omega - \mathbf{k} \cdot \mathbf{u}^*}{kv_e} \quad (57)$$

which is obtained by writing the rate of momentum transfer as

$$\mathbf{R}_e = -nm\nu_{11}^* \cdot (\mathbf{u}^* - \mathbf{u}_w) \quad (58)$$

as in (27). For transport along the magnetic field and an anisotropic spectrum the $l = 3, 5$ components may not be negligible, but a similar relation can be obtained from the solution for the longitudinal perturbation of the distribution function. Because friction and the contribution to wave growth is larger for low speed particles, the effective drift velocity for instability is reduced from the parallel drift velocity. From the transfer rate across the magnetic field, we obtain an effective drift velocity

$$\mathbf{u}_\perp^* = \mathbf{u}_\perp + \frac{\bar{\rho}_{u,\lambda}}{\Omega_e}\mathbf{e}_0 \times \frac{\mathbf{R}_e}{nm} - \frac{v_e^2}{\Omega_e}\mathbf{e}_0 \times \left(\bar{\rho}_{n,\lambda}\nabla lnn - \bar{\rho}_{T,\lambda}\nabla lnT_e + \frac{\nabla a_{-1}}{a_{-3}}\right) \quad (59)$$

The terms in addition to the mean velocity $u_\perp$ arise from a skewing of the distribution function due to the speed dependence of scattering. They are strongly dependent on the shape of the energy distribution $F(w)$, as reflected by the transport coefficients and directly by the last term in (59). a_{-1} is a form factor of F which again is normalized to unity for a Maxwellian, and $a_{-1} = 0.8832$ for the distribution (28) with $s = 5$. The transport coefficients are also expressed in terms of form factors and for $s = 5$ have the values $\bar{\rho}_{u,\lambda} = \bar{\rho}_{n,\lambda} = 0.982$, $\bar{\rho}_{T,\lambda} = 1.991$ [*Dum*, 1978b]. The second term in (59) introduces a rotation of the effective drift velocity, and hence of the resulting wave spectrum, which is proportional to the effective collision frequency. This interesting effect was observed in perpendicular shock experiments. The rotation angle was typically 20° with respect to the diamagnetic current [*Machalek and Nielsen*, 1973; *Muraoka et al.*, 1973]. It is quantitatively confirmed by computer simulation [*Dum et al.*, 1974; *Biskamp et al.*, 1975].

The additional effective drift related to the density and temperature gradients is comparable to the diamagnetic drift. Indeed, it could be shown for a perpendicular shock wave experiment that only these additional terms made ion sound instability possible over prolonged periods [*Söldner et al.*, 1977]. These measurements were also in excellent agreement with a hybrid code which used ion particles and described electrons by the anomalous transport theory sketched here. Although it would be very desirable, of course, to actually compute the ion sound wave spectrum, a local switch on-off condition $u_1 < u* < u_2$ for turbulence and anomalous transport was used instead. Here u_1 and u_2 are the critical drift velocities for ion sound instability evaluated for the initial (Maxwellian) ion distribution and the final ion distribution with an extended high energy tail, respectively. This marginal stability approach can be justified by the fact that electron relaxation and initial wave growth are very fast compared to ion tail buildup and the macroscopic time scales. The marginal stability condition $u_2 \approx u^*$ is well satisfied indeed at late times.

5. CONCLUSIONS

Our presentation should have made it clear that there can be no useful general theory of anomalous transport. In many cases, such as the acceleration of a selected group of particles by a turbulent wave spectrum, there may be no transport theory at all, but one must instead work directly with the kinetic equations for particle and wave distributions.

After identifying the relevant instabilities, the wave-particle interaction (speed and angle dependence of turbulent scattering) should be examined in detail for simplifying features which will allow a reduced description in terms of a set of parameters. This set will generally include the fluid variables, mass, momentum, energy, the fluctuation level and other features such as characteristic wave numbers etc. We have sketched this approach in some detail for ion sound turbulence. These fluctuations scatter electrons much like electron-ion collisions, but there is no mechanism similar to electron-electron collisions which would maintain a Maxwellian energy distribution. The energy distribution therefore must be determined from a reduced kinetic equation which in addition to local turbulent heating contains nonlocal speed dependent convection and diffusion terms. Fortunately in most cases evolution will rapidly become self-similar. The anisotropic part of the electron distribution is determined as a functional of the energy distribution and the perturbing forces by assuming that elastic scattering dominates. Velocity integration yields transport relations which strongly depend in magnitude as well as structure on the shape of the energy distribution, as expressed by a set of form factors. The fluctuation level and the shape of the wave spectrum, as expressed by characteristic wave numbers and factors describing the anisotropy of the spectrum, determine the effective collision frequencies. The distribution function determined from this transport theory also enters the quasilinear growth rate, which thus can be related to transport processes. Some of these methods, or similar methods should also be applicable to other instabilities in arriving at a set of generally coupled relations between various transport terms and the perturbing gradients, drifts etc. Even if one may be unable to close the loop by also determining the wave spectrum self-consistently, such relations should be still very useful in understanding many aspects of anomalous transport and their interrelationship.

We discussed some of the simplifications which arise for transport across a strong magnetic field, e.g. by drift waves. Some transport terms, may be directly expressed in terms of the wave spectrum and the dielectric properties. While such relations are very useful, they represent no shortcut to a self-consistent determination of the distribution functions responsible for anomalous transport. Certainly, they should not be applied blindly with the conventional model distributions used in linear instability analysis, even this may give a nonzero result.

Compared to collisional fluctuation spectra which can be expressed as functionals of the distribution functions, it is much more difficult to determine the evolution of turbulent wave spectra. Fortunately, this task can often be simplified by the generally large difference between microscopic and macroscopic scales in that only some average properties of the microscopic state are required for the solution of macroscopic equations. Finite difference schemes for fluid equations in effect amount to such averaging. A frequently occurring case is that the plasma approaches a state of marginal stability, possibly with relaxation oscillations about this state. This case can be modeled by turning on strong anomalous transport when instability thresholds are exceeded locally. As we have demonstrated for the case of ion sound turbulence and shock waves, a detailed understanding of the relation between various transport processes and wave growth may still be needed in this marginal stability approach.

In conclusion, yes it is possible and necessary to go beyond elementary estimates of effective collision frequencies and develop useful algorithms for anomalous transport if one starts with a careful examination of the microscopic processes and looks for simplifying features.

Acknowledgments. This work was partially supported by NASA Grant Nos. NAG5-2255 and NAGW 1532, by the Air Force Office of Scientific Research Contract No. F49620-93-1-0287 and the Phillips Laboratory Contract No. F19628-91-K-0043, while the author stayed at the MIT Center for Space Research. The author wishes to thank T. Chang for his hospitality and stimulating discussions.

REFERENCES

Balescu, R., Anomalous fluxes in the plateau regime for a weakly turbulent, magnetically confined plasma, *Phys. Fluids B, 2,* 2100, 1990.

Biskamp, D., R. Chodura, and C. T. Dum, Ion-sound spectrum and wave-electron interaction in perpendicular shocks, *Phys. Rev. Lett, 34,* 131-134, 1975.

Bohm, D., E. Burhop, and H. S. W. Massey, The use of probes for plasma exploration in a strong magnetic field, in *The Characteristics of Electrical Discharges in Magnetic Fields,* edited by A. Guthrie and R. K. Wakerling, pp. 13-76, Mc Graw Hill, New York, 1949.

Chen, F. F., *Introduction to Plasma Physics and Controlled Fusion,* Plenum, New York, 1984.

Connor, J. W., and H. R. Hastie, Survey of theories of anomalous transport, *Plasma Phys. and Controlled Fusion, 36,* 719-795, 1994.

Dum, C. T., Anomalous heating by ion sound turbulence, *Phys. Fluids, 21,* 945-955, 1978a.

Dum, C. T., Anomalous electron transport equations for ion sound and related turbulent spectra, *Phys. Fluids, 21,* 956-969, 1978b.

Dum, C. T., Electrostatic waves and anomalous transport in the solar wind, in *Proc. Solar Wind V, NASA Conf. Publ.,* vol. 2280, edited by M. Neugebauer, pp. 369-376, NASA, Washington, D. C., 1983.

Dum, C. T., Coupling of macroscopic and small scale phenomena, *Space Science Rev., 42,* 467-484, 1985.

Dum, C. T., Classical transport properties of plasmas, in *Physical Processes in Hot Cosmic Plasmas,* edited by W. Brinkmann, A. C. Fabian and F. Giovanelli, pp. 157-180, Kluwer Academic Publishers, Dordrecht, 1990.

Dum, C. T., R. Chodura, and D. Biskamp, Turbulent heating and quenching of the ion sound instability, *Phys. Rev. Lett, 32,* 1231-1234, 1974.

Dum, C. T., and T. H. Dupree, Nonlinear stabilization of high-frequency instabilities in a magnetic field, *Phys. Fluids, 13,* 2064, 1970.

Dupree, T. H., Nonlinear theory of drift wave turbulence and enhanced diffusion, *Phys. Fluids, 10,* 1049-1055, 1967.

Galeev, A. A., and L. I. Rudakov, Nonlinear theory of drift instability of an inhomogeneous plasma in a magnetic field, *Zh. Eksp. Teor. Fiz.,* Engl. Transl., *18,* 444-449, 1964.

Galeev, A. A., and R. Z. Sagdeev, Nonlinear Plasma Theory, in *Rev. of Plasma Physics,* vol. 7, edited by M. A. Leontovich, p. 1, Consultants Bureau, New York, 1979.

Gary, S. P., B. J. Anderson, R. E. Denton, S. A. Fuselier, and M. E. McKean, A limited closure relation for anisotropic plasmas from the Earth's magnetosheath, *Phys. Fluids, 1,* 1676-1683, 1994.

Horton, W., Drift wave turbulence and anomalous transport, in *Handbook of Plasma Physics, Basic Plasma Physics II,* vol. 2, edited by A. A. Galeev and R. N. Sudan, pp. 383-449, Elsevier, New York, 1984.

Itoh, S-I., Anomalous viscosity due to drift wave turbulence, *Phys. Fluids B, 4,* 796-803, 1992.

Kadomtsev, B. B., *Plasma Turbulence,* Academic, New York, 1965.

LaBelle, J., R. A. Treumann, Plasma waves at the dayside magnetopause, *Space Sci. Revs., 47,* 175-202, 1988.

Liewer, P. C., Measurements of microturbulence in tokamaks and comparison with theories of turbulence and anomalous transport, *Nuclear Fusion, 25,* 543-621, 1985.

Lysak, R. L., and C. T. Dum, Dynamics of magnetosphere-ionosphere coupling including turbulent transport, *J. Geophys. Res., 88,* 365-380, 1983.

Machalek, M. D., and P. Nielsen, Light-scattering measurements of turbulence in a normal shock, *Phys. Rev. Lett, 31,* 439-442, 1973.

Manheimer, W. M., and T. M. Antonsen, A theory of electron confinement in tokamaks, *Phys. Fluids, 22,* 957-970, 1979.

Manheimer, W. M., and J. Boris, Selfconsistent theory of a collisionless resistive shock, *Phys. Rev. Lett, 28,* 659-662, 1972.

Muraoka, K., S. Nakai, D. D. R. Summers, and J. W. M. Paul, Anisotropy of turbulence in a collisionless shock, *J. Plasma Phys., 10,* 135-140, 1973.

Muschietti, L., and C. T. Dum, Current driven ion cyclotron turbulence-Evolution of the electron distribution function and wave spectrum, *J. Geophys. Res., 95,* 173-185, 1990.

Ross, D. W., On standard forms of transport equations and fluxes, *Comments Plasma Phys. Controlled Fusion, 12,* 155-162, 1989.

Shaing, K. C., Neoclassical quasilinear transport theory of fluctuations in toroidal plasmas, *Phys. Fluids, 31,* 2249-2265, 1988.

Söldner, F., C. T. Dum, and K.-H. Steuer, Electron and ion heating in a high-voltage belt pinch, *Phys. Rev. Lett, 39,* 194-197, 1977.

Thorne, R. M., and B. T. Tsurutani, Wave particle interactions in the magnetopause boundary layer, in *Physics of Space Plasmas, SPI Conf. Proc. Reprint Series,* vol. 10, edited by T. Chang, G. B. Crew and J. R. Jasperse, pp. 119-150, Scientific Publishers, Cambridge, MA., 1991.

Treumann, R. A., J. LaBelle, and R. Pottelette, Plasma diffusion at the magnetopause: the case of lower hybrid drift waves, J. Geophys. Res., 96, 16009-16013, 1991.

Treumann, R. A., J. LaBelle, G. Haerendel, and R. Pottelette, Anomalous plasma diffusion and the magnetopause boundary layer, *IEEE Trans. Plasma Sci., 20,* 833-842, 1992.

Tsurutani, B. T., and R. M. Thorne, Diffusion processes in the magnetopause boundary layer, *Geophys. Res. Lett., 9,* 1247, 1982.

C. T. Dum, Max Planck Institut für Extraterrestrische Physik, P. O. Box 1603, D85740 Garching, Germany.

An Upper Bound for the Proton Temperature Anisotropy

S. Peter Gary

Los Alamos National Laboratory, Los Alamos, New Mexico

This tutorial describes recent research concerning the upper bound on the hot proton temperature anisotropy imposed by wave-particle scattering due to enhanced fluctuations from the electromagnetic proton cyclotron anisotropy instability. This upper bound, which has been observed in both the magnetosheath and the outer magnetosphere, represents a limited closure relation for the equations of anisotropic magnetohydrodynamics. Such a closure relation has the potential to improve the predictive capability of large-scale anisotropic models of the magnetosphere.

1. INTRODUCTION

Two questions addressed by tutorials at this Workshop are: (1) Where, when, and how are anomalous plasma effects relevant to space plasma transport? and (2) Can we develop useful expressions/algorithms for the anomalous transport coefficients?

First, for this tutorial I would like to redefine question (1). Although "anomalous" is usually understood to be equivalent to "non-collisional", in my opinion the term has a rather negative implication, suggesting lack of understanding of the underlying processes. On the contrary, we understand, at least in a qualitative sense, that microinstability growth often leads to plasma transport such as pitch-angle scattering or beam degradation. Therefore, when instabilities are the likely source of transport, it is more descriptive to use the term "wave-particle" in describing such transport. Because I am concerned only with instability-driven transport, I will follow such usage here.

The framework for wave-particle transport is a familiar part of collisionless plasma physics. A non-Maxwellian property, and in particular one which corresponds to an anisotropy, of the jth species velocity-space distribution function $f_j(\mathbf{v})$ represents potential instability growth or, in the jargon, "free energy". Examples of free energy include component/component relative drifts, a temperature anisotropy of a species, or a density gradient (which, in steady state, implies an anisotropic velocity distribution). If the free energy is sufficiently large, one or more associated microinstabilities will grow [*Gary*, 1993]; the consequence of such growth is enhanced magnetic and/or electric field fluctuations. These enhanced fluctuations, in turn, imply scattering of the particles and consequent wave-particle transport.

As Chris Dum emphasized in his tutorial, there is "no useful general theory of anomalous transport" in collisionless plasmas. A plasma can, and often does, bear several different kinds of free energy at once, thereby implying the potential to excite an even greater number of instabilities, each of which requires a different theoretical formulation to describe in a useful way the resulting transport. There are a few qualitative guidelines that can help sort out the physics; for example, the hypothesis [*Gary*, 1980] that the primary effect of wave-particle interactions due to a particular microinstability is to reduce the source of free energy driving that instability.

If this hypothesis is valid, then a useful procedure for the theoretical study of wave-particle transport would be to address a single source of free energy and a single, most likely, associated instability, and to determine how that instability acts to reduce that free energy. The resulting theory, although limited, may then become the basis for determination of secondary consequences as well as the incorporation of further kinds of free energy. For example, the primary consequence of the wave-particle scattering by fluctuations driven by the hot proton anisotropy instability discussed below is the

Cross-Scale Coupling in Space Plasmas
Geophysical Monograph 93
Copyright 1995 by the American Geophysical Union

reduction of that anisotropy. These fluctuations also yield heating of cool proton and heavy ion components, but this is a secondary effect which should addressed only after the primary consequence is well understood.

So my (incomplete) answer to question (1) is: Wave-particle transport may arise whenever $f_j(\mathbf{v})$ is sufficiently anisotropic and the resulting enhanced field fluctuations are sufficiently large. Detailed theoretical descriptions will be different for each source of free energy and for each associated instability, and the question of relevance must be addressed separately in each case.

If wave-particle transport coefficients are considered to be close analogues of the well-known collisional transport coefficients, then my answer to question (2) is negative. Although theoretical expressions for such analogues abound in the literature [*Gary*, 1980, and references therein], I do not believe that any of these expressions have demonstrated quantitative agreement with observations or have been widely accepted for use by the magnetospheric modeling community. This may be because global simulations have not attained the spatial resolution at which wave-particle effects become discernable. On the other hand, there may be a basic problem with the way wave-particle transport has been formulated. *Coroniti* [1985] has examined several magnetospheric applications of wave-particle resistivity and has concluded that it "fails to provide an adequate or complete description of the phenomenae" for reconnection and auroral field-aligned currents.

Why? One answer may be that the enhanced fluctuations, which constitute a fundamental factor in most theoretical expressions for wave-particle transport, are typically observed by spacecraft to be localized and/or sporadic. Particularly well-documented examples are the observations [*LaBelle et al.*, 1986] of patchy, bursty enhanced lower-hybrid-like enhanced electric field fluctuations along low altitude auroral magnetic field lines. *Vago et al.* [1992] further showed a one-to-one correspondence between these fluctuations and perpendicularly heated ions, and current theories [*Retterer et al.*, 1994, and references therein] have been successful in describing the strongly local, strongly time dependent character of the phenomenon. Given this insight, it is not surprising that efforts to cast a fundamentally small-scale, fast-time process in terms of a formalism intended to describe the slowly varying collisional properties of a plasma should fail.

So my answer to (2) is: Probably not. Wave-particle effects are observed to be bursty and inhomogeneous, and are not well represented by slowly varying transport expressions such as resistivity.

An alternate method for representing consequences of wave-particle transport in large-scale computer models may be in terms of free energy bounds. The concept is simple, and is based on the *Kennel and Petschek* [1966] idea of a instability-induced energetic particle flux limit.

The application of this idea to plasma properties and fluid codes was suggested also some time ago by *Manheimer and Boris* [1977]. If the free energy in question exceeds the threshold of the instability in question, that mode will grow, the associated field fluctuations will become enhanced, and wave-particle scattering will reduce that free energy to a threshold value. This happens on a time scale fast compared to any macroscopic changes, so that the local consequences of wave-particle scattering may be represented in a fluid code as a local bound on the macroscopic variable which corresponds to the free energy. If the free energy does not exceed the instability threshold there are no wave-particle consequences and the corresponding macroscopic variable follows its temporal evolution predicted by the conventional fluid model. The application of this upper bound to a large-scale model represents a limited closure relation for the equations of that anisotropic model, and may affect the predictive capability of such a model.

This upper bound procedure is not necessarily applicable to all free energies or to all macroscopic models. For example, suppose the free energy is a field-aligned current $\mathbf{J}_\parallel$ sustained by a field-aligned electric field $\mathbf{E}_\parallel$. Further suppose that a sudden fluctuation triggers an instability which limits $\mathbf{J}_\parallel$ in some small region of the plasma. If this current is constrained to flow along the background magnetic field $\mathbf{B}_o$, this limit on the parallel current must propagate along that field line and a local effect would soon have broad nonlocal implications. If the macroscopic model in question cannot address these nonlocal effects properly, the application of the upper bound will fail. On the other hand, if the free energy is sustained by electromagnetic forces acting across $\mathbf{B}_o$, such as magnetic compression or $\mathbf{E} \times \mathbf{B}_o$ convection, the consequences of wave-particle activity need not be communicated along the background field, and an upper bound on the anisotropy may be imposed locally in a large-scale modeling code without the complicating factor of nonlocal consequences. A second reason why an upper bound condition is usually simpler to implement than conventional wave-particle transport expressions is that the former is based on linear theory and does not require an estimate of the fluctuating field energy.

This tutorial describes a detailed application of this method to anisotropic space plasmas. We have used linear theory and computer simulations to examine the threshold condition of the electromagnetic proton cyclotron anisotropy instability (hereafter the "proton cyclotron instability"), and have compared it against observations in the magnetosheath and in the outer magnetosphere. In another tutorial from this Workshop, Joachim Birn describes how he has included the anisotropy upper bound in an MHD model of the magnetotail, and how it affects the resulting dynamics of the system.

The notation here is the same as that used by *Gary*

et al. [1994c]. In particular, $\beta_{\parallel j} \equiv 8\pi n_j T_j/B_o^2$ and γ_m denotes the maximum growth rate of an instability.

2. PROTON ANISOTROPY BOUND: MAGNETOSHEATH

The proton cyclotron instability is driven by the proton temperature anisotropy $T_{\perp p}/T_{\parallel p} > 1$ on the Alfvén/proton cyclotron branch of the linear Vlasov dispersion relation. The maximum growth rate γ_m typically arises at $kc/\omega_p \simeq 0.5$, where ω_p is the proton plasma frequency, and at propagation parallel or antiparallel to the background magnetic field, i.e., at $\mathbf{k} \times \mathbf{B}_o = 0$ [see, for example, *Gary*, 1993]. The associated enhanced fluctuations have two robust characteristics: they are predominantly magnetic and they are confined to $\omega_r < \Omega_p$ where Ω_p is the proton cyclotron frequency. I term these "proton-cyclotron-like" fluctuations.

If the protons can be described with a single, bi-Maxwellian distribution, linear theory predicts that the threshold condition for this instability can be written in the form

$$\frac{T_{\perp p}}{T_{\parallel p}} - 1 = \frac{S_p}{\beta_{\parallel p}^{\alpha_p}} \quad (1)$$

where S_p is a function of γ_m/Ω_p, but $\alpha_p \simeq 0.42$ for a broad range of maximum growth rates [*Gary et al.*, 1994a; *Gary and Lee*, 1994]. One-dimensional hybrid simulations of the proton cyclotron instability show ensembles of computations with late-time conditions which also satisfy Equation (1) with $\alpha_p \simeq 0.50$ [*Gary et al.*, 1994a], demonstrating that this condition corresponds to an upper bound on the temperature anisotropy and that the value of α_p is not greatly sensitive to the detailed mechanism of instability excitation.

In the terrestrial magnetosheath well downstream of the bow shock, the proton distribution function approximately corresponds to a single bi-Maxwellian. Observations in this region have shown not only the frequent presence of enhanced proton-cyclotron-like fluctuations [*Anderson and Fuselier*, 1993, and references therein], but also that, under highly compressed conditions, the proton temperature anisotropy well satisfies Equation (1) with $S_p = 0.85$ and $\alpha_p = 0.48$ [*Anderson et al.*, 1994]. Although the numerical values of the coefficients on the right-hand side of (1) vary with the choice of maximum linear growth rate at threshold and on the presence of other ionic species [e.g., *Gary et al.*, 1994b], these variations are relatively weak for magnetosheath parameters so that several different sheath observations have yielded similar upper bounds [*Hau et al.*, 1993; *Phan et al.*, 1994; *Fuselier et al.*, 1994]. These results show that the upper bound on $T_{\perp p}/T_{\parallel p}$ imposed by the proton cyclotron instability is a well established feature of the terrestrial magnetosheath. This bound has been used as a limited closure relation in a fluid model of proton temperatures in the magnetosheath and has successfully described the separate evolution of $T_{\perp p}$ and $T_{\parallel p}$ in a magnetosheath crossing [*Denton et al.*, 1994].

3. PROTON ANISOTROPY BOUND: OUTER MAGNETOSPHERE

The representation of the proton distribution function as a single bi-Maxwellian is not an appropriate approximation for many magnetospheric plasmas. Observations from Los Alamos magnetospheric plasma analyzers (MPAs) on geosynchronous satellites [*Bame et al.*, 1993] confirm many earlier observations that outer magnetospheric ions typically consist of two components [*McComas et al.*, 1993]. The hot component (which we denote by subscript h) typically is relatively tenuous ($n_h \lesssim 1$ cm^{-3}), has a temperature of several keV, and usually bears an anisotropy such that $T_{\perp h} > T_{\parallel h}$ [*Mauk and McPherron*, 1980; *Anderson and Hamilton*, 1993]. The cool ion component (subscript c) is observed to be much cooler (1 eV $\lesssim T_c \lesssim$ 10 eV) with a much wider range of densities (0.1 cm$^{-3} < n_c \lesssim$ 100 cm^{-3}) [*Reasoner et al.*, 1983; *Sojka and Wrenn*, 1985; *Moldwin et al.*, 1994]. The cool ions are generally regarded as having originated in the ionosphere where they are presumably created at still colder temperatures (of order 0.1 eV).

The presence of a cool proton component does not inhibit the growth of the proton cyclotron instability driven by the hot anisotropic ions; rather, under many circumstances a nonzero n_c actually enhances the maximum growth rate [*Cornwall and Schulz*, 1971] and lowers the threshold anisotropy of this growing mode. Thus proton-cyclotron-like fluctuations are observed throughout the magnetosphere, that is, within the plasmasphere [for example, *Kintner and Gurnett*, 1977], at geosynchronous orbit [e.g., *Young et al.*, 1981; *Fraser*, 1985], and in the outer magnetosphere [*Anderson et al.*, 1992a, and references therein]. Observations of enhanced low-frequency electric fluctuations have been used to infer the presence of proton-cyclotron-like magnetic fluctuations on auroral field lines as well [*Crew et al.*, 1990].

My co-workers and I have studied the growth of the proton cyclotron instability and the subsequent wave-particle interactions in the context of a relatively simple model which assumes a spatially homogeneous plasma, electromagnetic fluctuations propagating at $\mathbf{k} \times \mathbf{B}_o = 0$, and two bi-Maxwellian proton components representing the hot and cool ion components of the outer magnetosphere. Using linear theory and computer simulations, we have derived an upper bound for the hot proton temperature anisotropy which correlates well with Los

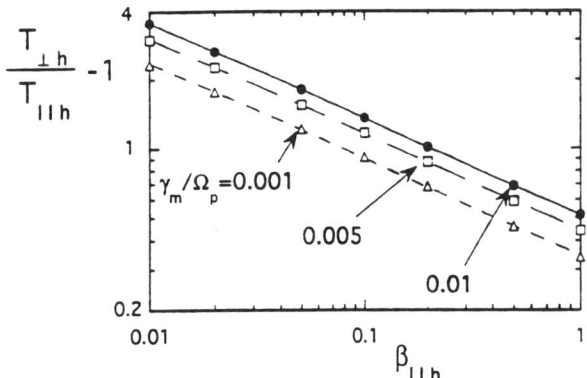

Fig. 1. The hot proton temperature anisotropy at three different values of the maximum growth rate of the proton cyclotron instability determined from linear Vlasov theory as a function of the parallel hot proton β. Values of the dimensionless plasma parameters are as indicated in Table 1 of *Gary et al.* [1994c]; here we choose $n_h/n_e = 0.50$. The individual points are computed from the linear dispersion equation; the corresponding lines are least-squares fits to the points and satisfy Equation (2) with $S'_h = 0.51$ and $\alpha_h = 0.42$ at $\gamma_m/\Omega_p = 0.01$, $S'_h = 0.44$ and $\alpha_h = 0.42$ at $\gamma_m/\Omega_p = 0.005$, and $S'_h = 0.34$ and $\alpha_h = 0.42$ at $\gamma_m/\Omega_p = 0.001$.

Alamos observations at geosynchronous orbit.

An instability threshold is determined not only by the values of the dimensionless variables describing the plasma, but also by the choice of maximum growth rate. Solutions of the linear Vlasov dispersion equation for the two-proton-component model [*Gary et al.*, 1994c] yield representative results for three different values of γ_m/Ω_p as illustrated in Figure 1. We use a least squares fitting procedure to obtain threshold expressions of the form

$$\frac{T_{\perp h}}{T_{\parallel h}} - 1 = \frac{S'_h}{\beta_{\parallel h}^{\alpha_h}} \quad (2)$$

where S'_h and α_h for each case are given in the figure caption. In this model, the threshold anisotropy is also a function of n_h/n_e [*Gary et al.*, 1994c]. We will not address the details of this functionality here, but only note that α_h is relatively insensitive to variations in the relative hot proton density and that the variations of S'_h with this parameter must be carefully considered when comparing theory against observations.

The next step in our research was to use this same two-component proton model in a fully nonlinear, self-consistent computer simulation of the proton cyclotron instability. Our computations used the one-dimensional hybrid code of *Winske and Omidi* [1993] under the assumptions of a homogeneous, periodic system and $\mathbf{k} \times \mathbf{B}_o = 0$. The parameters of the simulations described here are essentially the same as those given in *Gary et al.* [1994c].

Figure 2 shows results from a representative simulation which are similar to earlier such computations with two or more ionic components [*Cuperman and Sternlieb*, 1977; *Omura et al.*, 1985; *Tanaka*, 1985]. The anisotropic hot protons excite the proton cyclotron instability, so that the fluctuating magnetic field grows in time, eventually reaching a maximum or saturation value at $t = t_{max}$ (Figure 2b). These enhanced fluctuations are well above their background level, which we found empirically to scale roughly as $|\delta B|^2/B_o^2 \simeq$

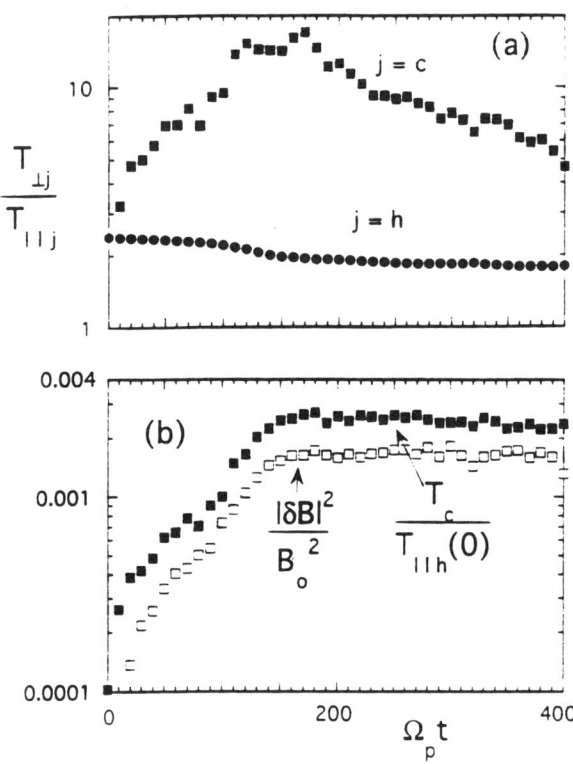

Fig. 2. Results as a function of time from a representative simulation of the proton cyclotron instability. The initial values of the dimensionless parameters are as given in Table 1 of *Gary et al.* [1994c] with $n_h/n_e = 0.10$, $\beta_{\parallel h} = 0.10$, and $T_{\perp h}/T_{\parallel h} = 6.07$, corresponding to an initial $\gamma_m = 0.10\Omega_p$. (a) The solid dots represent the temperature anisotropy of the hot proton component, the solid squares represent the temperature anisotropy of the cool proton component, and the crosses represent the parallel temperature of the hot protons normalized to the initial value of that parameter. (b) The square dots represent the parallel temperature of the cool protons normalized to the initial value of the parallel hot proton temperature, and the open squares represent the total fluctuating magnetic field energy density normalized to the energy density of the background magnetic field [From *Gary et al.*, 1994c].

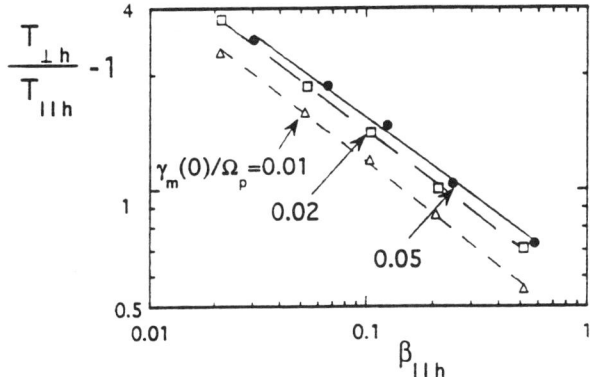

Fig. 3. Late-time values of the hot proton temperature anisotropy from simulations of the proton cyclotron instability as a function of the corresponding late-time $\beta_{\|h}$. Initial parameters are as given in Table 1 of *Gary et al.* [1994c] except that $T_c(0)/T_{\|h}(0) = 10^{-4}$; here $n_h = 0.50 n_e$. The simulations are grouped into three ensembles corresponding to $\gamma_m(0)/\Omega_p = 0.05$ (solid circles), $\gamma_m(0)/\Omega_p = 0.02$ (open squares), and $\gamma_m(0)/\Omega_p = 0.01$ (open triangles). The individual points are results from individual simulations; the corresponding lines are least-squares fits to the points and satisfy Equation (2) with $S'_h = 0.58$ and $\alpha_h = 0.42$ at $\gamma_m(0)/\Omega_p = 0.05$, $S'_h = 0.52$ and $\alpha_h = 0.44$ at $\gamma_m(0)/\Omega_p = 0.02$, and $S'_h = 0.42$ and $\alpha_h = 0.45$ at $\gamma_m(0)/\Omega_p = 0.01$.

$10^{-3}\beta_{\|h}^{1.4}$. The enhanced field fluctuations scatter the hot protons and cause $T_{\perp h}/T_{\|h}$ to become smaller with time (Figure 2a).

Our procedure has been to carry out a large number of simulations, to characterize each simulation in terms of its initial values, and to collect into an ensemble simulations in which only one initial dimensionless parameter changes. In Figure 3 we plot the hot proton temperature anisotropy at late times in each simulation as a function of the corresponding late-time $\beta_{\|h}$. The different symbols correspond to results from different ensembles of simulations characterized by different initial values of the growth rate. In each case the results well satisfy Equation (2), demonstrating once again that the simulations yield a threshold condition very similar to that derived from linear theory, and that this threshold corresponds to an upper bound on the hot proton temperature anisotropy.

Although both linear theory and simulations yield an anisotropy limit of the form of Equation (2) with similar values of α_h, the anisotropy/β relation is not completely specified until the value of S'_h is determined. This factor is clearly a function of γ_m/Ω_p which, in principle, can also be determined from theoretical considerations if the macroscopic forces driving the instability are well understood. Here, however, we choose to determine γ_m/Ω_p and S'_h empirically; that is we match plots of the form of Equation (2) against the data to determine the value of the maximum growth rate which provides the best fit to the observations.

To carry out such an empirical fit, we note that Figure 2 of *Gary et al.* [1994c] indicates that for $0.001 \leq n_h/n_e \leq 0.10$ the hot proton anisotropy at instability threshold is relatively independent of the relative hot proton density near $\gamma_m/\Omega_p \simeq 0.005$. So in Figure 4 we plot hot proton anisotropies from a ten day period as observed by the MPA instrument at geosynchronous orbit, corresponding to this same range of hot proton relative densities. Details of the data analysis and the magnetic field model used in determining $\beta_{\|h}$ are described in *Gary et al.* [1994c]. We also display in Figure 4 two lines corresponding to the predictions of linear theory at $n_h/n_e = 0.01$; the results clearly show that Equation (2) represents an upper bound on the data with most of the points bounded by the S'_h and α_h values corresponding to $\gamma_m/\Omega_p = 0.005$.

Figure 4 does not resemble the magnetosheath observations of *Anderson et al.* [1994] in which the anisotropies cluster near threshold. We interpret those sheath observations as corresponding to conditions under which large scale compressive forces constantly act to increase $T_{\perp p}/T_{\|p}$, continually exciting the instability and steadily maintaining the plasma near threshold. In contrast, the widely distributed points of Figure 4 indicate that the hot proton anisotropy at geosynchronous

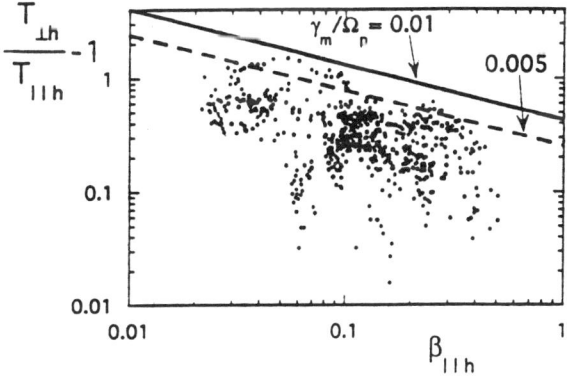

Fig. 4. The hot proton temperature anisotropy as a function of $\beta_{\|h}$ using the geosynchronous orbit data set described in *Gary et al.* [1994c] for $0.001 \leq n_h/n_e \leq 0.10$. The individual points correspond to data from January 21-31, 1994. The two lines represent solutions to the linear Vlasov dispersion equation with $n_h/n_e = 0.01$ at thresholds of the proton cyclotron instability. Both lines satisfy Equation (2) with $\alpha_h = 0.49$. The solid line corresponds to $S'_h = 0.42$, from the threshold condition $\gamma_m/\Omega_p = 0.01$, whereas the dashed line corresponds to $S'_h = 0.25$, from the threshold condition $\gamma_m/\Omega_p = 0.005$.

orbit is driven only sporadically, a conclusion in agreement with *Anderson et al.* [1992b] who observed enhanced proton-cyclotron-like fluctuations relatively infrequently near geosynchronous orbit.

4. COOL PROTON ENERGY GAIN: OUTER MAGNETOSPHERE

If proton-cyclotron-like enhanced fluctuations are often observed in the outer magnetosphere, and if these fluctuations impose an upper bound on the the hot proton anisotropy, they may also contribute to the heating of the cool ion components. Observations typically indicate that the cool helium and cool oxygen ion components gain more energy from wave-particle interactions than do the cool protons. The heating of the heavy ions is usually attributed to resonant wave-particle interactions at relatively low altitudes which often requires a nonlocal model for a theoretical analysis [e.g., *Roux et al.*, 1982; *Thorne and Horne*, 1994, and references therein]. For example, Tom Chang's tutorial at this Workshop describes the role of proton-cyclotron-like fluctuations in the heating of oxygen ions in the central plasma sheet region of the auroral zone.

However, observations near the magnetic equator in the outer magnetosphere indicate that transverse proton energization, although nonresonant and therefore relatively weak, is due to comparatively local wave-particle interactions [*Anderson and Fuselier*, 1994]. This result suggests that the temperature of the cool protons may be described in terms of local plasma parameters. Linear theory of the proton cyclotron instability is insensitive to the temperatures and anisotropies of the cool ion components if those temperatures are sufficiently small compared to $T_{\|h}$. Thus dispersion theory provides no information about T_c, and we have pursued this question using the same one-dimensional hybrid simulations we used to study the hot proton temperature anisotropy.

Figure 2 shows that the proton cyclotron instability not only heats cool protons, but imparts this energy in the directions perpendicular to $\mathbf{B}_o$ during the linear growth phase of the instability. This increse in perpendicular energy is due to a nonresonant wave-particle interaction and is therefore reversible, so we refer to T_c as an "apparent temperature". However, Figure 2a shows that the enhanced fluctuations yield pitch-angle scattering at later times, implying that, if the interaction time is sufficiently long, the process becomes partly irreversible and suggesting that it may play a role in plasmaspheric refilling [*Singh and Horwitz*, 1992, and references therein].

Observations from the MPA instrument at geosynchronous orbit indicate that n_h and $T_{\|h}$ show much less variation than the cool proton density and temper-

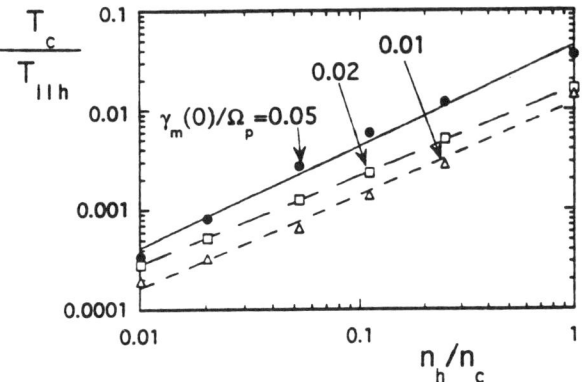

Fig. 5. The dimensionless cool proton temperature at t_{max} as a function of the hot/cool proton relative density from simulations of the proton cyclotron instability. Initial parameters are as given in Table 1 of *Gary et al.* [1994c] except that $T_c(0)/T_{\|h}(0) = 10^{-4}$; here $\beta_{\|h}(0) = 0.10$. The simulations are grouped into three ensembles corresponding to $\gamma_m(0)/\Omega_p = 0.05$ (solid circles), $\gamma_m(0)/\Omega_p = 0.02$ (open squares), and $\gamma_m(0)/\Omega_p = 0.01$ (open triangles). The individual points are results from individual simulations; the corresponding lines are least-squares fits to the points and satisfy Equation (3) with $S'_c = 0.044$ and $M_c = 1.01$ at $\gamma_m(0)/\Omega_p = 0.05$, $S'_c = 0.017$ and $M_c = 0.89$ at $\gamma_m(0)/\Omega_p = 0.02$, and $S'_c = 0.012$ and $M_c = 0.93$ at $\gamma_m(0)/\Omega_p = 0.01$ [After *Gary et al.*, 1995].

ature, so it is appropriate to use the former quantities to normalize the latter variables in our simulations. Thus in Figure 5 we show $T_c/T_{\|h}$ at t_{max} as a function of n_h/n_c for three simulation ensembles corresponding to three different values of $\gamma_m(0)/\Omega_p$. We find that the results are well fit by expressions of the form

$$\frac{T_c}{T_{\|h}} = S'_c \left(\frac{n_h}{n_c}\right)^{M_c} \quad (3)$$

where the corresponding values of S'_c and M_c are stated in the figure caption. If we assume that S'_c is relatively independent of $\beta_{\|h}$ and that $\gamma_m(0)/\Omega_p = 0.01$ provides a good correspondence between the simulation results and the observations of Figure 4, we obtain

$$\frac{T_c}{T_{\|h}} \simeq 0.010 \left(\frac{n_h}{n_c}\right)^{0.9} \quad (0.01 \leq n_h/n_c \leq 1.00) \quad (4)$$

If we assume $n_h/n_c = 0.01$, Equation (4) implies $T_c/T_{\|h} = 1.7 \times 10^{-4}$; using $T_{\|h} \simeq 6$ keV, we have $T_c \simeq 1$ eV, a result commensurate with observations in the outer plasmasphere [*Comfort et al.*, 1985] and at geosynchronous orbit with magnetospheric conditions when $n_h \ll n_c$ [*Moldwin et al.*, 1994]. In contrast, $n_h/n_c = 1.0$ implies $T_c \simeq 50$ eV, which is somewhat higher than the maximum value of cool proton temperatures usually observed in the outer magnetosphere [*Moldwin et al.*, 1994; *Anderson and Fuselier*, 1994]. Therefore, although Equation (4) predicts the same type of inverse correlation between n_c and T_c that is sometimes observed in the magnetosphere [*Comfort et al.*, 1985; *Sojka and Wrenn*, 1985; *Moldwin et al.*, 1994; *Gary et al.*, 1994c], it is clear that further studies are necessary to determine whether this equation represents a useful, quantitative description of the cool proton apparent temperature.

5. CONCLUSIONS

Equation (2) of the Introduction can be rephrased as follows: How can the consequences of wave-particle transport be best represented in large-scale computer models of the magnetosphere? I have argued that, for some free energies and some instabilities, this may not be as an analogue of the usual collisional transport coefficients. Rather, expressions for free energies at instability thresholds can provide upper bounds for some macroscopic variables which, in turn, may be implemented as limited closure relations in large-scale fluid models of space plasmas. These upper bounds have two important properties: they do not depend on the fluctuating field amplitudes, and they approximately represent the local, sporadic nature of wave-particle transport. These two properties imply that these bounds are simpler to implement than the usual wave-particle transport coefficients, and that they may be a more appropriate way of representing the inherently small-scale, fast-time aspects of plasma microphysics.

In this tutorial I have discussed the example of the proton cyclotron instability which theory and computations predict imposes an upper bound on the hot proton temperature anisotropy and which, in turn, shows good agreement with observations in the magnetosheath and at geosynchronous orbit. Similarly, with other coworkers I have shown that the whistler heat flux instability implies a generous, but properly scaled, upper bound for the electron heat flux as measured by the Ulysses spacecraft in the solar wind of the ecliptic plane [*Gary et al.*, 1994d].

But the theoretical and computational procedures which lead to upper bound expressions may not produce a useful result for every free energy or for every microinstability. For example, if more than one free energy is present, or if a particular type of free energy is not easily represented in terms of macroscopic parameters which are used in large-scale fluid models, a theoretical upper bound may not be readily applied to a macroscopic computation. Another case problematic case is that of instabilities driven by currents parallel to $\mathbf{B}_o$; the fundamentally local character of our formulation of the upper bound procedure may not provide useful results for this inherently nonlocal problem.

Another impediment to the widespread application of microscopically-derived upper bounds is that macroscopic computer models may not be ready to implement such bounds. For example, a limit on the proton temperature anisotropy is of no use in an isotropic MHD code which imposes $T_\perp = T_\parallel$ *ab initio*. I contend that the assumption of isotropy is not necessarily valid for representing the full range of large-scale magnetospheric phenomena, and that it would be useful to compare the predictions of an isotropic MHD code against the results of an anisotropic MHD code in which our anisotropy upper bound provides a limited closure relation for the $T_\perp$ equation [e.g., *Denton et al.*, 1994]. Comparison of the two results against each other and against observations might yield insight into when or if such a limited closure relation is important in the magnetosphere. Joachim Birn describes in another tutorial at this Workshop his first such comparisons through the use of his magnetotail model.

In summary I believe that the use of microscale theory and simulations to obtain free energy upper bounds from instability thresholds holds promise, in at least some cases, for providing limited closure relations for anisotropic MHD and other fluid models of macroscale plasma physics. Much research remains to be done to extend this procedure to growing modes beyond the present example of the proton cyclotron instability, and to establish whether the resulting limited closure relations can significantly improve the predictive capability of large-scale models of the magnetosphere. Should such a research program prove successful for the terrestrial magnetosphere, then it is likely that such upper bounds may also be useful in the description of other collisionless plasma systems, including the magnetospheres of other planets as well as the plasmas associated with astrophysical objects.

Acknowledgments. The research summarized here was carried out in collaboration with several coworkers. I especially thank Brian Anderson, Richard Denton, Steve Fuselier, Dave McComas, Michelle Thomsen, and Dan Winske for numerous discussions and substantial support. This work was performed under the auspices of the U.S. Department of Energy (DOE) and was supported by the DOE Office of Basic Energy Sciences, Division of Engineering and Geosciences, the SR&T Program of the National Aeronautics and Space Administration (NASA), and the Laboratory-Directed Research and Development Program of Los Alamos National Laboratory.

REFERENCES

Anderson, B. J., and S. A. Fuselier, Magnetic pulsations from 0.1 to 4.0 Hz and associated plasma properties in the Earth's subsolar magnetosheath and plasma depletion layer, *J. Geophys. Res.*, *98*, 1461, 1993.

Anderson, B. J., and S. A. Fuselier, Response of thermal ions to electromagnetic ion cyclotron waves, *J. Geophys. Res.*, *99*, 19,413, 1994.

Anderson, B. J., and D. C. Hamilton, Electromagnetic ion cyclotron waves stimulated by modest magnetospheric compressions, *J. Geophys. Res.*, *98*, 11,369, 1993.

Anderson, B. J., R. E. Erlandson, and L.J. Zanetti, A statistical study of Pc 1-2 magnetic pulsations in the equatorial magnetosphere, 1. Equatorial occurrence distributions, *J. Geophys. Res.*, *97*, 3075, 1992a.

Anderson, B. J., R. E. Erlandson, and L.J. Zanetti, A statistical study of Pc 1-2 magnetic pulsations in the equatorial magnetosphere, 2. Wave properties, *J. Geophys. Res.*, *97*, 3089, 1992b.

Anderson, B. J., S. A. Fuselier, S. P. Gary and R. E. Denton, Magnetic spectral signatures in the Earth's magnetosheath and plasma depletion layer, *J. Geophys. Res.*, *99*, 5877, 1994.

Bame, S. J., D. J. McComas, M. F. Thomsen, B. L. Barraclough, R. C. Elphic, J. P. Glore, J. T. Gosling, J. C. Chavez, E. P. Evans, and F. J. Wymer, Magnetospheric plasma analyzer for spacecraft with constrained resources, *Rev. Sci. Instrum.*, *64*, 1026, 1993.

Comfort, R. H., J. H. Waite, Jr., and C. R. Chappell, Thermal ion temperatures from the retarding ion mass spectrometer on DE 1, *J. Geophys. Res.*, *90*, 3475, 1985.

Cornwall, J. M., and M. Schulz, Electromagnetic ion-cyclotron instabilities in multicomponent magnetospheric plasmas, *J. Geophys. Res.*, *76*, 7791, 1971; Correction, *J. Geophys. Res.*, *78*, 6830, 1973.

Coroniti, F. V., Space plasma turbulent dissipation: reality or myth?, *Space Sci. Revs.*, *42*, 399, 1985.

Crew, G. B., T. Chang, J. M. Retterer, W. K. Peterson, D. A. Gurnett, and R. L. Huff, Ion cyclotron resonance heated conics: theory and observations, *J. Geophys. Res.*, *95*, 3959, 1990.

Cuperman, S., and A. Sternlieb, Numerical-experimental investigation of the enhancement of the electromagnetic ion cyclotron instability by cold plasma, *J. Geophys. Res.*, *82*, 181, 1977.

Denton, R. L., B. J. Anderson, S. P. Gary, and S. A. Fuselier, Bounded anisotropy fluid model, *J. Geophys. Res.*, *99*, 11,225, 1994.

Fraser, B. J., Observations of ion cyclotron waves near synchronous orbit and on the ground, *Space Sci. Revs.*, *42*, 357, 1985.

Fuselier, S. A., B. J. Anderson, S. P. Gary, and R. E. Denton, Inverse correlations between the ion temperature anisotropy and plasma beta in the Earth's quasi-parallel magnetosheath, *J. Geophys. Res.*, *99*, 14,931, 1994.

Gary, S. P., Wave-particle transport from electrostatic instabilities, *Phys. Fluids*, *23*, 1193, 1980.

Gary, S. P., *Theory of Space Plasma Microinstabilities*, Cambridge University Press, Cambridge, 1993.

Gary, S. P., and M. A. Lee, The ion cyclotron instability and the inverse correlation between proton anisotropy and proton beta, *J. Geophys. Res.*, *99*, 11,297, 1994.

Gary, S. P., M. E. McKean, D. Winske, B. J. Anderson, R. E. Denton, and S. A. Fuselier, The proton cyclotron instability and the anisotropy/β inverse correlation, *J. Geophys. Res.*, *99*, 5903, 1994a.

Gary, S. P., P. D. Convery, R. E. Denton, S. A. Fuselier, and B. J. Anderson, Proton and helium cyclotron anisotropy instability thresholds in the magnetosheath, *J. Geophys. Res.*, *99*, 5915, 1994b.

Gary, S. P., M. B. Moldwin, M. F. Thomsen, D. Winske, and D. J. McComas, Hot proton anisotropies and cool proton temperatures in the outer magnetosphere, *J. Geophys. Res.*, *99*, 23,603, 1994c.

Gary, S. P., E. E. Scime, J. L. Phillips, and W. C. Feldman, The whistler heat flux instability: Threshold conditions in the solar wind, *J. Geophys. Res.*, *99*, 23,391, 1994d.

Gary, S. P., M. F. Thomsen, L. Yin, and D. Winske, Electromagnetic proton cyclotron instability: interactions with magnetospheric protons, *J. Geophys. Res.*, submitted, 1995.

Hau, L.-N., T.-D. Phan, B. U. Ö. Sonnerup, and G. Paschmann, Double-polytropic closure in the magnetosheath, *Geophys. Res. Lett.*, *20*, 2255, 1993.

Kennel, C. F., and H. E. Petschek, Limit on stably trapped particle fluxes, *J. Geophys. Res.*, *71*, 1, 1966.

Kintner, P. M., and D. A. Gurnett, Observations of ion cyclotron waves within the plasmaphere by Hawkeye 1, *J. Geophys. Res.*, *82*, 2314, 1977.

LaBelle, J., P. M. Kintner, A. W. Yau, and B. A. Whalen, Large amplitude wave packets observed in the ionosphere in association with transverse ion acceleration, *J. Geophys. Res.*, *91*, 7113, 1986.

Manheimer, W., and J. P. Boris, Marginal stability analysis–A simpler approach to anomalous transport in plasmas, *Comments Plasma Phys. Controlled Fusion*, *3*, 15, 1977.

Mauk, B. H., and R. L. McPherron, An experimental test of the electromagnetic ion cyclotron instability within the earth's magnetosphere, *Phys. Fluids*, *23*, 2111, 1980.

McComas, D. J., S. J. Bame, B. L. Barraclough, J. R. Donart, R. C. Elphic, J. T. Gosling, M. B. Moldwin, K. R. Moore, and M. F. Thomsen, Magnetospheric plasma analyzer: Initial three-spacecraft observations from geosynchronous orbit, *J. Geophys. Res.*, *98*, 13,453, 1993.

Moldwin, M. B., M. F. Thomsen, S. J. Bame, D. J. McComas, and K. R. Moore, The structure and dynamics of the outer plasmasphere: A multiple geosynchronous satellite study, *J. Geophys. Res.*, *99*, 11,475, 1994.

Omura, Y., M. Ashour-Abdalla, R. Gendrin, and K. Quest, Heating of thermal helium in the equatorial magnetosphere: a simulation study, *J. Geophys. Res.*, *90*, 8281, 1985.

Phan, T.-D., G. Paschmann, W. Baumjohann, and N. Sckopke, The magnetosheath region adjacent to the dayside magnetopause: AMPTE/IRM observations, *J. Geophys. Res.*, *99*, 121, 1994.

Reasoner, D. L., P. D. Craven, and C. R. Chappell, Characteristics of low-energy plasma in the plasmasphere and plasma trough, *J. Geophys. Res.*, *88*, 7913, 1983.

Retterer, J. M., T. Chang, and J. R. Jasperse, Transversely accelerated ions in the topside ionosphere, *J. Geophys. Res.*, *99*, 13,189, 1994.

Roux, A., S. Perraut, J. L. Rauch, C. de Villedary, G. Kremser, A. Korth, and D. T. Young, Wave-particle interactions near Ω_{He^+} observed on board GEOS 1 and 2; 2. Generation of ion cyclotron waves and heating of He^+ ions, *J. Geophys. Res.*, *87*, 8174, 1982.

Singh, N., and J. L. Horwitz, Plasmasphere refilling: Recent observations and modeling, *J. Geophys. Res.*, *97*, 1049, 1992.

Sojka, J. J., and G. L. Wrenn, Refilling of geosynchronous flux tubes as observed at the equator by GEOS 2, *J. Geophys. Res.*, *90*, 6379, 1985.

Tanaka, M., Simulations of heavy ion heating by electromagnetic ion cyclotron waves driven by proton temperature anisotropies, *J. Geophys. Res.*, *90*, 6459, 1985.

Thorne, R. M., and R. B. Horne, Energy transfer between energetic ring current H^+ and O^+ by electromagnetic ion cyclotron waves, *J. Geophys. Res.*, *99*, 17,275, 1994.

Vago, J. L., P. M. Kintner, S. W. Chesney, R. L. Arnoldy, K. A. Lynch, T. E. Moore, and C. J. Pollock, Transverse ion acceleration by localized lower hybrid waves in the topside auroral ionosphere, *J. Geophys. Res.*, *97*, 16,935, 1992.

Winske, D., and N. Omidi, Hybrid codes: Methods and applications, in *Computer Space Plasma Physics: Simulation Techniques and Software*, edited by H. Matsumoto and Y. Omura, p. 103, Terra Scientific, Tokyo, 1993.

Young, D. T., S. Perraut, A. Roux, C. de Villedary, R. Gendrin, A. Korth, G. Kremser, and D. Jones, Wave-particle interactions near Ω_{He^+} observed on GEOS 1 and 2, 1. Propagation of ion cyclotron waves in He^+-rich plasma, *J. Geophys. Res.*, *86*, 6755, 1981.

S. P. Gary, M. S. D466, Los Alamos National Laboratory, Los Alamos, NM 87545 (Internet: pgary@lanl.gov)

Interrelationship of Local and Global Physics in the Low Altitude Ionosphere

G. Ganguli

Plasma Physics Division, Naval Research Laboratory, Washington, D.C.

Space plasma physics is characterized by temporal and spatial scale-sizes which are widely disparate. Many physical phenomena in the space plasma environment are highly dependent on the interrelationship of processes taking place on these vastly different scales. During evolution the macro-scale parameters such as currents, gradients in density, velocity, etc., can often exceed threshold values locally and may seed micro-scale phenomena such as waves and instabilities. The instabilities, in turn, can affect the macro-scale parameters via anomalous transport. This interrelationship offers a considerable challenge in analyzing and modelling space plasma phenomena. Ideally, it is desirable to model the entire spectrum of relevant scale sizes self-consistently. This however, is practically impossible since the local (micro-scale) processes are usually tied to ion or electron cyclotron frequencies and cyclotron radii whereas the global (macro-scale) processes may take place on several minutes or longer timescales and may be spread over several Earth radii. Therefore, the effects of local physics should be incorporated into global scale models via parameterizations. As a specific example we discuss recent observations which indicate low altitude (below 2000 kms) ion energization which is co-located with the global convection flow reversal region. In this region, the convection velocity V is generally small but spatial gradients in V can be large. As a result, Joule heating is small but effects due to dV/dx can be significant. We show that the observed magnitudes of velocity shear in the convection flow are sufficient to excite meso-scale low frequency waves, such as the Kelvin-Helmholtz mode. The Kelvin-Helmholtz modes can nonlinearly steepen and seed plasma waves in the range of ion cyclotron to lower hybrid frequencies which are a potential source for ion heating.

1. INTRODUCTION

The study of the interaction of the solar wind with Earth's magnetic field, the transfer of mass, energy, and momentum from the solar wind to the magnetosphere, their fate in the magnetosphere, and their ultimate dissipation in the ionosphere are the central theme of the space plasma research. This chain of events is highly complex and depends on an interplay between physical processes occurring at vastly disparate spatial and temporal scales. Higher resolution of modern in-situ probes is providing direct evidence of this interplay. At the largest scale the solar wind-magnetoplasma interaction is described by the MHD formalism [Siscoe, 1982; Hill, 1982]. This interaction with Earth's magnetic field creates a boundary layer, the magnetopause, whose scale size can be on the order of a few ion gyroradii (ρ_i) or smaller [Song, et al., 1990; Lee, 1991]. Owing to the short scale-size ($\sim O(\rho_i)$), the dynamics in the boundary layers must be resolved in greater detail and is best described by a kinetic formalism. Intense wave signatures have been observed in the magnetopause [Gurnett et al., 1979; Song et al., 1990]. Generation and propagation of these small-scale waves can affect the macroscopic steady-state structure of the magnetopause (and transport properties) either in a direct coherent fashion or via turbulence [Winske et al., 1991; Thomas and Winske, 1993]. In this example, the interaction of the large scale solar wind with the Earth's magnetic field seeds the smaller-scale processes within the boundary layer and self-consistently evolves with them. A similar interdependent cross-scale phenomenon occurs in the Earth's magnetotail leading to increased stress in the

Cross-Scale Coupling in Space Plasmas
Geophysical Monograph 93
This paper is not subject to U.S. copyright. Published in 1995
by the American Geophysical Union

plasma sheet boundary layer (PSBL), intense waves, relaxation of stress, and a resultant self-consistent steady-state structure of the PSBL which has been discussed [Ganguli et al., 1994a; Romero and Ganguli, 1994]. Thus, the entry of the solar wind mass, energy, and momentum into the magnetosphere involves an intricate interplay of physical processes occurring at highly diverse spatial and temporal scale sizes and is a microcosm of much of the physics prevalent elsewhere in space plasmas, including the ionosphere which we discuss in this article.

In nature, the interdependence of multi-scale phenomena is the norm rather than an exception. For example, turbulence in neutral gas is an excellent example of cross-scale phenomena which has been extensively studied over a number of years. Although plasmas are more complex than neutral gases, there are still similarities with regard to the description of turbulence. Indeed much of the foundations of plasma turbulence are similar to that of neutral gases [Sagdeev, 1966]. What makes a plasma system more complex is the possibility of a greater variety of small-scale collective processes. This is especially true of plasmas in a magnetic field such as the space plasma. As demonstrated by Romero and Ganguli [1994] for the case of a stressed PSBL, relaxation of a stressed plasma system is often accompanied by waves which can affect the steady-state structure and transport properties. An increase of stress during geomagnetically active periods followed by relaxation is fundamental to many plasma processes occurring in space, all of which collectively determine the global morphology and transport properties.

The importance of small-scale waves in plasma transport can be recognized by examining the equation that determines the evolution of the plasma,

$$Lf(r,v,t) = Cf(r,v,t), \qquad (1)$$

along with the Maxwell equations for self-consistent fields. Here $L \equiv \partial/\partial t + v \cdot \nabla + (F/m) \cdot \nabla v$, C is the collision operator, F is the total force, and f is the distribution function. The effects of collisions and self-consistent waves which enter through the operators C and L respectively are quite different. Collisions tend to relax and establish a local Maxwellian in a typical time scale, say τ. The self-consistent fields on the other hand are responsible for dispersion properties with a typical time scale $2\pi/\omega$. If $\omega\tau \gg 1$, then the collisions can be neglected and we recover the Vlasov equation,

$$Lf(r,v,t) = 0. \qquad (2)$$

Since the randomizing effects of the collisions are no longer present, it may appear that a Vlasov plasma framework is not suitable to describe relaxation processes which are irreversible. However, as mentioned earlier, relaxation processes in a collisionless plasma are accompanied by waves which can scatter plasma particles randomly and thereby introduce an effective collision frequency. The wave-generated 'anomalous' effects have been found to result in a severe degradation of plasma confinement in fusion devices. This problem continues to be a major concern in fusion plasmas and is responsible for the lack of success in realizing and harnessing power from fusion reactors. Since much of the space plasma environment is collisionless, and the relaxation in a collisionless plasma is dependent on collective effects, the study of waves and instabilities in space plasmas is indispensable.

The most challenging task in the study of the cross-scale coupling phenomena is the inclusion of plasma instabilities in large scale transport studies which are mostly numerical. The most general and self-consistent approach would employ a full three dimensional particle-in-cell code with resolution fine enough to account for the shortest time and space scales in the problem. Typically this would imply resolving ion and electron gyro frequencies and radii and possibly plasma frequencies over at least several Earth radii for temporal periods of hours. Given the state-of-the-art of computational hardware, this is beyond the current capabilities. The better approach is to rely on a simple fluid/MHD approach for the description of large-scale phenomena and include the important effects of instabilities as local plasma parameters exceed prescribed threshold conditions [Davidson and Krall, 1977; Papadopoulos, 1977; Manheimer and Boris, 1977]. This framework has been successfully applied to space plasma transport in a few cases [S. Ganguli and Palmadesso, 1987; Lysak and Hudson, 1987; Brown et al., 1991]. These studies are quite encouraging and reveal that anomalous transport can be crucial in some instances. For example, in the classical plasma outflow from the ionosphere to the magnetosphere, the ion gas is expected to experience an adiabatic cooling as it expands through the diverging geomagnetic flux tubes with $T_{||} > T_{\perp}$ [S. Ganguli and Palmadesso, 1987; Schunk and Watkins, 1981]. When the effects of the ion cyclotron waves in this region; such as anomalous heating and resistivity; were included in the transport code, S. Ganguli et al., [1988; 1991] found that there is plasma energization in the direction perpendicular to the geomagnetic field lines and the H^+ ion anisotropy is reversed to $T_{\perp} > T_{||}$. This energization process is consistent with the observations of Moore et al., [1986a,b].

So far most space applications of cross-scale coupling phenomena are one dimensional studies and are restricted to the magnetic field aligned direction. Important cross-scale phenomena can occur in the direction transverse to the magnetic field as well [Ganguli et al., 1994b; Huba, 1994]. Such effects may ellucidate several features of the plasma behaviour in the lower ionospheric region which are not yet well understood. One such example is ion energization at lower altitudes and its co-location in regions with velocity shear such as the convection flow reversal region and auroral arcs. In the following section we discuss this process in more detail.

2. ENERGIZATION AT LOW ALTITUDES

It was expected that the composition of the magnetospheric plasma is predominantly light ions of solar wind origin. However, recent observations of heavy ions previously thought to be gravitationally bound, have challenged this belief [Pollock et al., 1990; Moore, 1991, and references therein]. Recognition of the ionosphere as a source of magnetospheric plasma began with magnetospheric observations of heavy ions [Shelley et al., 1972; 1976]. There are now numerous observations of outflowing heavy O^+ ions. Lockwood et al. [1985] reported O^+ outflows known as the 'cleft ion fountain'. These heated ionospheric ions emerge from a restricted source region associated with the polar cleft topside ionosphere and are dispersed across the polar cap by the convection electric field [Horwitz, 1987]. It has been reported that the distribution function reveals perpendicular ion heating [Moore, 1984; Moore et al., 1986a,b; Moore, Heelis, private communications] which cannot be accounted for by frictional heating alone, thereby indicating the possible presence of wave heating via plasma instabilities. Other observations at higher (topside ionospheric) and lower (sounding rocket) altitudes of ion heating and conic acceleration are growing in number [Tsunoda et al., 1989, Moore et al., 1986, Whalen et al., 1978, Yau et al., 1983]. More recently, observations of large ion outflows resulting from transverse heating have also been made by the SMS on the Akebono spacecraft [Whalen et al., 1991].

The ion upflow events such as in the cleft ion fountain [Lockwood et al., 1985] appear to be driven by a process at low altitude (below 2000 kms) which heats the ions and initiates a slow upflow. Other processes operating at higher altitude (above 2000 kms) can provide additional energization and may accelerate the flow [S. Ganguli and Palmadesso, 1987; Lundin et al., 1990; Borovsky, 1984; Block and Falthammar, 1990; Temerin and Roth, 1986; Ball and Andre, 1991; Chang, 1993; Crew et al., 1990, and the references therein]. However, it is not yet clear what the energization mechanism in the source region below 2000 kms is. Energization in the source region is the first and perhaps the most critical step for ion migration to higher altitudes, where they can access other heating mechanisms mentioned above and then ultimately escape into the magnetosphere. At low altitudes, the cleft ion fountain heating is observed in high density, low velocity plasmas, where it is difficult to excite current driven instabilities with observed magnitude of the field aligned current. Thus, current driven instabilities [Kindel and Kennel, 1971; Satyanarayana et al., 1985] cannot be invoked to explain the transverse ion heating at these low altitudes. On the other hand, the ion outflow is often found to be correlated with the shear in the transverse velocity and localized in the region where the Joule heating rate is minimal (for example, Tsunoda et al., [1989], Lu et al., [1992], Pollock et al., [1990], and references therein). Evidence of transverse ion heating with velocity shear is also available [Moore et al., 1985, Marklund et al., 1994]. Therefore, the possible role of velocity shear in the energization of ions is an important and outstanding issue which we discuss in this article.

3. CORRELATION OF VELOCITY SHEAR WITH ION ENERGIZATION

The frictional heating of ions (Joule heating) in the F region ionosphere under conditions of strong convection has been discussed by several authors including Whealton and Woo [1971], St. Maurice and Schunk [1979], and more recently Kinzelin and Hubert [1992]. It is widely assumed that Joule heating is sufficient to account for the heating of heavy ions at low ionospheric altitudes [Loranc et al., 1991]. However, increasing number of observations suggest otherwise. A thorough analysis of the HILAT data by Tsunoda et al., [1989] led to the conclusion that there is a good correlation of ion outflow with the gradient in the convective flow and not with its magnitude. A similar study by Lu et al., [1992] using the DE 2 data reveals a remarkable anticorrelation of joule heating with ion outflow while at the same time points to a good correlation of outflow with the spatial gradient in the convection velocity which peaks in the reversal region. More recent evidence and a theoretical model provided in Ganguli et al., [1994b] supports these conclusions. Joule heating can produce highly nonthermal ion distributions when the convection velocity exceeds the neutral thermal speed and the ion-neutral collision frequency is between the ion gyrofrequency and the ion-ion collision frequency. The nonthermal features tend to be thermalized at altitudes above a few hundred km as the ion-neutral collision rate falls below the ion-ion collision rate. The heating is limited to thermal speeds on the order of the convection velocity. Moreover, the existing theory of Joule heating does not predict the production of a hot tail in the ion distribution function, as has often been reported from the observations. Joule heating clearly maximizes where the convection velocities are largest and can be important, but it evidently fails to explain the consistent correlation of ion energization and outflow with shear in the convective flow.

There are good physical reasons for the non thermal heating at the flow reversal region where the velocity shear is large. It has been shown in Ganguli et al., [1994b] that velocity shear enables the plasma to tap into the huge reservoir of free energy available in the convective flow. We review this argument in the following. Observations indicate that the typical ion flux in the upflow region is around 10^8 ions/cm^2-s, with an average energy of around 1 - 5 eV. Thus, the energy flux that needs to be imparted to the ions for upwelling is around 10^{-4} to 10^{-3} ergs/cm^2-s. An estimate of the total power available in the convective flow was considered to be P = $\Sigma_p E_\perp(x)^2$, where Σ_p is the height integrated conductivity and $E_\perp(x)$ is the electric field associated with the convective flows and it is sheared. If this electric field was uniform (as is commonly assumed), then a transformation to the frame moving with the convection velocity would imply that there is no free energy available. However, the shear in the convective flow velocity cannot be removed by such a transformation and it is this feature that can make the large reservoir of free energy accessible. Using $E_\perp \sim (10 - 50)$ mV/m and $\Sigma_p \sim 12$ mho, we see that the total power available is around 1.2 to 30 ergs/cm^2-s which is orders of magnitude larger than necessary. Note, however, that the estimate of the total power available is an underestimate. This is because the process we discuss is capable of tapping energy from the convection flow even in the absence of conductivity provided there is a velocity shear. A more appropriate estimate would be to use the flux tube integrated energy density in the convection flow as the energy available per unit area from the solar wind. This is effectively limitless for our purpose. Thus, even if a very small fraction of the total available energy can be dissipated by the instabilities to energize the ions, then ion upflow can easily be sustained. In the next section we show that coupling meso and micro scale processes due to velocity shear can achieve this. In the following discussions we will not repeat the details of the calculations of Ganguli et al., [1994b]. Instead we will emphasize the physical picture of the model.

4. COUPLING OF MESO AND MICRO SCALE PROCESSES DUE TO VELOCITY SHEAR

4.1 Meso Scale Dynamics

Velocity shear can change the wave dispersion characteristics of a magnetoplasma significantly [Ganguli et al., 1988a;1988b; Koepke et al., 1994]. In addition to giving rise to new wave modes, velocity shear can significantly modify other normal modes of a plasma, sometimes stabilizing [Ganguli et al., 1989b; Basu and Coppi; 1992] and sometimes destabilizing [Ganguli et al., 1989a; Ganguli and Palmadesso; 1988; Nishikawa et al., 1990] the wave. A classic instability due to velocity shear which has been studied for nearly a century is the Kelvin-Helmholtz (KH) instability [Rayleigh, 1896]. The KH instability is a low frequency mode ($\omega \ll \Omega_i$, the ion cyclotron frequency) with long wavelength ($k_y \rho_i \ll 1$, where k_y and ρ_i are the perpendicular wave vector and ion gyroradius respectively). Even a small magnitude of velocity shear can trigger the KH instability and therefore this instability is likely to be triggered in the flow reversal region (although other low frequency waves are possible as well).

Amplitudes of plasma waves once triggered are amplified at the expense of the available free energy, such as velocity shear, and evolve into a nonlinear stage. As ellucidated by Sagdeev, [1966], the instabilities and nonlinear evolution result in turbulence and steepening which lead to the reduction of the characteristic scale sizes. As the initial energy in the large scale convective flow is channeled into smaller scale sizes the spatial gradients increase leading to localized regions of stronger velocity shear. This is similar to the formation of eddies in hydrodynamic turbulence. This feature has often been seen in numerical simulations [e.g., see Keskinen et al., 1988; Seyler, 1990]. Another feature of the evolution is to generate steep density gradients and associated local velocity shear due to nonlinear convective effects. As shown in the simulations of the KH instability [see Fig. 9 in Ganguli et al., 1994b], the initial density gradient is convected by the waves and locally steepened. These steepened density structures remain quasi steady over the time scales associated with the low frequency wave. As the scale size of the density structures become comparable to or smaller than an ion gyrodiameter, an ambipolar electric field is self-consistently developed. This occurs because the ion gyroradius is much larger than an electron gyroradius and therefore a single electron can not charge neutralize an ion if they are both magnetized and their guiding centers are on the same field line. In a uniform or weakly inhomogeneous plasma (i.e., the density gradient scale $L_n \gg \rho_i$) there are always other electrons whose guiding centers are on different field lines available to charge neutralize the ion. If on the other hand L_n is short (i.e., $\leq 2\rho_i$), there will always be a charge imbalance due to the gyromotion of the plasma ions and the electrons. To maintain quasi-neutrality a localized ambipolar d.c. electric field will be self-consistently generated which effectively demagnetizes the ions and accelerates them towards the electrons. A detailed Vlasov formulation to compute this d.c. electric field and its spatial gradient has been developed by Romero et al. [1990]. Their results indicate that when $L_n \sim \rho_i$ a strongly sheared flow will be generated at the same scale.

From the above considerations we conclude that the conditions prevailing in the convective flow reversal region and in auroral arcs lead to meso scale instabilities such as the KH instability which grow and nonlinearly evolve to seed localized regions of strongly sheared flows. Moreover, there is good evidence of low frequency wave steepening in the ionosphere. These phenomena have been and continue to be extensively investigated both theoretically and observationally [Pfaff et al., 1987; Tsunoda, 1988; Huba, 1988; Kelley, 1989]. An important parameter which measures the magnitude of transverse velocity shear effects is the velocity shear frequency defined by $\omega_s = V_E/L$ where V_E is the peak flow transverse to the magnetic field while L is the scale size of this flow. From the simulations described in Ganguli et al., [1994b] we see that the initial large scale flow with $\omega_s \sim 0.1\Omega_i$ (where Ω_i is the ion gyrofrequency) can give rise to localized regions with $L_n \sim \rho_i$ within which $\omega_s \sim \Omega_i$ or larger. The dynamics of the plasma within these localized regions, which are of the order of ρ_i, can no longer be analyzed by the meso scale fluid code. A kinetic description will be necessary to study them. We discuss this in the following sections.

4.2 Micro Scale Dynamics

To analyze the micro scale dynamics at the eddy level we have to understand two broad catagories of plasma processes: (i) Weak shear ($\omega_s < \Omega_i$) effects, and (ii) strong shear ($\omega_s > \Omega_i$) effects. In the weak shear regime both ions and electrons are magnetized while in the strong shear limit the ions become unmagnetized. This can be easily seen from the equations of motion that includes an inhomogeneous electric field in the x direction transverse to the uniform magnetic field assumed in the z direction. For simplicity consider this electric field to be linear, i.e., $E = E'x$, where $E' = dE/dx$ is a constant. The equations of motion can be written as,

$$d^2x/dt^2 = (e/m)E'x + \Omega dy/dt \quad (3)$$

$$d^2y/dt^2 = -\Omega dx/dt. \quad (4)$$

From (3) and (4) it follows that

$$d^2x/dt^2 + \Omega_1^2 x = 0, \quad (5)$$

where $\Omega_1 = \eta^{1/2}\Omega$ is the effective gyrofrequency, $\eta = 1 + V_E'/\Omega$, and $V_E' = -eE'/m\Omega$. Since $V_E' = \omega_s$, we see that as shear increases to $|\omega_s| \sim \Omega$ the effective gyrofrequency goes to zero, i.e., the ions loose magnetization. For stronger shear cases ($|\omega_s| > \Omega$) which can happen when the scale size of the steepened layers is comparable to ρ_i, the ions may be treated as unmagnetized fluid. A more detailed calculation of orbits in inhomogeneous electric fields including higher order effects is given in Appendix-A of Ganguli et al., [1988a]. In the following we discuss the physics in the weak and strong shear regimes.

4.2.1 Weak Shear Regime ($\omega_s < \Omega_i$)

The KH instability is the classic mode in the weak shear regime which has been extensively studied for decades. This instability is a low frequency long wavelength mode and has been invoked in numerous space and laboratory applications. However, there is good evidence both in laboratory and space plasmas which indicates the possibility of higher frequency ($\omega \sim \Omega_i$) and shorter wavelength modes associated with velocity shear [Mozer et al., 1977; Temerin et al., 1981; Song et al., 1990; Jassby, 1970; Sato et al., 1986; Koepke et al., 1992]. Since the frequency of these waves is around Ω_i, they can be an efficient source for ion energization and therefore they are of much interest to space plasmas. To study these waves a general kinetic theory for electrostatic waves that includes an inhomogeneous d.c. electric field transverse to a uniform magnetic field is used [Ganguli et al., 1988a]. It is found that besides the classical KH modes, there is a new class of oscillations which are generated by the inhomogeneity in the energy density introduced by a localized d.c. electric field, and hence we call them the inhomogeneous energy density driven instability (IEDDI). Depending on the magnitude of velocity shear and background parameters, the frequency of the IEDDI can be anywhere from below the ion cyclotron frequency to above it and the wavelengths could be smaller than an iongyroradius to several ion gyroradii long. Detailed parametric study and applications of this instability are given elsewhere and hence are not repeated here [Ganguli et al., 1994b and references therein; Gavrichtchaka et al., 1995, (this volume)].

To get a physical picture of the IEDD mechanism we assume an idealized electric field shown in Fig. 1, where a uniform electric field E_0 is localized in region I defined by $|x| \leq L/2$ and zero elsewhere (in region II). The dispersion relation of the electrostatic ion Bernstein modes is

$$D(\omega) = 1 - \Gamma_0(b) - \sum_{n>0} \frac{2\omega^2}{\omega^2 - n^2\Omega^2}\Gamma_n(b), \quad (6)$$

where $k_\parallel \sim 0$ is assumed. The energy density of these modes is

$$U \propto \omega \frac{\partial D}{\partial \omega} = \omega \left(\sum_{n>0} \frac{4\omega\Gamma_n n^2\Omega^2}{(\omega^2 - n^2\Omega^2)^2} \right) = \omega^2 \sigma(\omega). \quad (7)$$

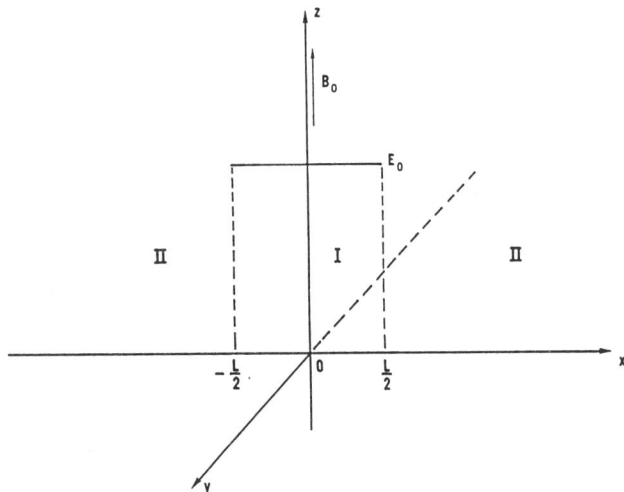

Fig. 1. A schematic representation of the piecewise continuous d.c. electric field profile.

Clearly, these are positive energy waves. Introduction of a uniform electric field in the x direction initiates an E X B drift in the y direction and consequently there is a Doppler shift in the frequency, i.e., $\omega \rightarrow \omega_1 = \omega - k_y V_E$. The energy density in the presence of the Doppler shift is, $U' \sim \omega \omega_1 \sigma(\omega_1)$, which can be negative provided $\omega \omega_1 < 0$ since $\sigma(\omega)$ is always positive. Now if we consider the localized field configuration as shown in Fig. 1 then, for sufficiently large E X B drift, the energy density in region I can become negative while it remains positive in region II. A nonlocal wavepacket can couple these two regions so that a flow of energy from region I into region II will enable the wave to grow. Based on this simple picture we can predict some gross features of the instability. For example, using the wave-kinetic description it is possible to obtain the energy balance condition for the system from which important scalings governing the growth rate can be predicted. The growth of the wave in region I implies a loss of energy from that region. By conservation of energy, this must be the result of convection of energy into region II and any local energy dissipation (S_-) or free energy release (S_+) processes in region I. The rate of growth of the total energy deficit in region I is proportional to the growth rate γ, the wave energy density U_I in region I, and the volume of region I, represented here by the extent in the x direction (L) of region I times a unit area $A_\perp$ in the plane perpendicular to x. The rate of convection through $A_\perp$ is just $V_G U_{II}$, where V_G is the group velocity in the x direction and U_{II} is the wave energy density in region II. We can then write the energy balance condition as

$$\gamma U_I L A_\perp \simeq (S_+ - S_- - V_G U_{II}) A_\perp , \qquad (8)$$

where S_+ and S_- represent the source and the sink in region I. First consider the case where there is no external source of free energy and since $k_{\parallel} \sim 0$, the natural dampings are absent as well. Therefore $S_+ = S_- = 0$. Now it is clear from (8) that if U_I is negative then γ can be positive and hence the growth of the wave is sustained by convection of energy into region II from region I. On the other hand if U_I is positive then the convection of energy out of region I would lead to a negative growth rate and therefore to damping of the waves. For $S_\pm = 0$, an important scaling can be predicted from (8), i.e., $\gamma/V_G \propto 1/L$ which with proper normalizations can be written as $\mathrm{Im}(k_x \rho_i) \propto \epsilon$ where $\epsilon = \rho_i/L$. In Fig. 6 of Ganguli et al., [1988a] this scaling law is confirmed. The instability is reactive in nature when $S_\pm = 0$. For the cases where there is a magnetic field aligned drift in addition to the transverse localized electric field, the term $S_\pm$ in (8) is not zero and can roughly be estimated (using the local theory) to be proportional to $LU_I \gamma_l$, where the local growth rate in the region I, $\gamma_l = -D_I/D_{R\omega}$, is evaluated at $\omega = \omega_r$. This term is dissipative in nature. Here D_R is the real part of the local dispersion relation as given in Eq. (6), and $D_{R\omega}$ is the ω derivative of D_R. In the frame moving with the electrons along the magnetic field with a drift speed V_d the ions get an additional Doppler shift equal to $k_\parallel V_d$ so that the net Doppler shifted frequency for the ions is $(\omega_1 - k_\parallel V_d)$. The D_I term, contributed by the electrons, is $\sim \pi^{1/2} \omega_1/(k_\parallel v_e) \propto \omega_1$. Assuming that the field aligned current is also localized within the region I we see that $D_{R\omega} = U_I/\omega_r = (\omega_1 - k_\parallel V_d)\sigma(\omega_1 - k_\parallel V_d)$. With these and noting that $U_{II} = \omega_r^2 \sigma(\omega_r)$, the energy condition becomes,

$$\gamma L A_\perp \omega_r (\omega_1 - k_\parallel V_d) \sigma(\omega_1 - k_\parallel V_d)$$
$$\simeq -\omega_1 \omega_r L A_\perp - V_G \omega_r \sigma(\omega_r) A_\perp . \qquad (9)$$

We have neglected the ion Landau and cyclotron dampings. First consider the case where the electric field is not strong enough to make $\omega_1 < 0$. We see that in the presence of a field aligned drift it is possible to have $(\omega_1 - k_\parallel V_d) < 0$ which can make the energy density in region I negative even when electric field alone is not sufficient to do so. This would be true even if V_d was uniform throughout in both regions I and II but subcritical by itself, i.e., $\omega > k_\parallel V_d$. As the condition $(\omega_1 - k_\parallel V_d) < 0$ is satisfied, the right hand side of (9) becomes positive indicating that there is growth. Conversely, it may be said that even for subcritical V_d, the presence of a d.c. electric field can lead to growth. If $\omega_1 < 0$

as well as $(\omega_1-k_\parallel V_d)<0$, convection will lead to growth while the first term in the right hand side will contribute to damping. In this case there can be growth for $V_d \sim 0$ provided the second term in Eq. (9) is larger than the first which now represents the electron Landau damping. This corresponds to the case where $R = k_y V_E/k_\parallel V_d >> 1$ discussed in Ganguli et al., [1989a] where $k_\parallel V_d$ can be neglected with respect to $k_y V_E$. In this limit the instability is purely reactive. Clearly, the latter case demands stronger shear while the combination of V_d and V_E case can be destabilized by much weaker shear. When both V_E and V_d are present the instability can be purely dissipative or partly dissipative and partly reactive depending on the magnitude of R. From the early studies of this mode it was found that velocity shear can increase the number of unstable roots [Ganguli et al., 1986]. Detailed characteristics of the various roots are now being investigated by Gavrichtchaka et al., [1995]. An interesting feature of the IEDD mechanism is that in the R << 1 limit, $K_y V_E$ cannot be neglected in the first term in the right hand side of (9) unless $\omega_r >> k_y V_E$ as well. This is important since even in the R << 1 limit which corresponds to weak velocity shear, the IEDD mechanism can influence wave propagation properties by moderating the wave phase speed ($\omega_1/k_\parallel$) parallel to the magnetic field and thereby affecting the Landau resonance condition. Specifically, it can enable a wave to satisfy the Landau resonance condition and draw energy from the field aligned drift even though it is not possible to satisfy this condition in the infinite homogeneous limit. This results in a reduction in threshold for the current driven ion cyclotron instability even for small magnitudes of velocity shear [Ganguli et al., 1989a; Ganguli and Palmadesso, 1988]. The frequencies, wave vectors, and other parameters in Eq. (9) must be consistent with the dispersion relation.

Past laboratory experiments in a Q-machine [Jassby, 1970; Sato et al., 1986; Van Niekerk et al., 1991] have indicated significant changes in the ion-cyclotron mode characteristics that were attributed to transverse electric fields present in the plasma. It was suggested that velocity shear may be responsible for these effects. Koepke et al., [1994] recently verified the existence of the IEDDI by measuring the identifying mode characteristics for laboratory conditions that include both the magnetic field aligned current and transverse velocity shear. Amatucci et al., [1994] demonstrated that this mode has a substantially lower threshold current density than the current driven ion cyclotron instability [Kindel and Kennel, 1971]. A comparison between the laboratory results and the theoretical predictions for experimentally relevant parameters shows very good consistency [Koepke et al.,

1995]. Additional experiments are underway to investigate the nature of the mode in greater detail.

The combination of V_d and V_E is a likely scenario in the ionospheric environment especially in the auroral arcs, shocks or double layers, and in the convective flow reversal region, where often subcritical field aligned drifts are reported but in the presence of velocity shear. Substantial velocity shear ($\omega_s \sim (0.1 - 0.2)\Omega_i$) in the low altitude ionosphere have been reported [Kelley and Carlson, 1977; Basu et al., 1988; Earle et al., 1989; Tsunoda et al., 1989]. Theoretical calculations indicate that ω_s in this range along with small field aligned drift, can support waves around the ion cyclotron frequency [Ganguli et al., 1989a; Gavrichtchaka et al., 1995 (this volume)]. These waves can lead to ion bulk heating [Ganguli et al., 1985] and contribute to anomalous viscosity and resistivity. In laboratory experiments at West Virginia University the value of ω_s was varied from 0 up to $0.5\Omega_i$, which is similar to the values reported from space observations. Theoretical calculations indicate that much larger ω_s is possible but their observation will require higher resolution detectors than are currently used. Recent observations from the Freja satellite indicate the existence of intense velocity shears associated with the black aurora where ω_s may be comparable to or larger than Ω_i [Marklund et al., 1994].

4.2.2 Strong Shear Regime ($\omega_s > \Omega_i$)

As the scale size of a steepened density layer becomes comparable to the ion gyroradius, ω_s can exceed Ω_i [Romero et al., 1990]. This may be ellucidated by a simple fluid picture, although the effects are somewhat minimized by the fluid description. Considering the ions to be unmagnetized its pressure balance is given by

$$eE - \frac{T_i}{n}\left(\frac{dn}{dx}\right) = 0 . \quad (10)$$

Electrons are magnetized and hence the electron momentum balance equation is

$$-eE - \frac{e\mathbf{V} \times \mathbf{B}}{c} - \frac{T_e}{n}\frac{dn}{dx} = 0 , \quad (11)$$

leading to

$$V_{\perp e} = -\frac{cE}{B} - \frac{cT_e}{eB}\left(\frac{1}{n}\frac{dn}{dx}\right) = -\frac{cE}{B}\left(1+\frac{T_e}{T_i}\right) . \quad (12)$$

Using a simple model for the steepened density structures, $n(x) = n_0(1+\delta+\tanh(x/L_n))$ where $\delta \ll 1$, we can express the shear frequency in terms of L_n,

$$\omega_s = \frac{1+\tau}{2\tau}\left(\frac{\rho_i}{L_n}\right)^2 \Omega_i \quad (13)$$

where $\tau = T_i/T_e$. For $\tau \sim 1$ the shear frequency can become comparable to Ω_i when the density scale size L_n approaches an ion gyroradius. Further reduction in L_n will rapidly make ω_s even larger. This effect is accentuated if electron temperature is larger than ion temperature.

The ions are unmagnetized in the strong shear limit while the electrons experience a sheared E X B flow. The free energy available in the sheared electron flow can support the electrostatic electron-ion-hybrid (EIH) modes [Ganguli et al., 1988b]. These are short wavelength oscillations around the lower hybrid frequency. The EIH modes prefer $k_\parallel \sim 0$ and $1/\rho_e \gg k_y \gg 1/\rho_i$. Therefore, this instability is fluid like and its eigenvalue condition is fairly simple to derive. From fluid equations it can be shown that the perturbed electron and ion densities are,

$$n_{1e}(x) = \frac{-\delta^2}{4\pi q_e}\left\{-\frac{d^2}{dx^2}+k_y^2 - \frac{k_y(V_E''(x) - \epsilon_n \Omega_e)}{(\omega - k_y V_E(x))}\right\}\phi_1(x), \quad (14)$$

$$n_{1i}(x) = \frac{1}{4\pi q_i}\frac{\omega_{pi}^2}{\omega^2}\left(-\frac{d^2}{dx^2}+k_y^2\right)\phi_1(x), \quad (15)$$

Here the subscript '1' implies perturbed quantities, ω_{pi} is the ion plasma frequency, $\epsilon_n = (dn/dx)/n = 1/L_n$, $\delta = \omega_{pe}/\Omega_e$, and $V_E'' = d^2 V_E/dx^2$. For details we refer to our previous publications [Ganguli et al., 1988a; Romero et al., 1992a]. Using the above expressions along with the Poisson equation we obtain,

$$\left(\frac{d^2}{dx^2}+k_y^2\right)\phi_1 - G(\omega)\frac{k_y(V_E''-\epsilon_n\Omega_e)}{(\omega-k_y V_E)}\phi_1 = 0, \quad (16)$$

where $G(\omega) = \delta^2/(\delta^2+1)(1-(\omega_{LH}/\omega)^2)$ and $\omega_{LH}^2 = \omega_{pi}^2/(1+\delta^2)$. The last term of Eq. 16 represents the free energy source. Clearly, there are two sources of free energy; (i) sheared electron flow ($V_E''(x)$ and $V_E(x)$), and (ii) the density gradient (ϵ_n). If $V_E'' = 0$ and V_E is uniform, then Eq. 16 reduces to the dispersion condition for the Lower Hybrid Drift (LHD) instability [Krall and Liewer, 1971]. On the other hand, if $\epsilon_n = 0$, then it reduces to the corresponding condition for the Electron-Ion-hybrid (EIH) instability [Ganguli et al., 1988b]. In general both sources can contribute. From numerical simulations it is found that if the velocity gradient is comparable to or larger than the wave frequency (ω_{LH} in this case) then the effects of velocity shear (and hence the EIH instability) can dominate [Romero et al., 1992b; Romero and Ganguli; 1993]. If the velocity shear is weak then the density gradient can lead to the LHD instability. Thus, the existence of a strong shear ($\omega_s \sim \omega_{LH}$) is not essential for lower hybrid waves as observed in the ionosphere [Kintner et al., 1992; Vago et al., 1992; Arnoldy et al., 1992; Lynch et al., 1994].

The EIH modes tap energy from the sheared electron flow by time averaging the flow. This is seen from the energy conservation condition for the EIH modes and explained in Ganguli et al., [1988b],

$$\frac{\partial}{\partial t}\int dx\left[\frac{|\tilde{E}|^2}{8\pi}+\frac{n_{0e}m_e}{2}\frac{|c\tilde{E}|^2}{B^2}+\right.$$

$$\left.\frac{n_{0i}m_i}{2}\frac{|e\tilde{E}|^2}{m_i^2\omega_r^2}+\frac{n_{0e}m_e}{2}|\tilde{x}|^2 V_E(x)V_E''(x)\right]=0, \quad (17)$$

where $\tilde{x} = \tilde{V}_x/(\omega_r - k_y V_E(x))$ is the displacement due to the wave-induced $\tilde{E}_y$ X B drift, $\tilde{V}_x = -c\tilde{E}_y/B_0$. Here $\tilde{E}_y = -ik_y\phi_1$ is the wave fluctuation in the y direction and $V_E(x)$ is the equilibrium electron flow due to the d.c. electric field. The first three terms are related to the a.c. electric field of the wave. The first term represents the electrostatic wave energy density in vacuum, the second term is the wave-induced kinetic energy of the electrons that are magnetized, and the third term is the wave-induced kinetic energy of the ions that are unmagnetized. The energy associated with the equilibrium flow $V_E(x)$ of the electrons is proportional to $V_E^2(x)$. At a given position x the wave fluctuations introduce a displacement $\tilde{x}$ which is proportional to the wave amplitude. Thus, when the waves are excited the mean flow at a given position x becomes $\langle V_E(x+\tilde{x})\rangle = V_E(x) + \langle \tilde{x}\rangle V_E'(x) + \langle \tilde{x}^2\rangle V_E''(x)/2 + ..$, where "$\langle\ \rangle$" implies time averaging and since $\tilde{x}$ is oscillatory, $\langle \tilde{x}\rangle = 0$. Consequently, when the waves are excited the difference in the energy associated with the equilibrium flow due to the d.c. electric field is proportional to $(\langle V_E(x+\tilde{x})\rangle^2 - V_E^2(x)) = |\tilde{x}|^2 V_E(x) V_E''(x)$, which is identical to the last term of Eq. (17). Thus, the last term of (17) represents the change in the energy associated with the equilibrium flow due to the d.c. electric field when the waves are present and the condition (17) represents the energy conservation of the system. The condition (17) indicates that reduction in the equilibrium flow energy at a given position x, which occurs due to the time averaging of the waves, is available as the free energy necessary for the growth of the instability. Note that the time averaging removes the first derivative and therefore the free energy is proportional to the second derivative of the d.c. electric field. Thus, the EIH instability is explicitly dependent on the second derivative of the electric field. The physics of the classical KH instability is identical to the EIH instability except that in that case the ions are magnetized and the electron contribution is negligible. If we ignore the ion contribution in in (17) and relabel the subscripts 'e' as 'i' then the condition (17) becomes pertinent to the KH instability.

Linear and nonlinear properties of the EIH instability have been investigated [Romero and Ganguli, 1993 and references therein]. It is found that the EIH instability has a larger perpendicular wavelength (larger phase velocity) than the LHD instability. This makes the EIH instability a better source for energizing the ions [Chang and Coppi, 1981]. The power spectra are broadband for both frequency and wavelength with peak around the lower hybrid frequency and $k_y L_E \sim 1$ where L_E is the scale size of the local 'eddy' flow. The anomalous viscosity and resistivity due to this instability have been parameterized in terms of velocity shear. For details we refer to our previous publications [Romero and Ganguli, 1993 and references therein]. The EIH instability may have been the source for the lower hybrid oscillation observed in the velocity shear region in the experiment of Yamada et al., [1977] and other experiments of laser produced plasma jets [Dimonte et al., 1991; Peyser et al., 1992; Mostovych et al., 1989].

4.2.3 Coupling of Meso amd Micro physics

From the above discussions we see that there is a similarity between the plasma processes due to velocity shear leading to non thermal heating in the convection flow reversal region and hydrodynamic turbulence. The nonlinear evolution of the large scale convection flow can result into small-scale eddy flows. Due to smaller scale size of the eddies the spatial gradients within these flows can become large and trigger collective effects. Depending on the magnitudes of these gradients different small scale dynamics may result in these structures. The present generation detectors may not always be able to resolve the details of the physics within an individual eddy layer, but their cumulative effects such as ion energization, co-location of ion energization with convection flow reversal region, wave signatures, features of the power spectral density, etc., are observable and are in good agreement with our model. We have parameterized the anomalous viscosity and resistivity due to the EIH instability, although further refinements are necessary. We have discussed the feed back prescription of the anomalous viscosity to the large scale flow in Appendix-B of Ganguli et al., [1994b]. Work is currently underway to parameterize the anomalous viscosity of the IEDDI and examine the effects of the feed back processes.

Perhaps the most impressive ionospheric evidence of such a coupling involving velocity shear is given by Kelley and Carlson, [1977]. They report the observation of an intense shear in plasma flow velocity at the edge of an auroral arc and associated with the shear are irregularities with two characteristic scale sizes. While the long wavelength irregularities could be explained by the KH instability, the origin of the shorter wavelength irregularities were not understood. This led Kelley and Carlson to conclude that "..*A velocity shear mechanism operating at wavelengths short in comparison with the shear scale length, such as those observed here, would be of significant geophysical importance.*". The wavelengths of the IEDDI and the EIH instabilities, which operate at the eddy level, are of the order of the eddy scale-size which are much smaller than the large-scale velocity shear. Therefore, our model is consistent with the Kelley and Carlson observations. A detailed study involving data and numerical simulations by Yamamoto et al., [1994] provides evidence of low frequency meso scale wave seeding high frequency small scale waves in the ionosphere as well. Also, high resolution data from the Freja satellite indicate the existence of intense velocity shears associated with black aurora [Marklund et al., 1994]. It is reported that there exists waves in the ion cyclotron and lower hybrid frequency ranges along with ion heating associated with the velocity shear. More recently, Moore et al., [1995] have analyzed the ARCS 4 rocket data and report the observation of vortical flows and associated ion heating when the rocket encountered an auroral arc. These observations are

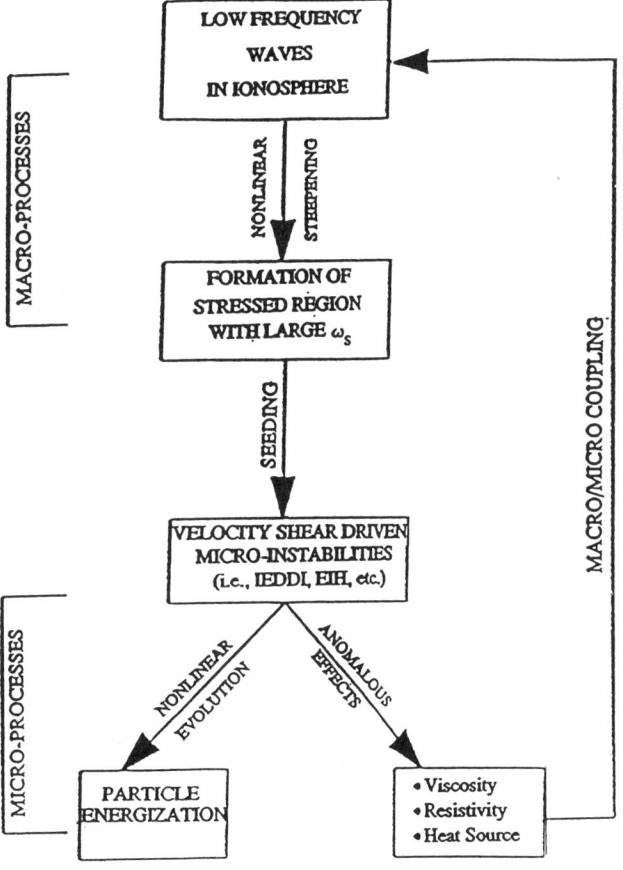

Fig. 2. A possible scenario.

consistent with our model predictions.

It is also interesting to note that in a particle simulation study of plasma-wall sheath, Theilhaber and Birdsall [1989] reported the formation of a large scale steady state vortex in the nonlinear stage due to the low frequency long wavelength KH instability. They also discovered that this vortex structure is associated with small scale high frequency ion cyclotron waves. Although their conditions are different from the ionosphere, the physics is similar to our model for the ionospheric plasmas and indicates that a KH vortex can be a source for the higher frequency ion cyclotron waves.

From our preliminary study a coherent picture of low altitude ion heating due to velocity shear is emerging. While meso-scale low frequency wave steepening leads to a stress build-up in localized layers which seeds the micro-instabilities, the micro-instabilities lead to stress relief through dissipation. In nature, these two opposing tendencies will lead to a balance which will prevent indefinite steepening of low frequency waves and result in a steady state. This also emphasizes the crucial role of micro-processes and the obvious need for coupling the micro- and the macro-physics models for a realistic representation of natural systems. Schematically the sequence of processes leading to energization is sketched in Fig. 2. The observed velocity shear in the ionosphere can excite meso-scale low frequency waves (such as the Kelvin-Helmholtz instability). The nonlinear evolution of low frequency waves leads to the formation of strongly sheared flow regions with large values of the shear frequency ω_s locally. With large ω_s, the nonlinearly developed low frequency waves can seed the high frequency velocity shear driven micro-istabilities. As the micro-instabilities grow they lead to ion energization and thereby initiate ion upflow. Other contributions by the micro-processes, such as viscosity, resistivity, and thermal conduction, affect the large-scale plasma modes and transport properties. This leads to small-scale-size structures. The process is iterated and a balance is reached.

5. DISCUSSION AND CONCLUSION

Our particle simulations of the EIH instability show that large reduction of the shear parameter are achieved within 20 lower hybrid times which is less than 2 ion gyroperiods [Romero and Ganguli, 1993]. In terms of typical ionospheric numbers ($\Omega_i \sim 300$ rad/s), in less than 0.04 s. Also, these shear layers are expected to be localized over regions on the order of an ion gyroradius. Thus, direct in-situ observation of strong velocity shears will be difficult unless experiments are specifically designed for it. Experiments with resolutions fine enough to reliably observe the physics within local structures of the order of an ion gyroradius are desirable. It is encouraging to note that density structures of the order of a few ρ_i capable of supporting strong shears have been reported in the nightside auroral ionosphere in the range of 500 - 1000 km [Kintner et al. 1992, Vago et al. 1992; Arnoldy et al., 1992; Lynch et al., 1994]. The origin of these structures are interpreted as due to a collapsing lower hybrid caviton [Chang, 1993 and references therein]. There is indeed good evidence in support of this elegant interpretation, although as Seyler [1994] points out the 'smoking gun', i.e. a collapsing soliton, has yet to be observed. From our analysis it appears that our model may also play a role in the formation of steep density gradients as a low frequency wave evolves and eddy flows develop. Local shear ($\omega_s < \Omega_i$) generated in the eddy flows can trigger the IEDDI which can heat the ions [Ganguli et al., 1985]. The heated ions are then subjected to a mirror force and, as suggested by Singh [1994], can easily be expelled from the region to form density cavities. As the cavities become deeper, the local shear becomes stronger. This can trigger the higher frequency ($\omega \sim \omega_{LH}$) EIH or the lower hybrid drift instability. This may explain the often reported association of lower hybrid turbulence with these density cavities. The appealing feature of this model is that the source of free energy is locally available in the auroral region where these observations are made. It is emphasized that the plausibility argument outlined above does not challenge the viability of the Chang theory of caviton collapse. In fact it is highly likely that both the processes are operating and more research is necessary to identfy the physical parameter that governs the dominance of either of the mechanism and the relationship of the mechanisms to each other.

Acknowledgments: Numerous stimulating discussions with Drs Tom Chang, Mark Koepke, Bill Amatucci, Tom Moore, Rod Heelis, Mike Keskinen, Craig Pollock, Hugo Romero, Kristina Lynch, Jim Horwitz, Nagendra Singh, Joel Fedder, Peter Palmadesso, Supriya Ganguli, Valeri Gavrichtchaka, Jim Carroll, and Pat Reiff are acknowledged. This work is supported by the Office of Naval Research.

REFERENCES

Amatucci, W.E., M.E. Koepke, J.J. Carroll III, and T.E. Sheridan, Geophys. Res. Lett., Observation of ion-cyclotron turbulence at small values of magnetic-field-aligned current, 21, 1595, 1994.

Arnoldy, R.L., K.A. Lynch, P.M. Kintner, J. Vago, S. Chesney, T.E. Moore, and C.J. Pollock, Bursts of transverse ion acceleration at rocket altitudes, Geophys. Res. Lett., 19, 413, 1992.

Ball, L and M. Andre, Heating of O^+ ions in the cusp/cleft: Double-cyclotron absorption versus cyclotron resonance, J. Geophys Res., 96, 1429, 1991.

Baron, M.J. and R.H. Wand, Joule heating in the high latitude ionosphere, J. Geophys. Res., 88, 4114, 1983.

Basu, B. and B. Coppi, Change of confinement properties and transition from absolute to non-normal mode microinstabilities, Phys. Fluids B, 4, 2817, 1992.

Basu Sunanda, Santimay Basu, E. MacKenzie, P.F. Fougere, W.R. Coley, N.C. Maynard, J.D. Winningham, M. Sugiura, W.B. Hanson, and W.R. Hoegy, Simultaneous density and electric field fluctuation spectra associated with velocity shears in the auroral oval, J. Geophys. Res., 93, 115, 1988.

Block, L.P. and C.-G. Falthammar, The role of magnetic-field-aligned electric fields in auroral acceleration, J. Geophys. Res., 95, 5877, 1990.

Borovsky, J.E., The production of ion conics by oblique double layers, J. Geophys. Res., 89, 2251, 1984.

Brown, D.G., G.R. Wilson, J.L. Horwitz, Self-consistent production of ion conics on return current region auroral field lines: A time-dependent semi-kinetic model, Geophys. Res. Lett., 18, 1841, 1991.

Chang, T. and B. Coppi, Lower hybrid acceleration and ion evolution in the supraauroral region, Geophys. res. Lett., 8, 1253, 1981.

Chang, T., Lower hybrid collapse caviton turbulence and charged particle energization in the topside auroral ionosphere and magnetosphere, Phys. Fluids B, 5, 2646, 1993.

Crew, G.B., T. Chang, J.M. Retterer, W.K. Peterson, D.A. Gurnett and R.L. Huff, Ion cyclotron resonance heated conics: theory and observations, J. Geophys. Res., 95, 3959, 1990.

Drummond, W.E. and M.N. Rosenbluth, Phys. Fluids., Anomalous diffusion arising from microinstabilities in a plasma, 5, 1507 (1962).

Dimonte, G. and L.G. Wiley, Dynamics of exploding plasmas in a magnetic field, Phys. Rev. Lett., 67, 1755, 1991.

Earle, G.D., M.C. Kelley, and G. Ganguli, Large velocity shears and associated electrostatic waves and turbulence in the auroral F-region, J. Geophys. Res., 94, 15321, 1989.

Ganguli, G., P. Palmadesso and Y.C. Lee, Electrostatic ion cyclotron instability due to a nonuniform electric field perpendicular to the external magnetic field, Geophys. Res. Lett., 12, 643, 1985.

Ganguli, G., Y.C. Lee and P.J. Palmadesso, A new mechanism for excitation of waves in a magnetoplasma I. Linear theory, in Ion Acceleration in the Magnetosphere and Ionosphere, Geophys. Monogr. Ser., 38, edited by T. Chang, p 297, AGU, Washington, D.C., 1986.

Ganguli, G. and P.J. Palmadesso, Electrostatic ion instabilities in the presence of parallel currents and transverse electric fields, Geophys. Res. Lett., 15, 103, 1988.

Ganguli, G., Y.C. Lee and P.J. Palmadesso, Kinetic theory for electrostatic waves due to transverse velocity shears, Phys. Fluids, 31, 823, 1988a.

Ganguli, G., Y.C. Lee and P.J. Palmadesso, Electron-Ion hybrid mode due to transverse velocity shear, Phys. Fluids., 31, 2753, 1988b.

Ganguli, G., Y.C. Lee, P.J. Palmadesso and S.L. Ossakow, Oscillations in a plasma with parallel currents and transverse velocity shears, in *Physics of Space Plasmas (1988)*, SPI Conference Proceedings and Reprint Series, edited by T. Chang, G.B. Crew and J.R. Jasperse, 8, pp. 231, Scintific Publishers, Inc., Cambridge, MA, 1989a.

Ganguli, G., Y.C. Lee, P.J. Palmadesso and S.L. Ossakow, D.C. electric field stabilization of plasma fluctuations due to a velocity shear in the parallel ion flow, Geophys. Res. Lett., 16, 735, 1989b.

Ganguli, G., H. Romero, and J. Fedder, Interaction between global MHD and kinetic processes in the magnetotail, Solar System Plasmas in Space and Time, Geophysical Monograph #84, J.L. Burch and J.H. Waite, Jr. eds, p. 135, 1994a.

Ganguli, G., M.J. Keskinen, H. Romero, R. Heelis, T. Moore, and C. Pollock, Coupling of microprocesses and macroprocesses due to velocity shear: An application to the low-altitude ionosphere, J. Geophys. Res., 99, 8873, 1994b.

Ganguli, S.B. and P.J. Palmadesso, Plasma transport in the auroral return current region, J. Geophys. Res., 92, 8673, 1987.

Ganguli, S.B. and P.J. Palmadesso, and H.G. Mitchell, Effects of electron heating on the current driven electrostatic ion cyclotron instability and plasma transport processes along auroral field lines, Geophys. Res. Lett., 15, 1291, 1988.

Ganguli, S.B., H.G. Mitchell, and P.J. Palmadesso, Effects of nonlinear kinetic electron heating on macroscopic plasma transport processes, in *Physics of Space Plasmas (1991)*, SPI Conference Proceedings and Reprint Series, edited by T. Chang, G.B. Crew and J.R. Jasperse, 11, pp. 197, Scintific Publishers, Inc., Cambridge, MA, 1991.

Gavrichtchaka, V., M.E. Koepke, J.J. Carroll III, and G. Ganguli, Frequency range and spectral width of waves associated with IEDD excitation mechanism, in Geophysical Monograph, (this volume), 1995.

Gurnett, D. A., R.R. Anderson, B.T. Tsurutani, E.J., Smith, G. Paschmann, G. Harendel, S.J. Bame, and C.T. Russell, Plasma wave turbulence at the magnetopause: Observations from ISEE 1 and 2, J. Geophys. Res., 84, 7043, 1979.

Heppner, J.P., M.C. Liebrecht, N.C. Maynard, and R.F. Pfaff, High-latitude distributions of plasma waves and spatial irregularities from DE 2 alternating current electric field observations, J. Geophys. Res., 98, 1629, 1993.

Hill, T.W., Solar-wind magnetosphere coupling, in *Solar-Terrestrial Physics Principles and Theoretical Foundations*, R.L. Carovillano and J.M. Forbes, eds, p. 261, (D. Reidel Publishing Company, Dordrecht, Boston, lancaster, 1982).

Horwitz, J.L., Parabolic heavy ion flow in the polar magnetosphere, J. Geophys. Res., 92, 175, 1987.

Horwitz, J.L., C.J. Pollock, T.E. Moore, W.K. Peterson, J.L. Burch, J.D. Winningham, J.D. Craven, L.A. Frank, and A. Persoon, The polar cap environment of outflowing O^+, J. Geophys Res., 97, 8361, 1992.

Huba, J.D., Hall dynamics of the Kelvin-Helmholtz instability, Phys. Rev. Lett., 72, 2033, 1994.

Huba, J.D., Theoretical simulation methods applied to high latitude ionospheric turbulence, in *Physics of Space Plasmas (1988)*, SPI Conference Proceedings and Reprint Series, edited by T. Chang, G.B. Crew and J.R. Jasperse, 8, pp. 49, Scintific Publishers, Inc., Cambridge, MA, 1989.

Jassby, D.L., Evolution and large-electric-field suppression of the transverse Kelvin-Helmholtz instability, Phys. Rev. Lett., 25, 1567, 1970.

Kelley, M.C., in The Earth's Ionosphere (Academic Press, London, 1989).

Kelley, M.C. and C.W. Carlson, Observation of intense velocity shear and associated electrostatic waves near auroral arc, J. Geophys. Res., 82, 2343, 1977.

Keskinen, M.J., H.G. Mitchell, J.A. Fedder, P. Satyanarayana, S.T. Zalezak and J.D. Huba, Nonlinear evolution of Kelvin-Helmholtz instability in the high latitude ionosphere, J. Geophys. Res., 93, 137, 1988.

Kintner, P.M., J.L. Vago, S.W. Chesney, R.L. Arnoldy, K.A. Lynch, C.J. Pollock, and T.E. Moore, Localized lower hybrid acceleration of ionospheric plasma, Phys. Rev. Lett., 68, 2448, 1992.

Kinzelin, E. and D. Hubert, Ion velocity distribution function in the upper auroral F region: 1. Phenomenological approach, J. Geophys. Res., 97, 4061, 1992.

Koepke, M.E., W.E. Amatucci, J.J. Carroll III, and T.E. Sheriden, Experimental verification of the inhomogeneous energy-density driven instability, Phys. Res. Lett., 72, 3355, 1994.

Koepke, M.E. and W.E. Amatucci, Electrostatic ion cyclotron wave experiments in the WVU Q-machine, IEEE Trans. Plasma Sci., 20, 631, 1992.

Koepke, M.E., J.J. Carroll III, W.E. Amatucci, V. Gavrichtchaka, and G. Ganguli, Velocity-shear-induced ion-cyclotron turbulence: Laboratory identification and space applications, Phys. Fluids B, (to appear), 1995.

Kindel, J.M. and C.F. Kennel, Topside current instabilities, J. Geophys. Res., 76, 3055, 1971.

Krall, N.A. and P.C. Liewer, Low frequency instabilities in magnetic pulses, Rev. A 4, 2094, 1971.

Lee, L.C., The Magnetopause: A tutorial review, in *Physics of Space Plasmas (1990)*, SPI Conference Reprint Series, # 10, T. Chang, G.B. Crew, and J. Jasperes, eds., (Scientific Publishers Inc., Cambridge, MA, 1991), p.33.

Lockwood, M, M.O. Chandler, J.L. Horwitz, J.R. Waite, Jr., T.E. Moore and C.R. Chappell, J. Geophys. Res., The cleft ion fountain, 90, 9736, 1985.

Loranc, M., W.B., Hanson, R.A. Heelis, and J.-P. St.-Maurice, A morphological study of vertical ionospheric flows in the high-latitude f region, J. Geophys. Res., 96, 3627, 1991.

Lu, G., P.H. Reiff, T.E. Moore, and R.A. Heelis, Upflowing ionospheric ions in the auroral region, J. Geophys. Res., 97, 16855, 1992.

Lundin, R., G. Gustafsson, A.I. Eriksson, and G. Marklund, On the importance of high-latitude low-frequency electric fluctuations for the escape of ionospheric ions, J. Geophys. Res., 95, 5905, 1990.

Lysak, R.L. and M.K. Hudson, Effects of double layers on magnetosphere-ionosphere coupling, Laser and Particle Beams (1987), 5, 351, 1987.

Manheimer, W and J.P. Boris, Marginal stability analysis - A simpler approach to anomalous transport in plasmas, Comments Plasma Phys. Cont. Fusion, 3, 15, 1977.

Marklund, G., L. Blomberg, C-G. Falthammar, and P-A. Lindqvist, On intense diverging electric fields associated with black aurora, Geophys. Res. Lett., 21, 1859, 1994.

Moore, T.E., Superthermal ionospheric outflows, Rev. Geophys. Space Phys., 22, 264, 1984.

Moore, T.E., M. Lockwood, M.O. Chandler, J.H. Waite, Jr., A. Persoon and M. Sugiura, Upwelling O^+ ion source characteristics, J. Geophys. Res., 91, 7019, 1986a.

Moore, T.E., J.H. Waite,Jr., M. Lockwood, and C.R. Chappell, Observation of coherent transverse ion acceleration, in Ion Acceleration in the Magnetosphere and Ionosphere, Geophys. Monogr. Ser., 38, edited by T. Chang, p 50, AGU, Washington, D.C., 1986b.

Moore, T.E., Origin of magnetospheric plasmas, Contribution in Solar-Planetary Relationship, p. 1039, US National Report 1987 - 1990, Twentieth General Assembly, IUGG, 1991.

Moore, T.E., M.O. Chandler, C.J. Pollock, D.L. Reasoner, R.L. Arnoldy, B. Austin, P.M. Kintner, and J. Bonnell, Plasma heating and flow in an auroral arc, J. Geophys. Res., (submitted), 1995.

Mozer, F.S., C.W. Carlson, M.K. Hudson, R.B. Torbert, B. Parady, J. Yatteau, and M.C. Kelley, Observation of paired electrostatic shocks in polar magnetosphere, Phys. Rev. Lett., 38, 292, 1977.

Nishikawa, K.-I., G. Ganguli, Y.C. Lee and P.J. Palmadesso, Simulation of ion-cyclotron-like modes in a magnetoplasma with transverse inhomogeneous electric field, Phys. Fluids., 31, 1568, 1988.

Nishikawa, K.-I., G. Ganguli, Y.C. Lee and P.J. Palmadesso, Simulation of electrostatic turbulence due to sheared flows parallel and transverse to the magnetic field, J. Geophys. Res, 95, 1029, 1990.

Papadopoulos, K., A review of anomalous resistivity for the ionosphere, Rev. Geophys. Spac. Phys., 15, 113, 1977.

Peyser, T.A., C.K. Manka, B.H. Ripin, and G. Ganguli, Electron-ion-hybrid instability in laser-produced plasma expansion across magnetic fields, Phys. Fluids B, 8, 2448, 1992.

Pfaff, R.F., M.C. Kelley, E. Kudeki, B.G. Fejer, and K.D. Baker, Electric field and plasma density measurements in the strongly driven daytime equatorial electrojet 2. Two -stream Waves, J. Geophys. Res., 92, 13597, 1987.

Pollock, C.J., M.O. Chandler, T.E. Moore, J.H. Waite, Jr., C.R. Chappell and D.A. Gurnett, J. Geophys. Res., A survey of upwelling ion event characteristics, 95, 18969, 1990.

Rayleigh, Lord, in Theory of Sound (MacMillan, London, 1896)(reprinted 1940), Vol. II, Chap. 21.

Romero, H., G. Ganguli, P.B. Dusenbery and P.J. Palmadesso, Equilibrium structure of the plasma sheet boundary layer-lobe interface, Geophys. Res. Lett., 17, 2313, 1990.

Romero, H., G. Ganguli, Y.C. Lee and P.J. Palmadesso, Electron-ion hybrid instabilities driven by velocity shear in a magnetized plasma, Phys. Fluids. B, 4, 1708, 1992a.

Romero, H., G. Ganguli, and Y.C., Lee, Ion acceleration and Coherent structures generated by lower hybrid shear-driven instabilities, Phys. Rev. Lett., 69, 3505, 1992b.

Romero, H. and G. Ganguli, Nonlinear evolution of a strongly sheared cross-field plasma flow, Phys Fluids B, 5, 3163, 1993.

Romero, H. and G. Ganguli, Relaxation of the stressed plasma sheet boundary layer, Geophys. Res. Lett., 21, 645, 1994.

Sagdeev, R.Z., Cooperative phenomenoa and shock waves in a collisionless plasma, in *Reviews of Plasma Physics (Leontovich, M.A., Ed.)*, 4, 23, 1966.

Sato, N., M. Nakamura, and R. Hatakeyama, Three-dimensional double layers inducing ion-cyclotron oscillations in a collisionless plasma, Phys. Rev. Lett., 57, 1227, 1986.

Satyanarayana, P., P.K. Chaturvedi, M.J. Keskinen, J.D. Huba, and S.L. Ossakow, Theory of current driven ion cyclotron instability in the bottomside ionosphere, J. Geophys. Res., 90, 12209, 1985

Schunk, R.W. and D.S. watkins, Electron temperature anisotropy in the polar wind, J. Geophys. Res., 86, 91, 1981.

Seyler, C.E., A mathematical model of the structure and

evolution of small-scale discrete auroral arcs, J. Geophys. Res., 95, 17199, 1990.

Seyler, C.E., Lower hybrid wave phenomena associated with density depletions, J. Geophys. Res., 99, 19513, 1994.

Shelley, E.G., R.G. Johnson and R.D. Sharp, Satellite observations of energetic heavy ions during a geomagnetic substorm, J. Geophys. Res., 77, 6104, 1972.

Shelley, E.G., R.D. Sharp and R.G. Johnson, Satellite observations of an ionospheric acceleration mechanism, Geophys Res. Lett., 3, 654, 1976.

Singh, N., Pondermotive versus mirror force in creation of the filamentary cavities in auroral plasma, Geophys. Res. Lett., 21, 257, 1994.

Siscoe, G.L., Solar system magnetohydrodynamics, in *Solar-Terrestrial Physics Principles and Theoretical Foundations*, R.L. Carovillano and J.M. Forbes, eds, p. 11, (D. Reidel Publishing Company, Dordrecht, Boston, lancaster, 1982).

St-Maurice, J.-P. and R.W. Schunk, Ion velocity distributions in high latitude ionosphere, Rev. of Geophys. and Sp. Sc., 17, 99, 1977.

Temerin, M. and I. Roth, Ion heating by waves with frequencies below the ion gyrofrequency, Geophys. Res. Lett., 13, 1109, 1986.

Temerin, M.C., C. Cattell, R.L. Lysak, M.K. Hudson, R.B. Torbert, F.S. Mozer, R.D. Sharp and P.M. Kintner, The small-scale structure of electrostatic shocks, J. Geophys. Res., 86, 11278, 1981.

Theilhaber, K. and C.K. Birdsall, Kelvin-Helmholtz vortex formation and particle transport in a cross-field plasma sheath, Phys. Rev. Lett., 62, 772, 1989.

Thomas, V.A. and D. Winske, Kinetic simulations of the Kelvin-Helmoltz instability at the magnetopause, J. Geophys. Res., 98, 11425, 1993.

Tsunoda, R.T., High latitude F region irregularities: A review and synthesis, Rev. Geophys., 26, 719, 1988.

Tsunoda, R.T., R.C. Livingston, J.F. Vickery, R.A. Heelis, W.B. Hanson, F.J. Rich and P.F. Bythrow, Dayside observations of thermal-ion upwellings at 800-km altitude: An ionospheric signature of cleft ion fountain, J. Geophys. Res., 94, 15277, 1989.

Vago, J.L., P.M. Kintner, S.W. Chesney, R.L. Arnoldy, K.A. Lynch, T.E. Moore, and C.J. Pollock, Transverse ion acceleration by lower hybrid waves in the topside auroral ionosphere, J. Geophys. Res., 97, 16935, 1992.

Van Niekerk, E.G., P.H. Krumm, and M.J. Alport, Electrostatic ion cyclotron waves driven by a radial electric field, Plasma Phys. Contr. Fusion, 33, 375, 1991.

Waite, J.H., Jr., T.E. Moore, M.O. Chandler, M. Lockwood, A. Persooon, and M. Sugiura, Ion energization in upwelling events, in Ion Acceleration in the Magnetosphere and Ionosphere, Geophys. Monogr. Ser., 38, edited by T. Chang, p 61, AGU, Washington, D.C., 1986.

Whalen, B.A., W. Bernstein and P.W. Daly, Geophys. Res. Lett., Low altitude acceleration of ionospheric ions, 5, 55, 1978.

Whalen, B.A., S. Watanabe, and A.W. Yau, Thermal and suprathermal ion observations in the low altitude transverse ion energization region, Geophys. Res. Lett., 18, 725, 1991.

Whealton, J.H. and S.B. Woo, Ion velocity distribution of a weakly ionized gas in a uniform electric field of arbitrary strength, Phys. Rev. A, 6, 2319, 1971.

Winske, D., S.P. Gary, and D.S. Lemmons, Diffusive transport at the magnetopause, in *Physics of Space Plasmas (1990)*, SPI Conference Reprint Series, # 10, T. Chang, G.B. Crew, and J. Jasperes, eds., (Scientific Publishers Inc., Cambridge, MA, 1991), p. 397.

Yamada, M. and D.K. Owens, Cross-field-current driven lower-hybrid instability and stochastic ion heating, Phys. Rev. Lett., 38, 1529, 1977.

Yamamoto. T., M. Ozaki, S. Inoue, K. Makita, and C.-I. Meng, Convective generation of "giant" undulations on the evening diffuse auroral boundary, J. Geophys. Res., 99, 19499, 1994.

Yau, A.W., B.A. Whalen, A.G. McNamara, P.J. Kellogg and W. Bernstein, J. Geophys. Res., Particle and wave observations of low-altitude ion ionospheric ion acceleration events, 88, 3411, 1983.

G. Ganguli, Code 6794, Plasma Physics Division, Naval Research Laboratory, Washington DC 20375

Microscale Effects From Global Hot Plasma Imagery

T. E. Moore and M.-C. Fok

Space Sciences Laboratory, NASA Marshall Space Flight Center, Huntsville, Alabama

J. D. Perez and J. P. Keady

Department of Physics, Auburn University, Auburn, Alabama

We have used a three-dimensional model of recovery phase storm hot plasmas to explore the signatures of pitch angle distributions (PADs) in global fast atom imagery of the magnetosphere. The model computes mass, energy, and position-dependent PADs based on drift effects, charge exchange losses, and Coulomb drag. The hot plasma PAD strongly influences both the storm current system carried by the hot plasma and its time evolution. In turn, the PAD is strongly influenced by plasma waves through pitch angle diffusion, a microscale effect. We report the first simulated neutral atom images that account for anisotropic PADs within the hot plasma. They exhibit spatial distribution features that correspond directly to the PADs along the lines of sight. We investigate the use of image brightness distributions along tangent-shell field lines to infer equatorial PADs. In tangent-shell regions with minimal spatial gradients, reasonably accurate PADs are inferred from simulated images. They demonstrate the importance of modeling PADs for image inversion and show that comparisons of models with real storm plasma images will reveal the global effects of these microscale processes.

INTRODUCTION

Microscale processes in a plasma are those having scales larger than the Debye length, but smaller than any other relevant scales in the plasma system or subsystems. The term also connotes "collective effects" as contrasted with collisional effects that occur on atomic collision scales. These microscale processes are thought to reduce the mean free path associated with the collisional effects, augmenting those weak effects in nearly collisionless space plasma systems. In such a case, microscale processes have the shortest time scales in the system.

There has been a long and honorable debate between those who advocate the importance of such microscale collective effects and those who are skeptical of their significance. The skeptics have argued convincingly that simple observation of microscale effects, which appear ubiquitous in space plasmas, is necessary but not sufficient to demonstrate that they have important consequences. They argue that all collisional effects must be fully evaluated and shown to be inadequate before a solid case can be made for significant microscale effects. The result has been a standoff in which the advocates have been led to state their results in terms of possibilities instead of certainties, e.g. "plasma waves *can* cause enhanced particle loss through diffusion into the loss cone." If space physics is to become a predictive science capable of supporting an operational space weather service, we must overcome this problem and reach the point where we know that microscale effects *will* produce specific consequences.

One approach that is being taken is to more rigorously and completely evaluate the actual effects of collisional processes in space plasmas. For example, *Fok et al.* [1995] have developed a fairly detailed model of the effects of charge exchange and Coulomb drag on the storm time, recovery phase hot plasmas of the inner magnetosphere. This model tracks the losses from the storm's hot plasma system during recovery and therefore leads to specific comparisons between the observed and modeled recovery phase plasma evolution. Preliminary indications from this comparison are that the modeled hot plasma decays more gradually in the early phase than the observed hot plasma does, as measured by the D_{st} index of the storm ring current. One hypothesis concerning this discrepancy is that plasma wave-driven pitch angle diffusion causes higher loss rates in the real world than in the model, which does not account for such microscale effects. In this paper, we identify a plan for evaluating this hypothesis in some detail, using observations that would be provided by a fast atom-imaging instrument, as proposed for the Magnetosphere Imager Mission being studied by Marshall Space Flight Center for the NASA Space Physics Division.

THREE-DIMENSIONAL HOT PLASMA MODEL AND HOT PLASMA IMAGING

The storm-time plasma model used herein is that of *Fok et al.* [1995], to which the reader is referred for additional details. In brief, the model solves a Boltzmann initial/boundary value problem with specified electric and magnetic fields, including the effects of charge exchange losses and Coulomb drag on energetic particles interacting with the hydrogen geocorona and the plasmasphere, respectively. Rather than tracking particle trajectories in detail, a bounce-averaged approach is taken. At present, a dipole magnetic field has been used for the region between 2.0 and 6.5 R_E. The circulation electric field is prescribed by the Volland-Stern model, and the plasmasphere is from the model of *Rasmussen et al.* [1993], with variations prescribed by the time series of K_p values for any specific storm interval. The initial and outer boundary conditions are derived from observations of storm main phase hot plasmas, i.e., those of *Hamilton et al.* [1988]. The main contribution of this storm plasma model lies in its attention to the pitch angle-dependent effects of charge exchange and Coulomb collisions. It computes the equatorial pitch angle distributions (PADs) as a function of ion species and location in the equatorial plane, leading to an effective three-dimensional description of the hot plasma distribution in space.

As compared with other recent simulations [e.g., *Roelof et al.*, 1985; *Williams et al.* 1992; *Moore et al.*, 1992] of fast neutral atom emission from the storm plasma region, the present work adds a realistic description of the spatial and temporal evolution of the hot plasma in three dimensions. The charge exchange reaction is essentially free of angular scattering, so the angular distribution of fast atoms emitted from each volume of space is essentially identical to the angular distribution of the parent ion population. At energies of tens of eV and higher, emitted fast atoms travel essentially in straight lines and form a basis for imaging the emissivity distribution from arbitrary vantage points. It is apparent that isotropic hot plasma produces a different flux brightness distribution than anisotropic hot plasma. The brightness distribution of an image is thus influenced as much by the (pitch) angular distribution of the hot plasma as it is by its spatial distribution.

It has been supposed that knowledge of the geocoronal hydrogen distribution allows it to be deconvolved from the fast atom images, yielding the spatial distribution of the hot plasma ions. However, as with most inverse problems, the solution is never unique. Selection of a credible solution from among the nonunique possibilities depends upon the availability and accuracy of other knowledge of the form of the hot plasma distribution. For example, the inversion is vastly improved when imagery is available from multiple complementary vantage points, permitting tomographic inversion. All such inversions proceed from forward modeling of the hot plasma distribution that must, at a minimum, include all parameters that significantly influence the images. The arguments above mandate that these parameters must in any case include a reasonable description of the possible hot plasma pitch angle distributions.

Acceleration and transport processes produce freshly injected storm plasmas ranging from nearly isotropic to "cigar-shaped," having magnetic field-aligned angular distributions. Charge exchange and Coulomb losses owing to interactions with the terrestrial neutral atmosphere remove ions most rapidly at field-aligned pitch angles, causing the equatorial distributions to become deficient in such particles, or "pancake" shaped. The ensuing loss of storm particles is very strongly influenced by the rate of pitch angle diffusion, which transports particles from long-lasting equatorial mirroring pitch angles to smaller angles, resulting in much more rapid loss. In turn, the rate of pitch angle diffusion is controlled by the amplitude of plasma waves that are driven by the pitch angle anisotropy and other microphysical features of the plasma. The actual form of the PAD is more isotropic than it would be in the absence of such waves, to a degree that is indicative of the intensity spectrum of the waves.

Clearly, proper accounting for the possible angular distributions within storm plasmas adds to the complexity of the image inversion process. On the other hand, no inversion that ignores angular distributions can conclusively identify spatial features of the hot plasma. Image inversions that properly account for PADs hold the promise to provide evidence of microscale, as well as macroscale, processes operative during storm recovery. In this paper, we assess the degree to which fast atom imagery can support the evaluation of microscale effects on storm recovery using an appropriate model. The three-dimensional drift-loss model provides the basis for simulating the flux of fast neutral atoms emitted as a result of the interaction of the storm-time plasmas with the hydrogen geocorona and plasmasphere, in the absence of pitch angle diffusion. However, the goal will be to identify means for determining, from real images, the true PADs of the hot plasma ions. To the degree that the inferred PADs are more isotropic than those predicted by this model, the results will indicate the presence of significant plasma wave intensities and resultant pitch angle diffusion.

To provide important global context, we begin with the simulation of polar views of fast atom emission from the storm recovery phase hot plasmas. A simple means for qualitative recovery of the integral ion flux is demonstrated using these images. However, storm plasma PAD features are expected to appear most clearly in the distribution of fast atom emissivity along selected field lines. Therefore, in subsequent sections we simulate equatorial views of the hot plasma in which equatorial and footpoint regions can be clearly distinguished. The same simple method for qualitative ion flux recovery is demonstrated on these images. Finally, a simple but quantitative method for inferring equatorial PADs is demonstrated. The results are discussed in terms of more advanced inversion techniques.

POLAR VIEWS

It is useful to gain a global perspective of the hot plasma distribution before focusing on the distribution along a particular field line. To gain such a global view of the hot storm plasmas, a view from high over the magnetic pole of the Earth is advantageous. Therefore we begin with development of polar views of the hot plasma.

Integral Hot Plasma Ion Flux

A useful step toward the production of simulated fast neutral atom images is the line-of-sight integration of the ion flux within a specified model distribution of magnetospheric hot plasma, without convolution with the geocoronal hydrogen distribution. Knowledge of the results absent the convolution are useful in the interpretation of fast atom images that incorporate the convolution. Plate 1 (left panel) shows the result of a line-of-sight integration of the ion flux from a polar vantage point, for protons of energy 1.7 keV, at a time 6 hours into the recovery phase of the major storm that occurred in February, 1986. In this integration, the ion flux within each step along the line-of-sight is taken from the relevant modeled local PAD according to the local pitch angle of the line-of-sight. Consequently, image brightness is controlled not only by the spatial distribution of ion flux, but also by the local PAD according to the predominant orientation of the line-of-sight relative to the local magnetic field.

Notable features of the 1.7 keV proton plasma that are evident from this image include the strong minimum of flux on the dayside, and the well-defined convecting cloud in the 0900LT sector (upper left of the left panel). We omit a similar image for the proton plasma at a higher energy of 100 keV, because the most notable feature would be the highly uniform distribution of flux in local time. This is a direct result of the very fast gradient and curvature drift of such energetic ions, and their relatively negligible convection drift speed.

Fast Neutral Atom Flux

When the cold atomic hydrogen geocoronal distribution is appropriately convolved into the polar images shown above, the differences (apart from absolute magnitudes), as shown in Plate 2, lie mainly in the radial distribution of emission, which responds to the strong peaking of geocoronal density close to the Earth. Because the ion fluxes are confined to magnetic field lines extending deep into the geocorona at high latitudes, the brightest emissions originate from the "footpoints" of the field lines, intersecting the Earth near the polar cap boundary. This produces the bright emission which appears inside the disk of the Earth indicated by the circle in these images.

We evaluate the fast neutral atom flux at 1-degree resolution for a hemispherical field of view, as shown in the left panel of Plate 2. In the right panel of Plate 2, we show the result of passing this flux through a simple instrument filter, specified by a pixel solid angle (4×4 degree in this case), an effective area (1 cm^2)), and an accumulation time (600 sec., assuming the image is obtained by a single row of pixels on a spinning spacecraft, and thus applying a pixel accumulation duty cycle of 4 degree of each 360-degree spin). The image is then expressed in counts for the accumulation, where integer counts have been selected at random from a range extending above and below the simulated value by one Poisson standard deviation, an approximate method for introducing counting noise.

Recovery Of Hot Plasma Ion Flux

Inspection of Plate 2 suggests that morphological features of the ion flux might be approximately recovered from the fast atom flux through a simple radial image rescaling to remove the radial dependence of the geocoronal hydrogen column density. Such a rescaling would implicitly assume that the geocoronal hydrogen and ion flux distributions along the line-of-sight are only weakly dependent on image pixel position, i.e., that they are factorable into separate line-of-sight integrations of the ion flux and the cold geocoronal density, forming separate ion flux and geocoronal images. To the extent that the ion flux is mainly confined to or largest at the equator, this assumption may be justifiable. With this simplifying assumption, the fast atom flux image F_{NA} can be approximately related to the integral ion flux image, F_i as follows:

$$F_{NA} = \sigma \int_{LOS} n_H f_i \, dl \approx \sigma \frac{1}{L} \int_{LOS} n_H \, dl \cdot \int_{LOS} f_i \, dl = \sigma \langle n_H \rangle \cdot F_i. \quad (1)$$

This equation applies separately to each image pixel, σ is the relevant cross section, L is the length of the line-of-

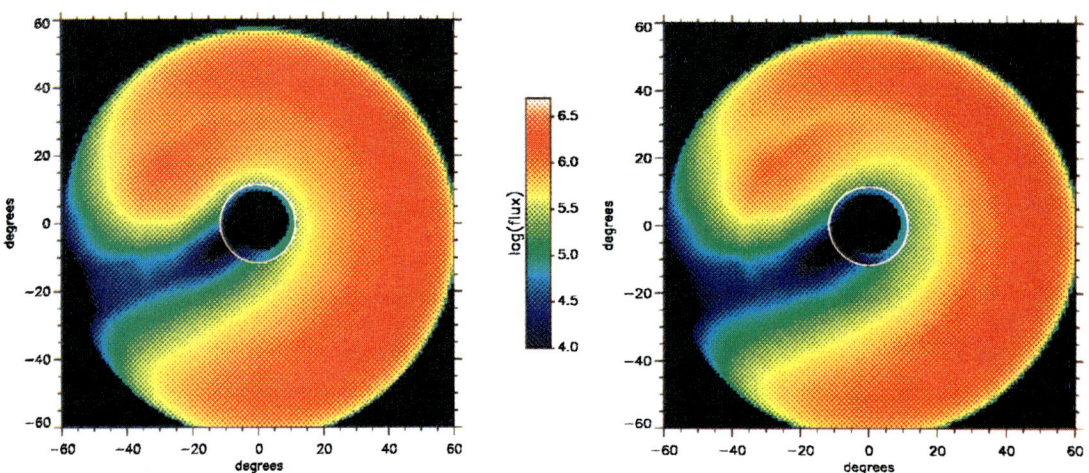

Plate 1. Comparison of model flux with flux as recovered from a simulated fast atom image. Left panel: the line-of-sight integral flux of 1.7 keV protons, as a two-dimensional hemispherical function of look angle, from a vantage point 5 R_E over the north pole, with the Sun toward the left. The three-dimensional ion flux is derived from the model of Fok et al. [1995], for an instant of time 6 hours into the recovery phase of a major magnetic storm (Feb. 1986 storm). Right panel: an estimate of the integral flux of 1.7 keV protons equivalent to the left panel, but derived through a simple recovery method from a simulated fast atom image of the same hot plasma population.

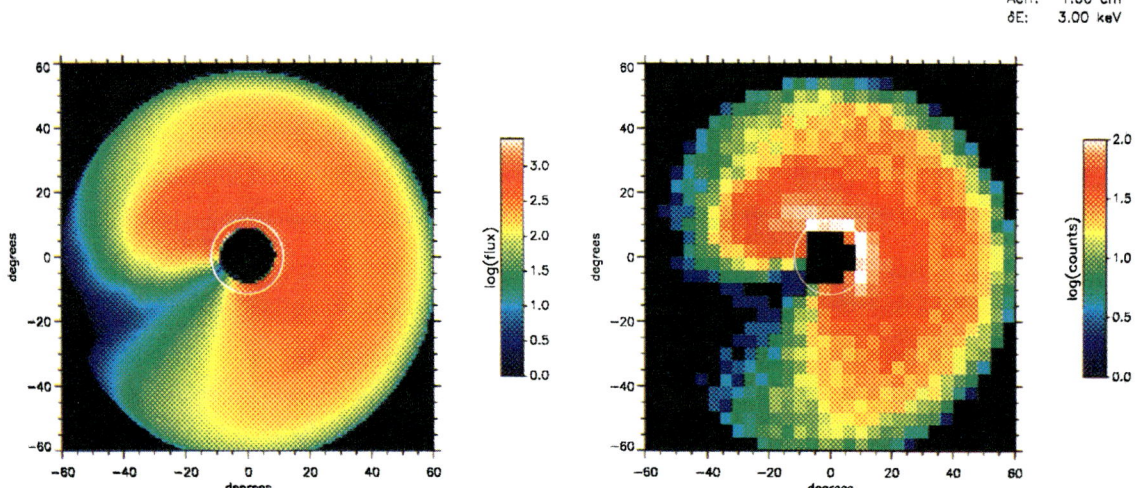

Plate 2. Comparison of the flux image of fast atom emission from the storm-time plasma, with the response to that flux of a hypothetical imaging instrument. Left panel: the fast atom image computed by the convolution of the ion flux distribution illustrated in Plate 1, left panel, with the geocoronal hydrogen density and the cross section for charge exchange between 1.7 keV protons and cold hydrogen, leading to a fast hydrogen atom and a cold proton. Right panel: the response of an imaging instrument to the flux distribution of the left panel, assuming an aperture of 1 cm², 4×4 degree pixels, and an energy bandwidth of 3 keV.

sight integration path, n_H is the geocoronal density, f_i is the ion flux in the viewing direction, and F_i is the ion integral flux. The geocoronal image $\langle n_H \rangle$ is implicitly defined by the equation. The integral ion flux can then be approximately recovered from the neutral atom image (F_{NA}), using the geocoronal hydrogen integral image, as follows:

$$F_i \approx F_{NA} \frac{1}{\sigma \langle n_H \rangle} \ . \qquad (2)$$

Plate 1 (right panel) shows the result for a polar view image of the magnetosphere at 1.7 keV. Recall that the left panel is the integral ion flux at 1 degree resolution. In the right panel is the quotient of the fast atom flux image at 1 degree resolution and the integral hydrogen density image, at the same resolution. It can be seen that the recovered ion flux is very similar to the model ion flux. Certainly, this technique provides a qualitatively credible reconstruction of the integral ion flux in this case.

This exercise is intended to demonstrate the plausibility of recovering meaningful ion flux distributions from fast atom imagery. It is clear that a more rigorous inversion will be required to quantitatively deconvolve the image, but this should not detract from the basic value of the method. Routine recovery of the images will yield information that will be useful in a browse mode where images must be selected for more rigorous inversion and analysis, or as a basis for initial guesses in an iterative inversion process.

EQUATORIAL VIEWS

Integral Hot Plasma Ion Flux

Complementary to the polar imagery of this storm plasma model, Plate 3 (left panel) shows the integral ion flux of protons with 1.7 keV energy at 6 hours recovery phase, but from a location at the equator near dusk local time and at a distance of 5 R_E. Notable features of the image include highly structured PADs that lead to strong spatial structuring of the ion flux, with contrasting behavior in the two halves of the plate. The vantage point and storm phase have been chosen so that the left (dayside) half of the left panel image reflects a region with "pancaked" PADs leading to equatorial peaking of the brightness, whereas the right (nightside) half of the image reflects a region with somewhat field-aligned PADs, leading to brightness maxima at the footpoints of the field lines in some locations. Since these are hemispherical all-sky images, distances are difficult to judge. The projections of $L=2, 4$, and 6 field lines are shown superposed on the image as a spatial reference system.

For comparison, Plate 4 (left panel) shows the integral ion flux from the same vantage point, but for protons of 100 keV energy, some 36 hours into the recovery phase. In this case, the ion flux is characterized by "pancaked" PADs on both day and night sides of the image with only modest spatial structuring of the ion flux or PAD. These characteristics reflect the late stage of the recovery phase, and the uniformity in local time that is expected for energetic ion populations with large drift speeds.

Fast Neutral Atom Flux

As in the case of polar views, the ion flux features are still generally discernible in the equatorial view of fast atom flux shown in the left side of Plates 5 (1.7 keV) and 6 (100 keV). However, these features remain severely distorted by the cold hydrogen density distribution, which enhances the emissivity in regions close to the Earth. This is true of both the 1.7 keV neutral atom images and the 100 keV images. The response of a hypothetical imaging instrument to these fluxes is shown in the right hand panels of the same two plates. Here it can be seen that a very high quality image is produced at 1.7 keV, but that the 100 keV image, in the hot tail of the proton energy distribution, is of marginal quality for the same aperture and exposure time, even though a much wider energy passband has been used.

Recovery of Hot Plasma Ion Flux

When we attempt the same ion flux recovery of the equatorial 1.7 keV and 100 keV fast atom images using the integral hydrogen density image, we find that this method is somewhat inaccurate within a substantial region of the inner magnetosphere for these equatorial views. The results of the operation for the 1.7 keV case are shown in the right panels of Plate 3 for direct comparison with the true model flux image. The differences between true and recovered ion flux result from the highly structured nature of the modeled plasma region, wherein local time structures appear as asymmetric depth variations along the lines of sight that make up the image pixels.

In the case of the equatorial 100 keV fast atom image, shown in the right hand panel of Plate 4, the simple radial image adjustment is reasonably accurate except in the innermost region, where the difference in line-of-sight distributions of ions and cold atoms becomes apparent. The improved accuracy at 100 keV stems from the greater uniformity of the energetic proton flux with local time and radius. In either case, it is clear that the equatorial fast atom images, when adjusted for the radial dependence of the hydrogen density, allow for qualitative discrimination between regions with "pancaked" and field-aligned PADs. In the following sections, we investigate the degree to which this can be used to quantitatively estimate the PADs in a specified region of space, using this simple image inversion method.

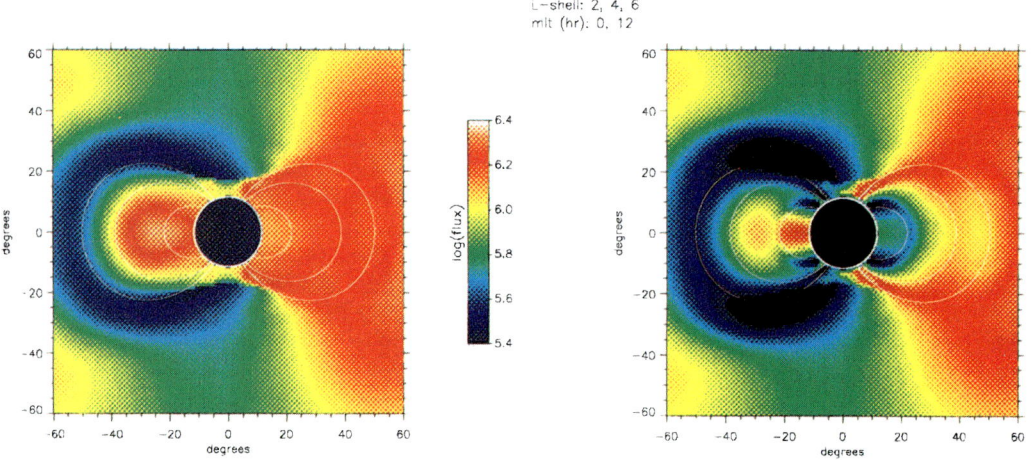

Plate 3. Similar to Plate 1, but for a vantage point located at the magnetic equatorial plane, at 1800 local time. Magnetic field lines at $L = 2, 4,$ and 6, at noon and midnight local times, have been superposed as a spatial reference system.

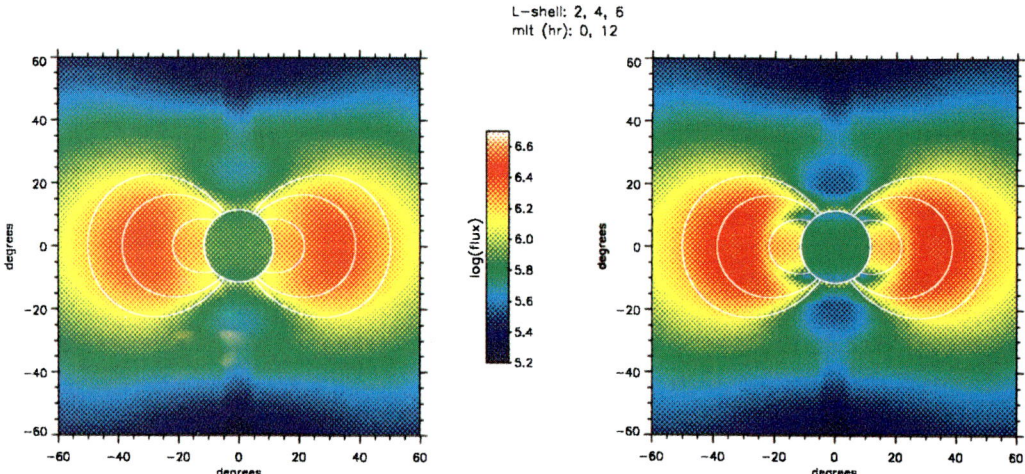

Plate 4. Similar to Plate 1, but for 100 keV protons at 36 hours into the storm recovery and a vantage point like that of Plate 3, located at the magnetic equatorial plane, at 1800 local time. Magnetic field lines at $L = 2, 4,$ and 6, at noon and midnight local times, have been superposed as a spatial reference system.

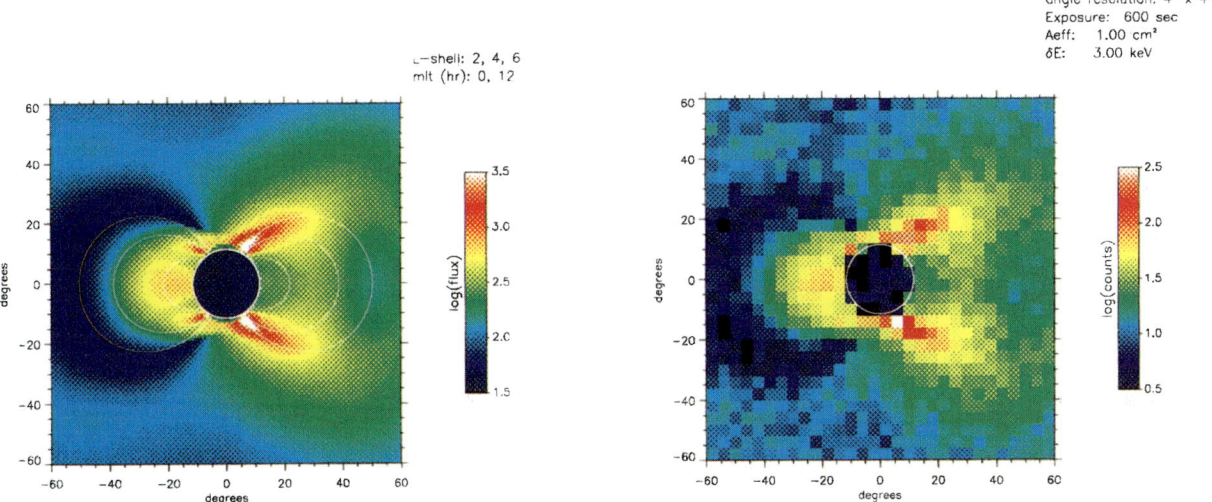

Plate 5. Similar to Plate 2, but for a vantage point located at the magnetic equatorial plane, 1800 local time. Magnetic field lines at $L = 2, 4$, and 6 at noon and midnight local times, have been superposed as a spatial reference system.

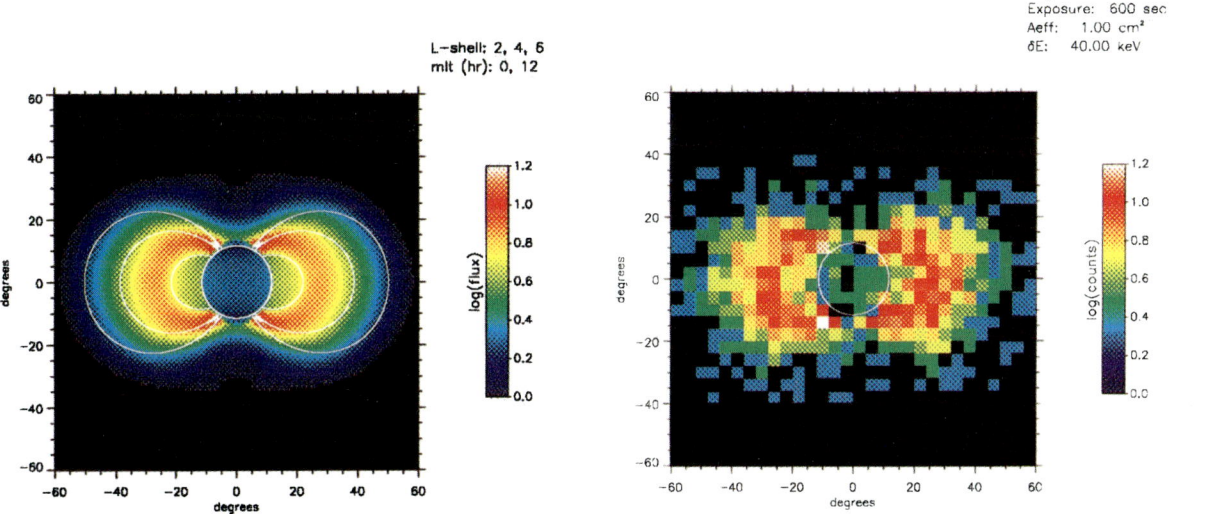

Plate 6. Similar to Plate 2, but for 100 keV protons 36 hours into the storm recovery and a vantage point like that of Plate 3, located at the magnetic equatorial plane, 1800 local time. Magnetic field lines at $L = 2, 4$, and 6 at noon and midnight local times have been superposed as a spatial reference system.

Tangent-Shell Field Line

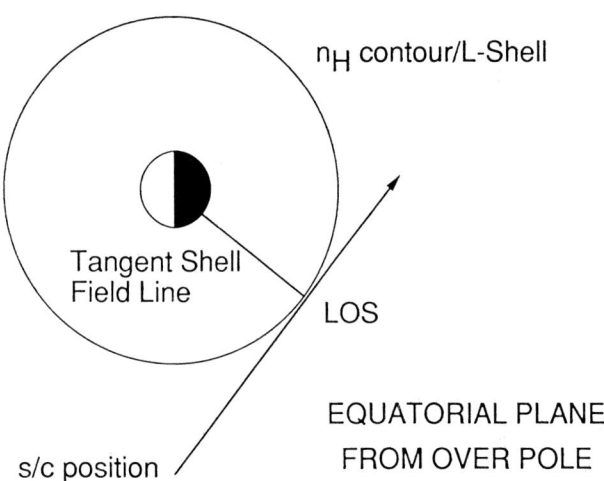

Fig. 1. Schematic illustration of the tangent-shell field line meridian, a concept needed for the interpretation of wide angle global imagery of fast atoms.

QUANTITATIVE PADS FROM IMAGES

Tangent-Shell Field Line Meridian

In discussing the images above, we have sometimes referred to them as if they represented the ion flux in the plane normal to the central line-of-sight. Since they are actually line-of-sight convolutions, this is not strictly true. They are more accurately described as best representing the ion flux within a "tangent-shell field line" meridian, which we define by the field line that is contained in a magnetic L-shell tangent to the current line-of-sight, and located at the point of tangency. This concept is closely related to the concept of an atmospheric limb, as illustrated in Figure 1. For example, an imager with a vantage point at 5 R_E at local dusk, viewing at an angle of 45 degree. from the line to the Earth, toward the nightshade, has a tangent-shell field line lying at a local time of approximately 2100 hours, with an L value of approximately 5.7.

Neutral atom images tend to be dominated by plasma conditions near the tangent-shell field line meridian, because this is the region along the line-of-sight where the neutral hydrogen density is strongly peaked. For a uniform distribution of ion flux, it is clear that the vast majority of emission along a particular line-of-sight originates near the tangent-shell field line. This is the basis of the method described in the following section, for associating an inferred PAD with a particular spatial region. The method will be compromised by strong departures from a uniform distribution of storm-time plasma, in which plasma regions far from the tangent-shell field line have dominant contributions to the line-of-sight convolution for a given pixel region.

Spatially Uniform Storm Plasma

As a simple demonstration of this technique, we use a spatially uniform storm plasma to compute normalized brightness dependencies upon geomagnetic latitude and L value of field lines, and upon the PAD exponent, n, where the PAD is assumed to be well-described by the function $\sin^n(\alpha_e)$, with α_e the equatorial pitch angle, and n a real exponent. The procedure is to compute the fast neutral atom image based on the uniform storm plasma with specified n's. Tangent-shell field lines are then overlaid on the images, and a set of brightness versus latitude curves is developed for each value of n. The set of curves then constitutes a renormalizable calibration between brightness curves and PADs. Plate 7 (left panel) shows the computed flux image for a uniform storm plasma spatial distribution with PAD index, $n=1$. The right panel of Plate 7 shows the set of flux curves computed for the indicated $L=4$ field line and the indicated set of PAD indexes, n. The curves can be readily seen to incorporate the cold geocoronal hydrogen distribution, having footpoint brightness that in general are larger than equatorial brightnesses, except for strongly "pancaked" PADs, i.e., those with large positive n.

PAD RESULTS

The superposition of specific fast atom image brightness curve upon the family of curves shown in Plate 7 allows inference of the PAD in the tangent-shell region that dominates the image. More generally, we would envision the parameterization of the curves and a fitting process whereby the observed image brightness would lead to an inferred set of PAD parameters. As a demonstration of this technique, Figure 2 shows two case studies from the images previously shown in Plates 5 and 6.

In the left panel of Figure 2, the 1.7 keV image data from dayside and nightside, along the $L=4$ field line, are superposed on the family of curves from Plate 7. It can be seen that there are notable differences between the shape of the data and the curves, but that qualitatively, the difference in pitch angle distribution is clearly evident. The equatorial (center) parts of the data traces reflect the true PADs significantly better than the higher latitude parts of the data traces, because of increasing contributions from regions well away from the tangent-shell meridian, in the inner parts of the images.

In the right panel of Figure 2, the 100 keV image data from the nightside are superposed on the same set of curves. Here there is remarkably good agreement between the shape of the modeled curves and that of the simulated data trace from the fast atom image. Moreover, the interpolated value of the PAD index n that is inferred from the position of the simulated data on the set of curves, is in

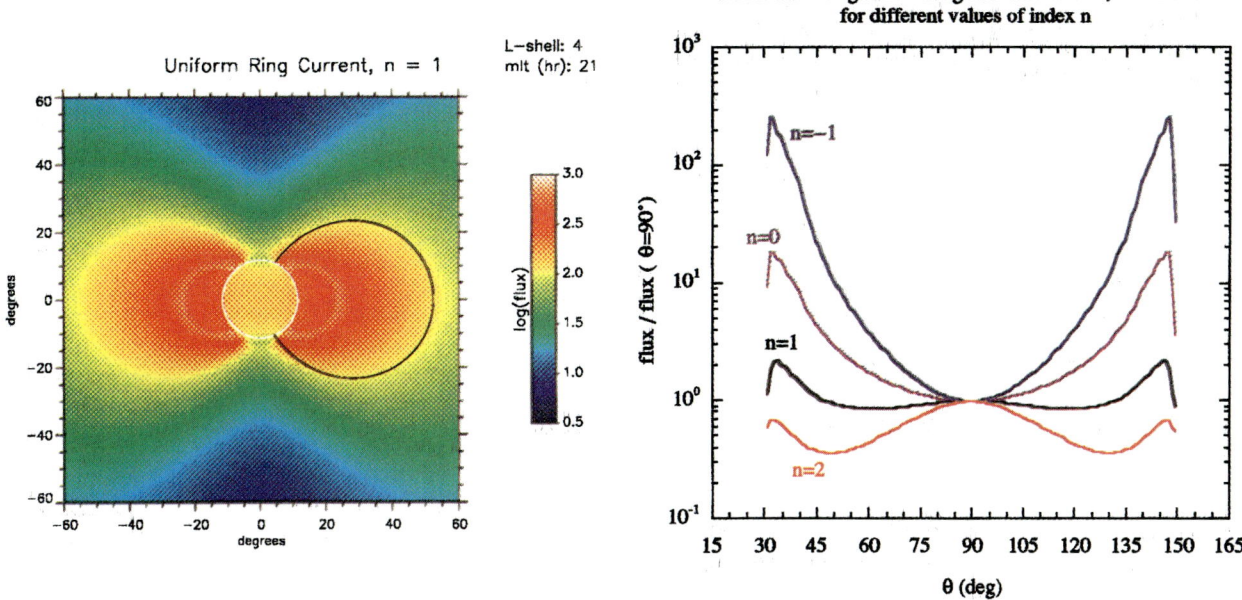

Plate 7. Left panel: the fast atom flux image generated by a uniform distribution of flux with a pitch angle distribution given by $\sin^n(\alpha_e)$, where $n=1$ throughout the modeled region. Right panel: the flux image brightness as a function of magnetic latitude along the $L=4$ field line indicated in the left panel, for the indicated values of n in the (uniform) PAD.

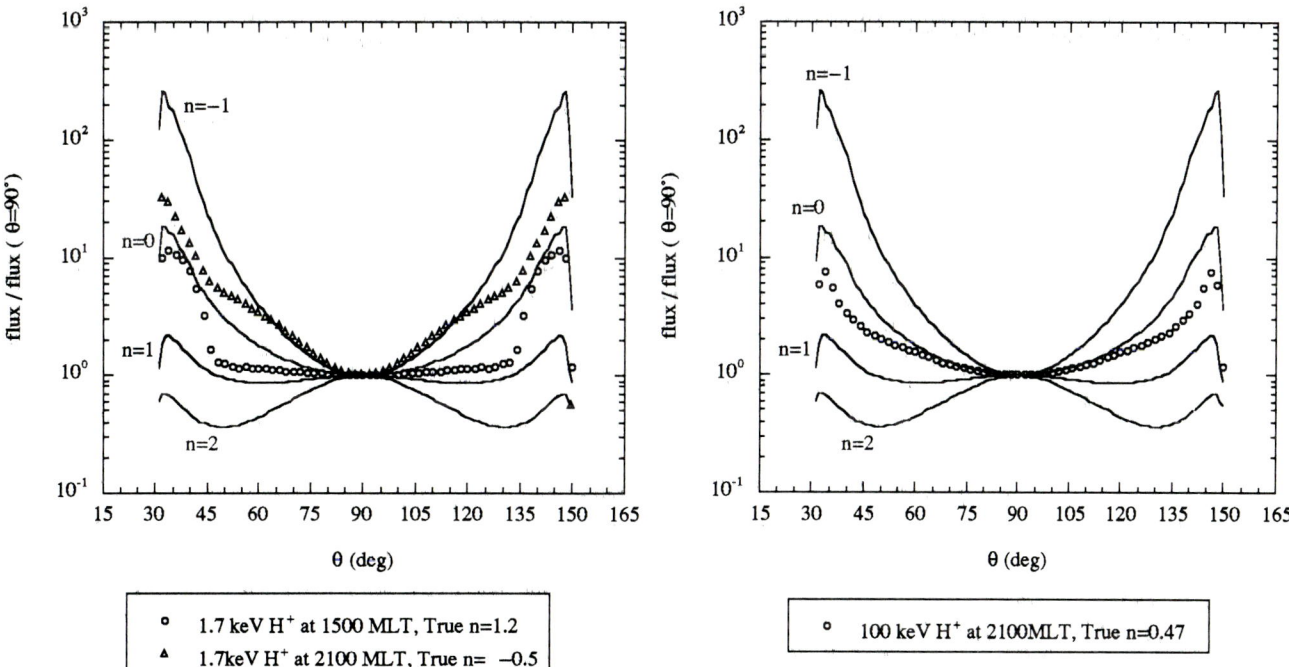

Fig. 2. Comparison of simulated image data with the tangent-shell field line flux curves of Plate 7. Left panel: case of 1.7 keV, 6 hours recovery phase. Right panel: case of 100 keV, 36 hours recovery phase.

good agreement (to one significant digit) with the true PAD in the tangent shell meridian region. The 100 keV case is clearly much more successful than the 1.7 keV case, because the nature of the 100 keV plasma distribution is much closer to that of the hot plasma used to generate the image flux curves; that is, more uniform in local time and radius.

SUMMARY/CONCLUSIONS

We have used a newly developed three-dimensional model of the recovery phase storm plasma to explore the signatures of pitch angle distributions in fast atom imagery of hot plasmas. We find that fast atom images are significantly influenced by the spatio-temporal characteristics of pitch angle distributions included in this model. Said differently, we have shown that global fast atom images contain detailed information about the pitch angle distribution of ions within the imaged regions. In the process, it has become clear that fast atom images cannot be accurately deconvolved without attention to PAD effects, since they can otherwise masquerade as spatial distribution effects.

When PAD effects are properly treated, magnetospheric hot plasma imaging promises a significant new capability to sense the global effects of microphysical processes. This would be done by monitoring the global distribution of plasma wave intensities through their diffusive effects on PADs. We have shown that the assumption of spatial uniformity provides a useful tool for quickly estimating PADs and identifying morphological features of the PAD in space. PAD inferences using this tool are best for high energy ions that are relatively uniform in local time, on L-shells where radial gradients are weak. However, useful qualitative inferences can be made almost trivially using this method, and gross parameters of the ion flux distribution can be estimated as inputs to a more rigorous iterative inversion procedure.

The limited successes of the simple techniques illustrated here suggest an optimistic view of the prospects for quantitative image inversion. Sophisticated deconvolution methods, such as a maximum entropy analysis and Bayesian statistics, are now being brought to bear on the problem, and very encouraging results are being obtained. This optimism is bolstered by the empirical fact that readily interpreted features of the images are subject to straight-forward modeling inferences.

Acknowledgments. The authors have benefited from stimulating discussions with D. L. Gallagher, who has contributed to the coding of the line-of-sight integration, and applied it to the simulation of images of cold He^+ plasma in scattered solar 30.4 nm EUV light. We are also indebted to M. A. Sloan of Computer Sciences Corp., who has contributed programming discipline and documentation to the analysis and display of these results. M.-C. Fok was a Resident Research Associate of the National Research Council at Marshall Space Flight Center during this work, which was also was supported by Marshall Space Flight Center and by the NASA Space Physics Division under UPN 433-90-00. Work at Auburn University was supported by grant NAG8-147 from the NASA JOVE program, and by grant NGT-51046 from the NASA/Marshall Space Flight Center Graduate Student Researcher Program.

REFERENCES

Fok, M.-C., T. E. Moore, J. U. Kozyra, G. C. Ho, and D. C. Hamilton, A three-dimensional ring current decay model, *J. Geophys. Res. 100*, 9619, 1995.

Hamilton, D. C., G. Gloeckler, F. M. Ipavich, W. Studemann, B. Wilken, and G. Kremser, Ring current development during the great geomagnetic storm of February, 1986, *J. Geophys. Res. 93*, 14,343, 1988.

Moore, K. R., D. J. McComas, H. O. Funsten, and M. F. Thomsen, Low energy neutral atoms in the Earth's magnetosphere modeling, in *Instrumentation for Magnetospheric Imagery*, ed. by S. Charkrabarti, p. 51, Proc. SPIE, p. 1744. (also submitted to Optical Science), 1992.

Rasmussen, C. E., S. M. Guiter, and S. G. Thomas, Two-dimensional model of the plasmasphere: refilling time constants, *Planet. Space Sci. 41*, p. 35, 1993.

Roelof, E. C., D. G. Mitchell, and D. J. Williams, Energetic neutral atoms (~50 keV) from the ring current: IMP 7/8 and ISEE 1, *J. Geophys. Res. 90*, p. 10,991, 1985.

Williams, D. J., E. C. Roelof, and D. G. Mitchell, Global Magnetospheric Imaging, *Rev. Geophys. 30(3)*, 183, 1992.

T. E. Moore and M.-C. Fok, Space Sciences Laboratory, NASA Marshall Space Flight Center, Huntsville, AL 35812.

J. D. Perez and J. P. Keady, Auburn University, Auburn, AL 36849.

Development of Non-Maxwellian Velocity Distributions as a Consequence of Nonlocal Coulomb Collisions

G. R. Wilson

Department of Physics and Center for Space Plasma and Aeronomic Research, University of Alabama, Huntsville

Coulomb collisions are the only microscale process present, to some degree, in every space plasma transport situation. The most rigorous formalism commonly used to describe these collisions is the Fokker-Planck equation. One of its properties is that the Maxwellian velocity distribution is its only time-independent solution. This however does not mean that Coulomb collisions cannot produce non-Maxwellian distributions. During the course of the relaxation process such non-Maxwellian distributions can develop when different species or different populations of the same species are flowing at supersonic speeds relative to each other or have different temperatures. The details of the relaxation process, described in the Fokker-Planck formalism, are unimportant for collision dominated or collisionless plasmas but are very important for plasmas that are in transition between these two states. In such cases the effects of collisions can become nonlocal when collision time scales and length scales become comparable to transport time and length scales. When this happens the nonMaxwellian velocity distributions may produce plasma flow results significantly different from those obtained from transport theory which assumes Maxwellian or near Maxwellian velocity distributions. In this paper we illustrate these concepts through examples of local collisional relaxation and collisional relaxation occurring together with field aligned transport.

1. INTRODUCTION

There exists two extreme situations in geospace plasmas. In the first, the collision time scales and length scales are much smaller than the time and length scales of interest so that the plasma can be considered to be collision dominated. Under these conditions the plasma will relax to Maxwellian so quickly that one can consider it to always be in that state. For such conditions the details of the collisional relaxation process are unimportant and a simple model, like the BGK collision model [*Bhatnagar et al.*, 1954], is sufficient for most modeling purposes. In the second extreme situation, the collisional time and length scales are so large compared to the time and length scales of interest that collisions can be ignored altogether.

Between these two extremes the collision process produces only a partial relaxation over time and length scales of interest. Coulomb collisions can then no longer be considered to be a truly microscopic process as it can in the collision dominated case. Particles still collide only with other particles within their local Debye spheres but the change in velocity experienced in crossing a Debye length is very small. A significant change in velocity only occurs after crossing many Debye lengths, or a macroscopic distance. Under such conditions collisions are said to be nonlocal since the velocity distribution at one location can evolve via collisions that occur at a distant location. For these situations a collision formalism which is faithful to the details of the relaxation process is needed in order to properly model the system.

This realization has led to the development of a number of kinetic transport models which incorporate the effects of Coulomb collisions through algorithms which mimic the behavior of the Fokker-Planck equation. These include studies of the transport of plasma on dipolar field lines [*Wilson et al.*, 1992; *Miller et al.*, 1993] using the Fokker-Planck collision algorithm of *Takizuka and Abe* [1977], and studies of the outflow of ionospheric plasma through the transition region using the *Takizuka and Abe* [1977] collision method [*Wilson*, 1992] and using a Monte Carlo method based directly on the friction and diffusion coefficients that come from the Fokker-Planck equation [*Barghouthi et al.*, 1993]. Kinetic models used for the study of ring current decay have been developed which include the Coulomb drag force (using the first term in the Fokker-Planck equation) [*Fok et al.*, 1993] and the full Fokker-Planck equation [*Jordanova et al.*, 1994]. Work has also been done on developing more efficient collision algorithms which can allow for different particle weighting among species [*Miller and Combi*, 1994], and remove the need to pair particles by using a grid-based collision force [*Jones et al.*, 1994].

2. APPLICATION OF FOKKER-PLANCK THEORY IN KINETIC SIMULATIONS

The Landau form of the Fokker-Planck equation, appropriate for describing Coulomb collisions in a plasma, is

$$\frac{\partial f_r}{\partial t} = -\nabla_{\mathbf{v}} \cdot (\mathbf{A}_r f_r) + 1/2 \nabla_{\mathbf{v}} \nabla_{\mathbf{v}} : (\mathbf{B}_r f_r) \quad (1)$$

where

$$\mathbf{A}_r = \sum_s \Gamma_{rs} \frac{m_r + m_s}{m_s} \frac{\partial}{\partial \mathbf{v}} \int \frac{f_s(\mathbf{v}')}{|\mathbf{v}-\mathbf{v}'|} d\mathbf{v}' \quad (2)$$

and

$$B_{r,ij} = \sum_s \Gamma_{rs} \frac{\partial^2}{\partial v_i \partial v_j} \int f_s(\mathbf{v}') |\mathbf{v}-\mathbf{v}'| d\mathbf{v}' \quad (3)$$

where e_s is species charge, m_s is species mass, Λ is the plasma parameter and

$$\Gamma_{rs} = \frac{\ln \Lambda}{4\pi} \left(\frac{e_r e_s}{\epsilon_0 m_r} \right)^2$$

The details of the derivation of this equation can be found in *Clemmow and Dougherty* [1969] and *Nicholson* [1983]. There are a number of assumptions made in the derivation of equation (1) including: (1) the two particle correlation function g is small compared to f^2, (2) g relaxes on a much smaller time scale than f, (3) the plasma is spatially homogeneous on Debye length scales, (4) all species distributions are stable, and (5) all gyro radii are large compared to the Debye length.

There is, however, no assumption about the form of the velocity distribution which at this point is still arbitrary.

The desire here is to use equation (1), or something that behaves very much like it, in a mesoscale kinetic plasma simulation. In such simulations one or more species are described through the collective dynamics of simulation macro-particles which act as phase space tracers. This approach is ideally suited for modeling collisionless and stable systems where particles do not interact with each other locally but only respond to externally applied fields and interact collectively at large length scales through a polarization electric field. Such models do not implicitly include local collective effects because the spatial resolution in the models is typically not sufficient to resolve them. Coulomb collisions can be added to such a model only through the use of an explicit procedure. One such approach is the method of *Takizuka and Abe* [1977]. (Another is demonstrated in the work of *Barghouthi et al.* [1993].)

Unlike the technique used by *Barghouthi et al.* [1993], which uses $\mathbf{A}_r$ and $\mathbf{B}_r$ directly to calculate the slowing and deflection of particles, or the method of *Casanova et al.* [1991] which descritizes the Fokker-Planck equation, the method of *Takizuka and Abe* [1977] uses a particle pairing technique. For collisions between particles of the same species (self-collisions) the procedure is to randomly pair simulation particles located in the same spatial cell and to compute the deflection of their relative velocity $\vec{U}$ according to that which would result from the cumulative effect of many small angle collisions experienced over the course of a time step. In the models where we have employed this technique we do not follow the full gyromotion of the simulation ions (i.e., their gyrophases are unknown) so we assume a random phase difference between the two collision partners for purposes of finding $\vec{U}$. The variable $\delta = \tan(\Theta/2)$ (Θ is the angular deflection of $\vec{U}$) is chosen randomly from a Gaussian distribution with zero mean and a variance $\langle \delta^2 \rangle$ given by

$$\langle \delta^2 \rangle = \frac{e^4 n_s \ln \Lambda}{2\pi \epsilon_0 m_s^2 |\vec{U}|^3} dt \quad (4)$$

where n_s is the species density in the cell, and m_s is the species mass. The resulting change in the relative velocity $\vec{U}$ is applied to both particles (see *Takizuka and Abe* [1997] for the equations used to perform the velocity coordinate transformations and changes to particle velocities) so that between them, momentum and energy is conserved. Since these quantities are conserved between each colliding pair, they are conserved for the entire population.

A slightly different procedure is used for the collisions between particles of species r and another species (s) which is present but not modeled in the simulation. In this case the species r particle is paired with a particle of species s whose velocity is randomly generated from an assumed velocity distribution. A relative velocity $\vec{U}$ is found and its deflection calculated, as for self collisions, but in this case the expression for $\langle \delta^2 \rangle$ is given by

$$\langle \delta^2 \rangle = \frac{e^4 n_s \ln \Lambda}{8\pi\epsilon_0 \mu^2 |\vec{U}|^3} dt \quad (5)$$

where n_s is the density of species s at the position of the simulation particle and μ is the reduced mass. The change in U is applied to the velocity of the species r particle while changes to the velocity of the species s particle are ignored. This approach can be used in situations where the density and velocity distribution of species s can be assumed to be unaffected by collisions with species r. The paper by *Takizuka and Abe* [1977] describes how collisions between particles of different species, which are both included in the model, can be performed when the different species particle weights (number of real particles represented by the macro-particle) are the same. If the weights are different then the method of *Miller and Combi* [1994] could be used.

In the Fokker-Planck equation the coefficient of dynamic friction, $\mathbf{A}_r$, and the velocity diffusion coefficient $\mathbf{B}_r$ both involve integrations over the velocity distribution of the collision partner species s. The particle pairing process is an approximation to this integration which is a good approximation when the time step is small compared to the collision time. In that case the velocity distribution does not change rapidly over many timesteps and the random selection of collision partners is equivalent to a Monte-Carlo integration.

Our collision technique, based on the method of *Takizuka and Abe* [1977], has been extensively tested. The particle pairing method allows explicit conservation of both momentum and energy. In the model, these quantities remain close, within round-off, to their original values, even after very long times (i.e. there is no energy or momentum drift). When the relaxation process is finished all final velocity distributions are Maxwellian over the entire velocity range where the distribution can be reliably determined. Tests of the relaxation of anisotropic distributions, slowing of streaming distributions, and the energy exchange between species with different temperatures were all performed. It was found that initially the rates of isotropization, slowing and energy exchange matched those produced by analytic theory. At later times there is a departure that can be accounted for by the nonMaxwellian distributions which develop during relaxation. The analytic theory assumes that all distributions remain Maxwellian through out the process. As a final test we simulated the relaxation of a shell distribution which was studied by *MacDonald et al.* [1957] who integrated the Fokker-Planck equation directly. We found that the simulated distribution function matched closely those of *MacDonald et al.* [1957] at each time during the process. In particular we also observed the overshoot of the distribution at low velocities and the long time required to fill the high speed tail.

3. COLLISIONAL RELAXATION IN THE ABSENCE OF TRANSPORT

In this section we show local calculations, which ignore transport, so as to illustrate how a species velocity distribution can be driven away from Maxwellian during the relaxation process. As a first example assume that species r is a minor species and that all other species have nondrifting Maxwellian velocity distributions with thermal speeds $v_s = \sqrt{2K_b T_s/m_s}$. With these assumptions there is no preferred direction in the plasma so that $\mathbf{A}$ becomes

$$\mathbf{A} = \sum_s 2\Gamma_{rs} \frac{m_r + m_s}{m_s} \frac{n_s}{v_s^2} G(v/v_s) \frac{\mathbf{v}}{v} \quad (6)$$

and $\mathbf{B}$ can be expressed in terms of a perpendicular (to $\mathbf{v}$) component

$$B_\perp = \sum_s n_s \Gamma_{rs} \frac{1}{v} \left[\Phi(v/v_s) - G(v/v_s) \right] \quad (7)$$

and a parallel component

$$B_\parallel = \sum_s 2n_s \Gamma_{rs} \frac{G(v/v_s)}{v} \quad (8)$$

where

$$G(x) = \frac{\Phi(x) - x\Phi'(x)}{2x^2} \quad (9)$$

and $\Phi(x)$ is the error function [*Hinton*, 1983]. From these three terms we can define the stopping frequency

$$\nu_s = \mathbf{A} \cdot \mathbf{v}/v^2 \quad (10)$$

the deflection frequency

$$\nu_d = 2B_\perp/v^2 \quad (11)$$

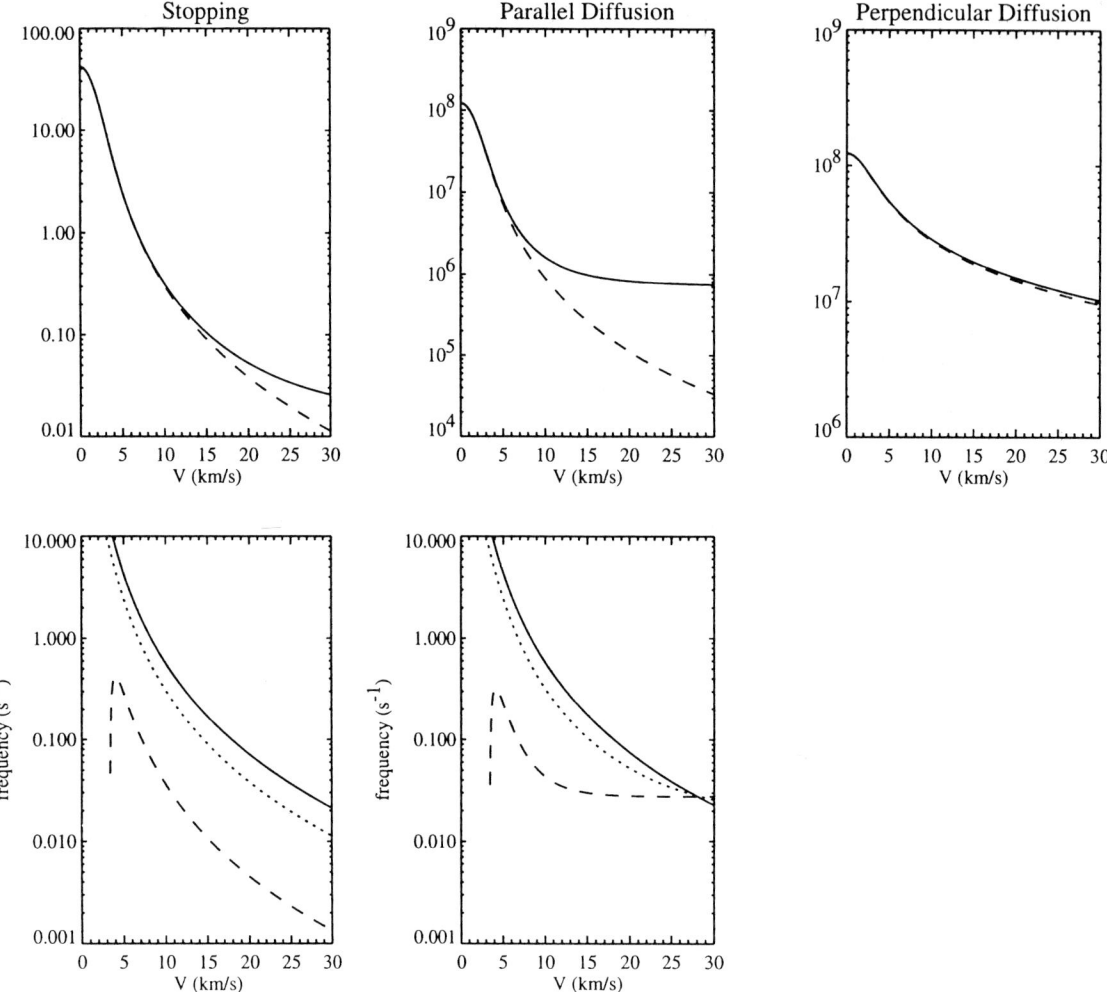

Fig. 1. The top three panels are plots of the stopping frequency (equation (10)), parallel diffusion coefficient (equation (11)), and perpendicular diffusion coefficient (equation (12)) for H^+–O^+ collisions only (dashed curves) and for H^+–O^+ and H^+–e^- collisions (solid curves). The background plasma consists of Maxwellian O^+ ions and electrons with a density of 10^5 cm^{-3} and a temperature of 3000 K. The bottom two panels show the stopping frequency (dotted curves), deflection frequency (solid curves), and the energy exchange frequency (dashed curves) for H^+–O^+ collisions only and for H^+–O^+ and H^+–e^- collisions.

and the energy exchange frequency

$$\nu_e = 2\nu_s - 2(B_\perp + 1/2 B_\parallel)/v^2 \qquad (12)$$

Figures 1 and 2 show simulation results for a case where there is an H^+ beam flowing through an O^+ and electron background. Initially the hydrogen ions have a flow speed of 20 km/s, a temperature of 100 K and a density which is very small compared to the oxygen ion density. The O^+ ions and the electrons are both assumed to have a density of 10^5 cm^{-3}, a temperature of 3000 K, and zero drift speed (typical density, slightly high temperature for the topside ionosphere [*Wahlund et al.*, 1992]).

The top three panels in figure 1 show the values of ν_s, $B_\perp$, and $B_\parallel$ as functions of velocity. The dashed lines indicate the values for collisions with the O^+ ions only while the solid lines are for collisions with both the O^+ ions and the electrons. The last two panels in figure 1 show the stopping, deflection, and energy exchange frequencies, as functions of velocity, for H^+–O^+ collisions only (1.d) and for H^+–O^+ and H^+–e^- collisions (1.e). As can be seen in this figure, the stopping (dotted curve), deflection (solid curve) and energy exchange (dashed curve) frequencies are strong functions of velocity. They also can have significantly different values at a given velocity. The main effect of adding collisions with electrons is to increase the energy exchange frequency at speeds above 10 km/s for this specific situation.

Figure 2 shows a time sequence of plots of the H^+ velocity distribution as a function of $v_\parallel$ and $v_\perp$ where parallel and perpendicular here refer to the initial H^+ drift velocity vector. The plots in the left hand column are for a case where only H^+–O^+ collisions are considered. The case shown in the right column adds the H^+–e^- collisions. The time of each plot is indicated to the right. The three numbers above each plot give the time multiplied by the stopping, deflection, and energy exchange frequencies evaluated at $v = 20$ km/s. Because the deflection frequency is so much larger than the energy exchange frequency, when only H^+–O^+ collisions are considered, the initial effect of collisions is to scatter the direction of the hydrogen ion velocity vectors without changing their magnitude much. After about 1 deflection time a shell distribution is formed. The interior of the shell begins to fill through the formation of a Maxwellian that is well separated from the shell surface. As time advances the number of ions in the Maxwellian grows at the expense of the shell until the distribution is a Maxwellian with an isotropic shoulder. At later times the shoulder diminishes as the H^+ ions come to thermal equilibrium with the O^+ ions. When H^+–e^- collisions are added, the deflection frequency is still larger than the energy exchange frequency, but not by as much as before. The shell distribution still forms but it fills more rapidly. Complete equilibration with the background is achieved in about 60 s which is a significantly shorter time interval than when H^+–e^- collisions are ignored.

Figures 3–5 show results for a case, different from the previous example. Here we start with an H^+ beam flowing through an H^+ and electron background. The beam has a velocity of 440 km/s (1 keV), a temperature of 100 K, and a density which is small compared to the background H^+. The background has a density of 10^3 cm^{-3}, a temperature of 1 eV and a drift speed of zero (Typical plasmasphere values [*Comfort et al.*, 1985]). Collisions with both ions and electrons are included in this case. Figure 3 shows, in the same format as figure 1, the stopping, deflection, and energy exchange frequencies with and without electron collisions. Above a speed of about 50 km/s collisions with electrons have a dominate effect where they greatly increase the energy exchange and stopping frequencies.

Figure 4 shows a series of H^+ velocity distribution plots at the times indicated in the upper left corner of each plot. The three numbers above each plot are the time multiplied by the stopping, deflection, and energy exchange frequencies at 440 km/s. Each of these plots

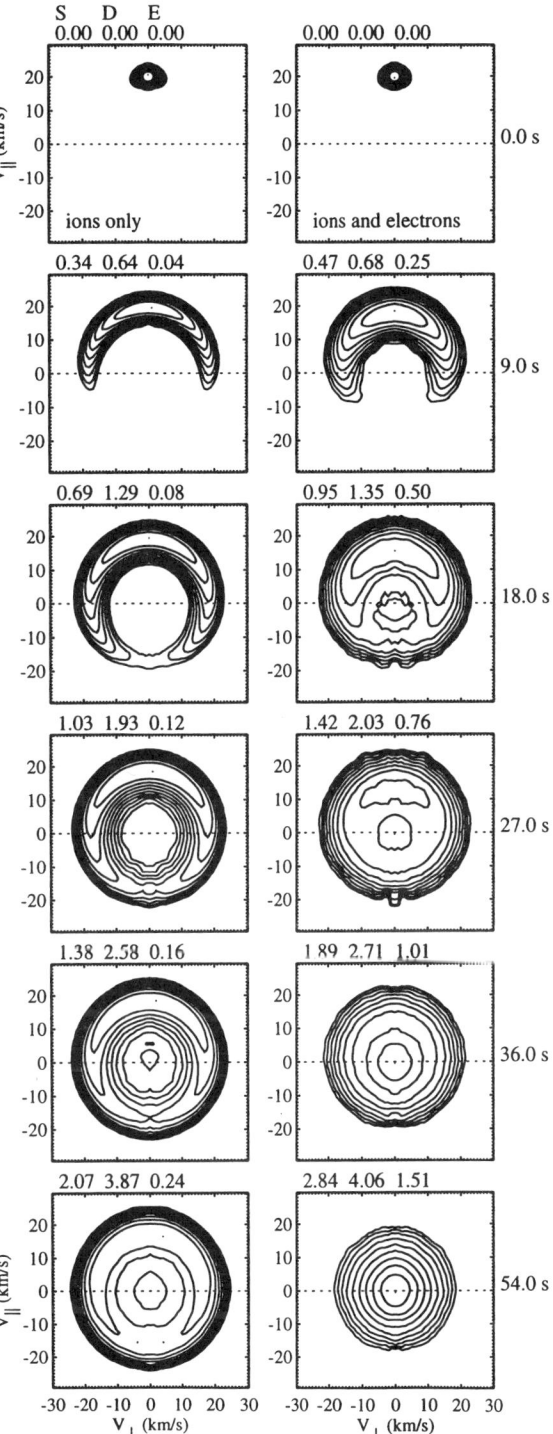

Fig. 2. Velocity distribution plots for an H^+ beam colliding with an O^+ background plasma or an O^+ and electron plasma. The background parameters are the same as figure 1. The three numbers above each plot are the time multiplied by the stopping, deflection, and energy exchange frequencies evaluated at 20 km/s. The contour line separation is $e^{-1/2}$

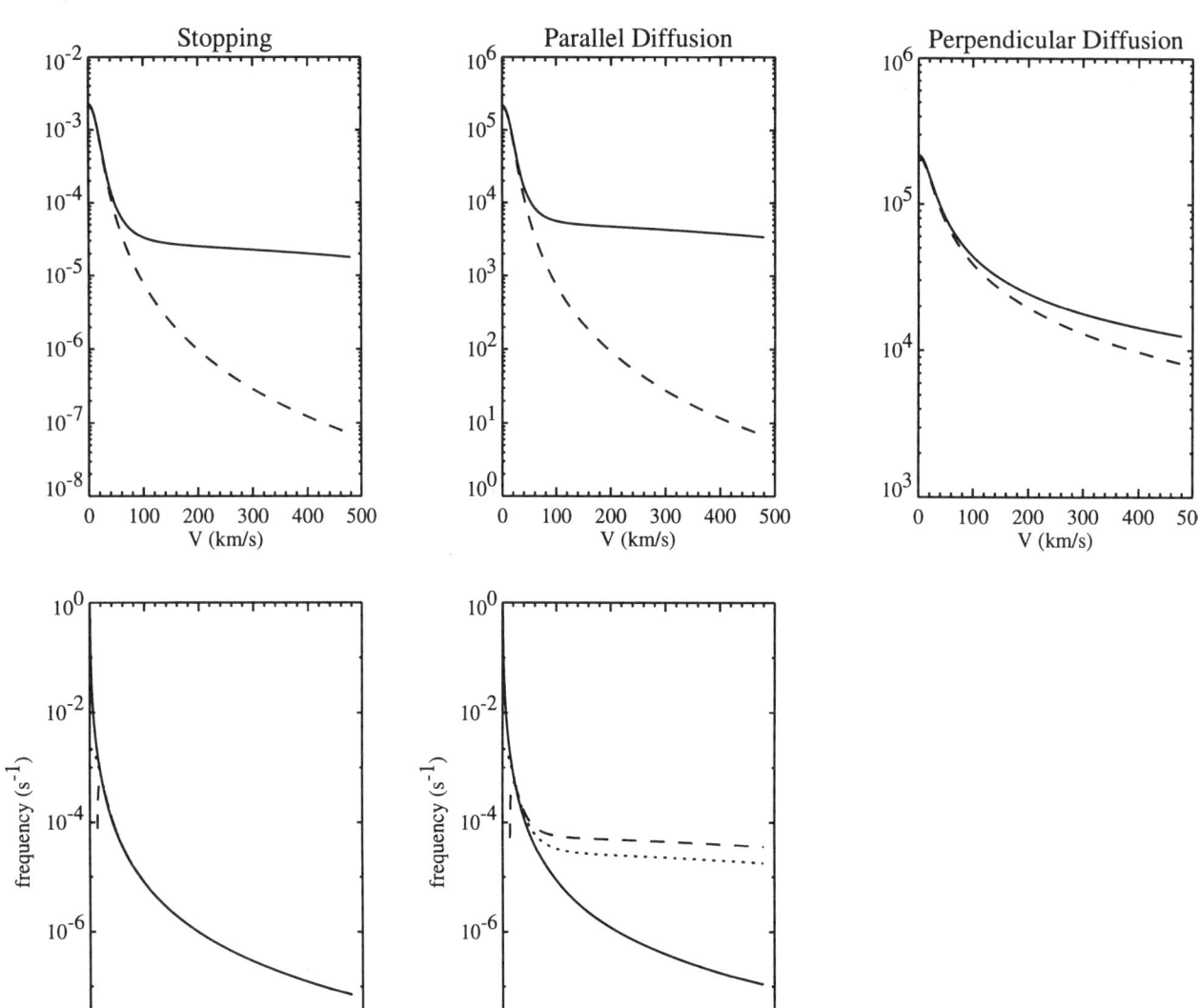

Fig. 3. Plots of collision coefficient and frequencies for an H^+ and electron background with a density of 10^3 cm^{-3} and temperature of 11600 K (1 eV). The format of this figure is the same as figure 1.

was prepared so that the $v_\parallel = 0$ value corresponds to the drift speed of the ions at that instant. We are then viewing the velocity distribution through a plot window which slides down the parallel velocity axis as the distribution slows. The actual drift speed of the beam ions is shown in the lower left corner of each plot.

Because the energy exchange frequency is the largest of the three frequencies, the dominant effect of collisions is the exchange of energy between the beam H^+ ions and the background electrons. This results in a relatively rapid slowing of the beam. After about 30 hours the ion speeds have dropped to the point where H^+–H^+ collisions start to become significant and a sig-

nificant amount of deflection becomes apparent. Near 40 hours a group of ions begins to congregate around the velocity space location of the background ions to which they are now strongly collisionally coupled. In the remaining hours the original beam decreases until it is merely an upward extended tail on the main body of the ion velocity distribution. Figure 5 shows the beam drift speed, temperature anisotropy, mean energy, and parallel and perpendicular temperatures as functions of time. During the first 35 hours H^+ electron collisions dominate while during the remaining hours H^+–H^+ collisions dominate.

As a final example I show two cases of temperature

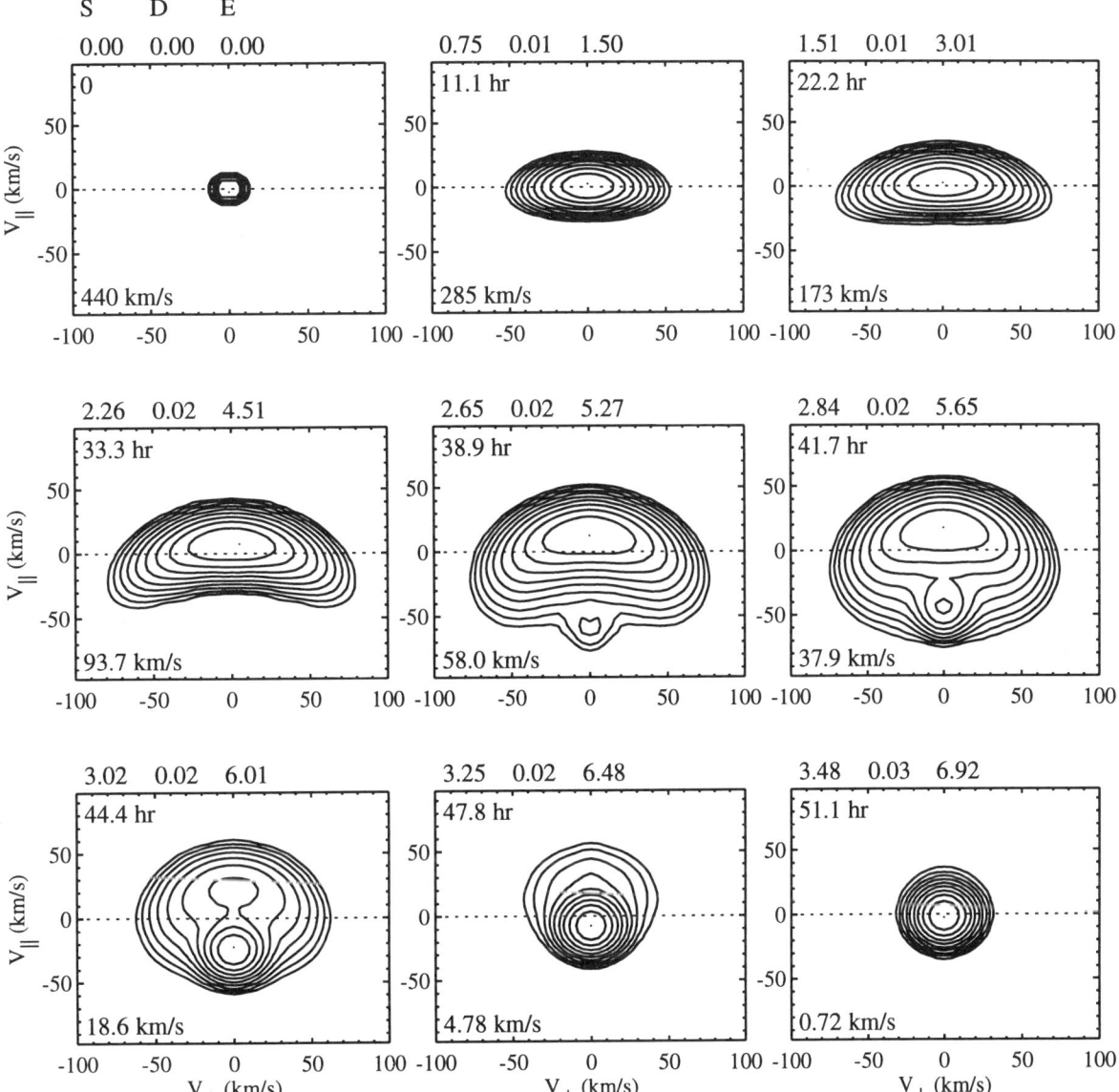

Fig. 4. Velocity plots for a 1 keV H^+ beam relaxing in an H^+ background plasma with parameters as indicated in figure 3. The time of each plot is indicated in the upper left corner. Each plot window was moved along the $v_\parallel$ axis so as to keep the distribution in the window center. The drift speed at each time is indicated in the lower left corner of each plot. The contour line separation is $e^{-1/2}$

equilibration between H^+ and O^+ ions. In each case both species start with Maxwellian velocity distributions and have equal densities. The total ion density is 1000 cm^{-3}. Unlike the previous examples, self collisions are included in this case but collisions with electrons are ignored. Figure 6 shows velocity distributions for both species, at an intermediate time during the relaxation process, plotted as the logarithm of f versus energy. The top two plots (at $t = 2.1 t_c$, where $t_c = 138$ s)

show the H^+ and O^+ distributions for a case where the initial H^+ temperature was 2000 K and the initial O^+ temperature was 10,000 K. The bottom two panels (at $t = 3.1 t_c$, where $t_c = 1022$ s) show the distributions for a case where the initial temperatures are reversed. In each plot the solid line is the distribution function while the over plotted dashed line is a Maxwellian with the same temperature so that one can see the degree of departure from Maxwellian.

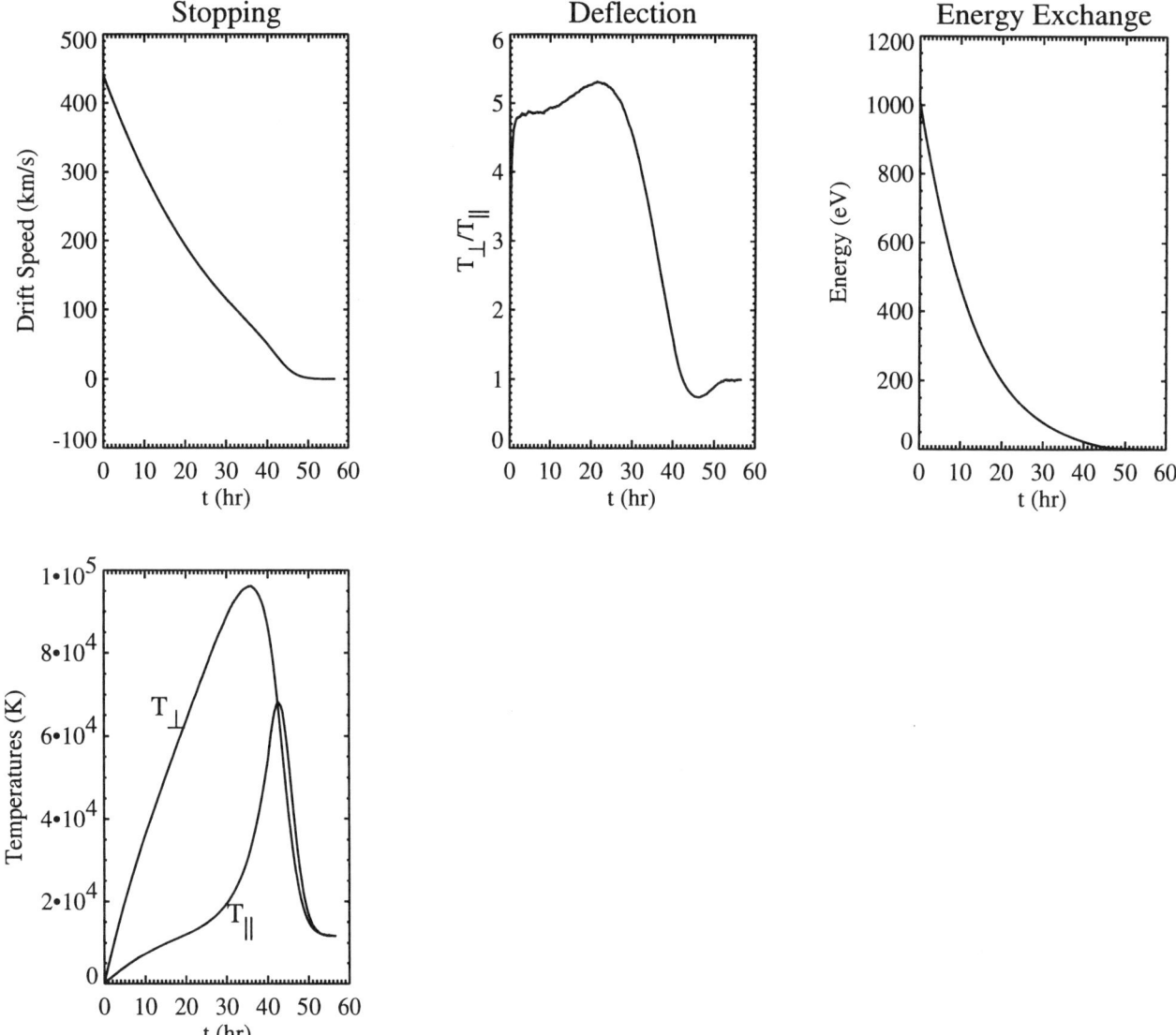

Fig. 5. Plots of the drift speed, temperature anisotropy, mean energy, and parallel and perpendicular temperatures for the case shown in figure 4.

One can see in these plots that the H^+ velocity distribution deviates significantly from Maxwellian while the O^+ distribution shows very little departure. The reason for the hydrogen departure is that over the energy range from about 0.5 to 5.0 eV the H^+ energy exchange frequency for H^+–O^+ collisions decreases by a factor of about 20. The result is that the H^+ ions in the high energy tail of the distribution will exchange energy with the oxygen ions more slowly then those in the core of the distribution. The core of the distribution relaxes faster than the tail leading to the bending of the distribution curve. On the other hand, the oxygen energy exchange frequency decreases by a factor of about 1.4 over this same energy range. Because of this smaller variation the bending of the oxygen distribution curve is much smaller than for hydrogen and is not apparent in the plots. The departure of the H^+ velocity distribution from Maxwellian occurs in spite of the effect of self collisions.

4. COLLISIONAL RELAXATION WITH TRANSPORT

As the above examples illustrate, during the relaxation process velocity distributions can deviate from Maxwellian even when they start as such. These de-

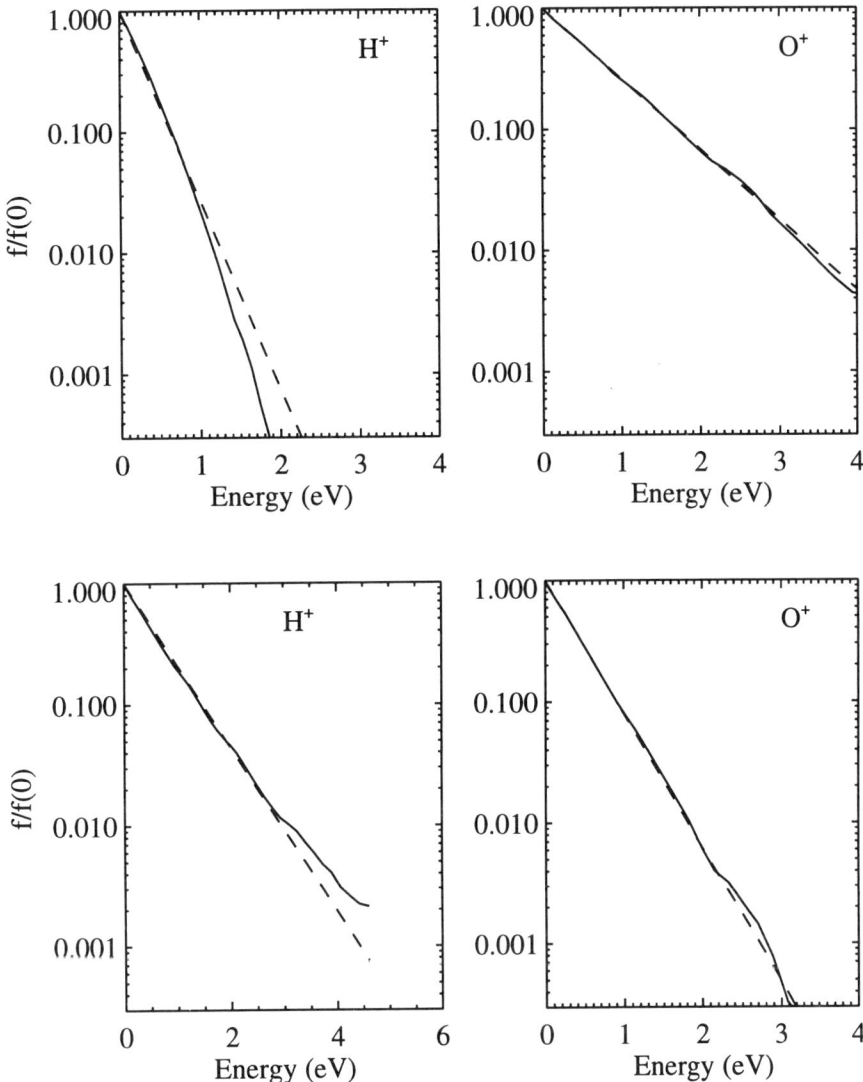

Fig. 6. Plots of the H$^+$ and O$^+$ velocity distributions (log f versus energy) at an intermediate point in the temperature relaxation process. In the top two panels $T_o(\text{H}^+) = 2000$ K while $T_o(\text{O}^+) = 10000$ K. In the bottom two panels the initial temperatures are reversed. The over plotted dashed curve is a Maxwellian with same temperature.

parts are the result of the often strong velocity dependence of the stopping, deflection, and energy exchange frequencies and the fact that under different conditions these three quantities can have significantly different values with one often dominating the others. Mass differences also play a role so that different species will depart from Maxwellian to different degrees. The densities, temperatures, and flow speeds used in the previous examples are typical values for geospace plasmas suggesting that these effects will be apparent there. However, the question remains as to the role that transport effects will play when combined with the collisional re-

laxation process. The previous examples were all local calculations which ignored transport. In this section we discuss examples which include transport.

4.1. Ion Outflow through the Transition Region

When light ions such as H$^+$ and He$^+$ flow out from the ionosphere they go from a collision dominated to a collisionless region. During this process collisions change from being local to nonlocal as the mean free path and collision time increase. When the Knudsen number (mean free path/O$^+$ scale height) goes through one the average light ion is on the verge of becoming

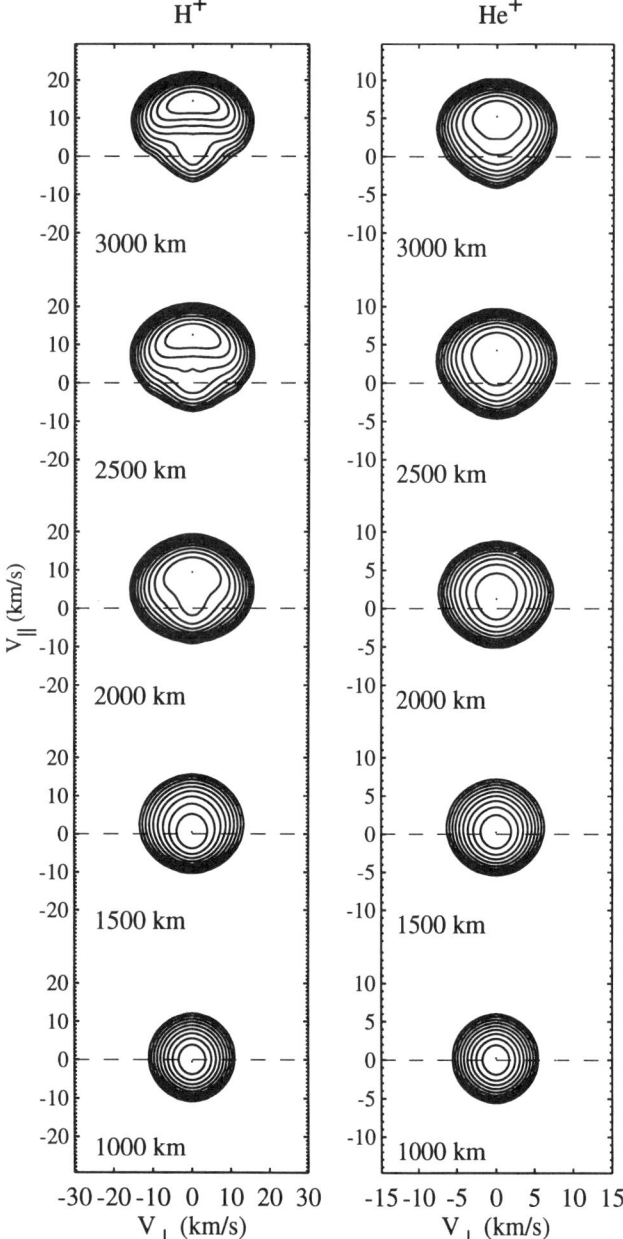

Fig. 7. Steady state H$^+$ and He$^+$ velocity distributions at the indicated altitudes. In this case the O$^+$ density decreased from 5×10^4 cm^{-3} at 1000 km to 700 cm^{-3} at 3000 km. Over the entire altitude range O$^+$ is the dominant ion species. The contour line separation is $e^{-1/2}$

not. This can lead to significant skewing of the velocity distribution. Also, the accelerating ions will still experience some collisions which will primarily deflect the ion's velocity away from the field line direction.

To see these effects we have performed simulations in which light ion particles are injected upward on a magnetic field line starting at low altitude in the collision dominated region. During each time step the particles are paired with an O$^+$ ion and their velocity deflected by the method described in section 2. Between deflections the ions are moved along the field line a distance of $v_\parallel \Delta t$ and are accelerated by the large scale forces (gravity, polarization electric, mirror). These processes are repeated each time step for each particle until steady-state is achieved. The results are then time averaged. Figure 7 shows steady state velocity distributions for H$^+$ and He$^+$ ions, at the indicated altitudes, for such a case. In the altitude range of figure 7 O$^+$ is the dominate ion (Figure 8) so that self collisions are a secondary collisional effect, but they have been included.

For each species shown in Figure 7, the velocity distributions near 1000 km are nearly Maxwellian with a very small drift. At higher altitudes a significant departure from Maxwellian, more apparent in H$^+$ then He$^+$, can be seen. One feature apparent in the H$^+$ velocity distributions, particularly at 2500 and 3000 km, is the inverted bowl shape that the core of this distribution forms. This feature forms for the same reason that the shell distribution forms for the H$^+$ ions in figure 2. In this case the H$^+$ ions accelerating up the field line tend to form a beam but they are scattering because deflection dominates energy exchange in the collisions with

collisionless. However, because of the $1/v^4$ dependence of the collision cross section some ions will already be collisionless and some will still be collision dominated. The collisionless ions will be free to accelerate up the field line in response to the outward pointing polarization electric field. The collision dominated ions will

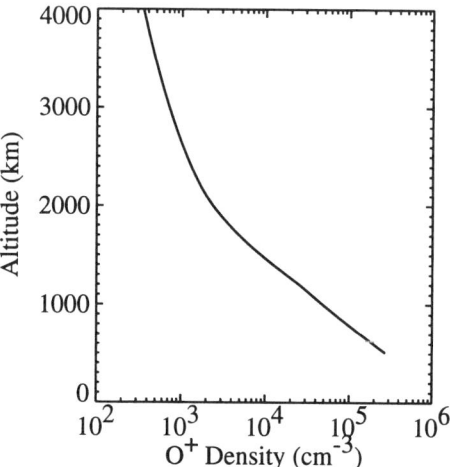

Fig. 8. The O$^+$ density versus altitude profile used for the case of Figure 7.

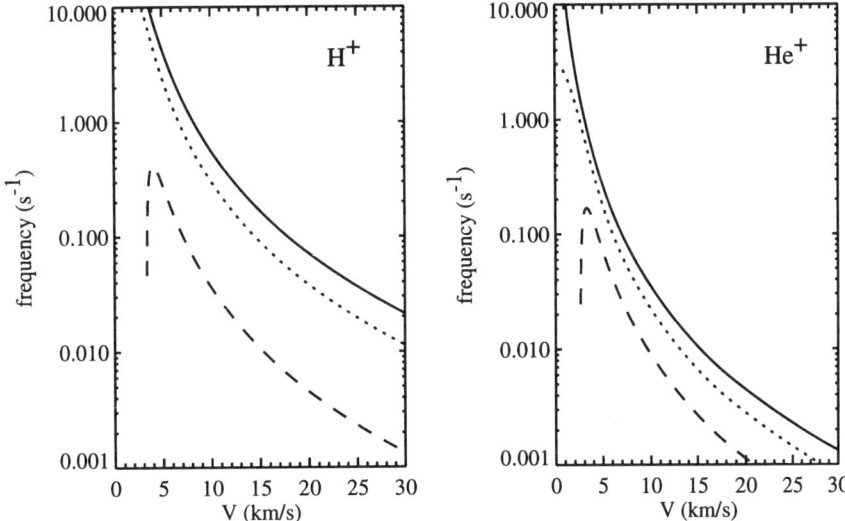

Fig. 9. Plots of the stopping (dotted curve), deflection (solid curve), and energy exchange (dashed curve) for H^+–O^+ collisions and He^+–O^+ The background O^+ ions had a density of 10^5 cm^{-3} and a temperature of 3000 K.

O^+. Also visible in the H^+ velocity distributions is a downward extending tail that goes to negative velocities. These are ions which are still collision dominated because of their close proximity to the O^+ ions in velocity space. This feature forms for reasons very similar to those which account for the downward tail seen in the distributions in figure 4.

The Helium ions also accelerate up the field line but their velocity distributions have significantly different features from the H^+ ions. They do not develop the inverted bowl like feature. The reason can be seen from figure 9 which shows the stopping (dotted curve), deflection (solid curve), and energy exchange (dashed curve) for both H^+ and He^+ ions colliding with O^+ ions whose density is 10^5 cm^{-3} and temperature is 3000 K. In the velocity range of 5–15 km/s for H^+ and 3–10 km/s for He^+ one can see that there is a bigger difference between the deflection and energy exchange frequencies for the H^+–O^+ collisions then there is for the He^+–O^+ collisions. This means that deflection is not such a dominate process for the He^+ ions as it is for the H^+ ions. Energy exchange is more important meaning that the He^+ ions will not gain as much energy in acceleration by the polarization electric as the H^+ ions because they will lose more of it in collisions with O^+.

One might think, looking at Figure 2, that collisions with the electrons, which were not included in these calculations, should have been. We redid the calculations including collisions with the electrons and saw very little change. The reason is that at the altitudes where collisions with O^+ and the electrons are significant most of the H^+ and He^+ ions have speeds small enough so that they interact collisionally only with the O^+ ions. At higher altitudes, where these ions have larger speeds, the background plasma density is low enough so that collisions do not play a major role and other factors, such as the polarization electric and magnetic mirror forces, dominate.

4.2. Refilling of Depleted Flux Tubes

It has long been realized that in order for depleted plasmaspheric flux tubes to refill following a magnetic storm, some process must act to scatter upflowing ionospheric particles from the source cone onto trapped trajectories. Since the initial densities are very low it has been assumed that Coulomb collisions could not be responsible but that some type of plasma wave was needed [*Schulz and Koons*, 1972]. However, we have demonstrated that Coulomb collisions are fully adequate to initiate flux tube refilling [*Wilson et al.*, 1992].

Our simulations show how collisions can initiate the refilling process even when the flux tube begins completely empty. Figure 10 shows the equatorial H^+ density, as a function of time, for an L= 6 flux tube connected to two ionospheric plasma reservoirs with densities of 100 cm^{-3} and temperatures of 6000 K. This H^+ density is lower that what is typically seen at 2000 km in the topside ionosphere while the temperature is above average for the same location. These values were cho-

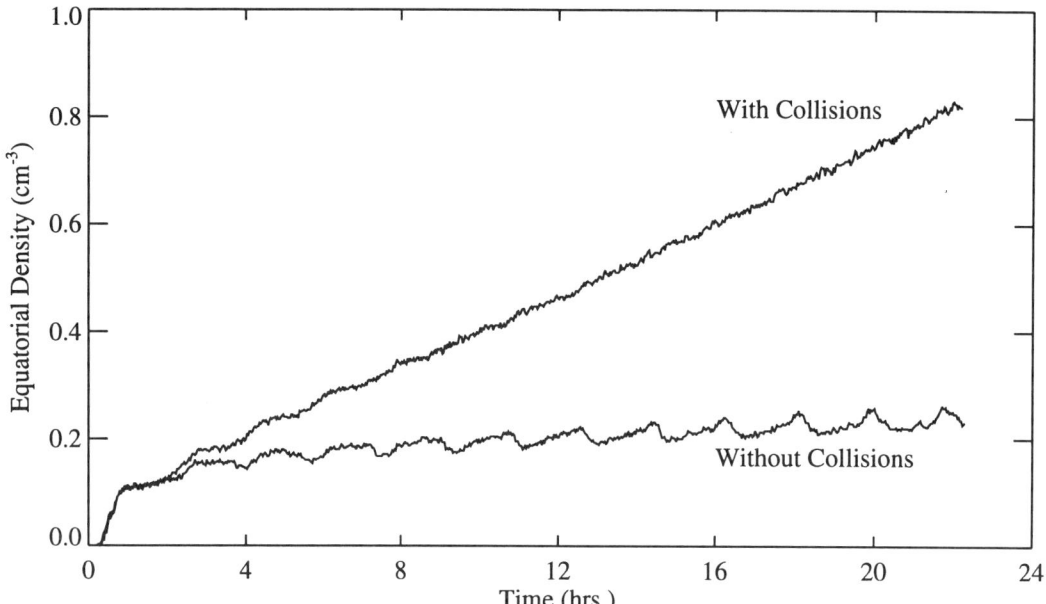

Fig. 10. Plot of the equatorial H$^+$ density for a refilling case with self collisions turned on and off. For this case L= 6 and the plasma reservoir at each base of the flux tube has a density of 100 cm^{-3} and a temperature of 6000 K.

sen so as to minimize the effects of collisions. One can see in Figure 10 that when collisions are turned on the density at the magnetic equator, of the L= 6 flux tube, increases above the levels that result without collisions operating. Clearly collisions make a difference, and the difference shows up at about two hours into the simulation. This demonstrates that the collisions which cause this difference occur at low altitudes because it takes about this length of time for the ions which first leave the ionosphere to travel to the conjugate end of the flux tube and then back to the equator. These low altitude collisions are effective at returning ions into the flux tube because as the ions approach the ionosphere their pitch angles approach 90° and only require a small angular deflection in order to be removed from the lose cone. Also, early in the refilling process there is a steep density gradient in the topside ionosphere which creates a strong upward directed electric field. As the down going ions flow against this electric field they lose energy, their speeds decrease, and their collision frequency goes up.

Figure 11 shows the velocity distribution of the H$^+$ ions at the equator for a case done on a L= 4.5 flux tube coupled into a realistic ionosphere. At early times one can see that the flows from the two conjugate ionospheres are different because of the different conditions in the respective northern and southern ionospheres. After a few hours ions flowing from each source have had time to reach the conjugate ionosphere, reflect via collisions, and return to the equator. Since each source now has ions flowing in both directions at the equator the two beams appear symmetric. They are still however very narrow in the perpendicular velocity direction due to magnetic folding which reduces the ion's perpendicular velocity as it flows from the point of its last collision (in the ionosphere) to the equator. In the following hours the two beams spread in both the perpendicular and parallel velocity directions until they merge. This spread occurs because the region where ions experience enough collisions to be reflected back towards the equator is moving toward the equator from both ionospheres. They therefore travel a shorter distance between their last collision and the equator and, as a consequence, experience less magnetic folding. The spread can also be attributed to a range of latitudes at which particles are reflected back to the equator.

In these simulations the refilling process occurs from the bottom up with high densities first developing near the ionosphere and then moving upward toward the magnetic equator. This is in contrast to earlier ideas of the refilling process [*Banks et al.*, 1971] where refilling occurs from the top downward with a high density region first developing near the equator and then spreading downward behind a pair of shock waves. The evolution of the equatorial velocity distribution seen in figure 11 is an example of the nonlocal nature of Coulomb

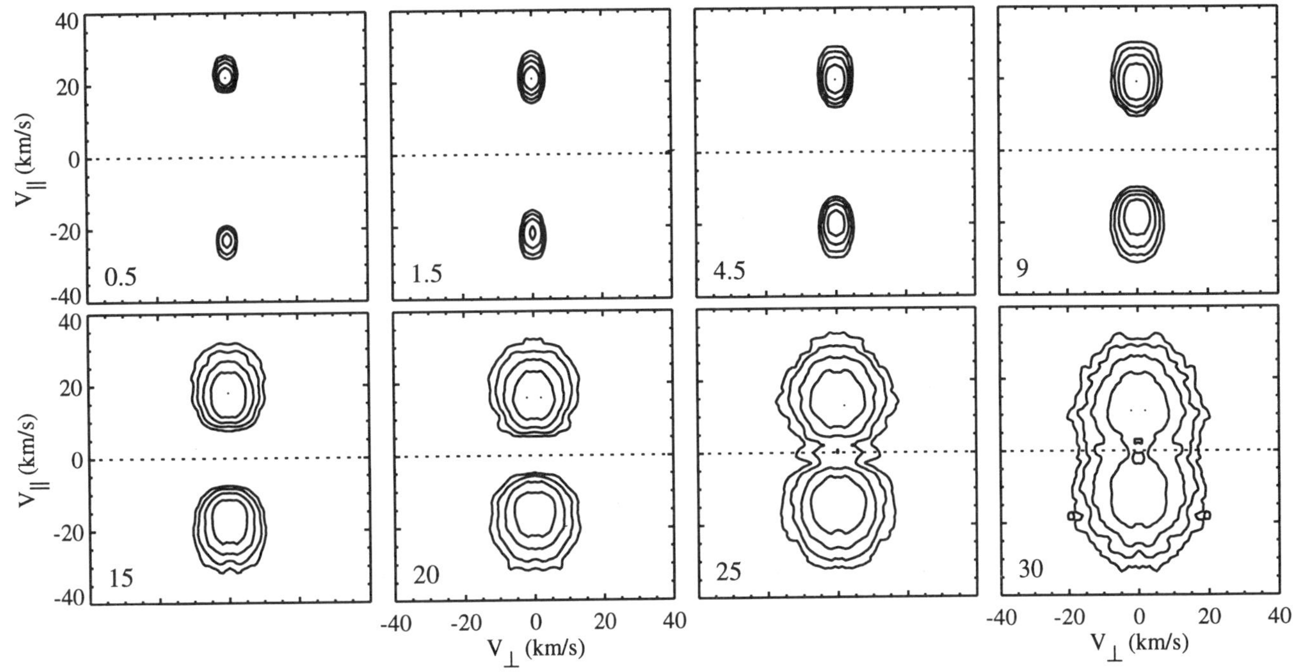

Fig. 11. Plots of the equatorial H$^+$ velocity distribution for a case done on an L= 4.5 flux tube connected to a realistic ionosphere. The times in each plot indicate the elapsed time (in hours) since refilling of the initially empty flux tube began. The contour line separation is e^{-1}.

collisions in this refilling situation. Most of the collisions which shape the equatorial velocity distribution do not occur at the equator but over an extended distance along the flux tube.

5. SUMMARY

In near geospace there can exist a situation where a plasma species is between the states of being collision dominated and being collisionless, because of large density gradients or time varying plasma densities. In such situations the collision time scales and length scales can become comparable to time and length scales of interest to transport processes. When this occurs the details of the collision process can no longer be ignored. A sophisticated description of collisions based on something like the Fokker-Planck equation is needed. It has been demonstrated, in this paper, that during the process of collisional relaxation highly nonMaxwellian velocity distributions can develop. When the collisional relaxation process extends over macroscopic time and length scales these nonMaxwellian velocity distributions must be taken into account in the modeling process. Such nonMaxwellian velocity distributions develop because of processes occurring in local collisions and distant (nonlocal) collisions in regions connected by transport to the point of interest. One consequence of these nonMaxwellian velocity distributions can be significant differences in the heat flows found from a fluid type transport model using BGK derived collision terms and a kinetic model using a Fokker-Planck collision algorithm. Transport models are, of course, not restricted to the use of the BGK collision model and can use more sophisticated, velocity dependent collision terms derived from the Fokker-Planck equation [*Schunk*, 1977], however, such models are unable to describe the transport consequences of the highly nonMaxwellian velocity distributions.

Acknowledgments. This work was supported by NASA grant NAGW-3470, and NSF grant ATM8911799, to the University of Alabama in Huntsville.

REFERENCES

Banks, P. M., A. F. Nagy, and W. I. Axford, Dynamical behavior of thermal protons in the mid-latitude ionosphere and magnetosphere, *Planet. Space Sci., 19*, 1053, 1971.

Barghouthi, I. A., A. R. Barakat, and R. W. Schunk, Monte Carlo study of the transition region in the polar wind: An improved collision model, *J. Geophys. Res., 98*, 17583, 1993.

Bhatnagar, P. L., E. P. Gross, and M. Krook, A model for collision processes in gases, *Phys. Rev.*, *94*, 511, 1954.

Casanova, M., O. Larroche, and J.-P. Matte, Kinetic simulation of a collisional shock wave in a plasma, *Phys. Rev. Lett.*, *67*, 2143, 1991.

Clemmow, P. C., and J. P. Dougherty, *Electrodynamics of Particles and Plasmas*, Addison-Wesley Publishing Company, Inc., Redwood City, CA, 1969.

Comfort, R. H., J. H. Waite, Jr., and C. R. Chappell, Thermal ion temperatures for the retarding ion mass spectrometer on DE 1, *J. Geophys. Res.*, *90*, 3475, 1985.

Fok, M.-C., J. U. Kozyra, A. F. Nagy, C. E. Rasmussen, and G. V. Khazanov, Decay of equatorial ring current ions and associated aeronomical consequences, *J. Geophys. Res.*, *98*, 19381, 1993.

Hinton, F. L., Collisional transport in plasma, in *Basic Plasma Physics I*, eds. A. A. Galeev and R. N. Sudan, North-Holland Publishing Co., Amsterdam, p. 147, 1983.

Jones, M. E., D. S. Lemons, R. J. Mason, V. A. Thomas, and D. Winske, A grid-based inter-particle collision model for PIC codes, submitted *J. Comput. Phys.*, 1994.

Jordanova, V. K., J. U. Kozyra, G. V. Khazanov, and A. F. Nagy, The effect of collisional processes on the pitch angle distributions of the ring current ions, *EOS* trans. AGU supp., p. 545, November 1, 1994.

Miller, R. H., and M. R. Combi, A Coulomb collision algorithm for weighted particle simulations, *Geophys. Res. Lett.*, *21*, 1735, 1994.

Miller, R. H., C. E. Rasmussen, T. I. Gombosi, G. V. Khazanov, and D. Winske, Kinetic simulation of plasma flows in the inner magnetosphere, *J. Geophys. Res.*, *98*, 19301, 1993.

Nicholson, D. R., *Introduction to Plasma Theory*, John Wiley & Sons, New York, NY, 1983.

Schulz, M., and H. C. Koons, Thermalization of colliding ion streams beyond the plasmasphere, *J. Geophys. Res.*, *77*, 248, 1972.

Schunk, R. W. Mathematical structure of transport equations for multispecies flows, *Rev. Geophys. Space Phys.*, *15*, 429, 1977.

Takizuka, T., and H. Abe, A binary collision model for plasma simulation with a particle code, *J. Comput. Phys*, *25*, 205, 1977.

Wahlund, J.-E., H. J. Opgenoorth, I. Häggström, K. J. Winser, and G. O. L. Jones, EISCAT observations of topside ionospheric ion outflows during auroral activity: Revisited, *J. Geophys. Res.*, *97*, 3019, 1992.

Wilson, G. R, J. L. Horwitz, and J. Lin, A semikinetic model for early stage plasmasphere refilling, 1: Effects of Coulomb collisions, *J. Geophys. Res.*, *97*, 1109, 1992.

Wilson, G. R, Semikinetic modeling of the outflow of ionospheric plasma through the topside collisional to collisionless transition region, *J. Geophys. Res.*, *97*, 10,551, 1992.

Wilson, G. R, J. L. Horwitz, and J. Lin, Semikinetic modeling of plasma flow on outer plasmaspheric field lines, *Adv. Space Res.*, *13*, (4)107, 1993.

G. R. Wilson, Department of Physics and Center for Space Plasma and Aeronomic Research, University of Alabama in Huntsville, Huntsville, Alabama.

Modeling of Spatial and Temporal Scales in Turbulent Flows and Their Relevance to Space Plasma Transport

Ram K. Avva and Ashok K. Singhal

CFD Research Corporation, Huntsville, Alabama

In this article, general features of turbulence and a description of its energy spectrum are introduced. Various length and time scales observed in turbulent flows are delineated. Current state of the art in modeling gas dynamics turbulence is reviewed. Popular turbulence models, their strengths and weaknesses are outlined. Interaction of turbulence and chemistry in reacting flows and selected modeling techniques are discussed. Finally the adequacy of conventional turbulence models for application to plasma turbulence is addressed.

1. INTRODUCTION

1.1. *Characteristics of Turbulence*

The phenomenon of turbulence in fluid flows has been observed for millennia. *Von Karman* [1937] was one of the first to attempt to define turbulence from a scientific point of view. He defined turbulence as an irregular motion which in general makes its appearance in fluids, gaseous or liquid, when they flow past solid surfaces or even when neighboring streams of fluid flow over one another. Later, *Hinze* [1975] brought in the concept of scales into the definition of turbulence. According to him, turbulence is an irregular condition of flow in which various quantities show a random variation with time and space over a wide range of scales, so that statistically distinct averages can be discerned. To put it more succinctly, turbulence is always transient and three-dimensional; it is neither completely deterministic nor stochastic, *i.e.* turbulent flows contain both organized and random motions. In other words, turbulent flow consists of lumps of fluid of disparate sizes, commonly referred to as eddies, which fluctuate with a wide range of frequencies. The large scale eddies derive energy from the mean motion and pass it on to smaller scales through a complex process of vortex stretching. This energy cascade continues till the energy is dissipated in the form of heat at the smallest possible eddies (known as Kolmogorov scales) due the action of molecular viscosity. The disparity between the largest and the smallest scales increases with Reynolds number.

A detailed discussion of the fundamentals of turbulence can be found in several text books, including *Tennekes and Lumley* [1972], and *Hinze* [1975].

1.2. *Scales in Turbulent Flows*

Over the years, various attempts have been made to divide the turbulent spectrum into different ranges for (i) to understand the turbulence structure from a fundamental point of view, and (ii) to facilitate in the development of turbulence models. A generally accepted classification is given below. In the notation employed below, the letter τ denotes time scale while L, λ etc. denote length scales. q^2 is twice the turbulent kinetic energy and ε is the dissipation rate of turbulence.

1.2.1. *Large scale.* This is the order of the size of the object responsible for producing large scale eddies, *e.g.* channel width or bluff body size. Large scales, in general, are not dependent on fluid molecular viscosity but very much flow dependent. This scale is also of the same order of size as the commonly used mixing length used in a number of turbulence models.

$$L \sim \frac{q^3}{\varepsilon} \qquad \tau \sim \frac{q^2}{\varepsilon} \qquad (1)$$

1.2.2 *Taylor micro-scale.* This is a length scale of intermediate range and falls between the large scales defined above and the kolmogorov scales defined next. Taylor micro-scale has little significance in turbulence modeling.

$$\lambda \sim \left(\frac{q^2 \nu}{\varepsilon}\right)^{\frac{1}{2}} \qquad \tau \sim \left(\frac{q^2 \nu}{\varepsilon^2}\right)^{\frac{1}{3}} \qquad (2)$$

1.2.3. *Kolmogorov scale.* This represents the smallest possible eddies where viscous dissipation takes place. Small scales do not depend on the flow geometry and inertia and hence are more universal.

$$\eta \sim \left(\frac{\nu^3}{\varepsilon}\right)^{\frac{1}{4}} \qquad \tau \sim \left(\frac{\nu}{\varepsilon}\right)^{\frac{1}{2}} \qquad (3)$$

1.2.4 *Molecular scale.* This is akin to the mean free path of molecules in kinetic theory of gases. Product of the mean free path and mean molecular velocity fluctuation is defined as molecular viscosity. In the defining relations below, c is the speed of sound.

$$\ell \sim \frac{\nu}{c} \qquad \tau \sim \frac{\nu}{c^2} \qquad (4)$$

In a fully turbulent flow, the following inequality generally holds among the different length scales defined above.

$$L \gg \lambda \gg \eta \gg \ell \qquad (5)$$

Of the scales described above, L and η are most relevant to turbulence modelers. They represent the largest and the smallest realizable length scales in a turbulent flow. Their ratio is a good measure of the width or disparity of the turbulent spectrum.

$$\frac{L}{\eta} = R_T^{\frac{3}{4}} \qquad R_T = \frac{q^4}{\varepsilon \nu} \qquad (6)$$

R_T is a turbulent Reynolds number and is generally 20-100 times smaller than the bulk flow Reynolds number.

Figure 1, reproduced from *Chapman* [1979], shows the energy spectra for a wide class of turbulent flows. Energy distribution among scales at large wave numbers, *i.e.* small scales, is practically identical for all flows. On the other hand, scales at smaller waver numbers, *i.e.* large scales, have disparate energy distribution. This characteristic of turbulent flows plays an important role in turbulence modeling which is discussed in the next section.

2. TURBULENCE MODELING

2.1 *Hierarchy of Turbulence Models*

The inherently complex characteristics of turbulence continues to pose an insurmountable challenge to turbulence modelers since the time of Osborne Reynolds. No turbulence model exists today that can adequately model the effect of all scales on the mean flow. Nevertheless a hierarchy of turbulence models have been developed over the years to model turbulence. A classification scheme was given by *Ferziger* [1987].

(1) Correlations
(2) Integral methods
(3) One-point Closure schemes
(4) Two-point closure schemes
(5) Large-eddy simulations
(6) Full Turbulence simulation.

The correlations are simple algebraic relations between various flow parameters, *e.g.* skin friction as a function of Reynolds number. Each correlation is applicable to only a particular and limited class of flows. Integral methods are developed mostly for boundary layers and free shear layers. Thin shear layer equations are integrated normal to the direction of the mean flow to yield ordinary differential equations which can be easily solved for. However integral methods are not suited for complex 3D flows.

In one-point closure schemes, all flow variables are decomposed into mean and fluctuating components. Navier-Stokes equations are averaged over time or ensembles to yield mean flow equations. Due to this averaging process, information about individual scales is lost. In fact, at any given point in the flow filed, only one representative time and length scales can be delineated. The nature of the turbulent scales, in particular the larger scales, is

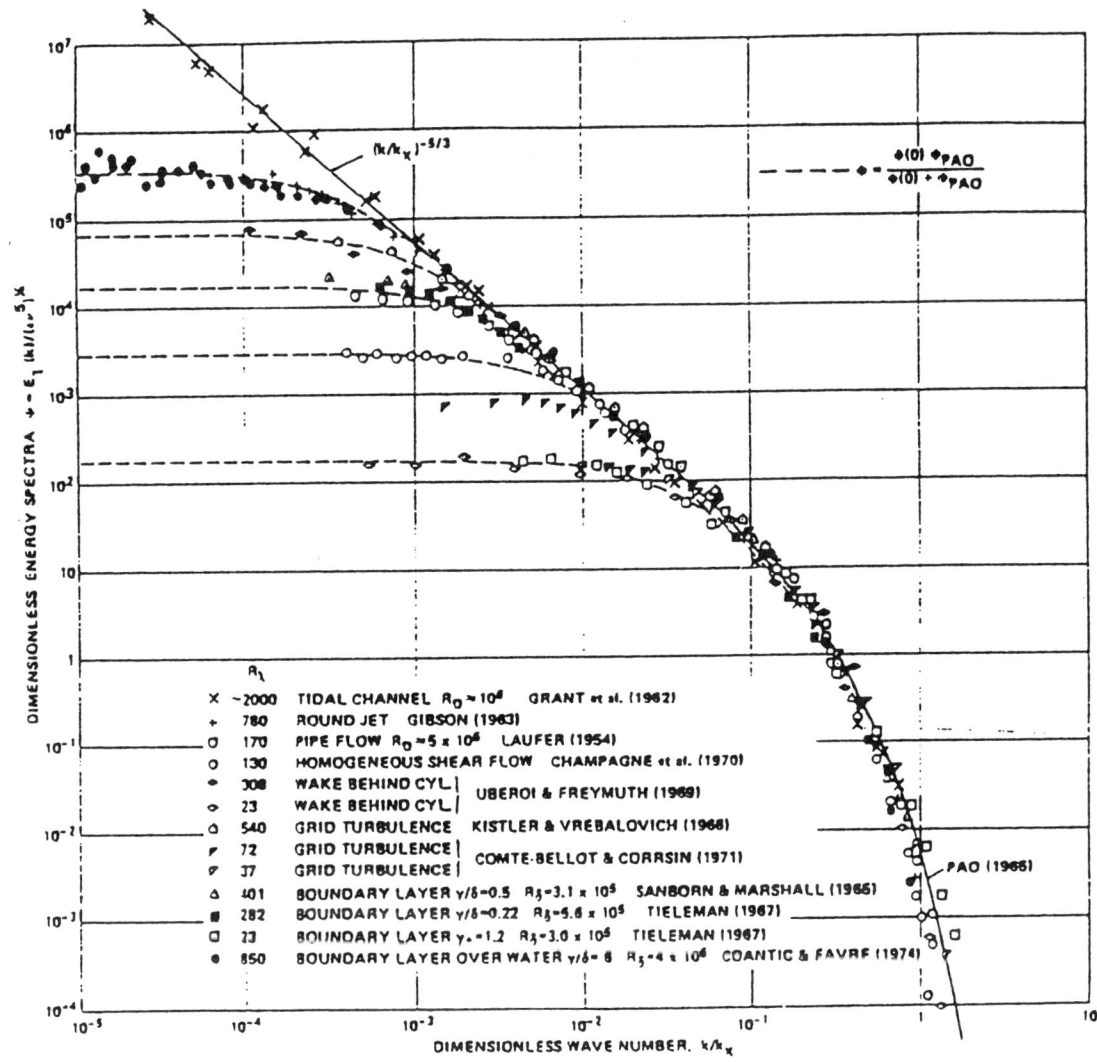

Fig. 1. Energy spectra in turbulent flows (from *Chapman*, 1979).

strongly affected by flow geometry, boundaries, and various other factors. It is precisely because of this reason, the one-point closure models tuned to one particular class of flows may not be accurate for another class of flows. Models based on one-point closure are dealt with in greater detail in the next section.

In two-point closure schemes, NS equations are transformed into spectral space so that the length scale information is easily delineated. They have so far been applied to simple homogenous flows only and have limited use in engineering applications. Interested readers may find more details in *Lesieur* [1987].

The Large Eddy Simulation (LES) and the Direct Numerical Simulation (DNS) are used to provide information about individual scales. In LES, the unsteady NS equations are averaged over space so that motion of large eddies is captured. However, information about subgrid scales, i. e. scales smaller than the grid size, is lost in the spatial averaging process. The effect of these small scales is modeled via subgrid scale models. It continues to be a matter of controversy till today whether the smaller scales are more universal in nature. Thus the accuracy of LES is limited by the accuracy of the subgrid scale models used. In DNS, this drawback is circumvented by solving for all scales. However, it requires very fine grids to resolve the smallest of scales. The need for very fine grids and the need to obtain the flow field for long periods of time to delineate meaningful statistical averages, both LES and DNS are computationally very expensive. For these reasons, they have been primarily used as research tools and not as engineering tools till today.

With respect to the classification scheme presented above, one should note:

(1) As one moves downwards in the list, each level requires more computational resources but less modeling than those above it.

(2) Each level computes more details of a flow than those above it, albeit at increased cost.

(3) The range of flows that may be simulated with a single model becomes broader as the level increases.

(4) The numerical methods used at each level differ.

(5) The amount of detail in the laboratory data required to provide initial and/or boundary conditions increases significantly with increasing level. Higher-level simulations require some data for which no suitable measurement techniques are available at the moment.

In all levels of turbulence modeling, the accuracy of spatial and temporal discretization plays a crucial role in determining the accuracy of the predictions. CFD codes employing one-point closure models which are commonly used for engineering purposes, should employ spatial and temporal schemes of order of accuracy 2 or higher to eliminate any false diffusion. To adequately resolve small scales and to accurately predict higher order turbulence quantities, LES and DNS need to employ higher-order spatial and temporal schemes or spectral methods. Spatial differencing schemes should preferably be of order 4 or higher while the temporal schemes should be of order 2-4.

2.2 Turbulence Models in Gas Dynamics

Most of the turbulence models used in Computational Fluid Dynamics (CFD) analyses fall under the category of one-point closure schemes. In these schemes, the Navier-Stokes equations are usually averaged over time or ensemble of statistically equivalent flows to yield averaged equations. In the averaging process, a flow quantity ϕ is decomposed in to mean and fluctuating parts. The following two types of averaging are generally used.

Reynolds (or time) Averaging:

$$\phi = \bar{\phi} + \phi' \text{ where } \bar{\phi} = (1/T) \int_{t_o}^{t_o+T} \phi \, dt \qquad (7)$$

Favre (or density) Averaging:

$$\phi = \tilde{\phi} + \phi'' \text{ where } \tilde{\phi} = \overline{\rho \phi}/\bar{\rho} \qquad (8)$$

When Navier-Stokes equations are density averaged, we obtain the Favre-averaged Navier-Stokes (FANS) equations given below. [For detailed derivation, see *Cebeci and Smith*, 1974.]

$$\frac{\partial}{\partial t}\left(\bar{\rho}\tilde{u}_j\right) + \frac{\partial}{\partial x_j}\left(\bar{\rho}\tilde{u}_i\tilde{u}_j\right) = -\frac{\partial \bar{p}}{\partial x_i} + \frac{\partial}{\partial x_j}\left[\bar{\mu}\left(\frac{\partial \tilde{u}_i}{\partial x_j} + \frac{\partial \tilde{u}_j}{\partial x_i} - \frac{2}{3}\frac{\partial \tilde{u}_m}{\partial x_m}\delta_{ij}\right)\right] + \frac{\partial}{\partial x_j}\left(-\bar{\rho}\widetilde{u_i''u_j''}\right) \qquad (9)$$

The FANS equations contain less information than the full NS equations, but have additional unknown terms $-\bar{\rho}\widetilde{u_i''u_j''}$ called the Reynolds stresses. These correlations between the fluctuating components arise in the averaging process, and need to be modeled to achieve closure of the FANS equations. Various models are used for modeling $u_i u_j$. These include:

(1) Mixing-length models
(2) One-equation models
(3) Two-equation models
(4) Algebraic Stress models
(5) Stress-transport models

The first 3 classes of models employ the generalized Boussinesq eddy viscosity concept in which the Reynolds stress $-\bar{\rho}\widetilde{u_i''u_j''}$ is treated as a linear function of the mean strain rate

$$-\bar{\rho}\widetilde{u_i''u_j''} = \mu_t\left(\frac{\partial \tilde{u}_i}{\partial x_j} + \frac{\partial \tilde{u}_j}{\partial x_i} - \frac{2}{3}\frac{\partial \tilde{u}_m}{\partial x_m}\delta_{ij}\right) - \frac{2}{3}\bar{\rho}k\delta_{ij} \qquad (10)$$

Here μ_t is known as the turbulent eddy viscosity and k is half the trace of the Reynolds stress tensor.

$$k = \frac{1}{2}\widetilde{u_k'' u_k''} \qquad (11)$$

By substituting Equation (10) in Equation (9), we obtain the modeled FANS equations.

$$\frac{\partial}{\partial t}(\bar{\rho}\tilde{u}_j) + \frac{\partial}{\partial x_j}(\bar{\rho}\tilde{u}_i\tilde{u}_j) = -\frac{\partial \bar{p}}{\partial x_i} + \frac{\partial}{\partial x_j}\left[(\bar{\mu}+\mu_t)\left(\frac{\partial \tilde{u}_i}{\partial x_j}+\frac{\partial \tilde{u}_j}{\partial x_i}-\frac{2}{3}\frac{\partial \tilde{u}_m}{\partial x_m}\delta_{ij}\right)\right] - \frac{2}{3}\frac{\partial}{\partial x_i}(\bar{\rho}k) \qquad (12)$$

Following the kinetic theory of gases, the eddy viscosity is generally modeled as the product of a velocity scale q and a length scale ℓ.

$$\mu_t = C\bar{\rho}q\ell \qquad (13)$$

where C is a constant of proportionality. Various models differ in the way q and ℓ are estimated and each of the following sections describes a turbulence model. In the description of models, the overbar for μ and ρ, and tilde for u, v, *etc.* will be dropped for convenience.

2.2.1. *Mixing length models.* In these models, both q and l are expressed as algebraic functions of mean flow field. These models have been very popular in computing boundary layer flows. Examples can be found in *Cebeci and Smith* [1974], *Baldwin and Lomax* [1978]. As no transport equations are solved for q or l, these models are inadequate for predicting complex flows with large separation, swirl etc.

2.2.2 *One-equation models.* As the name implies, these models employ a transport equation for the evolution of q and an algebraic equation for l. Examples are *Wolfshtein* [1969], *Norris and Reynolds* [1975]. Because the length scale is algebraically prescribed, one-equation models are only marginally better than the mixing-length models. They are mostly replaced by the more versatile two-equation models.

2.2.3 *Two-equation models.* Evolution of both the velocity and length scales is computed via transport equations. Models of this category are by far the most popular in engineering applications as they have a fairly wide range of applicability and yet economical for use with current day computers. Majority of these models employ an equation for k, the turbulent kinetic energy and ε, the dissipation rate. Some of the popular models are listed below.

- Standard k-ε model [*Launder & Spalding*, 1974]
- Low-Re k-ε model [*Chien*, 1982]
- Multiscale k-ε model [*Kim & Chen*, 1988]
- RNG-based k-ε model [*Yakhot et al.* 1993]
- 2-layer k-ε model [*Rodi*, 1991]
- k-ω model [*Wilcox*, 1991]

As these models assume one eddy viscosity to model the whole Reynolds stress tensor, they are inaccurate for predicting flows with strong anisotropy. For example, two-equation models do not adequately predict stress-induced secondary flows. Also they do not satisfactorily capture the effects of extra strains such as curvature, rotation, pressure gradient etc.

2.2.4. *Algebraic stress models.* These models are developed to overcome the deficiencies of two-equation models in particular the assumption of isotropic eddy viscosity. In stead of expressing the Reynolds stresses by Equation (10), these models obtain the individual stresses via algebraic relations which are functions of k, ε, and P, the turbulence production rate. k and ε are obtained from transport equations. Some examples are *Gibson* [1978] and *Demuren and Rodi* [1984]. These models have shown some success in predicting the effects of extra strains such as curvature, swirl, and rotation. However, they are inadequate for predicting complex flows in which the evolution of the stress tensor plays an important role.

2.2.5. *Stress transport models.* These models employ a transport equations for each component of the Reynolds stress tensor. These equations involve higher order correlations such as $u_i u_j u_k$, which need to be modeled. Higher order correlations are very difficult to measure accurately. They are also more difficult to model accurately. The computational CPU requirements also increase several fold with these models. However, they appear to be the models of near future. *Launder* [1989] demonstrated the superiority of these models for a number of complex industrial flows.

Interested reader may find excellent reviews of one-point closure models in *Reynolds* [1976], *Rodi*

[1980], *Lakshminarayana* [1987], and *Wilcox* [1993].

2.3. Turbulence/Chemistry Interactions

Conventional turbulence models discussed above were developed for nonreacting flows and cannot properly model mean reaction rates. A major difficulty with reacting flows is that reaction rates, which are highly nonlinear functions, must be averaged. The reaction rates depend on the kinetics mechanism, which differs greatly for different fuels. The effects of heat release also modify the turbulent flow field. These effects are not considered in conventional turbulence modeling.

One approach for turbulent combustion is the eddy breakup model. The reaction is assumed to be fast enough that it is controlled by turbulent mixing. The time scale from the turbulence model therefore replaces the time scale of the reaction. Another approach is to assume the fluctuations are not important and to use the average values of temperatures and composition to calculate reaction rates.

The joint probability density function (PDF) of composition must be used to account for the effects of temperature and composition fluctuations on finite-rate kinetics. The most promising approach to solve turbulent flows with finite-rate chemistry is the PDF transport method [*Pope*, 1990], which removes the need to model effects of turbulent fluctuations on finite-rate reactions and can, in principle, be used with any desired kinetics mechanism. PDF transport models have been successfully used to calculate a number of turbulent flames.

The PDF transport method solves the transport equation for a joint PDF of at least the composition variables. Velocity, energy dissipation rate, or other variables can be included as well. The composition PDF is sufficient to remove the need to model mean reaction rates. The mean value of any function of composition, such as temperature or density, can be determined from the composition PDF. The effects of convection and molecular mixing must be modeled in the composition PDF approach. The joint velocity-composition PDF removes the need for gradient transport models, but requires modeling of viscous terms and the pressure gradient. A simpler method than solving a PDF transport equation is to prescribe a parametric form of the PDF. Transport equations for the parameters are solved by conventional means. The prescribed PDF method is only appropriate when the heat-release chemistry can be assumed to be very fast, although slow reactions can be considered if they can be decoupled from the heat release reactions. In this case the concentrations, temperature, and density can be related to a single scalar variable, the mixture fraction. A variety of shapes can be prescribed for the PDF of one variable. Multidimensional PDF's are required for general finite-rate kinetics. The prescription of a multidimensional PDF is not practical. The number of parameters increases dramatically with the number of variables (unless assumptions are made about independence of variables) and the prescribed PDF must have nonzero probability only for allowable compositions.

The composition PDF contains no information about the velocity field and must be supplemented with a flow solver that includes conventional turbulence modeling. The velocity-composition PDF does not require any external specification of the velocity field, but requires special treatment of the mean pressure gradient. The mean pressure field is either solved by a conventional flow solver [*Correa and Pope*, 1992 and *Roekaerts*, 1991] or with a solution of a Poisson equation [*Anand, Pope, and Mongia*, 1990] for elliptic flows.

The main disadvantage of PDF methods is that finite difference methods cannot be used to solve the PDF transport equation because the CPU time increases exponentially with number of independent variables [*Pope*, 1985]. The PDF transport equation can be more efficiently solved by Monte Carlo methods.

2.4. Current Status of Turbulence Modeling in Gas Dynamics

The state of the art in turbulence modeling may be summarized as below.

(1) Since the time of Osborne Reynolds (1883), thousands of mathematicians, scientists, and engineers have studied various aspects of incompressible fluid turbulence resulting in substantial experience base.

(2) A vast number of turbulence modeling methodologies have evolved in the past century ranging from Moody's charts to Full Turbulence Simulation.

(3) Turbulence models have been successfully applied in the design of all kinds of flow devices ranging from water pipes to space shuttle.

(4) Considerably less effort has been devoted in understanding and modeling turbulent heat transfer. Majority of the models use Reynolds analogy concept while a few employ a T'^2 equation. Examples are *Plumb and Kennedy* [1977], and *To and Humphrey* [1986]. Models with transport equations for $u'T'$ have yet to be developed.

(5) Effect of compressibility (density fluctuations) on turbulence is only being addressed within the last 15 years. See *Zeman* [1989] and *Sarkar and Lakshmanan* [1991]. These models are yet to be tested for complex industrial flows.

(6) Modeling of turbulent reactive flows continues to be one of the hardest problems in gas dynamics community. Accounting for the effect of density, velocity, and concentration fluctuations on reaction rates and heat release terms has so far been an insurmountable. Conventional Boussinesq eddy diffusivity models proved to be futile. Stochastic methods such as Monte Carlo PDF techniques are slowly gaining success.

(7) Besides the acquirement of conventional wisdom (bubbles augment turbulence while particles attenuate turbulence), very little focussed effort has been directed in modeling turbulence in multiphase flows. Most multiphase turbulence models are nothing but incompressible turbulence models modified on the basis of heuristic arguments. Only recently, two-time averaging and full turbulence simulation techniques are being employed to understand multiphase turbulence.

(8) LES and FTS techniques could not replace the conventional turbulence models as envisaged some 25 years ago. Nor could they make the wind tunnel obsolete.

3. MODELING OF PLASMA TURBULENCE

Any engineer or scientist interested in analyzing turbulent plasma flows will face the following question: Are turbulence models developed for gas dynamics adequate for plasma turbulence? Unfortunately, lack of sufficient experience with plasma flows makes it difficult for the authors of this article to answer this question with any reasonable confidence. Nevertheless, the authors would venture to make the following suggestions based on their experience with gas dynamics.

(1) Gas dynamics turbulence theory and models are generally applicable to dense plasma flows in which continuum hypothesis holds good.

(2) In addition to velocity and density fluctuations, electric (and/or magnetic) fluctuations need to be taken into account for modeling plasma turbulence. Cross correlations such as $e'u'$ need to be modeled to achieve closure.

(3) In dense plasma where macroscopic eddies are responsible for plasma turbulence, phenomenological models such as mixing-length [e.g. *Baldwin and Lomax*, 1978] models may be applied. In dilute plasma where microscopic (discrete) particle behavior is important, conventional turbulence models may not be appropriate.

(4) Turbulence in gas dynamics flows is mostly generated at walls and around shear layers (e.g. mixing layers). Thus conventional turbulence modeling techniques may be useful in modeling confined dense plasma flows as in MHD reactors.

(5) In Space Plasma, wall generated turbulence is probably insignificant compared to the turbulence generated by electro-magnetic field fluctuations. Modeling techniques used for reacting flows may provide a better starting point for modeling space plasma turbulence.

(6) In addition to the turbulence theory of fluid dynamics, the linear/non-linear

perturbation and instability theory of wave interference and oscillation can provide useful clues for modeling shock waves and turbulence in cosmic plasmas.

(7) In ordinary fluid dynamics (OFD), energy cascade occurs from large eddies (long-wavelength modes) to small eddies (short-wavelength modes). The opposite occurs in plasma flows due to wave-wave interactions. Short-wavelengths waves tend to coalesce into large-wavelength waves which are less energetic. Plasma turbulence models should be consistent with this phenomenon.

(8) Plasma Turbulence also leads to an important phenomenon called "anomalous resistivity" in which electrons are slowed down by collisions with random electric field fluctuations, rather than with ions. Plasma turbulence models should also take this into account.

Fundamental aspects of plasma turbulence and different modeling strategies are addressed by *Cap* [1982], *Chen* [1984], *Chen* [1987], *Kadomtsev* [1985], and *Tsytovich* [1977].

Acknowledgements. The assistance of our colleagues Drs. Andy Leonard and Ben Yu is greatly appreciated. Thanks are due to Ms. Marni Kent for her efforts in the preparation of this manuscript.

REFERENCES

Anand, M.S., Pope, S.B., and Mongia, H.C., Pressure Algorithm for Elliptic Flow Calculations with the PDF Method, *CFD Symposium on Aeropropulsion*, NASA Lewis Research Center, 1990.

Baldwin, B.S. and Lomax, H., Thin Layer Approximation and Algebraic Model for Separated Flows, *Paper No. AIAA-78-257*, 1978.

Cap, F.F., *Handbook on Plasma Instabilities*, Academic Press, New York, 1982.

Cebeci, T. and Smith, A. M. O., *Analysis of Turbulent Boundary Layers*, Academic Press, New York, 1974.

Chapman, D. R., Computational Aerodynamics Development and Outlook, *AIAA Journal*, Vol. 17, pp. 1293-1313, 1979.

Chen, F.F., *Introduction to Plasma Physics and Controlled Fusion*, Plenum Press, NY, 1984.

Chen, L., *Waves and Instabilities In Plasmas*, World Scientific, New Jersey, 1987.

Chien, K. Y., Predictions of Channel and Boundary-Layer Flows with Low-Reynolds-Number Turbulence Model, *AIAA Journal*, Vol. 20, pp. 33-38, 1982.

Correa, S. M. and Pope, S. B., Comparison of a Monte Carlo PDF/Finite Volume Mean Flow Model with Bluff-Body Raman Data, *Twenty-Fourth Symp. (Intl.) on Combust.* 279-285, 1992.

Demuren, A. O. and Rodi, W., Calculation of Turbulence-driven Secondary Motion in Non-circular Ducts, *J. of Fluid Mechanics*, Vol. 140, pp. 180-222, 1984.

Ferziger, J. H., Simulation of Incompressible Turbulent Flows, *Journal of Computational Physics*, Vol. 69, pp. 1-48, 1987.

Gibson, M. M., An Algebraic Stress and Heat-Flux Model for Turbulent Shear Flow with Streamline Curvature, *Int. J. Heat and Mass Transfer*, Vol. 21, pp. 1609-1617, 1978.

Hinze, J. O., *Turbulence*, McGraw Hill Inc., New York, 1975.

Kadomstev, B.B., *Plasma Turbulence*, Academic Press, New York, 1985.

Kim, S. W. and C. P. Chen, A Multiple Time-Scale Turbulence Model Based on Variable Partitioning of Turbulent Kinetic Energy Spectrum, *Paper No. AIAA-88-1771*, 1988.

Lakshminarayana, B., Turbulence Modeling for Complex Shear Flows, *AIAA J.*, Vol. 24, No. 12, pp. 1900-1917, 1987.

Launder, B. E., Second-Moment Closure and Its Use in Modelling Turbulent Industrial Flows, *Int. J. of Num. Meth. in Fluids*, Vol. 9, pp. 963-985, 1989.

Launder, B.E., and Spalding, D.B., The Numerical Computation of Turbulent Flows, *Comp. Methods Appl. Mech. Eng.*, Vol. 3, pp. 269-289, 1974.

Lesieur, M., *Turbulence in Fluids: Stochastic and Numerical Modeling*, Martinus Nijhoff Publ., Dordrecht, Netherlands, 1987.

Norris, L. H. and Reynolds, W. C., Turbulent Channel Flow with a Moving Wavy Boundary, *Rep. No. FM-10*, Thermosciences Div., Dept. of Mech. Engr., Stanford University, Stanford, CA, 1975.

Plumb, O. A. and Kennedy, L. A., Application of a k-ε Model to Natural Convection From a Vertical Isothermal Surface, *Trans. of ASME, J. of Heat Transfer*, Vol. 99, pp. 79-85, 1977.

Pope, S. B., PDF Methods for Turbulent Reactive Flows, *Prog. Energy Combust. Sci.* Vol. 11, 119-192, 1985.

Pope, S. B., Computations of Turbulent Combustion: Progress and Challenges, *Twenty-Third Symp. (Intl.) on Combust.* 591-612, 1990.

Reynolds, W. C., Computation of Turbulent Flows, *Annual Review of Fluid Mechanics*, Vol. 8, pp. 183-208, 1976.

Rodi, W., *Turbulence Models and Their Application in Hydraulics*, IAHR Publ., DELFT, The Netherlands, 1980.

Rodi, W., Experience with Two-Layer Models Combining the k-ε Model with a One-equation Model Near the Wall, *AIAA-91-0216*, 1991.

Roekaerts, D., Use of a Monte Carlo PDF Method in a Study of the Influence of Turbulent Fluctuations on Selectivity in a Jet-Stirred Reactor, *Appl. Sci. Research* Vol. 48, 271-300, 1991.

Sarkar, S. and Lakshmanan, B., Application of a Reynolds Stress Turbulence Model to the Compressible Shear Layer, *AIAA J.*, Vol. 29, No. 5, pp. 743-749, 1991.

Tennekes, H. and Lumley, J. L., *A First Course in Turbulence*, MIT Press, Cambridge, Massachusetts, 1972.

To, W. M. and Humphrey, J. A. C., Numerical Simulation of Buoyant, Turbulent Flow, *Int. J. Heat Mass Transfer*, Vol. 29, No. 4, pp. 573-610, 1986.

Tsytovich, V. N., *Theory of Turbulent Plasma*, Consultants Bureau, New York, 1977.

Von Karman, T., Turbulence, Twenty-fifth Wilbur Wright Memorial Lecture, *Journal of Royal Aeronautical Society*, Vol. 41, pp. 1109, 1937.

Wilcox, D. C., A Half Century Historical Review of the k-ω Model, *AIAA-91-0615*, 1991.

Wilcox, D.C., *Turbulence Modeling for CFD*, DCW Industries, La Canada, California, 1993.

Wolfshtein, M., The Velocity and Temperature Distribution in One-dimensional Flow with Turbulence Augmentation and Pressure Gradient, *International Journal of Heat and Mass Transfer*, Vol. 12, pp. 301-318, 1969.

Yakhot, V., Orszag, S. A., Thangam, S., Gatski, T. B. and Speziale, C. G., Development of Turbulence Models for Shear Flows by a Double-Expansion Technique, *Physics of Fluids*, Vol. 4, pp. 1510-1520, 1992.

Zeman, O., Dilatation Dissipation: The Concept and Application in Modeling Compressible Mixing Layers, *CTR Manuscript 100*, Center for Turbulence Research, Stanford, CA, 1989.

Ram K. Avva and Ashok K. Singhal, CFD Research Corporation, 3325 Triana Blvd., Huntsville, Alabama 35812.

Regional Particle Simulations and Global Two-Fluid Modelling of the Magnetospheric Current System

R. M. Winglee

Geophysics Program, University of Washington, Seattle, Washington

Global MHD modelling can provide the overall topology of the magnetosphere. However, at critical boundary layers, such as the magnetopause, meso-scale processes can be important in controlling the coupling of energy and momentum from the solar wind into the magnetosphere. Regional particle simulations are used to show the importance of some meso-scale processes, arising from the differential penetration of ions and electrons across the magnetopause and the generation of field-aligned currents that map into the noon and midnight auroral sectors. These processes are then incorporated into the global context through a newly developed modified two fluid treatment. Using both 2-D and 3-D simulations, it is shown that not only are the particle effects seen in the regional modelling recovered in the global modelling, but that realistic mapping of both field lines and currents can be attained.

1. INTRODUCTION

Magnetohydrodynamic (MHD) simulations are presently one of the methods means for providing global or macro-scale modelling of the magnetosphere. Such simulations have been successful in predicting the overall characteristics of the magnetosphere under varying solar wind conditions. Current efforts are focussed on either high-resolution two-dimensional simulations of specific regions of the magnetosphere [e.g., *Wei et al.*, 1990; *Shi et al.* 1991] or global three-dimensional simulations of the magnetosphere [e.g., *Ogino*, 1986; *Fedder and Lyon*, 1987; *Watanabe and Sato*, 1990; *Hesse and Birn*, 1991; *Usadi et al.*, 1993].

One of the limitations of the above global modelling is that the grid spacing is typically a few thousand kilometers at best. As such, the global simulations have problems treating the dynamics of discrete magnetospheric boundary layers which are in turn important in determining the energy and plasma transport between different regions. A key example is the magnetopause. Its position may vary by several earth radii, depending on the solar wind dynamic pressure and strength and direction of the interplanetary magnetic field (IMF) [*Sibeck et al.*,1991; *Roelof and Sibeck*, 1993; *Petrinec and Russell*, 1993], but it remains relatively thin, typically between about 400 km to 1000 km [*Berchem and Russell*, 1982]. The magnetopause is also highly structured [e.g., *Song et al.*, 1990, 1993].

The above motion of the magnetopause under varying solar wind conditions can in principle be described by the above global MHD simulations. However, these global MHD simulations do not predict the observed structure of the magnetopause which are presumably driven by dynamical processes relevant to meso- and/or micro-scale lengths. Thus, there is a need to incorporate the coupling between global processes that predict the overall positions and plasma conditions with regional (meso-scale) or local (micro-scale) processes that determine the structure and plasma interactions between different plasma populations. The importance of the coupling of macro-scale to meso- and microscale processes is not limited to just the magnetopause but is also relevant to other regions where there are thin boundary layers, including the cusp, the nightside auroral region and the plasma sheet boundary layer.

Only recently have there been attempts to evaluate

meso- and micro-scale processes and show how they can modify global processes. In an initial application to magnetic reconnection in the magnetotail [*Winglee and Steinolfson*, 1993], particle simulations were used to identify the types of conditions needed to drive magnetic reconnection in the presence of a normal magnetic field, and to determine the effective plasma resistivity. It was shown that kinetic effects lead to the generation of field-aligned currents and the ejection of much more compact and faster plasmoids than predicted in resistive MHD. The differences in plasmoid ejection were then confirmed using global 2-D MHD simulations using the the effective plasma resistivity determined from the particle simulations [*Steinolfson and Winglee*, 1993].

The kinetic effects that lead to the generation of the field-aligned currents and effective plasma resistivity are due to meso-scale processes at boundary layers, arising from the differential penetration of ions and electrons across the boundary layers. These processes are reviewed in section 2. To date it has been difficult to incorporate these processes in global models since the particle simulations from which they are derived cannot presently provide full global models. This problem has been recently overcome by *Winglee* [1994] where a modified two-fluid treatment has been developed that is able to provide full global modelling, incorporating the field-aligned currents and the effective plasma resistivity seen in the particle simulations without the restrictions of artificial mass ratios and limited spatial and temporal scales that plague particle simulations.

The initial 2-D results from the modified two-fluid simulations are also reviewed in section 2. These simulations confirm the relevant physics. In order to make detailed predictions for the mapping of currents and magnetic fields from the ionosphere out to the magnetosphere, 3-D two-fluid simulations are presented in section 3. This 3-D mapping provides important insight into the locations where coupling between global and meso-scale processes is important to the transport of solar wind energy and plasma to the magnetosphere.

2. PARTICLE EFFECTS AT THE MAGNETOPAUSE

2.1 A Physical Picture

The importance of incorporating particle effects at the magnetopause is illustrated in Fig. 1. As the shocked solar wind (magnetosheath) plasma impinges on the magnetopause, the dayside magnetic field is compressed producing the Chapman-Ferraro currents in the dawn-dusk direction. At this stage there are no field-aligned currents that map from the noon sector into the dayside auroral region. However, the Chapman-Ferraro currents are not the only currents generated in this region. These additional currents arise because the solar wind ions, with their large gyro-radii are able to penetrate on average further than the electrons. This differential penetration leads to a net accumulation of charge in the outer regions of the magnetopause and a net positive charge in the inner regions. Because these field lines are tied to the ionosphere, quasi-neutrality can be maintained by pulling ionospheric electrons up and into the inner regions of the magnetopause. The magnetosheath electrons in the outer regions are then free to stream along the field lines into the cusp. Some of these electrons may mirror and continue to convect into the nightside magnetosphere while others may precipitate into the dayside ionosphere, maintaining the charge neutrality of the ionosphere.

As a result, a circulatory current system can be set up around the magnetopause with a portion of the current mapping directly along the field lines into the dayside ionosphere as indicated in Fig. 1. The Chapman-Ferraro currents are incorporated in global MHD models. However, the above additional field-aligned currents that arise from the differential penetration of the solar wind ions and electrons across the magnetopause are not. As such the magnetopause is a critical region where the coupling of global influences with meso- and micro-scale processes are important.

2.2. Regional Particle Simulations

The generation of these additional plasma currents in relation to the MHD current system has been recently investigated by *Winglee* [1994] using regional particle simulations. These initial simulations were two-dimensional and focussed on the noon-midnight meridian and incorporate a magnetic dipole field for the initial terrestrial magnetic field. An ambient plasma is load throughout the system down to some minimum radius from the magnetic dipole (equivalent to typically about $2 - 3\ R_e$ where R_e is the radius of the Earth). The density of this ambient plasma is assumed to decrease inversely with the cube of the radial distance such that the density near the estimated position of the magnetopause is about a tenth of that near the inner boundary. The magnetic field of a two-dimensional magnetic field decreases with the square of the radial distance. Thus, the Alfvén speed decreases as the square root of the radial distance, being equivalent to about 500 km/s on the magnetospheric side of the magnetopause. This magnetic field strength is

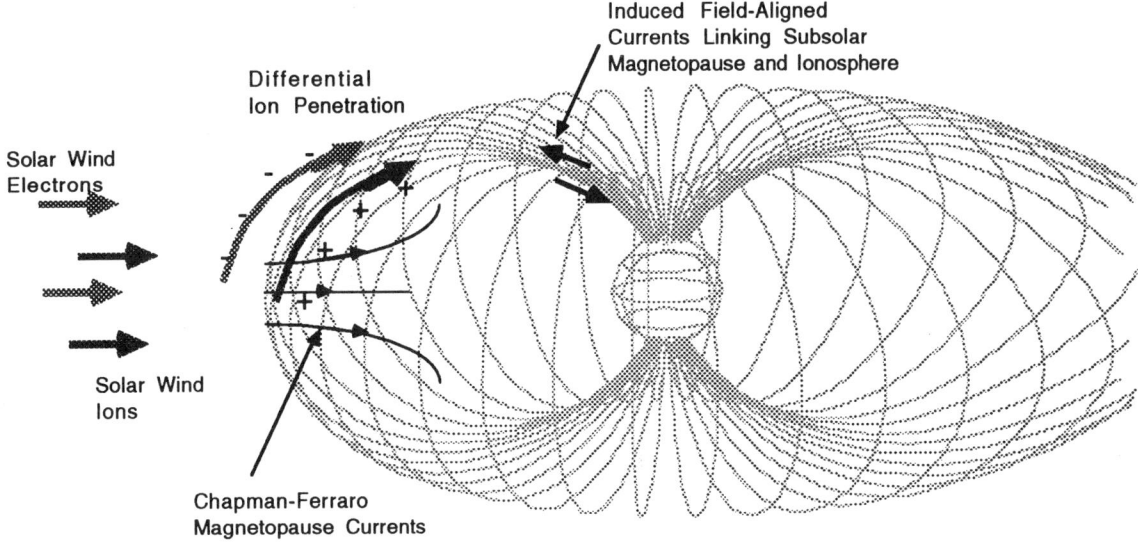

Fig. 1. Schematic showing the interaction of the solar wind with the dayside magnetosphere. The ram pressure of the solar wind compresses the dayside magnetic field producing the Chapman-Ferraro currents in the dawn-dusk direction. The differential penetration of solar wind ions and electrons can produce additional currents that are field-aligned and enter the dayside ionosphere.

equivalent to an electron cyclotron frequency in the vicinity of the magnetopause $\Omega_{e,mp}$ that ranges from 0 to 0.2 of the magnetosheath plasma frequency $\omega_{e,sh}$. The temperature of the ambient plasma increases with radial distance such that pressure balance is maintained with the normalization and the ion thermal speed near the estimated position of the magnetopause is approximately equal to the local Alfvén speed. The initial electron temperature is assumed to be equal to the ion temperature.

Solar wind is continuously injected into the system so that the magnetopause self-consistently forms through its own efforts to balance plasma and magnetic pressures. The inflowing plasma is assumed to have a drift speed $v_{b,sh}$ (which is equivalent to 350 km/s) which is twice its thermal speed $v_{T,sh}$, and its density is taken as 10 times the initial magnetospheric density near the magnetopause. A total of 350,000 multi-charged particles are used to represent the different plasma populations. The average particle per cell is fractionally less than unity since much of the simulation system is essentially empty space where the magnetosheath plasma is unable to penetrate. However, in the vicinity of the magnetosheath there are typically 2 - 3 particles per cell.

There are no boundary conditions imposed on the fields because they are exactly specified by the plasma and dipole currents in the system. Instead, boundary conditions on the particle dynamics must be imposed. Any particle crossing the boundary is reflected back outwards. In addition, a collisional operator is applied to a region extending fractionally out from the inner boundary and this operator scatters all particles, irrespective of its origin, according to a specified collisional cross-section that mimics the properties of the actual ionosphere [*Winglee*, 1990; *Winglee and Steinolfson*, 1993]. These collisions provide the plasma at low altitudes with a Pederson conductivity sufficient to close any field-aligned currents, plus they act as a sink for the energy of any precipitating, energetic particles. In this way, the present particle simulations directly incorporate crucial physics tied to the dynamics of the ionosphere.

Such regional particle simulations can only be considered a first step to resolving the coupling between global influences and meso- and micro-scale processes since the particle simulations must use an artificial ion-to-electron mass ratio (16 in the present case) to maximize effective length and time scales. Even with this approximation realistic global length scales cannot be modelled. On a grid of 384 × 192 only a few hundred ion gyro-radii can be resolved along the length of the magnetopause. Nevertheless, they are important as they show the relevant physics. As shown here, these processes once identified can be incorporated into global modelling using more accurate fluid treatments than ideal MHD.

As an example of the comparison between global (MHD) and kinetic processes affecting the magnetopause, Fig. 2 shows the derived plasma flow (left

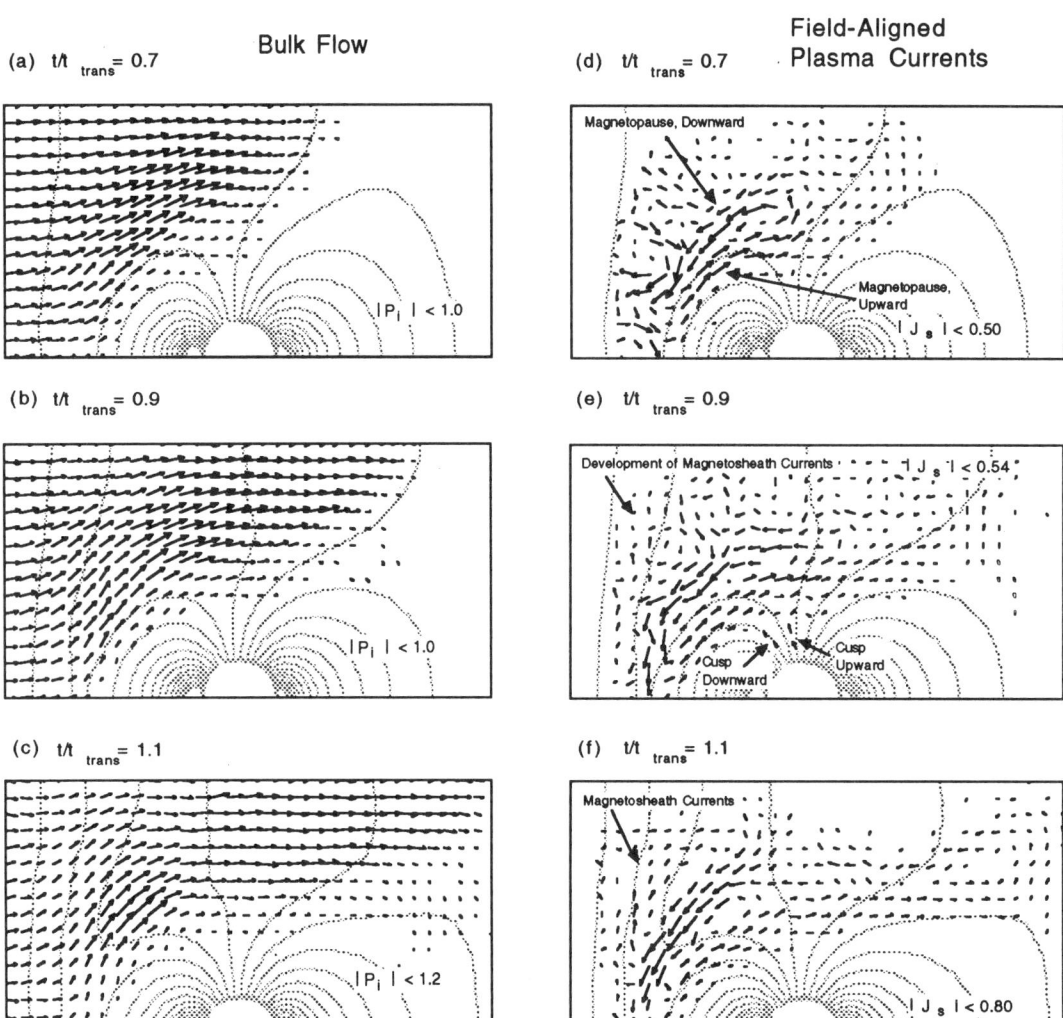

Fig. 2. A regional particle simulation of magnetopause interactions in the noon-midnight meridian. The magnetic field is indicated by the dotted lines. The arrows in the left hand panels indicate the evolution of the bulk flow while the arrows in the right hand panels indicate the field-aligned currents. The bulk flow is essentially the same as seen in MHD models with the solar wind plasma being deflected around the magnetopause. However, contrary to MHD, there are significant field-aligned currents present.

hand side) and induced currents in the noon-night meridian (right hand side). Time scales are relative to a transit time across the system. Similar to global MHD simulations, the injected plasma moves into the system and compresses the dayside magnetosphere (Fig. 2a). When the plasma eventually encounters sufficiently strong magnetic field, it is deflected over the polar cap and a well defined magnetopause is seen (Fig. 2b). With the injection of southward IMF, dayside reconnection occurs leading to the peeling away of some of the dayside magnetic field and the generation of accelerated plasma flows.

The flows are similar to that seen in global MHD. However, for this configuration MHD predicts there should be no currents seen in the plane of the simulation system. As indicated by the right hand side of Fig. 2, there are instead intense plasma currents flowing around the magnetopause, similar to the schematic in Fig. 1. The downward currents at the magnetopause coincide in a region where there is predominantly magnetosheath plasma mixed with a small concentration of magnetospheric plasma and may correspond to the sheath transition region identified by *Song et al.* [1990, 1993]. In the upward current regions the relative concentration of the two plasma populations are reversed, and as such may correspond to the boundary layer regions associated with the magnetopause. These currents are seen to intensify in association with dayside reconnection during the southward turning of the IMF (see *Winglee* [1994] for more details).

The important point to the above particle simulations is that there are crucial kinetic or particle effects that are vital to the dynamics of discrete regions such as the magnetopause. The next step is to determine whether these regional or local processes affect the global dynamics and, if so, a method is required to incorporate the above effects into a more realistic global model.

2.3. Global 2-D Two-Fluid Simulations

The particle results can be reconciled with fluid models by going to higher order forms of Ohm's law and using the fact that there are other electric fields at boundary layers besides the convective electric field incorporated in ideal MHD. By achieving this reconciliation between particle and fluid models, one is then able to incorporate meso-scale processes into the global environment. This reconciliation is achieved by reexamining the generalized Ohm's law from two-fluid theory (as opposed to the idealized Ohm's law of MHD). If the currents are assumed to be slowly varying but allow the plasma to have finite spatial gradients, the generalized form of Ohm's law can be approximated by

$$\boldsymbol{E} + \frac{\boldsymbol{V} \times \boldsymbol{B}}{c} \simeq \eta \boldsymbol{J} + (\frac{\boldsymbol{J} \times \boldsymbol{B}}{ecn_e} - \frac{1}{en_e}\nabla P_e) \quad (1)$$

where η is the collisional plasma resistivity, and n_e and P_e are the electron plasma number density and pressure, respectively. The MHD approach is to further neglect the terms in the brackets in (1). This approximation forces a replacement of the collisional plasma resistivity with an anomalous or artificial resistivity in order to produce dissipation of the magnetic field on timescales comparable to the observed timescale of substorms.

An explicit solution to (1) can be attained by splitting the currents and associated fields into their global and boundary layer (meso-scale) contributions. Specifically, the global currents are primarily the ideal MHD currents supported by the convective electric field, both of which are orthogonal to the noon-midnight meridian for the present case. For simplicity, these quantities are referred to as body currents or fields since they are associated with global processes. The boundary layer or plasma currents and associated fields are related to meso- or micro-scale processes at the surface of the object and hereafter referred to as surface currents or fields. Unlike the body currents and fields, the surface currents and fields are in the plane of the magnetic field and fluid flow for the present case.

With this decomposition, (1) reduces to

$$\boldsymbol{E}_b + \frac{\boldsymbol{V} \times \boldsymbol{B}_b}{c} = \eta \boldsymbol{J}_b + \frac{\boldsymbol{J}_s \times \boldsymbol{B}_b}{ecn_e} \quad (2a)$$

$$\boldsymbol{E}_s + \frac{\boldsymbol{V} \times \boldsymbol{B}_s}{c} = \frac{\boldsymbol{J}_b \times \boldsymbol{B}_b}{ecn_e} - \frac{1}{en_e}\nabla P_e \quad (2b)$$

where (2a) approximates the generalized Ohm's law in the plane orthogonal to the fluid flow and the body magnetic field, while (2b) represents the component in the plane of the body magnetic field. In deriving (2a) and (2b), we have used the fact that $\boldsymbol{J}_b \times \boldsymbol{B}_s$ is equal to zero since the vectors are parallel by definition, and $\boldsymbol{J}_s \times \boldsymbol{B}_s$ is assumed second order and therefore neglected.

Note also that the inclusion of the plasma surface currents in (2a) leads to an effective resistivity on the body (MHD) currents. This effective resistivity is given by

$$\eta_{eff} = \frac{(\boldsymbol{J}_s \times \boldsymbol{B}_b)_y}{ecn_e(\boldsymbol{J}_b)_y} \quad (3)$$

and has the feature that it increases with increasing magnetic field strength. This result has been noted in particle simulations of the magnetotail by *Winglee and Steinolfson* [1993] and is contrary to many of the forms assumed in resistive MHD simulations. The above form for the effective resistivity is important as it spreads the dissipation rate more uniformly across a current sheet, and, as such, enables better storage of energy in the magnetotail during the growth phase of a substorm than predicted by resistive MHD [*Steinolfson and Winglee*, 1993].

An analytic solution to (2b) for the form of the surface currents can be attained by noting that $\boldsymbol{E}_s$ is basically an electrostatic, ambipolar electric field whose potential across the region is approximately equal to the energy of the plasma flowing across the region, i.e.,

$$\boldsymbol{E}_s \simeq \frac{1}{en_e}\nabla(\rho V^2/2 + P) \quad (4)$$

Assuming that the plasma is in approximate equilibrium (or alternatively slowly varying), so that $\nabla P \simeq \boldsymbol{J} \times \boldsymbol{B}/c$, the substitution of (4) into (2b) yields,

$$\boldsymbol{B}_s = \frac{\boldsymbol{J}_b}{en_e V^2}\frac{\boldsymbol{V}\cdot\boldsymbol{B}}{}(\frac{\rho V^2/2 + P_e}{P}) \quad (5)$$

and where the corresponding surface current is given by $\boldsymbol{J}_s = \nabla \times \boldsymbol{B}_s$.

Equations (2a) and (5) represent a higher order solution to the generalized Ohm's law than that given by ideal MHD. They can be solved in conjunction with the standard fluid equations for conservation of

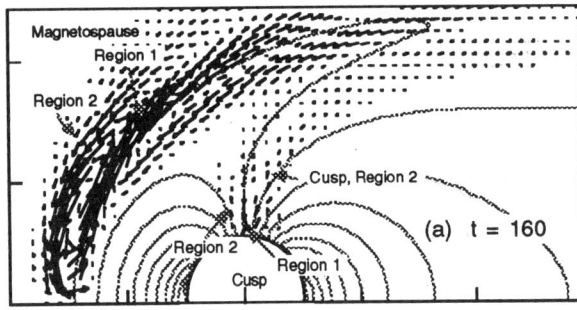
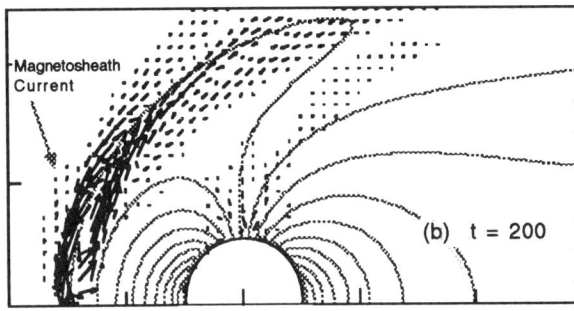
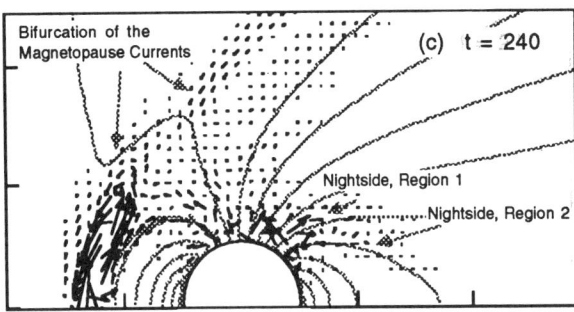
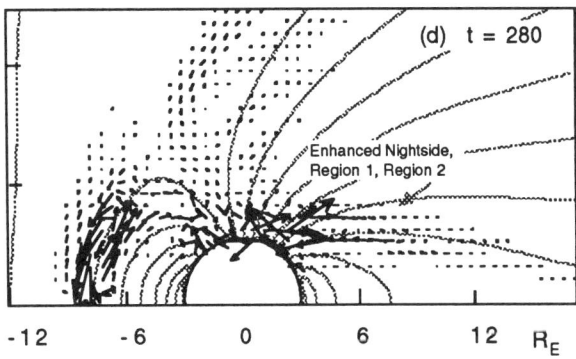

Fig. 3. The evolution of magnetic field and field-aligned currents derived from the MPD prescription in a 2-D simulation of the noon-midnight meridian. The current system is able to reproduce those seen in the particle simulations in Fig. 2.

mass, momentum and energy (for the exact equations see *Winglee* [1994]). These equations are hereafter referred to as the magnetoplasma dynamic (MPD) equations as they incorporate at least some of the physics associated with the differential penetration of ions and electrons across discrete magnetospheric boundary layers. The above approximations eliminate the need for assuming an artificial mass ratio which so often limits particle simulations. However, they are perturbative solutions and may underestimate the strength of the currents, particularly in very narrow boundary regions.

The MPD equations are solved using a two-step Lax-Wendroff differencing scheme. The full simulation system covers a region 30 R_E on the dayside and flanks, and 80 R_E on the nightside on a 375 × 125 rectangular grid, with the inner boundary is set at $3R_E$. An inner radius about the earth is set at 3 R_e. The solar wind flows in from the left-hand boundary at a speed of 350 km/s and the time unit in the figures corresponds to about a minute. The initial magnetospheric plasma density is assumed to fall off as the cube of the radial distance from the earth's center out to a radius of 12 R_e after which it is held constant at 0.2 of the solar wind density. The magnetospheric inner density is set at 5 times the solar wind density. The magnetic field is set such that the Alfvén speed near the earth center is equivalent to about 1000 km/s.

The boundary conditions at the inner radius are that the tangential velocity is assumed to be conserved, while the normal component is set to zero. The tangential magnetic field is set to zero which corresponds to conducting boundary conditions. The energy and density of the ambient plasma at the inner boundary are allowed to floated excepted that the minimum density is not allowed to fall below a tenth of its initial value (otherwise the Alfvén speed becomes very large and the computations slow down).

In order to show that the MPD equations can reproduce the surface current system seen in the particle simulations, Figure 3 shows the evolution of the magnetic field and surface currents in the near-earth region as the IMF is turned southward in a 2-D simulation of the noon-midnight meridian. However, for clarity and ease of comparison with Fig. 2, only the inner portion of the simulation box is shown. The magnetic field is drawn in as the dotted lines and the inner radius of the simulation by the semi-circle in the lower edge. Arrows are plotted only at the points where the magnitude of the current is above 10% of the maximum current.

The fluctuations in Figure 3 are less than in the particle analogue in Figure 2 since the latter involves the integration of the particle distributions which contain thermal and wave variations. As a result, the thresholding on the arrows is half that of the particle simulations and the penetration of the currents into the auroral regions can be more easily seen in Figure 3, even at early times before there is significant reconnection (Figures 3a and 3b).

On comparing with Figure 2, it is seen that the MPD equations are indeed able to reproduce the overall pattern of the surface currents seen in the particle simulations, with downward pointing currents on the magnetosheath side of the magnetopause (denoted as region 1 currents) and upward directed surface currents on the magnetospheric side. The magnitude of the current on the magnetospheric side appears to be fractionally larger which is opposite to that seen in the particle simulations. During southward IMF (Figures 3c and 3d), the magnetopause surface currents are seen to bifurcate with the more intense region being confined to the magnetopause and a weaker, spatially separate region being confined to high latitudes in the mantle. These currents, particularly those that are connected to the ionosphere, are seen to increase during southward turning of the IMF. There are also nightside field-aligned currents which that show an even stronger intensification with the southward turning of the IMF and substorm onset.

3. 3-D TWO-FLUID GLOBAL SIMULATIONS

The above results are important in establishing particle effects that are fundamental to the dynamics of boundary layers into the global context. However, 2-D modelling cannot provide realistic modelling of the magnetosphere. In this section, the global two-fluid modelling is extended into 3-D. In order to test the modelling, a comparison is made with the semi-empirical model of *Tsyganenko* [1989], hereafter referred to as T89M. Since the T89M model is a static model, a meaningful comparison requires the fluid modelling be in a near-steady equilibrium rather than for disturbed magnetospheric conditions, such as during a substorm where the field-lines and plasma are moving around rapidly.

Thus, the solar wind conditions for the present case study are taken so that the magnetopshere resembles a $Kp = 0$ configuration. Specifically, the solar wind dynamic pressure is assumed to be 1 nPa, with a density of 3 cm^{-3} and a speed of 450 km/s. The IMF is assumed to be northward at 3.6 nT for the first two hours as the system tries to come into equilibrium with the inflowing solar wind. After this initialization period the IMF is turned to zero and the following results are after another 30 minutes have passed, at which time the system is only involving slowly.

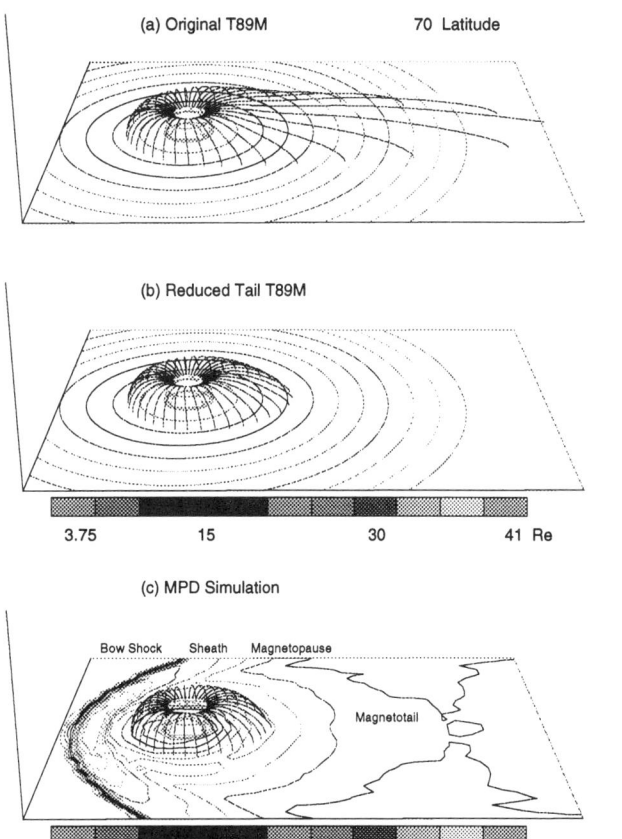

Fig. 4. Mapping of the magnetic field lines at 70° for $Kp = 0$, using (a) the T89 model, (b) the T89M model where the tail current is reduced by 30% and (c) 3-D MPD models.

The algorithms for the 3-D two-fluid modelling are essentially the same as described in the previous section. The main difference is that the grid resolution is slightly coarser at 0.625 R_E, and the inner radius of the system is set at 3.75 R_E. The grid size is $160 \times 89 \times 48$ covering a region 26 R_E on the dayside, 30 R_E on the flanks, and 86 R_E on the nightside. The density around the inner radius is set at plasmaspheric levels of 200 cm^{-3} and decreases radially with distance as r^{-5}. The corresponding equatorial Alfvén speed at the inner radius is 1100 km/s.

As an example of the mapping of the field lines, Figure 4a shows the mapping of fields derived from the T89M model for a latitude of 70° and $Kp = 0$. This latitude is of interest because it typifies those of the nightside auroral region. The sun is to the left and the concentric circles in the $x - y$ plane indicate distance with the spacing being equal to $3.75 R_E$. It is seen that the T89M predicts very stretched field lines even for low Kp. However,

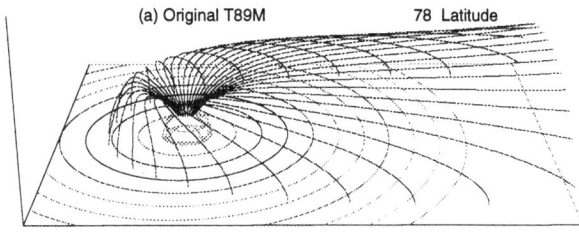
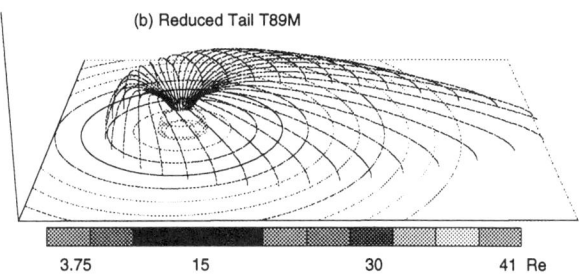
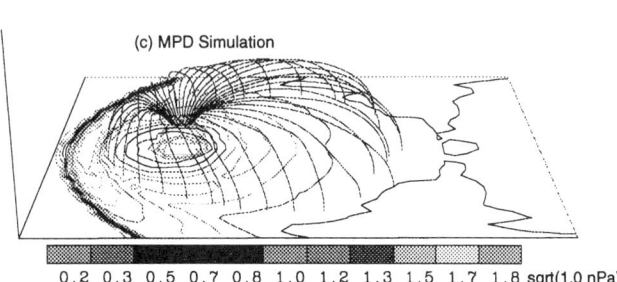

Fig. 5. As in Fig. 4, except that the latitude is 78° which is close to the last closed field line in the dayside.

Huang and Frank [1994] have shown that the T89M model underpredicts the strength of B_z in the tail. This discrepancy can be reduced by decreasing the strength of the tail current by about 30% [*Elsen and Winglee*, 1995]. The predicted mapping for the reduced tail current model is shown in Figure 4b. The reduction in the tail current clearly pulls the field lines in much closer to the Earth.

The results from the simulation model are shown in Figure 4c. The format differs in that the contours in the $x-y$ plane are now for the square root of the plasma pressure (a square root is taken to give higher dynamic range). These pressure contours give a ready indicator of critical boundary layers, including the bow shock, magnetosheath, magnetopause, and magnetotail. For the present conditions, the subsolar bow shock and magnetopause are at 16.5 R_E and 12.5 R_E, respectively. It is seen that the mapping of the dayside field lines is in excellent agreement with both Figures 4a and 4b while the nightside mapping of field lines most closely resembles the reduced tail current model.

As a further example of the mapping of the magnetic field lines, Figure 5 shows the mapping of the field lines at 78° latitude for the same models in Figure 4. This latitude is predicted to be close to the last closed field-line by the T89M model. It is seen that the modelling is consistent with this prediction. The mapping of the nightside field lines more closely resembles the reduced tail current model (Fig. 5b) rather than the very stretched mapping of T89M (Fig. 5a). This difference in the nightside mapping is due to the presence of some northward IMF as the system was allowed to come into equilibrium. Introduction of southward IMF would produce a more stretched tail but the system would be much more dynamic.

The important point is that the mapping of the field lines is fairly consistent with existing static models. The next problem is to determine the mapping of the currents. As in section 2, these currents can be decomposed into the components that would be derived from MHD and those additional currents arising from kinetic (particle) effects at the boundary layers (i.e. from meso-scale processes). The mapping of these two current systems out from 78° latitude is illustrated in Figure 6. Contours of the square root of the plasma pressure are again used to indicate position. Since the visualization of the mapping of the MHD currents in 3-D is difficult, two perspectives are given: (a) the mapping just in the dusk sector and (b) a landscape view showing the mapping between the dawn and dusk sectors.

The dayside MHD currents at this high latitude have the sense of region 0 currents [e.g., *Iijima and Potemra*, 1976] in that they are away from the ionosphere on the dawn side, cross over the noon-midnight meridian through the high-latitude boundary layer and flow down into the dusk ionosphere. Similarly, on nightside there are weak currents from the dawn side that cross through the plasma sheet boundary layer and flow into the dusk auroral region. In other words, the ionospheric MHD currents are strongest in the dawn and dusk sectors where there is strong vorticity induced by the solar wind flow along the flanks. However, there are no MHD currents that directly map from the ionosphere out to anywhere near the subsolar magnetopause nor into the magnetotail. It is in these latter regions where particles that precipitate into the auroral regions are believed to originate. The lack of mapping of ionospheric currents into these critical magnetospheric regions is consistent with other MHD models [*Ogino*, 1986].

The corresponding mapping of the plasma surface currents is illustrated in Figure 6c. Only the cutaway view for the dusk sector is shown because the plasma currents are more truly field-aligned and

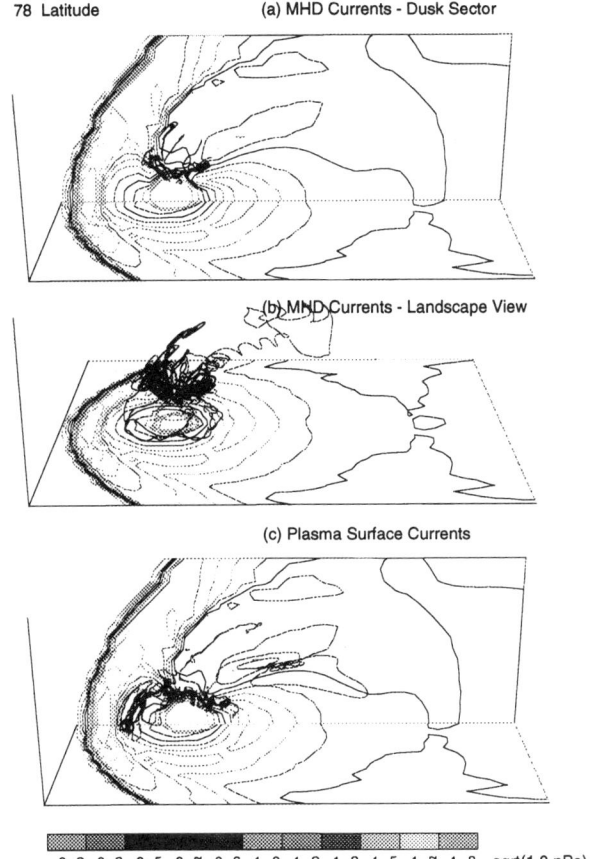

Fig. 6. The mapping of the MHD currents (solid lines in the top two panels) and the plasma surface currents starting at 78° latitude. In the top and bottom panels the contours indicating the boundaries in both the noon-midnight meridian and in the dusk equatorial region. Only the plasma surface currents directly map to the subsolar magnetopause.

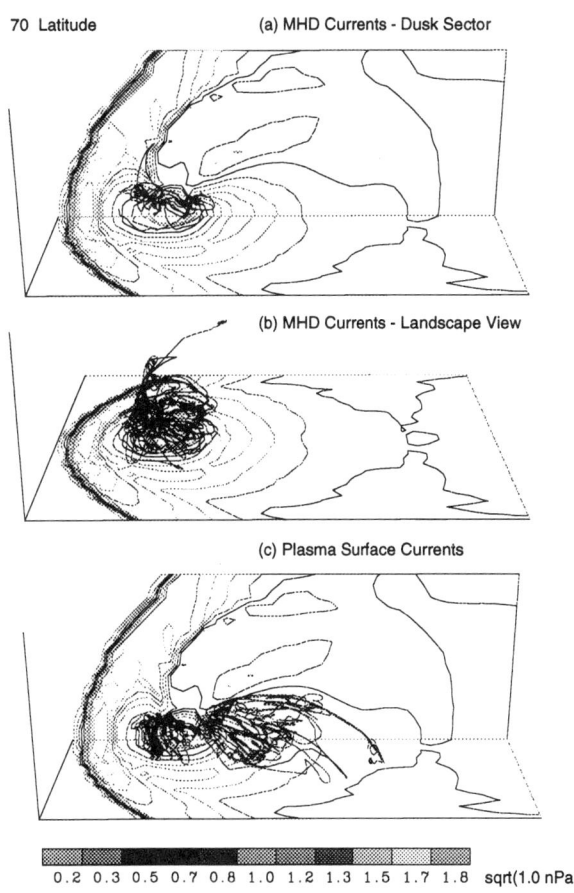

Fig. 7. As in Fig. 6, except that the latitude is 70°. The plasma surface currents at this latitude have direct access to the magnetotail while the MHD currents do not.

in general do not cross the noon-midnight meridian. The plasma currents for this high latitude are only significant on the dayside and map out to the earthward side of subsolar magnetopause. They then loop back, forming a similar circulatory pattern, as in Figures 2 and 3. They also have only a limited extent in local time from about 1000 to 1400 local time.

The corresponding mapping of the currents at 70° is shown in Figure 7. The MHD currents (Figs. 7a and 7b) have the opposite sense of those in Figure 6 and correspond to the region 1 current system. The dayside component of these currents crosses the noon-midnight meridian through the plasma mantle, rather than the high latitude boundary layer. However, there is again no direct linkage with either the subsolar magnetopause or the magnetotail. On the other hand, the plasma surface currents in the nightside map out into the near-to-middle tail region (with the same sense as dusk side region 1), some of which then curls back towards the nightside ionosphere at lower latitudes (having the same sense as dusk side region 2). The dayside portion of these currents also have the sense of region 1 out of the ionosphere, then map out into the inner magnetosphere and curl back to lower latitudes in the ionosphere in the same sense as region 2 currents.

4. SUMMARY

The above results show that the global MHD current system includes important components to the magnetospheric current system, including the magnetopause and magnetotail current sheets and the dawn and dusk components of the region 0 and 1 current systems. However, there is no direct linkage between the MHD currents and the subsolar magnetopause nor the near-to-middle magnetotail where reconnection is believed to play an important part in the acceleration of particles that may precipitate into the noon and midnight sectors of the auroral regions.

The reason for this discrepancy is that much of the particle acceleration can be expected to occur on small scales where spatial structures of boundary layers are critical to the particle dynamics. Localized (or meso-scale) particle simulations show that across such boundary layers, the differential penetration of ions leads to ambipolar (or space-charge) electric fields that can lead to enhanced acceleration both across and along the magnetic field. The latter acceleration gives rise to field-aligned currents not incorporated in MHD. These additional plasma currents provide a direct linkage to reconnection sites during the southward turning of the IMF. MHD being a single fluid treatment cannot make such a connection.

These meso-scale currents are incorporated into the global context through the newly developed modified two-fluid approach called magnetoplasma dynamics (MPD) as developed here and in *Winglee* [1994]. It has been shown that this MPD modelling can produce realistic mapping of both magnetic field lines and currents and provide a much more comprehensive understanding of the magnetosphere than single fluid MHD. A full parameter study of the mapping of field lines in 3-D is to be presented in *Winglee and Elsen* [1995].

Acknowledgments. This work was supported by National Science Foundation grant ATM-9321665 to the University of Washington. The simulations were supported by the Cray Y-MP at the Artic Research Computing Center at the University of Alaska, Fairbanks, and the Cray C-90 at the San Diego Supercomputing Center.

REFERENCES

Berchem, J., and C. T. Russell, The thickness of the magnetopause current layer: ISEE 1 and 2 observations *J. Geophys. Res.*, 87, 2108, 1982.

Elsen, R. K., and R. M. Winglee, Mapping of magnetic field lines and auroral currents in the noon-midnight meridian, *J. Geophys. Res.*, submitted, 1995.

Fedder, J. A., and J. G. Lyon, The solar wind-magnetosphere-ionosphere current voltage relationship, *Geophys. Res. Lett.*, 14, 880, 1987.

Hesse, M., and J. Birn, Magnetospheric-ionospheric coupling during plasmoid evolution: First results, *J. Geophys. Res.*, 96, 11513, 1991.

Huang, C. Y., and L. A. Frank, Magnitude of B_z in the neutral sheet of the magnetotail, *J. Geophys. Res.*, 99, 73, 1994.

Ogino, T., A three-dimensional MUD simulations of the interaction of the solar wind with the Earth's magnetosphere: The generation of field-aligned currents, *J. Geophys. Res.*, 91, 6791, 1986.

Petrinec, S. M., and C. T. Russell, External and internal influences on the size of the dayside terrestrial magnetosphere *Geophys. Res. Lett.*, 20, 339, 1993.

Roelof, E. C., and D. G. Sibeck, Magnetospheric shape as bivariate function of interplanetary magnetic field B_z and solar wind dynamic pressure *J. Geophys. Res.*, 98, 21,421, 1993.

Shi, Y., C. C. Wu, and L. C. Lee, Magnetic field reconnection patterns at the dayside magnetopause : An MHD simulation study, *J. Geophys. Res.*, 96, 17,627, 1991.

Sibeck, D. G., R. E. Lopez and E. C. Roelof, Solar wind control of the magnetopause shape, location and motion, *J. Geophys. Res.*, 96, 5489, 1991.

Song, P., R. C. Elphic, C. T. Russell, J. T. Gosling and C. A. Cattell, Structure and properties of the subsolar magnetopause for northward IMF: ISEE Observations *J. Geophys. Res.*, 95, 6375, 1990.

Song, P., C. T. Russell, R. J. Fitzenreiter, J. T. Gosling, M. F. Thomsen, D. G. Mitchell, S. A. Fuselier, G. K. Parks, R. R. Anderson, and D. Hubert, Structure and properties of the subsolar magnetopause for northward interplanetary magnetic field: Multiple-instrument particle observations *J. Geophys. Res.*, 98, 11,319, 1993.

Steinolfson, R. S., and R. M. Winglee, Energy Storage and Dissipation in the Magnetotail during Substorms: 1 MHD Simulations, *J. Geophys. Res.*, 98, 7537, 1993.

Tsyganenko, N. A., *Planet. Space Sci.*, 37, 5, 1989.

Usadi, A., A. Kageyama, K. Watanabe and T. Sato, A global simulation of the magnetosphere with a long tail: Southward and northward interplanetary magnetic field, *J. Geophys. Res.*, 98, 7503, 1993.

Wei, C. G., L. C. Lee, and A. L. La Belle-Hamer, A simulation study of the vortex structure in the low-latitude boundary layer *J. Geophys. Res.*, 95, 20,793, 1990.

Winglee, R. M., 1994, Non-MHD Influences on the Magnetospheric Current System, *J. Geophys. Res.*, 99, 13,437, 1994.

Winglee, R. M., and R. K. Elsen, Mapping of magnetic field lines, auroral currents and magnetopause utilizing a 3-D modified two-fluid model, *J. Geophys. Res.*, submitted, 1995.

Winglee, R. M., and R. S. Steinolfson, Energy Storage and Dissipation in the Magnetotail during Substorms: 1 Particle Simulations, *J. Geophys. Res.*, 98, 7519, 1993.

R. M. Winglee, University of Washington, Geophysics Program, Box 351650 Seattle, WA 98195-1650.

Frequency Range and Spectral Width of Waves Associated With Transverse-Velocity Shear

V. Gavrishchaka, M. E. Koepke, J. J. Carroll III, and W. E. Amatucci

Department of Physics, West Virginia University, Morgantown, West Virginia

G. Ganguli

Plasma Physics Division, Naval Research Laboratory, Washington, D.C.

Theory and laboratory experiment are combined to investigate processes driven by inhomogeneities believed to couple phenomena in different classes of length and time scales. The focus here is on inhomogeneity with a scale length comparable to or larger than an ion gyroradius, on cyclotron and sub-cyclotron fluctuations, and on clarifying the role played by a recently discovered class of instabilities driven by transverse-velocity shear (Ganguli et al., *Phys. Fluids* 31, 823, 1988; Koepke et al., *Phys. Rev. Lett.* 72, 3355, 1994). Predictions from the general nonlocal eigenvalue condition are shown to be consistent with results from a laboratory experiment. The emphasis, in this paper, involves an investigation of the roots of the eigenvalue condition for space-relevant parameters. It is found that for each set of parameters there exist multiple roots with comparable growth rates. In some cases, the roots extend to zero frequency. The predicted behavior of these roots is used to interpret the qualitative aspects of the wave-spectrum behavior observed in the laboratory and in particle-in-cell simulations, such as broadening, shifting, and increasing spikiness at large values of transverse-velocity shear.

INTRODUCTION

Recently, several authors have discussed the role of transverse-velocity-shear-driven instabilities in understanding processes driven by inhomogeneities associated with space plasma domains where a sheared flow exists, *e.g.*, convective-flow-reversal regions in the ionosphere, magnetospheric boundary layers, and shocks [Ganguli et al., 1994; Amatucci et al., 1994; Yamamoto et al., 1994; Pritchett, 1993; Miura, 1987; Keskinen, 1988; Kintner, 1976]. The well known Kelvin-Helmholtz (KH) mechanism [Chandrasekhar, 1961] and the relatively new inhomogeneous energy-density driven (IEDD) and electron-ion hybrid (EIH) mechanisms [Ganguli et al., 1985, Ganguli et al., 1988a] are associated with these instabilities and, together, cover a wide range of length and time scales, from much larger to much smaller than those associated with ion gyromotion.

The low-altitude ion energization model described by Ganguli [this issue] may be a good example of coupling of phenomena across different scale sizes. The model depends on shear in the transverse velocity to trigger the meso-scale waves such as the Kelvin-Helmholtz instability. It also depends on the existence of smaller-scale, shear-driven instability mechanisms, such as the IEDD and EIH instabilities. The KH instability can be excited by weak shear, a condition identified with a small but finite value of ω_s/ω_{ci}, where ω_{ci} is the ion gyro-frequency and the shear frequency ω_s ($\equiv dv_E/dx$) is estimated to be the peak value of the transverse-velocity profile along x divided by the profile's inhomogeneity scale size. Values of ω_s/ω_{ci} above 0.1 have been measured in space [Kelly and Carlson, 1977; Basu et al., 1988, Earle et al., 1989; Wahlund et al., 1994]. The FREJA satellite, with capabilities to at least partially resolve fine structure in the electric field profile, has observed electric field structures in the dayside auroral region with magnitudes as large as 1 V/m and spatial widths as narrow as 100 m [G. Marklund, private communication, 1995], indicating intense transverse-velocity shears. In terms of ω_s, values as high as 400 s^{-1} are indicated, a significant fraction of the ion gyrofrequency (maximum values of $\omega_s/\omega_{ci} \approx 0.2$ for hydrogen

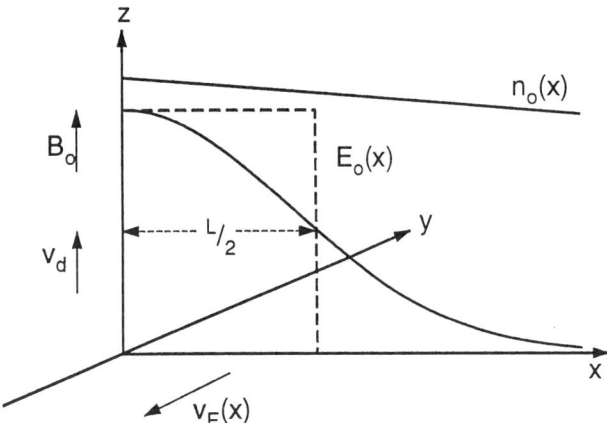

Fig. 1. Properties of the steady-state used for the general eigenvalue condition. The directions of magnetic field, relative drift velocity of electrons, and E × B velocity, as well the profiles of density and inhomogeneous electric field, are shown. The dashed and solid lines correspond to the sharp and smooth $v_E(x)$ profiles, respectively, as described in the text. In this paper, $dn/dx = 0$.

and $\omega_s/\omega_{ci} \approx 2$ for oxygen). These conditions, which can be accompanied by magnetic-field-aligned current, may excite the IEDD instability. Values of ω_s/ω_{ci} of order unity have been reported in laboratory investigations [Yamada and Owens, 1977] and are attainable in our experiment.

The growth of the KH instability can proceed to a nonlinear stage during which vortices in electrostatic potential and plasma flow develop [*e.g.*, Yamamoto et al., 1994]. The evolution of these vortices is expected to create between each vortex large, localized gradients of potential and flow that exceed the excitation threshold of the IEDD and EIH microinstabilities. Cross-scale coupling occurs because waves and particles interacting on a small-scale will have large-scale effects on the plasma, such as ion energization, spectral distortion, and morphological changes. A number of space observations appear to support this model [Tsunoda et al., 1989; Lu et al., 1992; Kelley and Carlson, 1977; Basu et al., 1988; Earle et al. 1989; Yamamoto et al., 1994]. Although observations with improved resolution are desirable for characterizing the dynamics in detail, the steepening of low-frequency waves is observed in the ionosphere and continues to be a topic of interest [Pfaff et al., 1987; Tsunoda et al., 1988, Huba 1988, Kelley, 1989]. Since the micro-scale physics is important to the energization model, a laboratory experiment was performed to verify the existence of the IEDD excitation mechanism and to study the nature of the waves associated with it [Koepke and Amatucci, 1992]. Specific theoretical predictions, based on experimentally relevant parameters, have been compared to the experi-mental results and shown to be consistent [Koepke et al., 1995]. In this paper are described results of a detailed parametric analysis of the dispersion relation for space-relevant parameters. The cases being considered can be categorized as weak shear, $\omega_s < \omega_{ci}$. The existence and nature of the multiple roots provide an explanation of the spiky, broadband IEDD wave spectrum seen in particle-in-cell simulations [Nishikawa et al., 1988] and similar to that seen experimentally in space and in the laboratory. Predictions of the general characteristics of the wave spectrum are qualitatively compared with laboratory observations. The consistency between theory and experiment, discussed in the sections that follow, adds an important level of validity to the space predictions.

GENERAL EIGENVALUE CONDITION

The kinetic theory for the IEDD instability mechanism has been developed [Ganguli et al., 1988b, 1989; Ganguli and Palmadesso 1988] for cases with and without a magnetic-field-aligned current. For weak shear conditions, the KH and IEDD instabilities can both exist, although the two mechanisms are different and the KH instability requires a very small value of k_z. This paper concentrates on the IEDD mechanism, which requires a larger k_z. The description of the steady-state used in the general model is shown in Figure 1 and includes a uniform magnetic field B_0 along z, a density gradient dn/dx, a relative drift $v_d(x) \hat{z}$ between the electrons and ions, and an E×B flow $v_E(x)$ due to a localized, transverse, dc electric field $E_x(x)$. The parameter L characterizes the localized nature of the inhomogeneous $v_E(x)$ profile. The general eigenvalue condition for this model is given by Ganguli et al. [1988b,1994],

$$\left(\rho_i^2 \frac{d^2}{dx^2} - \frac{1+\sum F_{ni}(x)\Gamma_n(b)+\tau(1+F_{0e}(x))}{\sum F_{ni}(x)\Gamma_n'(b)} \right)\Psi(x) = 0, \quad (1)$$

where $F_{n\alpha} = A_{n\alpha} - B_{n\alpha}$, $A_{n\alpha}(x) = [(\omega_1 + \omega_{2\alpha} - \omega_\alpha^*)/\sqrt{2}\,|k_z|v_\alpha]$ $Z\{(\omega_1 - \omega_{2\alpha} - n\omega_{ci})/\sqrt{2}\,|k_z|v_\alpha\}$, $B_{n\alpha}(x) = -(\omega_{3\alpha}/\sqrt{2}\,|k_z|v_\alpha)$ $Z'\{(\omega_1 - \omega_{2\alpha} - n\omega_{ci})/\sqrt{2}\,|k_z|v_\alpha\}$, $\Gamma_n(b) = I_n(b)\exp(-b)$, I_n are the modified Bessel functions, Z is the plasma dispersion function, $b = (k_y\rho_i)^2$, $\omega_1 = \omega - k_y v_E(x) - k_z v_{d\alpha}$, $\omega_{2\alpha} = k_y v_E''(x)\rho_\alpha^2/2$, $\omega_{3\alpha} = k_y v_d'(x)\rho_\alpha$, ω_α^* is the diamagnetic drift frequency, v_α is the thermal velocity, $v_{d\alpha}$ is the parallel drift velocity, ρ_i is the ion gyroradius, $\tau = T_i/T_e$, α indicates the species (electron or ion), the apostrophe indicates the derivative with respect to the argument, and Ψ is the fluctuating electrostatic potential. Eq. (1) is solved numerically for the case $v_{dt} = 0$, $\omega_{2\alpha} = 0$, $\omega_{3\alpha} = 0$, and $\omega_\alpha^* \approx 0$ for eigenvalues using parameters relevant to the laboratory experiment [$(k_y\rho_i)^2 = 0.15$, $k_z/k_y = 0.1$, ε ($= \rho_i/L$) $= 0.35$, $v_{de}/v_i = v_d/v_i = 30$, $v_E/v_i \leq 1.5$, $m_i/m_e = 7\times10^4$, $T_i/T_e = 1$] or space [$(k_y\rho_i)^2 \approx 0.3$, $k_z/k_y \approx 0.1$, $\varepsilon = 0.025$, $v_d/v_i = 30$, $v_E/v_i \leq 1.5$, $m_i/m_e = 3\times10^4$, $T_i/T_e \approx 1$]. Note that $dn/dx = 0$ is used throughout this paper. The coupling of the IEDD mechanism with the electrostatic drift mechanism is the subject of a future paper.

SHARP-BOUNDARY MODEL OF THE IEDD INSTABILITY MECHANISM

The general character of the roots of Eq. (1) can be understood intuitively using a nonlocal sharp-boundary model introduced a decade ago [Ganguli et al., 1985] wherein two

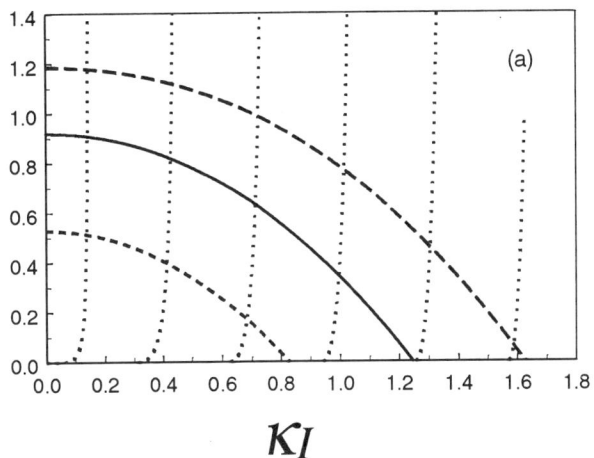
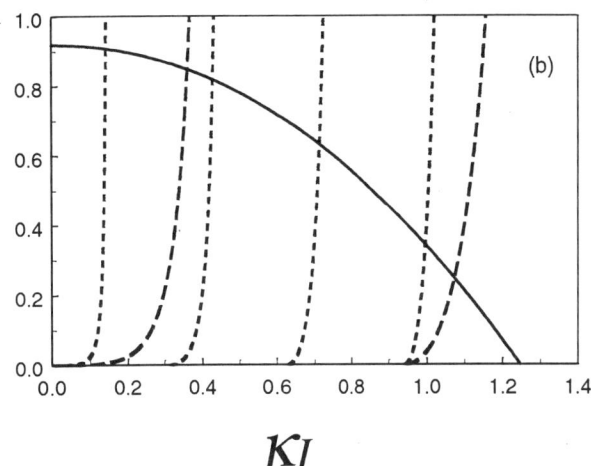

Fig. 2. Intersections of Y_1 (left-hand side of nonlocal dispersion relation) vs. κ_I (vertical lines) and Y_2 (right-hand side of nonlocal dispersion relation) vs. κ_I (quarter circles) give the roots. In (a) v_E is varied while $\varepsilon = 0.05$ is fixed: 1.5 (outer quarter circle), 1.0 (middle), 0.5 (inner quarter circle). In (b) ε is varied while v_E is fixed: $0.5v_i$, (four vertical lines), $1.5v_i$, (two vertical lines).

regions are identified on opposite sides of an infinitesimally thin transverse-velocity-shear layer located at $x = \pm L/2$. In region I, the field, $E_I = E_0$, is large enough to make the energy density negative. Surrounding region I is region II, with no field, $E_{II} = 0$. No magnetic-field-aligned current is included in this simple model. The mode is described in each region with a separate differential equation, $\left(\partial^2/\partial\xi^2 + \kappa_{I,II}^2\right)\Psi_{I,II}\{\xi\} = 0$, where $\xi = x/\rho_i$, $(\kappa^2)_{I,II} = (Q/A)_{I,II}$, $Q = D(\omega_1, k)$, $A = -\sum_{n>0}[\omega_1^2 \Gamma_n'/(\omega_1^2 - n^2\omega_{ci}^2)]$, $(\omega_1)_I = \omega - k_y E_0/B_0$, and $(\omega_1)_{II} = \omega$. The extended (i.e., nonlocal) IEDD wave packet couples fluctuations on opposite sides of the point where ω_1 changes sign, and grows as energy is transferred across this velocity-shear layer. Even solutions are considered, $\Psi_I = \Psi_{I0}\cos(\kappa_I\xi)$, $\Psi_{II} = \Psi_{II0}\exp(i\kappa_{II}\xi + i\delta)$, and logarithmic derivatives are required to be continuous at $\xi = \pm L/2\rho_i$. This yields the sharp-boundary dispersion relation $-\kappa_I\tan(\kappa_I/2\varepsilon) = i\kappa_{II}$.

Although the sharp-boundary model elucidates the basic mechanism of the IEDD instability and provides a nonlocal dispersion relation in the form of an algebraic equation that can be solved easily for complex frequency, it is not appropriate for the case of a smooth $v_E(x)$ profile. The general kinetic theory of the IEDD instability results in an integral equation that for $\varepsilon < 1$, can be expressed as a second-order differential equation, as given in Eq. (1), which is solved numerically using $v_E(x)/(v_E)_{max} = \text{sech}^2(x/L)$ to obtain the results in this paper.

PROPERTIES OF THE MULTIPLE ROOTS

To conveniently obtain roots of the nonlocal dispersion relation associated with the sharp-boundary model, consider only the real part of $-\kappa_I\tan(\kappa_I/2\varepsilon) = i\kappa_{II}$. Use only the first-order term of the large-argument expansion of the real part of the ion Z-function and neglect the electron Z-function. In the small-b limit, consider only terms with $n = -1, 0, +1$. Then ω and κ_{II} can be expressed in terms of κ_I. With these approximations, the real part of the nonlocal dispersion relation becomes $Y_1(\kappa_I) = Y_2(\kappa_I)$, where $Y_1(\kappa_I) = \kappa_I^2\tan^2(\kappa_I/2\varepsilon)$ and $Y_2(\kappa_I) = 2(\Gamma_1 M_0 - N)/(\Gamma_0' + \Gamma_1' M_0)$. Here we use $\tan(\kappa_I/2\varepsilon) > 0$, $N = 1 + \tau - \Gamma_0$, $M_0 = 2\omega^2/(\omega^2 - 1)$, $M_1 = (N - \kappa_I^2\Gamma_0'/2)/(\kappa_I^2\Gamma_1'/2 + \Gamma_1)$, and $\omega = k_y v_E + (M_1 - 2/M_1)^{-1/2}$. Each intersection of the curves $Y_1(\kappa_I)$ vs κ_I and $Y_2(\kappa_I)$ vs κ_I corresponds to a root of the nonlocal dispersion relation. The quasiperiodicity of function Y_1 results in several discrete roots ω_n, each corresponding to a real part ω_{Rn} and an imaginary part γ_n, for each set of parameters (b, u, etc.). A similar situation appears in quantum mechanics with the energy levels of a one-dimensional, square potential well. In the plasma case, however, the potential Q is more complicated and the eigenvalues are complex numbers. The behavior of these roots as the parameters v_E and ε are varied is shown in Figure 2. The number of roots for a given set of parameters are seen to increase for larger v_E and for smaller ε, consistent with experimentally obtained IEDD wave spectra [Amatucci et al., 1994] which indicate the presence of several spikes for values of v_E and ε comparable to those determined from space observations, and in contrast with current-driven electrostatic ion-cyclotron wave spectra which show a single spectral feature.

Although the sharp-boundary model elucidates the basic mechanism of the IEDD instability and provides a nonlocal dis-persion relation in the form of an algebraic equation that can be solved easily for complex frequency, it is not appropriate for the more realistic case of a smooth $v_E(x)$ profile. The general kinetic theory of the IEDD instability results in an integral equation that for $\varepsilon < 1$, can be expressed as a second-order differential equation, as in Eq. (1), which is solved numerically using $v_E(x) = v_0\text{sech}^2(x/L)$ to obtain the results in this paper.

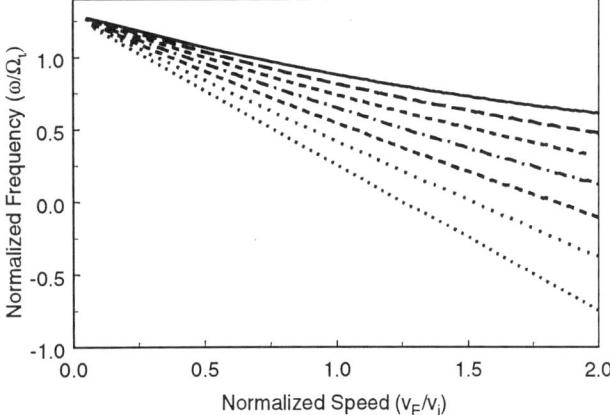

Fig. 3. Shifting and spreading of multiple roots of general eigenvalue condition as v_E is varied. Here $\varepsilon = 0.025$, $v_d = 30 v_i$, $\tau = 0.5$, $\mu = 29392$.

RESULTS FOR A SMOOTH $v_E(x)$ PROFILE

The existence and character of the multiple roots, as illustrated with the sharp-boundary model, is recovered in roots obtained using a smooth profile in the general eigenvalue condition. Figure 3 shows the v_E dependence of several unstable roots calculated for conditions relevant to geospace. Frequency is normalized to the ion-cyclotron frequency and velocity is normalized to the ion thermal velocity. The property that all roots shift and spread apart for larger v_E has implications for the wave spectrum, as shown in Figure 4. Near $v_E = 0$ there is a range over which the number of roots increases with v_E. This range depends on many parameters. For laboratory conditions, the roots also shift and spread apart, but less dramatically and without extending so far below ω_{ci}. Particle-in-cell simulations [Nishikawa et al., 1988] also find that the IEDD wave spectrum becomes spikier and broader as v_E increases.

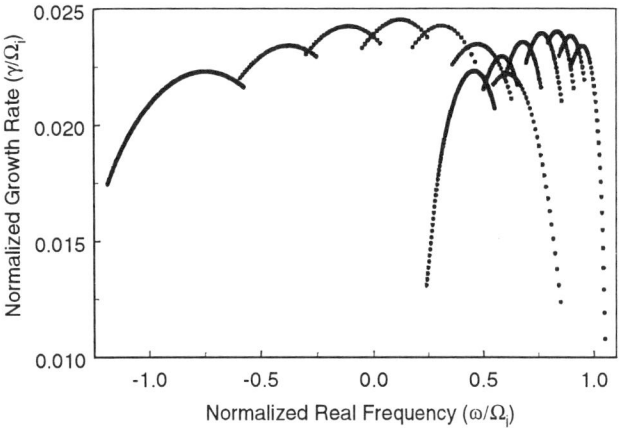

Fig. 4. Spectrum of multiple roots of general dispersion relation. Here $\varepsilon = 0.025$, $v_d = 30 v_i$, $\tau = 0.5$, $\mu = 29392$.

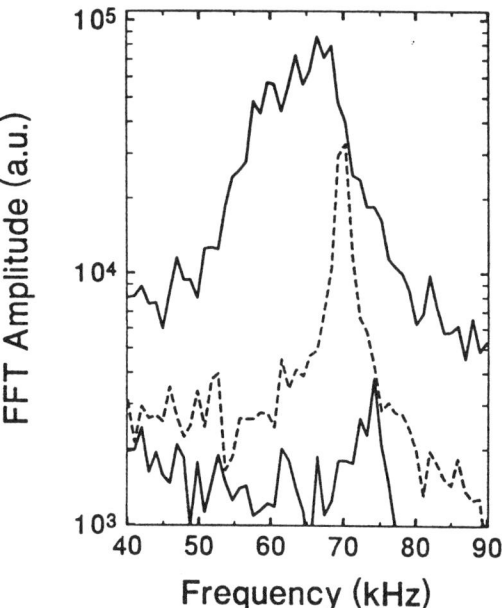

Fig. 5. Experimentally measured IEDD wave spectra (solid lines) and CDEIC wave spectra (dashed line). Small and large IEDD wave spectra correspond to small and large values of v_E, respectively.

Figure 5 shows the shifting and broadening in the spectra acquired in the laboratory experiment for small and large v_E. For small v_E, IEDD waves are single peaked and narrowband (FWHM ≈ 2 kHz, or 2.7%). For large v_E, IEDD waves are spiky and broadband (FWHM ≈ 12 kHz, or 17%). It is not certain if the number of spectral spikes increases from a small to a large number as v_E increases or if several spikes overlap at small v_E and become resolved and spread out at larger v_E. Frequencies below the potassium-cyclotron frequency are occasionally seen for these laboratory IEDD waves.

DISCUSSION

The existence of multiple roots of the IEDD eigenvalue condition is demonstrated to be significant in predicting and interpreting experimentally obtained wave spectra. The dramatic shift of these roots as the shear is increased enables IEDD waves to couple across scale sizes. The broadband, spiky, shifted wave spectra observed in the laboratory provide important credibility to the theory which predicts shifts of approximately $\Delta\omega/\omega \approx 0.1$ for experimentally relevant parameters. Koepke et al. [1995] discuss other aspects of the comparison between theory and experiment more quantitatively. One could hope that the consistency between laboratory experiment and theory should hold for space-relevant parameters because the value of shear is smaller. These results encourage the further investigation of the behavior of IEDD roots, which is presently underway.

From the experimental point of view, the properties of the multiple roots help one see how they may be responsible for the spiky, broadband IEDD wave spectrum observed in the laboratory. These results indicate that shear in the transverse plasma flow can play an important destabilizing role and significantly affect the spectrum and characteristics of the waves.

The capability of IEDD waves for cross-scale coupling appears to be significantly greater than that of CDEIC waves by virtue of the wider frequency range associated with the former. The observed v_E dependence of the IEDD waves indicates shifting and broadening, consistent with the predictions presented here. This behavior is in contrast with the v_E-independent frequency of current-driven electrostatic ion-cyclotron waves [Motley and D'Angelo, 1963; Drummond and Rosenbluth, 1962]. The broader spectrum and lower magnetic-field-aligned current threshold for IEDD wave excitation [Amatucci et al., 1994] also increases the likelihood that IEDD waves participate in wave-particle processes coupling geospace phenomena associated with different length and time scales.

Acknowledgments. This work was supported by the National Science Foundation and the Office of Naval Research.

REFERENCES

Amatucci, W. E., M. E. Koepke, J. J. Carroll III, and T. E. Sheridan, Observation of ion-cyclotron turbulence at small values of magnetic-field-aligned current, Geophys. Res. Lett., 21, 1595, 1994.

Basu, S., S. Basu, E. MacKenzie, P. F. Fourgere, W. R. Coley, N. C. Maynard, J. D. Winningham, M. Sugiura, W. B. Hanson, and W. R. Hoegy, Simultaneous density and electric field fluctuation spectra associated with velocity shears in the auroral oval, *J. Geophys. Res.*, 93, 115, 1988.

Kelley, M. C., and C. W. Carlson, Observation of intense velocity shear and associated electrostatic waves near an auroral arc, J. Geophys. Res., 82, 2343, 1977.

Chandrasekar, *Hydrodynamic and Hydromagnetic Stability*, Oxford Univ. Press, New York, chap. 11, 1961.

Drummond, W. E., and M. N. Rosenbluth, "Anomalous diffusion arising from microinstabilities in a plasma", *Phys. Fluids*, 5, 1507, 1962.

Earle, G. D., M. C. Kelley, and G. Ganguli, Large velocity shears and associated electrostatic waves and turbulence in the auroral F region, *J. Geophys. Res.*, 94, 15321, 1989.

Ganguli, G., Y. C. Lee and P. J. Palmadesso, Electrostatic ion-cyclotron instability caused by a nonuniform electric field perpendicular to the external magnetic field, *Phys. Fluids*, 28, 761, 1985.

Ganguli, G., Y. C. Lee, and P. J. Palmadesso, Electron-ion hybrid mode due to transverse velocity shear, Phys. Fluids, 31, 2753, 1988a.

Ganguli, G., and Y. C. Lee, Kinetic theory for electrostatic waves due to transverse velocity shears, Phys. Fluids, 31, 823, 1988b.

Ganguli, G., and P. J. Palmadesso, Electrostatic ion instabilities in the presence of parallel currents and transverse electric fields, *Geophys. Res. Lett.*, 15, 103, 1988.

Ganguli, G., Y. C. Lee, P. Palmadesso, and S. L. Ossakow, Oscillations in a plasma with parallel currents and transverse velocity shears, in *Physics of Space Plasmas* (1988), SPI Conf. Proc. and Reprint Series, No. 8, T. Chang, G. B. Crew, and J. R. Jasperse, eds., Scientific Publishers, Cambridge, MA, p. 231, 1989.

Ganguli, G., M. J. Keskinen, H. Romero, R. Heelis, T. Moore, and C. Pollock, Coupling of microprocesses and macroprocesses due to velocity shear: An application to the low-altitude ionosphere, J. Geophys. Res., 99, 8873, 1994.

Ganguli, G., Interrelationship of local and global physics in the low-altitude ionosphere, this issue.

Keskinen, M. J., H. G. Mitchell, J. A. Fedder, P. Satyanarayana, S. T. Zalezak, and J. D. Huba, Nonlinear evolution of Kelvin-Helmholtz instability in the high latitude ionosphere, J. Geophys. Res., 93, 137, 1988.

Kindel, J. M., and C. F. Kennel, Topside current instabilities, *J. Geophys. Res.*, 76, 3055-3078, 1971.

Kintner, P. M., Observations of velocity-shear-driven plasma turbulence, J. Geophys. Res., 81, 5114, 1976.

Koepke, M. E., and W. E. Amatucci, Electrostatic ion-cyclotron wave experiments in the WVU Q machine, *IEEE Trans. Plasma Sci.*, 20, 631, 1992.

Koepke, M. E., W. E. Amatucci, J. J. Carroll III, and T. E. Sheridan, Experimental verification of the inhomogeneous energy-density driven instability, *Phys. Rev. Lett.*, 72, 3355, 1994.

Koepke, M. E., W. E. Amatucci, J. J. Carroll III, V. Gavrishchaka, and G. Ganguli, Velocity-shear-induced ion-cyclotron turbulence: Laboratory identification and space applications, Phys. Plasmas, 2, 2523, 1995.

Lu, G., P. H. Reiff, T. E. Moore, and R. A. Heelis, Upflowing ionospheric ions in the auroral region, J. Geophys. Res., 97, 16855, 1992.

Miura, A., Simulation of Kelvin-Helmholtz instability at the magnetospheric boundary, J. Geophys. Res., 92, 3195, 1987.

Nishikawa, K.-I., G. Ganguli, Y. C. Lee, and P. J. Palmadesso, Simulation of ion-cyclotron-like modes in a magnetoplasma with transverse inhomogeneous electric field, *Phys. Fluids*, 31, 1568, 1988.

Pfaff, R. F., M. C. Kelley, E. Kudeki, B. J. Fejer, K. D. Baker, Electric field and plasma density measurements in the strongly driven daytime equatorial electrojet 2. Two-stream waves, J. Geophys. Res., 92, 13597, 1987 1987.

Pritchett, P. L., Simulation of collisionless electrostatic velocity-shear-driven instabilities, Phys. Fluids, B5, 3770, 1993.

Tsunoda, R. T., R. C. Livingston, J. F. Vickrey, R. A. Heelis, W. B. Hanson, F. J. Rich, and P. F. Bythrow, Dayside observations of thermal-ion upwellings at 800-km altitude: An ionospheric signature of the cleft ion fountain, J. Geophys. Res., 94, 15277, 1989.

Wahlund, J.-E., P. Louarn, D. Chust, H. de Feraudy, A. Roux, B Holback, P.-O. Dovner, and G. Holmgren, On ion-acoustic turbulence and the nonlinear evolution of kinetic Alfven waves in aurora, Geophys. Res. Lett., 21, 1831, 1994.

Yamamoto, T., M. Ozaki, S. Inoue, K. Makita, and C.-I. Meng, Convective generation of giant undulations on the evening diffuse auroral boundary, J. Geophys. Res., 99, 19499, 1994.

Yamada, M., and D. K. Owens, Cross-field-current driven lower-hybrid instability and stochastic ion heating, Phys. Rev. Lett.., 38, 1529, 1977.

V. Gavrishchaka (gavrish@wvnvms.wvnet.edu),
M. E. Koepke (koepke@wvnvms.wvnet.edu) and
J. J. Carroll III, (carroll@wvnvms.wvnet.edu), Department of Physics, West Virginia University, Morgantown, WV 26506-6315
W. E. Amatucci (amatucci@ppd.nrl.mil.gov) and
G. Ganguli (gang@ppd.nrl.mil.gov), Plasma Physics Division, Code 6700, Naval Research Laboratory, Washington, DC 20375.

Micro/Mesoscale Coupling in the Auroral Region: Observations

J. L. Burch

Southwest Research Institute, San Antonio, Texas

Mesoscale phenomena in the Earth's auroral zone are in many cases easily detectable with single spacecraft. However, the microscale processes, which are often responsible for the mesoscale phenomena we observe and sometimes result from them, are difficult to study experimentally or theoretically because of the fine spatial and temporal scales within which they operate. The study of coupling between micro- and mesoscale processes is proceeding through a combination of experiment, theory, and numerical simulation; but the unique identification of causal relationships has not been possible with the existing data sets. Four examples of mesoscale phenomena: inverted-V events, ion conics, auroral kilometric radiation, and suprathermal electron bursts, are discussed in this paper in terms of the ongoing search for the microscale processes associated with them. New data with much higher spatial and temporal resolution than ever before obtained from orbital spacecraft have recently become available from the Freja mission and are planned for the FAST mission. With these high-resolution data sets, the unique identification of plasma processes responsible for some of the auroral mesoscale phenomena will finally be possible.

INTRODUCTION

The Earth's magnetosphere has been characterized by numerous spacecraft-borne measurements as a collection of mesoscale phenomena that are strongly coupled. The coupling processes produce dynamical phenomena such as magnetospheric substorms, which transport energy and plasma efficiently and macroscopically throughout the magnetospheric system. The many mesoscale phenomena, such as the cusp, flux transfer events, the magnetopause, inverted-V events, and systems of field-aligned currents, collectively define the magnetosphere and its channels of mass, momentum and energy transport [*Burch et al.*, 1995]. One might consider the mesoscale as containing most of the individual magnetospheric phenomena that have been observed and characterized, while the smaller-scale processes that produce the phenomenology of the mesoscale, and through which the mesoscale phenomena interact one with another, occupy a microscale, in which processes have in general been inferred rather than directly observed. In global models the microscale processes must be parameterized rather than explicitly included. In part, this limitation follows from inadequate computing power, but more importantly, it arises because the resolution of the plasma measurements in space and time have not been sufficient to characterize these processes fully.

The low to mid-altitude auroral regions exhibit many examples of coupling between mesoscale and microscale processes, and the resulting mesoscale phenomena can often be characterized as anomalous transport. Some examples of the production of anomalous transport in the auroral oval are listed in Table 1. Figure 1 shows graphically the space-time regimes that loosely define the microscale and mesoscale phenomena that are important in the low-altitude magnetosphere. As shown in Figure 1, the auroral arc lies at the boundary between the mesoscale and microscale in the auroral oval. Individual arcs, having latitudinal widths as small as 100 m [*Borovsky*, 1995], are readily observed with ground-based optical techniques and with sounding rockets, but their connection with the larger-scale inverted-V events, which are routinely observed by low-altitude polar-orbiting satellites, is still not known and is in fact a major puzzle for auroral physicists.

Also near the microscale-mesoscale boundary are phenomena such as ion conics, suprathermal electron bursts, solitary kinetic Alfvén waves, and lower hybrid cavitons, which can be examined rather completely at the resolution now available from Freja, as shown in Figure 1. Microscale processes operating at the electron gyroscale, such as auroral kilometric radiation (AKR) and electron conics, will require even higher time resolution for their

Table 1. Examples of the production of anomalous transport in the auroral oval.

Observed Mesoscale Phenomenon	Induced Microscale Processes	Observed Mesoscale Plasma Transport
Alfvén Wave	Cyclotron Resonance, Anomalous Resistivity, Landau Damping, Double Layers, Ion Acoustic Turbulence	Electron and Ion Acceleration
Auroral Electron Beams	Nonlinear LH Wave Interactions Producing Caviton Turbulence and Particle Heating	Energization of Ion Conics and Counter-streaming Electrons

complete specification (see Figure 1), although the growth rates for associated instabilities (e.g., the cyclotron maser) are within the resolution obtainable by FAST, which should sample the AKR source region.

The remainder of this paper will concentrate on the observational aspects of recent progress that has been made on the production of mesoscale plasma transport by microscale processes in the auroral region. Specifically, the role of non-MHD Alfvén waves in the acceleration of auroral electrons, the transverse acceleration of ionospheric ions, and the processes responsible for AKR and suprathermal electron bursts are discussed. The ongoing mission of the Freja high-data-rate low-altitude auroral spacecraft and the upcoming mission of the FAST high-data-rate mid-altitude auroral spacecraft promise to yield rapid progress in these areas, and their potential contributions to auroral physics will be discussed.

ROLE OF ALFVÉN WAVES IN THE ACCELERATION OF AURORAL PRIMARY ELECTRONS

There has been much progress over the past two decades in determining the altitude range of the auroral acceleration region [*Reiff et al.*, 1993] and in identifying phenomena such as electrostatic shocks and density cavities that are characteristics of the regions. The electron and ion distribution functions have been found to be consistent with parallel electric fields; however, the source of these electric fields and their time behavior have not been identified as yet. It is generally agreed that the parallel potential drops arise in response to strong field-aligned currents, but the possible roles of phenomena such as anomalous resistivity, double layers, and lower hybrid waves with strong parallel electric field components have not been tested in an unambiguous way because of experimental limitations.

It is known that disturbances are transmitted throughout the magnetosphere by field-aligned currents and the associated Alfvén waves, which are responsible for the strongly coupled nature of the magnetosphere. Generalized theories of magnetosphere-ionosphere coupling by Alfvén waves have become well developed over time since their introduction by *Goertz and Boswell* 1979], most notably by the analytical theory and simulation results of R. L. Lysak and co-workers [*e.g., Lysak,* 1985]. Although the possible role of Alfvén waves in auroral electron acceleration has not been tested fully, the recent high-resolution electromagnetic field data of the Freja satellite [*Louarn et al.*, 1994] have provided new evidence of their potential importance.

As hydromagnetic phenomena, Alfvén waves do not have the intrinsic capability of supporting parallel electric fields. However, the leading edge of an Alfvénic disturbance can develop a shock wave and a parallel electric field as it enters the ionosphere where

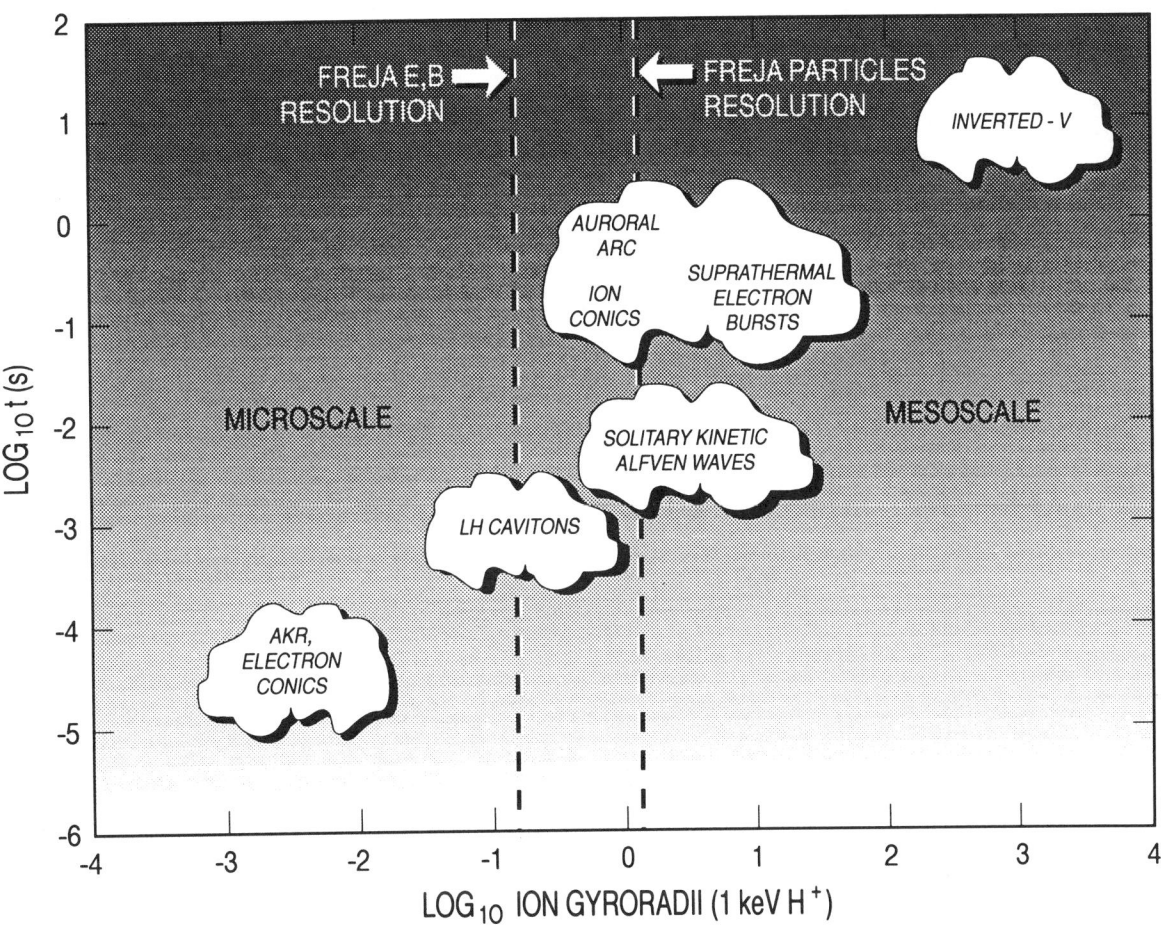

Fig. 1. Some important phenomena in the low-altitude polar magnetosphere located according to their approximate temporal and spatial scales, with the spatial dimension measured in gyroradii of 1 keV hydrogen ions. Also noted are the Freja spacecraft spatial resolutions for fields and particles and approximate identifications of the microscale and the mesoscale.

it is dissipated [see Figure 4 of *Goertz and Boswell*, 1979]. Also, as pointed out by *Hasegawa and Mima* [1978], a kinetic (i.e., non-MHD) Alfvén wave will evolve whenever finite Larmor radius effects become important, and such waves will contain a parallel electric field component. A similar situation will exist in the cold plasma limit when electron inertial effects become important, and in this situation the term inertial Alfvén waves is often used to describe these phenomena [*Temerin et al.*, 1993]. Generally, below altitudes of about five Earth radii the electron thermal velocity is less than the Alfvén velocity so the electron inertia effects will dominate, while the finite ion Larmor radius effects will dominate higher up. In both cases there are several possible paths to turbulent interactions and the rather rapid dissipation of the Alfvén waves during which electromagnetic energy is transformed into the mechanical energy of accelerated electrons. For example, the parallel electric field component could couple into ion acoustic waves; Landau damping by electrons or ions could occur if the particles are moving at the phase velocity of the wave; or resonant ion cyclotron interactions could occur if the ion velocity Doppler shifts the wave into cyclotron resonance. As shown in Table 1, all of these possibilities would lead to dissipation, particle acceleration, and turbulence, which *Lysak and Carlson* [1981] have shown will cause the Alfvén wave to support a parallel electric field.

The Alfvén velocity is lower at the Freja altitude (apogee 1760 km) than at the one Earth radius typical altitude of the auroral acceleration region because of the rapid increase of plasma density toward lower altitudes. Therefore, Alfvén wave phenomena should be easier to observe at the lower altitude [*Louarn et al.*, 1994]. Because of the high Alfvén speeds at the S3-3 altitude it is possible that the phenomena identified as electrostatic shocks in that data set [*Mozer et al.*, 1977] might actually have been electromagnetic Alfvén waves.

The high data rate of Freja has allowed it to detect what most

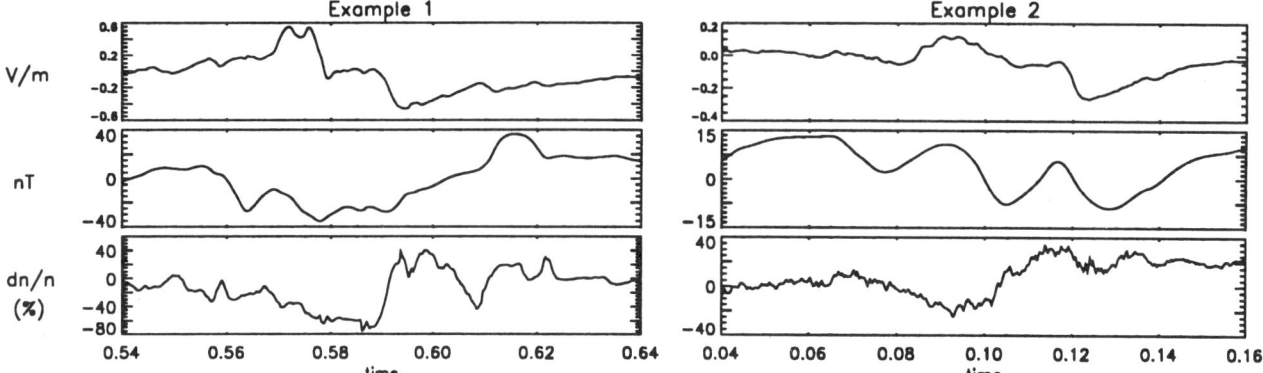

Fig. 2. Two examples of inertial Alfvén waves (indicated by the large electric-field fluctuations centered at 0.58 s in example 1 and at 0.10 s in example 2) observed by the Freja spacecraft [*Louarn et al.*, 1994].

probably are solitary kinetic Alfvén waves (or inertial Alfvén waves) [*Louarn et al.*, 1994] along with ion acoustic turbulence [*Wahlund et al.*, 1994] at 1700 km altitude in the auroral oval. Two examples of solitary kinetic Alfvén waves observed by Freja are shown in Figure 2. The key aspect of these observations that identifies them as Alfvén waves is the fact that $\Delta E/\Delta B$ is approximately equal to the Alfvén velocity ($\sim 5 \times 10^6$ m/s) as is expected for normally propagating Alfvén waves. The electron inertial length at this altitude is ~ 5 km/$\sqrt{m_e}$, which is traversed by the Freja spacecraft in about 0.02 seconds. Reference to Figure 2 shows that this time interval is approximately equal to the time duration of the detected Alfvén waves, which suggests that they are inertial Alfvén waves.

TRANSVERSE ION ACCELERATION

Because of the difficulty in making plasma and field measurements at the microscale, the unique identification of auroral microscale processes has in many cases not been possible. For example, fluctuating double layers have been postulated to be responsible for several different phenomena in the auroral region, including electron conics [*André and Eliasson*, 1992], Knight's trapped electrons [*Knight*, 1973; *Louarn et al.*, 1990], counter-streaming electrons [*Sharp et al.*, 1980], and co-located upward-moving ion and electron distributions [*Hultqvist et al.*, 1991]. Similarly, lower hybrid waves have been invoked to explain ion conics [*Kintner et al.*, 1992], suprathermal electron bursts [*Robinson et al.*, 1989], and the acceleration of auroral primary electrons [*Bingham et al.*, 1984].

Theories for the production of ion conics, in particular, have been difficult to confirm because of the limited altitude range of the individual measurements. For example, *Temerin* [1986] showed that the "elevated conic" distribution functions (see Figure 3 herein and Figure 1 of *Temerin*, 1986) could be produced by parallel heating over an extended altitude range as well as by the generally accepted model of parallel acceleration combined with transverse acceleration within a restricted altitude range, with the conic shape being produced by the magnetic mirror force as the distribution moves upward in altitude. Further uncertainty has been introduced by the inability in many experiments to identify the wave modes accompanying the ion conics.

Chang [1993] has noted that the auroral electron beams can excite lower hybrid waves, which can produce transverse acceleration of ionospheric ions. However, the lower hybrid wavelengths are generally much larger than the ion gyroradius. Chang has proposed that nonlinear wave interactions near the lower hybrid frequency through a modulational instability, such as the collapse of waves into soliton (caviton) turbulence, could play a key role in the acceleration of ionospheric ions as well as electrons. The resulting caviton turbulence would be characterized by shorter wavelengths, which could match the ion motion and result in resonant acceleration.

Recent high-resolution multiprobe wave measurements by the Freja spacecraft have focused new attention on the lower hybrid cavitons such as observed by *Kintner et al.* [1992] on the TOPAZ 3 sounding rocket. Kintner et al. observed conics at 90° pitch angles, indicating measurements in the transverse acceleration region, within deep density cavities that were filled with lower hybrid waves. The Freja lower hybrid caviton measurements have not as yet been associated with transverse ion acceleration. This result might be attributable to several factors, including (1) altitude differences (e.g., the Freja altitude of 1700 km is a few hundred kilometers above the TOPAZ 3 apogee); (2) a possible requirement for preheating of the plasma before transverse acceleration becomes effective; and (3) a requirement for the conversion of the long wavelength lower hybrid waves to the shorter wavelengths required for resonant interactions at the ion gyroscale. This latter requirement could be satisfied by the caviton collapse suggested by *Chang* [1993] or by other conversion channels such as mode conversion to the ion Bernstein mode as suggested by *Koskinen* [1985].

THE SOURCE OF FREE ENERGY FOR AURORAL KILOMETRIC RADIATION

It is generally agreed that auroral kilometric radiation (AKR) is generated by the cyclotron maser instability in auroral density cavities. According to *Wu and Lee* [1979] the most likely source of free energy for the generation of AKR is the loss-cone distribution within a region of depleted plasma density where the ratio of electron plasma frequency to gyrofrequency (f_p/f_g) is less than 0.3. This distribution provides the requirement of the cyclotron maser instability that $df/dv_\perp > 0$, and the theory has had success in explaining several observations. However, the source of free energy for the instability has not been uniquely identified. Velocity-space density gradients in the auroral electron distribution, including those at the edge of the loss cone and at the low-energy side of the auroral beam, have been investigated as possible sources [*Omidi and Gurnett*, 1982; *Lin et al*, 1986; *Ungstrup et al.*, 1990]. Recent attention has focused on trapped electrons in Knight's forbidden region [*Knight*, 1973] with energies less than those of the auroral primary electrons [*Louarn et al.*, 1990; *Menietti et al.*, 1993]. One difficulty in finding the source of free energy for AKR has been the small number of measurements that are available within the source region where the AKR frequencies extend down to or slightly below the electron cyclotron frequency. Another problem is the lack of electron measurements with high enough time resolution to observe the electron density gradient during the growth time of AKR.

Omidi and Gurnett [1982] have demonstrated that f(v) distributions obtained by the S3-3 satellite within a source region of AKR seem to provide the large growth rate necessary to explain the observations with a cosmic noise background wave source and typical ray path lengths of 100 km. This analysis points out the fundamental problem of measurement time resolution in trying to resolve the source of free energy for AKR. That is, the AKR growth rate is calculated to be $\leq 10^4 \mathrm{s}^{-1}$, and no satellite has been capable of obtaining a complete distribution in phase space on this time scale. Typically the best satellites require several seconds to obtain a complete distribution in phase space. During this time, the waves within an AKR source region would have grown and already altered the existing plasma distribution. This point was also made by *Melrose* [1986]. The FAST spacecraft, with apogee near 4000 km, should sample the lower-altitude reaches of the AKR source region with burst-mode electron distribution function measurements having high enough temporal resolution to resolve this problem.

Observations of electron distributions in the AKR source region have recently been reported by investigators of the Viking satellite. *Ungstrup et al.* [1990] used Viking data within an AKR source region to search for the source of free energy. Ungstrup et al. favor the loss-cone because the loss-cone gradients were steeper just outside the source region than within it, supposing

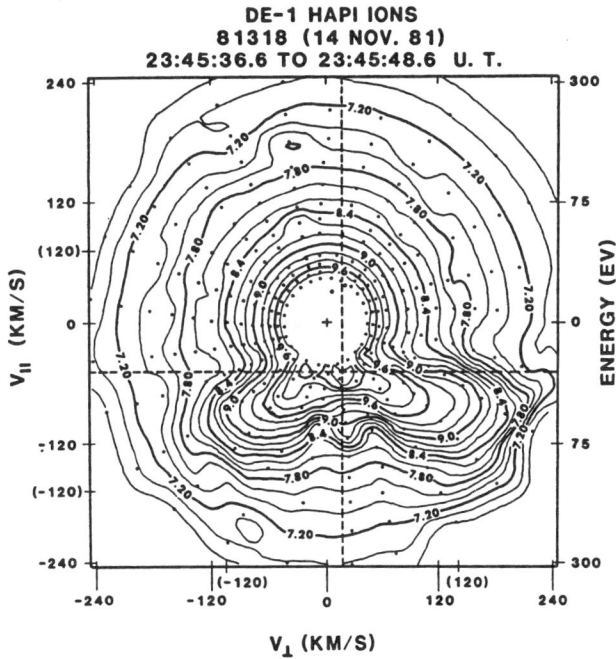

Fig. 3. Contour plot of an elevated conic ion distribution function measured by DE-1. Black dots denote where the measurements were made. [*Winningham and Burch*, 1984]

that the gradient was smoothed out by the cyclotron maser instability.

Louarn et al. [1990] made simultaneous measurements of electromagnetic fields and particle distributions measured during a crossing of an AKR source region. Louarn et al. have determined that trapped electrons may play a more important role in the generation of AKR than loss-cone distributions--a conclusion that was based on an instability analysis of $df/dv_\perp > 0$ in the "forbidden" region of velocity space [*Knight*, 1973], which lies outside the loss cone and between the ionospheric and magnetospheric electron populations. *Louarn et al.* [1990] suggested fluctuating double layers as a process for moving electrons into the forbidden region. Although a plateau was actually observed along the $v_\perp$ axis, it was supposed by *Louarn et al.* [1990] that the velocity space gradient necessary for wave growth was smoothed out by the cyclotron maser instability in a time less than the measurement cycle of the Viking plasma instrument, thus underscoring the measurement time-scale problem pointed out above.

Menietti et al. [1993] analyzed DE-1 particle and wave data taken near and within an AKR source region in the nightside auroral region. They compared their results to those reported by Viking and also to recent numerical simulation studies of *Winglee and Pritchett* [1986]. A mechanism suggested by *Winglee and Pritchett* [1986], which could be responsible for producing a distribution with $df/dv_\perp > 0$ outside the forbidden region, is

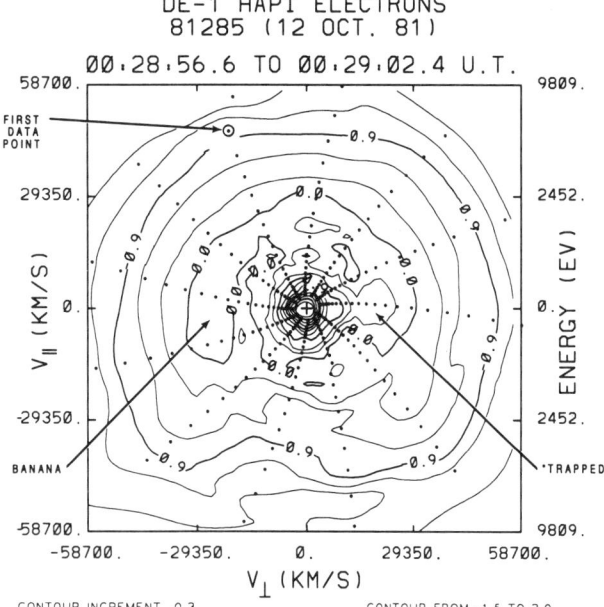

Fig. 4. Contour plot of an electron distribution function measured by DE-1 within a nightside AKR source region [*Menietti et al.*, 1993]. The data were taken at the black dots, beginning at the top of the figure and proceeding counterclockwise, spiraling successively in from high energies to low.

supported by the DE-1 data. In this case the cyclotron maser instability itself leads to population of the forbidden region as well as to the AKR wave growth. An example of this type of distribution is shown in the contour plot in Figure 4. The small black dots in Figure 4 indicate the data points used in the contour plot; the spiral structure is produced by the energy scan of the electron detector (from high to low energy) and the spin of the spacecraft, which is counterclockwise in Figure 4 beginning near the top of the figure. In Figure 4 the distribution function in the $-v_\perp$ half plane is clearly of a different character from that in the $+v_\perp$ half plane, which was sampled about three seconds later. The former distribution resembles the distribution proposed by *Winglee and Pritchett* [1986] to result from the spreading of a downward-directed beam in velocity space as it moves to lower altitudes, resulting in positive velocity-space gradients along both the parallel and perpendicular velocity axes.

The simulation studies of *Winglee and Pritchett* [1986] demonstrated how the competition between the bump-in-tail instability and the cyclotron maser instability can produce diffusion along both the parallel and perpendicular directions, with the resulting distribution depending on the initial parameters such as the electron density and the wave propagation angle. The electron distribution in the left-hand portion of Figure 4 resembles the "banana" distribution of *Winglee and Pritchett* [1986], which has $df/dv_\perp > 0$. In the right portion of Figure 4, the positive velocity space gradient along the perpendicular axis is absent, and the "forbidden" region is populated with electrons as reported by *Louarn et al.* [1990]. The distribution in Figure 4 may indicate a temporal evolution of the distribution function as AKR is generated and the forbidden region is populated, but much more experimental evidence with data acquired at higher rates will be necessary to confirm this possibility.

ACCELERATION MECHANISMS FOR SUPRATHERMAL ELECTRON BURSTS

The acceleration of auroral primary electrons produces inverted-V events with the characteristics of acceleration by parallel potential differences. There also is a second class of auroral acceleration mechanisms involving nonadiabatic, or diffusive, processes that lead to burst-type electron distributions. The electron bursts are typically field-aligned over a broad range of energies and are often associated with conical ion distributions [*Klumpar*, 1979]. Electron distributions with designations such as field-aligned bursts, counterstreaming electrons, suprathermal bursts, and edge precipitation belong to this category. The burst-type distributions have been identified as the primary charge carriers for major Birkeland current elements such as the morning-side region-1 currents [*Burch et al.*, 1983], the currents associated with polar-cap theta arcs [*Menietti and Burch*, 1987], and the evening-side region-2 currents [*Klumpar and Heikkila.*, 1982]. The burst distributions have also been associated with the flickering aurora because of the periodicities they have exhibited with periods in the range of a few seconds [*McFadden et al.*, 1987]. An example of an upward-directed suprathermal electron burst associated with a downward field-aligned current is shown in Figure 5 from *Marshall et al.* [1988].

After the initial discovery of field-aligned electron bursts by *Hoffman and Evans* [1968] and further analysis by *Arnoldy* [1974], *Johnstone and Winningham* [1982] investigated them with the ISIS 2 data and referred to the distributions as s-uprathermal bursts. Although it is not known for sure whether the electron bursts are primarily spatial or temporal effects, Johnstone and Winningham noted mean time durations of about two and six seconds for bursts observed by satellites and sounding rockets, respectively, which would suggest that the satellite duration reflects mainly a spatial effect with a scale size of about 14 km. The suprathermal bursts observed at low altitudes with ISIS were primarily directed downward, although some upward bursts were also reported. Measurements at higher altitudes within and above the auroral acceleration regions have now shown an essentially equal probability of observing upward and downward bursts, which are associated with downward and upward Birkeland currents, respectively [*Marshall et al.*, 1991]. Often, however, the upward and downward bursts are observed in very close proximity to one another or together as counterstreaming electrons [*Sharp et al.*, 1980] with their relative intensities determining whether the net Birkeland current is upward or downward [*Marshall et al.*, 1991].

Johnstone and Winningham [1982] noted in the suprathermal bursts a strong field alignment over a broad range of energies and a tendency for the bursts to occur in the central plasma sheet (CPS) region or in the boundary plasma sheet (BPS), but in the regions surrounding inverted-V structures rather than within them. Nonetheless, they considered the suprathermal burst process to be closely related to the inverted-V mechanism because the upper energy limit of the bursts seemed to rise and then fall again with latitude or, when an inverted V was present, to fall off with increasing distance from the edge of a nearby inverted V. In an earlier sounding rocket experiment, however, *Arnoldy* [1974] reported field-aligned electron burst distributions that seemed to fill in the energy regime below the peak in the primary auroral electron energy spectrum. Similarly, *Burch et al.* [1979] and *McFadden et al.* [1987] observed field-aligned electron bursts within inverted-V structures with the field-aligned electrons having a wide energy range extending up to, but not above, the primary electron beam energy.

The close association that has been observed between field-aligned electron burst distributions and both Birkeland current systems and inverted-V events suggests that the bursts are another example of phenomena that result from the coupling of mesoscale phenomena into microscale processes in the upper ionosphere. The mechanism responsible for the electron bursts has not yet been determined, although it has been noted that the bursts are often accompanied by upward-moving ion conics and that the field-aligned motion of the suprathermal electrons and the cyclotron motion of the conic ions have similar time scales. This notion has led to several models of the electron bursts, all of which involve waves in the VLF frequency range that are generated within an auroral acceleration region [*Temerin et al.*, 1986; *Lotko*, 1986; *McFadden et al.*, 1987]. Another class of models for the field-aligned electron bursts involves flickering or moving double layers [*Sharp et al.*, 1980; *Hultqvist*, 1991].

The role of the suprathermal bursts in carrying field-aligned currents suggests a spatial dependence that would tend to belie their bursty nature, at least in the vicinity of the large-scale region-1 currents. It also suggests a close link with the Alfvén waves associated with the field-aligned currents themselves. Whereas, attention has so far been focused on the importance of the non-MHD Alfvén waves observed by Freja [*Louarn et al.*, 1994] for the acceleration of auroral electrons in inverted-V events, it is possible that these waves can also produce suprathermal electron bursts in other regions of field-aligned currents. The model of *Goertz and Boswell* [1979] demonstrates the development of upward and downward parallel electric field components at the leading edges of inertial Alfvén waves, for field-aligned currents that are upward and downward, respectively. These field-aligned components are produced in the

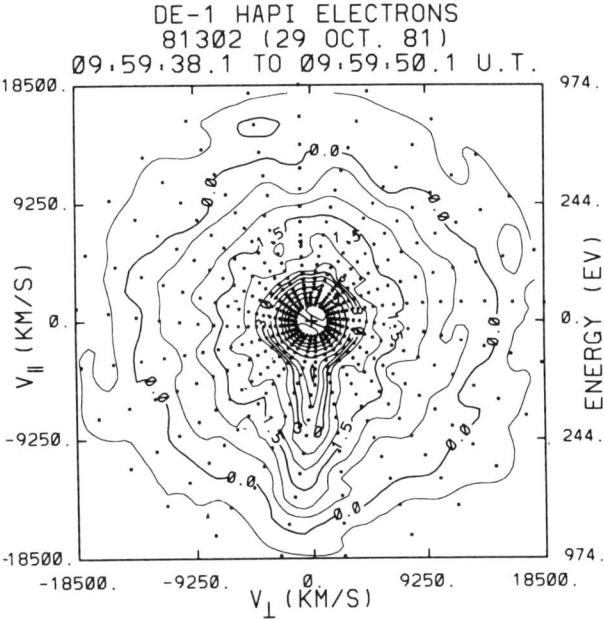

Fig. 5. Contour plot of a suprathermal burst electron distribution function associated with a downward field-aligned current measured by DE-1. [*Marshall et al.*, 1988]

region where the Alfvén waves are being dissipated, which apparently can extend from the auroral acceleration region (~4000 - 8000 km) down to the Freja altitude of 1700 km. Field-aligned acceleration of cold electrons over an extended altitude range would tend to produce a strongly field-aligned distribution over a broad energy range since electrons with sources both near to and distant from the spacecraft, which have undergone lesser and greater periods of acceleration, respectively, would be observed together.

CONCLUSIONS

Mesoscale phenomena in auroral plasmas are well known (inverted-Vs, AKR, ion and electron conics, suprathermal electron bursts, field-aligned current sheets), but microscale phenomena have been very difficult to isolate. It is to be expected that the high-resolution data of Freja and FAST will solve several problems related to: (1) the respective roles of Alfvén waves and earthward-streaming ion beams in powering the aurora, (2) the roles of lower hybrid cavities and ion cyclotron waves in ionospheric ion acceleration, (3) the relative importance of upper hybrid waves and stochastic parallel acceleration in producing electron conics, (4) the transport and acceleration of cold electrons to produce suprathermal bursts, and (5) the relative importance of quasi-static and periodic parallel electric fields in the acceleration of auroral primary electrons.

The acquisition of these new high-resolution space plasma data sets should finally allow meaningful plasma physics to be done in the Earth's low-altitude polar magnetosphere. Lack of high-resolution data in the past has led to the inability to test directly a number of theories concerning a wide range of magnetospheric plasmas. Examples of theories that have not been directly tested are strong pitch-angle diffusion by whistler-mode waves [*Kennel and Petschek,* 1966], anomalous resistivity producing parallel electric fields in the topside ionosphere [*Kindel and Kennel,* 1971], ion tearing mode instability of the magnetotail neutral sheet [*Schindler,* 1974], and reconnection at the dayside magnetopause [*Dungey,* 1961]. High-resolution data such as obtained by Freja and FAST in the auroral regions will forge stronger links between experiment and theory in the future, hopefully allowing more rapid progress toward understanding of magnetospheric plasma phenomena throughout the magnetosphere than has been possible up to now.

REFERENCES

André, M., and L. Eliasson, Electron acceleration by low frequency electric field fluctuations: electron conics, *Geophys. Res. Lett., 19,* 1073-1076, 1991.

Arnoldy, R. L., Auroral particle currents and Birkeland currents, *Rev. Geophys. Space Phys.,12,* 217-231,1974.

Bingham, R. D., D. A. Bryant, and D. S. Hall, A wave model of the aurora, *Geophys. Res. Lett, 11,* 327-330, 1984.

Borovsky, J. E., Fine-scale structures in auroral arcs: an unexplained phenomenon, in *Space Plasmas: Coupling Between Small and Medium Scale Processes,* edited by M. Ashour-Abdalla,T. Chang, and P. Dusenbery, 255-267, Geophysical Monograph 86, American Geophysical Union, Washington, D. C., 1995.

Burch, J. L., S. A. Fields, and R. A. Heelis, Polar cap electron acceleration regions, *J. Geophys. Res., 84,* 5863-5874, 1979.

Burch, J. L., P. H. Reiff, and M. Sugiura, Upward electron beams measured by DE-1: a primary source of dayside region-1 Birkeland currents, *Geophys. Res. Lett., 10,* 753-756, 1983.

Burch, J. L., C. S. Lin, J. D. Menietti, and R. M. Winglee, Coupling between mesoscale and microscale processes in the cusp and auroral plasmas, in *Space Plasmas: Coupling Between Small and Medium Scale Processes,* edited by M. Ashour-Abdalla, T. Chang, and P. Dusenbery, 269-283, Geophysical Monograph 86, American Geophysical Union, Washington, D. C., 1995.

Chang, T., Lower-hybrid collapse, caviton turbulence, and charged particle energization in the topside auroral ionosphere and magnetosphere, *Phys. Fluids B, 5,* 2646-2656, 1993.

Dungey, J. W., Interplanetary field and auroral zones, *Phys. Rev. Lett., 6,* 47, 1961.

Goertz, C., and R. Boswell, Magnetosphere-ionosphere coupling, *J. Geophys. Res., 84,* 7239-7246, 1979.

Hasegawa, A., and K. Mima, Anomalous transport produced by kinetic Alfvén wave turbulence, *J. Geophys. Res., 83,* 1117-1123, 1978.

Hoffman, R. A., and D. S. Evans, Field-aligned electron bursts at high latitudes observed by OGO 4, *J. Geophys. Res., 73,* 6201-6214, 1968.

Hultqvist, B., H. Vo, R. Lundin, B. Aparacio, P.-A. Lindqvist, F. Gustafsson, and B. Holback, On the upward acceleration of electrons and ions by low-frequency electric field fluctuations observed by Viking, *J. Geophys. Res., 96,* 11,609-11,615, 1991.

Johnstone, A. D., and J. D. Winningham, Satellite observations of suprathermal bursts, *J. Geophys. Res., 87,* 2321-2329, 1982.

Kennel, C. F, and H. E. Petschek, Limit on stably trapped particle fluxes, *J. Geophys. Res., 71,* 1-28, 1966.

Kindel, J. M., and C. F. Kennel, Topside current instabilities, *J. Geophys. Res., 76,* 3055-3078, 1971.

Kintner, P. M., J. Vago, S. Chesney, R. L. Arnoldy, K. A. Lynch, C. J. Pollock, and T. Moore, Localized lower hybrid acceleration of ionospheric plasma, *Phys. Rev. Lett., 68,* 2448, 1992.

Klumpar, D. M., Transversely accelerated ions: an ionospheric source of hot magnetospheric ions, *J. Geophys. Res., 84,* 4229-4237, 1979.

Klumpar, D. M., and W. J. Heikkila, Electrons in the ionospheric source cone: evidence for runaway electrons as carriers of downward Birkeland currents, *Geophys. Res. Lett., 9,* 873-876, 1982.

Knight, S., Parallel electric fields, *Planet. Space Sci., 21,* 741-750, 1973.

Koskinen, H. E. J., Lower hybrid parametric processes on auroral field lines in the topside ionosphere, *J. Geophys. Res., 90,* 8361-8369, 1985.

Lin, C. S., J. L. Burch, C. Gurgiolo, and C. S. Wu, DE 1 observations of hole electron distribution functions and the cyclotron maser resonance, *Ann. Geophys. 4,* 33-39, 1986.

Lotko, W., Diffusive acceleration of auroral primaries, *J. Geophys. Res., 91,* 191, 1986.

Louarn, P., A. Roux, H. de Feraudy, D. Le Queau, M. Andre, and L. Matson, Trapped electrons as a free energy source for the auroral kilometric radiation, *J. Geophys. Res., 95,* 5983-5996, 1990.

Louarn, P., J. E. Wahlund, T. Chust, H. de Feraudy, A. Roux, B. Holback, P. O. Dovner, A. I. Eriksson, and G. Holmgren, Observation of kinetic Alfvén waves by the Freja spacecraft, *Geophys. Res. Lett., 21,* 1847-1850, 1994.

Lysak, R. L., Auroral electrodynamics with current and voltage generators, *J. Geophys. Res., 90,* 4178-4190, 1985.

Lysak, R. L., and C. W. Carlson, The effect of microscopic turbulence on magnetosphere-ionosphere coupling, *Geophys. Res. Lett., 8,* 269-272, 1981.

Marshall, J. A., J. L. Burch, J. R. Kan, and J. A. Slavin, DE-1 observations of return current regions in the nightside auroral oval, *J. Geophys. Res., 93,* 14,542-14,548, 1988.

Marshall, J. A., J. L. Burch, J. R. Kan, P.H. Reiff, and J. A. Slavin, Sources of field-aligned currents in the auroral plasma, *Geophys. Res. Lett., 18,* 45-48, 1991.

McFadden, J. P., C. W. Carlson, M. H. Boehm, and T. J. Hallinan, Field-aligned electron flux oscillations that produce flickering aurora, *J. Geophys. Res., 92,* 11,133-11,148, 1987.

Melrose, D. B., A phase-bunching mechanism for fine structures in auroral kilometric radiation and Jovian decametric radiation, *J. Geophys. Res., 91,* 7970-7980, 1986.

Menietti, J. D., and J. L. Burch, DE 1 observations of theta aurora plasma source regions and Birkeland current charge carriers, *J. Geophys. Res., 92,* 7503-7518, 1987.

Menietti, J. D., J. L. Burch, R. M. Winglee, and D. A. Gurnett, DE-1 particle and wave observations in auroral kilometric radiation (AKR) source regions, *J. Geophys. Res., 98,* 5865-5880, 1993.

Mozer, F. S., C. W. Carlson, M. K. Hudson, R. B. Torbert, B. Parady, J. Yatteau, and M. C. Kelley, Observations of paired electrostatic shocks in the polar magnetosphere, *Phys. Rev. Lett., 38,* 292, 1977.

Omidi, N., and D. A. Gurnett, Growth rate calculations of auroral kilometric radiation using the relativistic resonance condition, *J. Geophys. Res., 87,* 2377-2384, 1982.

Reiff, P. H., G. Lu, J. L. Burch, J. D. Winningham, L. A. Frank, J. D. Craven, W. K. Peterson, and R. A. Heelis, On the high- and low-altitude limits of the auroral electric field region, in *Auroral Plasma Dynamics,* edited by R. L. Lysak, 143-154, Geophysical Monograph 80, American Geophysical Union, Washington, D. C., 1993.

Robinson, R. M., J. D. Winningham, J. R. Sharber, J. L. Burch, and R. Heelis, Plasma and field properties of suprathermal electron bursts, *J. Geophys. Res., 94,* 12,031-12 1989.

Sharp, R. D., E. G. Shelley, R. G. Johnson, and A. G. Ghielmetti, Counterstreaming electron beams at altitudes of 1 R_E over the auroral zone, *J. Geophys. Res., 85,* 92-100, 1980.

Schindler, K., A theory of the substorm mechanism, *J. Geophys. Res., 79,* 2803-2810, 1974.

Temerin, M., Evidence for a large bulk ion conic heating region, *Geophys. Res. Lett., 13,* 1059-1062, 1986.

Temerin, M., C. Carlson, and J. P. McFadden, The acceleration of electrons by electromagnetic ion cyclotron waves, in *Auroral Plasma Dynamics,* edited by R. L. Lysak, 155-162, Geophysical Monograph 80, American Geophysical Union, Washington, D. C., 1993.

Ungstrup, E., A. Bahnsen, H. K. Wong, M. André and L. Matson, Energy source and generation mechanism for auroral kilometric radiation, *J. Geophys. Res., 95,* 5973-5981, 1990.

Wahlund, J.-E., P. Louarn, T. Chust, H. de Feraudy, A. Roux, B. Holback, P.-O. Dovner, and G. Holmgren, On ion acoustic turbulence and the nonlinear evolution of kinetic Alfvén waves in aurora, *Geophys. Res. Lett., 21,* 1831-1834, 1994.

Winglee, R. M., and P. L. Pritchett, The generation of low-frequency electrostatic waves in association with auroral kilometric radiation, *J. Geophys. Res., 91,* 13,531-13,541, 1986.

Winningham, J. D., and J. L. Burch, Observations of large scale ion conic generation with DE-1, in *Physics of Space Plasmas (1982-4),* edited by J. Belcher, H. Bridge, T. Chang, B. Coppi, and J. Jasperse, 137-158, Scientific Publishers, Inc., Cambridge, MA, 1984.

Wu, C. S., and L. C. Lee, A theory of the terrestrial kilometric radiation, *Astrophys. J., 230,* 621, 1979.

J. L. Burch, Southwest Research Institute, San Antonio, TX 78228

Semikinetic Simulation of Effects of Ionization by Precipitating Auroral Electrons on Ionospheric Plasma Transport

D. G. Brown, P. G. Richards, J. L. Horwitz, and G. R. Wilson

Department of Physics and Center for Space Plasma and Aeronomic Research, University of Alabama, Huntsville

Precipitating auroral electrons have two primary effects upon the ionosphere: Ionization of neutral atoms and heating of thermal electrons. Increased ionization will raise both the ion and electron densities, while electron heating will modify the ambipolar electric field. In this paper we concentrate on the effects of only the ionization upon transport of ionospheric ions. The ionospheric plasma is modelled by a time-dependent, semikinetic, transport model. We follow kinetic ion gyrocenters which move subject to the macroscopic forces of gravity, ambipolar electric force, and magnetic mirror force. They are also subject to such microscopic effects as Coulomb collisions, chemical reactions, and collisions with neutrals. Once the ionospheric plasma has reached an equilibrium with the neutral atmosphere (generated by the MSIS-86 model), an impulse of precipitating electrons is applied to the simulated region. The rate of additional O^+ production due to the precipitating electrons is calculated with a 2-stream energy loss code, and is used to generate new O^+ test particles at the appropriate locations. We find that, as well as enhancing the O^+ density, there is an increase in H^+ production, the latter is a result of the enhanced O^+ density which increases the rate of H^+ production by the accidentally resonant H–O^+ charge exchange reaction. These density enhancements result in an increase in both the O^+ and H^+ upgoing fluxes.

1. INTRODUCTION

Auroral electron precipitation may result in ionization (which increases both ion and electron densities) and electron heating (which enhances the ambipolar electric field). Together these effects may increase the outflow of ions from the ionosphere. Such precipitation may be characterized by Maxwellian energy distributions with energies ranging from a few hundred eV to a few tens of KeV [e.g., *Lilensten et al.*, 1990, *Sandahl et al.*, 1990]. However, only precipitation with energies on the order of hundreds of electron volts generate ionization at sufficiently high altitudes to significantly modify ionospheric outflows [*Roble and Rees*, 1977]. Thus, in this paper, we focus on such low energy auroral electron precipitation.

Richards (this monograph), studies the response of the ionosphere to one hour of soft electron precipitation using the FLIP (Field Line Interhemispheric Plasma) model. We, however, adopt a generalized semikinetic approach for the ionospheric plasma. This approach builds on earlier work with such models in the polar wind, return current, and plasmaspheric regions [e.g., *Wilson et al.*, 1990, *Brown et al.*, 1991, *Ho et al.* 1993, *Lin et al.*, 1994] and the transition region model of *Wilson* [1992]. Though computationally intensive, the kinetic treatment of the ionospheric ions allows examination of ion velocity space distributions as the plasma reacts to various stimuli. Here we present results from the combination of our ionospheric plasma model with electron precipitation induced ionization. Though the effects of ionization alone upon the velocity space distributions are minor these initial results represent a step towards more completely modelling auroral phenomena

TABLE 1. Parameters describing the location and conditions used as inputs for the MSIS-86 neutral atmosphere model.

Parameter	Value
Magnetic Latitude	70°
Magnetic Longitude	180°
Day	183
Local Time	12 hours
F10.7	90
F10.7A	90
Ap Index	18

in situations where a kinetic approach for the ion species is necessary.

2. MODEL

We model the transport of ionospheric plasma, in one dimension, along a high latitude field line. The lower boundary is located at an altitude of 200 km. Although electron precipitation of the energies used in these simulations does result in significant O^+ production at this altitude, O^+ outflow is inhibited by interactions with the neutral atmosphere and so we assume photochemical equilibrium in determining the O^+ density at this boundary. We will see later that this assumption is justified. The upper boundary of the simulation is located at 3 R_E. Particles passing beyond this boundary are assumed to escape, however, in the results presented here we concentrate on altitudes below 4000 km. The model consists of four components: a background neutral atmosphere; H^+ and O^+ ionospheric ions; ionospheric electrons; and precipitating suprathermal auroral electrons.

The neutral atmosphere is generated with the MSIS-86 model [*Hedin*, 1987]. The parameters describing the location and conditions used as inputs to the neutral atmosphere model are contained in Table 1. The model returns density profiles for N_2, O_2, O, and H neutral species, which are maintained as constant for the duration of each simulation.

The ionospheric ions are treated as kinetic gyrocenters, which move subject to the macroscopic forces of the ambipolar electric field, gravity, and magnetic mirror force. They also are subject to microscopic effects: Coulomb self-collisions; H^+–O^+ Coulomb collisions; reactions and collisions with the components of the neutral atmosphere; H–O^+ accidentally resonant charge exchange; and, O^+ production as a result of the precipitating auroral electrons. The techniques used to model these micro- and mesoscale effects on the ionospheric plasma are much the same as those of *Wilson* [1992, 1994]. However, the model now extends over a larger altitude range. As in *Wilson* [1994] the ionospheric electrons are treated as a neutralizing, Boltzmann distributed fluid, however, in this set of simulations the electrons are not isothermal. The electron temperature increases from 2555 K to 5000 K between the altitudes of 500 km and 1 R_E (Figure 1), and this profile is held constant throughout the simulations. The outflows for a lower electron temperature would, of course, be somewhat reduced.

The flux tube is initiated with an estimate of the O^+ density profile. This profile, and that for H^+, then evolve subject to macroscopic forces, chemical reactions, and collisions until a new equilibrium is obtained. A brief impulse of suprathermal auroral electron precipitation then begins. The precipitating electrons are modeled by the two-stream energy loss code of *Richards and Torr* [1990] which takes as its inputs parameters describing the precipitating electrons, the neutral atmosphere, and the ionospheric plasma parameters. The two-stream model returns the O^+ production rate due to the electron precipitation as a function of altitude, which is then used to modify the rate of production of O^+ simulation particles within the semikinetic model.

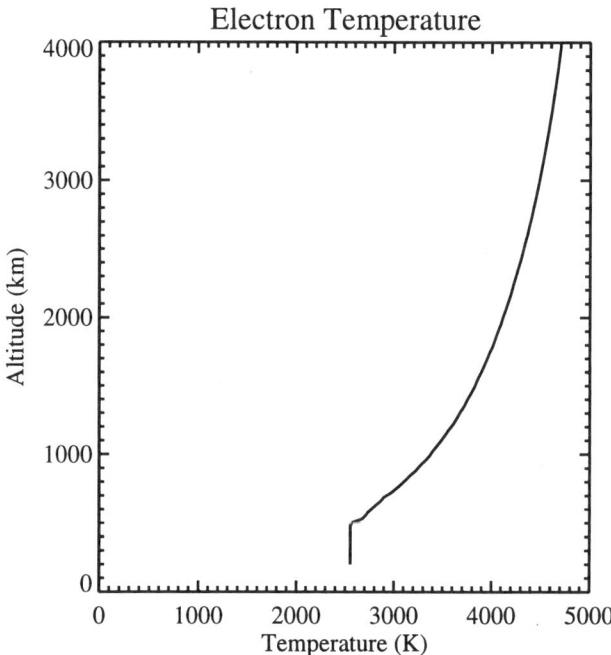

Fig. 1 Electron temperature as a function of altitude from 200 to 8000 km. The temperature is constant above 8000 km.

3. RESULTS

We examine the response of the ionospheric plasma to two separate temporal variations and energy spectra of precipitating suprathermal electron impulses. In both cases the energy flux is 5 ergs cm^{-2} s^{-1} and the electrons have a Gaussian energy distribution. We concentrate on soft electron precipitation which causes ionization at sufficiently high altitudes to effect ion outflow from the ionosphere. In the first case the characteristic energy is 500 eV, with a duration of 640 seconds, while in the second case we reduce the characteristic energy to 100 eV and the duration to 280 seconds. The resulting rates of O$^+$ ion production are shown in Figure 2, and these remain essentially constant during the period of precipitation (being most sensitive to the neutral atmosphere altitude profile which is held constant). The peak in O$^+$ production moves upward for lower energy precipitation, due to the shorter stopping distance for lower energy electrons, which is a consequence of the peak around 100 eV in the dependence of the ionization cross-section on electron energy (*Richards and Torr* [1990], Figure 2).

The changes in ion density and velocity, as well as flux, before, during, and after the period of 500 eV electron precipitation are illustrated in Figure 3. In order to better illustrate the changes in O$^+$ density we plot the ratio of the O$^+$ density at a particular time to the initial density. Similarly, to bring out the change in the H$^+$ drift velocity, we plot the difference between the H$^+$ velocity profile at a specific time and the initial profile. The O$^+$ density increases by a factor of 1.8 around 200 km (at 400 seconds). This density enhancement spreads up the flux tube at a velocity of about 2 kms^{-1}, reaching an altitude of 4000 km by 2000 seconds. There is a corresponding increase in the upgoing O$^+$ flux, from about 10^8 to about 3×10^8. This flux enhancement persists more than 10 minutes after the precipitation has ceased. We also note an enhancement in the H$^+$ density. This is a result of the increased upflow of O$^+$, which alters the equilibrium in the H–O$^+$ charge exchange reaction. The increased production of H$^+$ from non-drifting H atoms also acts to reduce the net drift velocity of H$^+$, however, the net effect is a slight increase in the upward flux of H$^+$ ions. The change in the production of H$^+$ occurs primarily between 500 and 800 km, with variations at higher altitudes being due to transport.

Figure 4 illustrates the changes in ion density and velocity, as well as flux, before, during, and after the period of 100 eV electron precipitation. The O$^+$ den-

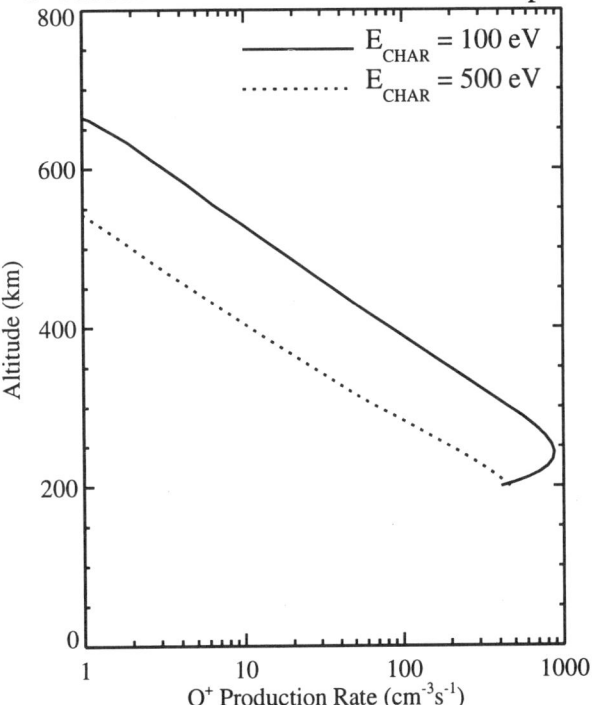

Fig. 2 Production rate of O$^+$, as a function of altitude, due to precipitating auroral electrons with characteristic energies of 100 eV (dotted) and 500 eV (solid).

sity increases by a factor of 2.6 around 250 km (at 400 seconds). Despite strong O$^+$ production reaching below the lower boundary of the model, we see no significant change in the O$^+$ flux below 250 km, justifying our assumption of photochemical equilibrium at the lower boundary. There is an upgoing pulse in the O$^+$ drift velocity which flows upward from about 300 km. This is a result of an electric field enhancement which is caused the modification of the O$^+$ density profile by the precipitation. As this pulse travels upward, the higher energy particles travel more quickly causing the pulse to disperse, and by 2000 seconds we also observe a return flow as the less energetic ions begin to fall back toward the Earth. The change in the O$^+$ velocity space distribution within this pulse is illustrated in the top two panels of Figure 5. The right hand panel shows the pulse of accelerated O$^+$, flowing through the initial background, and corresponds to an increase in the upward, parallel, heat flux.

As in the previous case, the increased upflow of O$^+$ ions shifts the equilibrium of the H–O$^+$ accidentally resonant charge exchange reaction towards H$^+$, leading to an enhancement in the H$^+$ density and flux, but a drop in its drift velocity. Again, the decrease in the

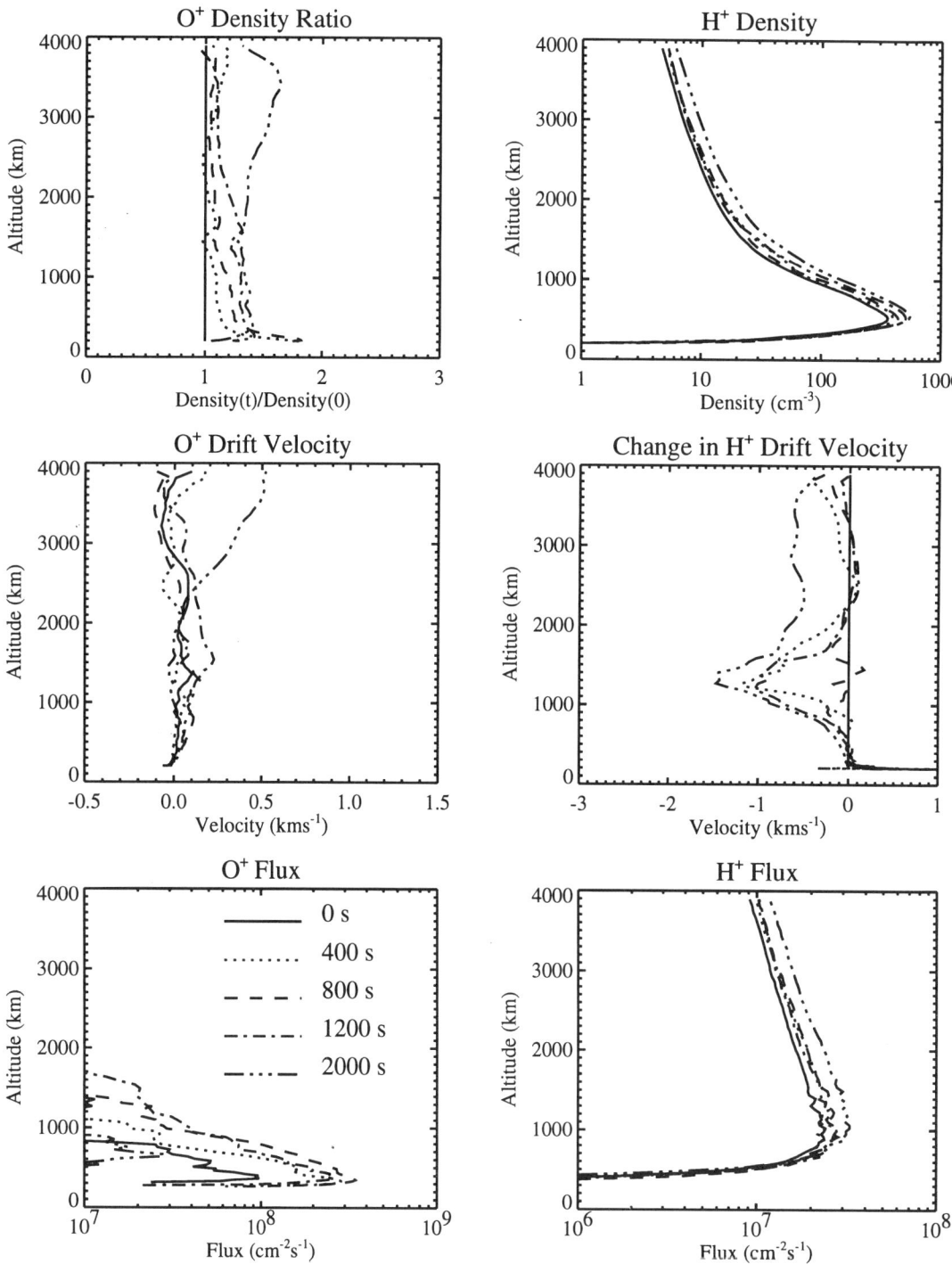

Fig. 3 Altitude profiles of O^+ and H^+ parameters for the 500 eV auroral electron precipitation case. The panels show the ratio of the O^+ density at time t its initial density (top left), H^+ density (top right), drift velocity of O^+ (center left), change in H^+ drift velocity (center right), and value of the ion fluxes (bottom panels) at the times indicated in the lower left panel.

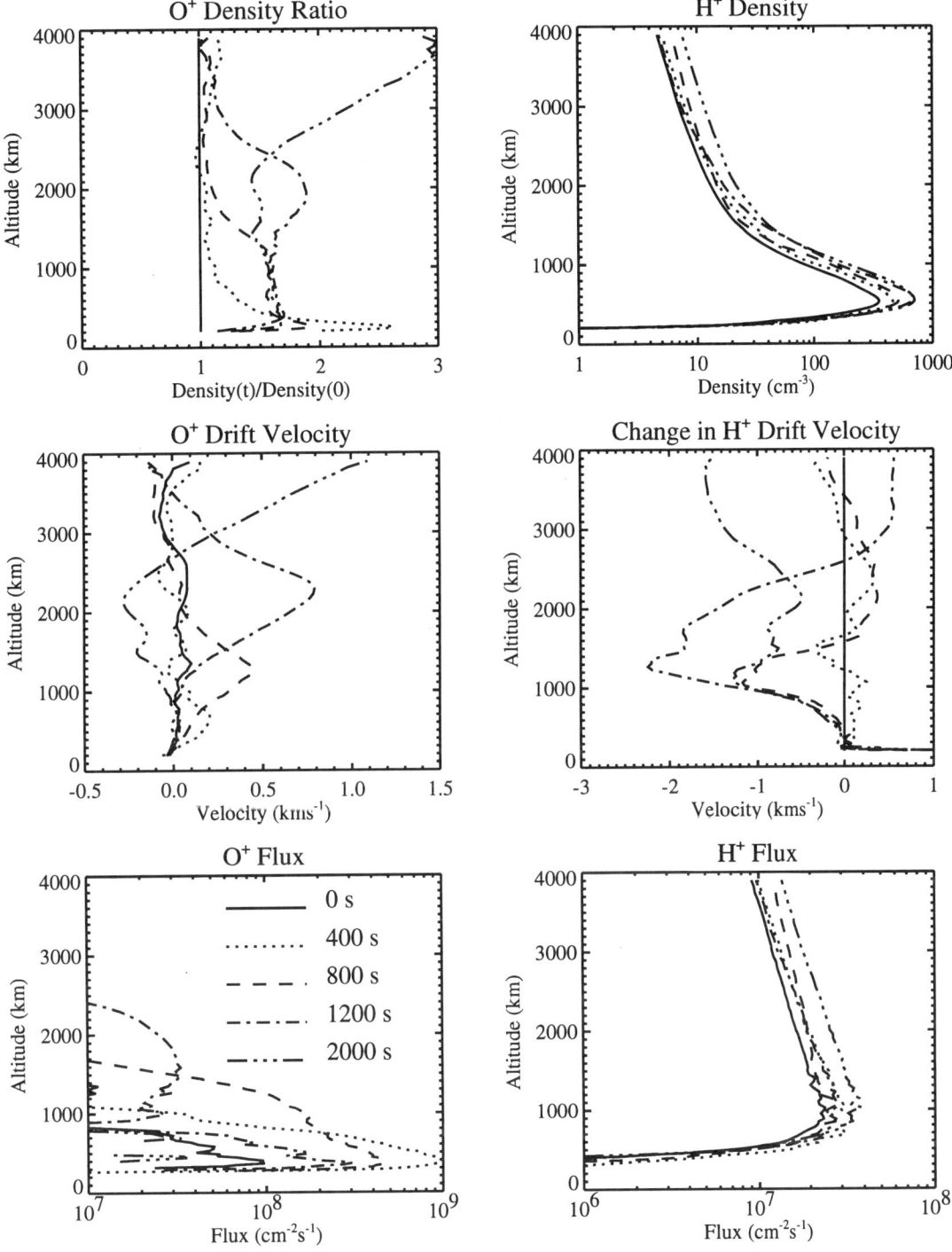

Fig. 4 Altitude profiles of O^+ and H^+ parameters for the 100 eV auroral electron precipitation case. The panels show the ratio of the O^+ density at time t its initial density (top left), H^+ density (top right), drift velocity of O^+ (center left), change in H^+ drift velocity (center right), and value of the ion fluxes (bottom panels) at the times indicated in the lower left panel.

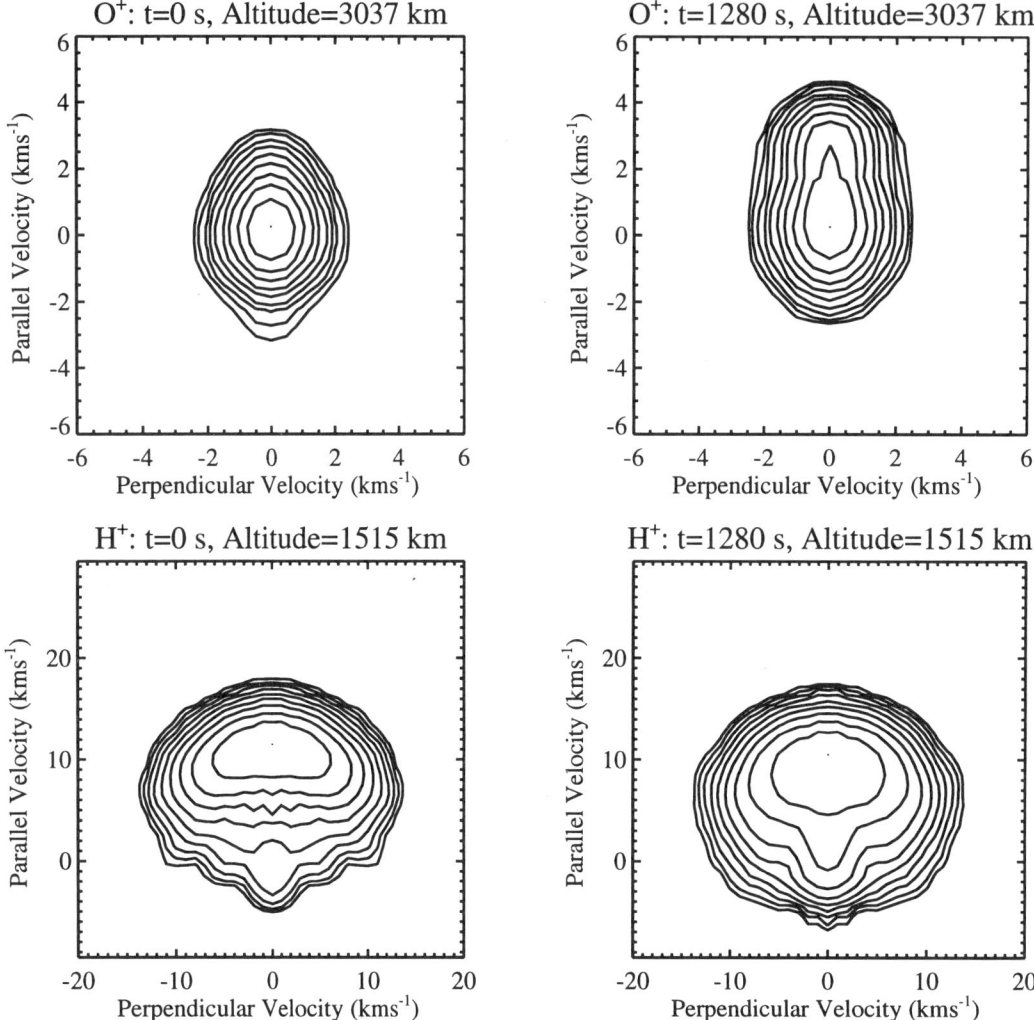

Fig. 5 Velocity space distributions for O^+ (top panels) and H^+ (bottom panels) before (left panels) and after (right panels) the period of precipitation, at the times and altitudes indicated.

H^+ drift velocity is due to the increased conversion of non-drifting, neutral H to H^+ ions. The resulting modification of the H^+ velocity space distribution is shown in the lower two panels of Figure 5. The right hand panel illustrates the decrease in upward, parallel, heat flux which results from the generation of non-drifting H^+ ions, and self-collisions.

4. DISCUSSION AND CONCLUSIONS

We find that O^+ ionization, resulting from precipitating, suprathermal, auroral electrons primarily effects the flux and lowest order moments of the ion populations. The O^+ density increases locally due to the increased production rate, which results in an electric field enhancement (via a change in the density gradient). This leads to enhanced O^+ upflow (and a corresponding increase in its flux). The increase in O^+ concentration shifts the equilibrium of the H–O^+ accidentally resonant charge exchange reaction, resulting in loss of slower O^+ ions, and enhanced H^+ density and flux. This reaction also locally reduces the H^+ drift velocity by creating H^+ ions from the non-drifting atomic hydrogen background.

We have seen that the altitude distribution and rate of O^+ production depend on the energy of the incident electrons, with more significant effects on ionospheric plasma transport occurring for lower energy precipitation. However, in our simulations we have used the neutral atmosphere for a specific date, time, and location.

The density profiles of the various neutral atmosphere components vary with time and solar activity, and consequently the production rates due to precipitation will also be modified by the neutral atmosphere selected [cf. *Germany et al., 1990*].

In future work we plan to include the heating effects of precipitating electrons on the thermal electron population, by using a fluid description for the ionospheric electrons. This will also allow us to model the effects of instabilities such as the current-driven ion cyclotron instability as in our earlier work [*Brown et al., 1991*].

Acknowledgments. This work was supported by NASA grant NAG-1554 and NSF grant ATM-9402310 to the University of Alabama in Huntsville. The authors thank the reviewer for useful comments and questions.

REFERENCES

Brown, D. G., G. R. Wilson, J. L. Horwitz, and D. L. Gallagher, 'Self-consistent' production of ion conics on return current region auroral field lines: A time-dependent, semikinetic model, *Geophys. Res. Lett.*, 18, 1841, 1991.

Germany, G., M. R. Torr, P. G. Richards, and D. G. Torr, The dependence of modeled OI 1356 and N_2 LBH auroral emissions on the neutral atmosphere, *J. Geophys. Res.*, 95, 7725, 1990.

Hedin, A. E., MSIS-86 thermosphere model, *J. Geophys. Res.*, 92, 4649, 1987.

Ho, C. W., J. L Horwitz, N. Singh, and G. R. Wilson, Plasma expansion and evolution of density perturbations in the polar wind: Comparison of semikinetic and transport models, *J. Geophys. Res.*, 98, 13581, 1993.

Lilensten, J., D. Fontaine, W. Kofman, L. Eliasson, C. Lathuillere, and E. S. Oran, Electron energy budget in the high-latitude ionosphere during viking/EISCAT coordinated measurements, *J. Geophys. Res.*, 95, 6081, 1990.

Lin, J., J. L. Horwitz, G. R. Wilson, and D. G. Brown, Equatorial heating and hemispheric decoupling effects on inner magnetospheric core plasma evolution, *J. Geophys. Res.*, 99, 5727, 1994.

Richards, P. G., and D. G. Torr, Auroral modeling of the 3371 Å emission rate: dependence on characteristic electron energy, *J. Geophys. Res.*, 95, 10337, 1990.

Roble, R. G., and M. H. Rees, Time-dependent studies of the aurora: Effects of particle precipitation on the dynamic morphology of ionospheric and atmospheric properties, *Planet. Space Sci.*, 25, 991, 1977.

Sandahl, I., L. Eliasson, A. Pellinen-Wannberg, G. Rostoker, L. P. Block, R. E. Erlandson, E. Friis-Christensen, B. Jacobsen, H. Lühr, and J. S. Murphree, Distriburion of auroral precipitation at midnight during a magnetic storm, *J. Geophys. Res.*, 95, 6051, 1990.

Wilson, G. R., Semikinetic modeling of the outflow of ionospheric plasma through the topside collisional to collisionless transition region, *J. Geophys. Res.*, 97, 10555, 1992.

Wilson, G. R., Semikinetic modeling of O^+ upflows resulting from E×B convection heating in the high latitude F region ionosphere, *J. Geophys. Res.*, 99, 17453, 1994.

Wilson, G. R., C. W. Ho, J. L. Horwitz, N. Singh, and T. E. Moore, A new kinetic model for time-dependent polar plasma outflow: Initial results, *Geophys. Res. Lett.*, 17, 263, 1990.

D. G. Brown, P. G. Richards, J. L. Horwitz, and G. R. Wilson, Department of Physics and Center for Space Plasma and Aeronomic Research, University of Alabama in Huntsville, Huntsville, Alabama 35899.

High Latitude Outflow of Centrifugally Accelerated Ions through the Collisional/Collisionless Transition Region

C. W. Ho, J. L. Horwitz, G. R. Wilson, and D. G. Brown

Department of Physics and Center for Space Plasma and Aeronomic Research, University of Alabama, Huntsville

In this paper, we show how centrifugal acceleration and frictional heating due to ionospheric convection affects O^+ and H^+ transport along open field lines from a collisional to a collisionless region (200 - 6000 km). The model we use is a generalized semikinetic model similar to that of *Wilson* [1994] who used it to study the effects of convection frictional heating of O^+ ions in the 300 - 1100 km altitude polar region. The present study is complementary to the work of *Horwitz et al.* [1994] on the effects of centrifugal acceleration of the collisionless polar wind. We found that in the altitude range considered here, centrifugal acceleration makes very little direct contribution to the overall upward force on the H^+ ions, but has a pronounced effect on them through the centrifugally accelerated O^+ ions. While the effect of centrifugal acceleration on the O^+ ions is macroscopic in nature, i.e., acceleration increases the bulk flow speeds and thus cools the ions adiabatically, the effect of centrifugal acceleration on the H^+ ions is to a large extent dominated by microscopic processes – mainly H^+-O^+ collisions. Because the collisional mean free path of a H^+ ion decreases rapidly with the decrease in the relative speed between the H^+ ion and the average O^+ flow speed, when the upward O^+ flow speed is increased by centrifugal acceleration, a normally collisionless high speed H^+ ion can now exchange energy with the O^+ ions. The result is an overall decrease in the H^+ flow velocity and a large upward or downward H^+ heatflux. The H^+ heatflux is upward or downward depending on whether the majority of the H^+ ions are held back by the collisional drag force that comes from the O^+ ions. During strong convection, when there is enhanced O^+ ion loss, the drop in O^+ density decreases the collisional mean free path of the H^+ ions. The majority of H^+ ions therefore stream upwards and leave a tail of collisional ions, resulting in a large downward heatflux. We therefore conclude that in moving through the collisional/collisionless transition region, the velocity space characteristics of the H^+ ions in the polar region are greatly dependent on the effects of centrifugal acceleration on the O^+ ions and the loss of O^+ ions by enhanced chemical reactions with the neutrals.

1. INTRODUCTION

It has been shown that $\mathbf{E} \times \mathbf{B}$ convection of ionospheric plasmas in the high latitude region produces a field-aligned upward centrifugal acceleration in the convection frame of reference [*Cladis*, 1986; *Horwitz et al.*, 1994]. This effect had also been considered in the single-particle-based model of ion trajectories [*Horwitz et al.*, 1987; *Delcourt et al.*, 1989], fluid simulation of ion transport in the large-scale magnetosphere [*swift*, 1990], and the semikinetic model of plasmasphere refilling [*Miller et al.*, 1993]. Recently, in studying the outflow of F-region O^+ ions due to convection frictional heating, *Wilson* [1994] used a generalized semikinetic model which included centrifugal acceleration but the effect was not stressed. The first detailed study of the effect of centrifugal acceleration on the ionospheric outflow of high latitude plasmas was conducted by *Horwitz et al.* [1994] using a collisionless generalized semikinetic model. *Horwitz et al.* found that centrifugal acceleration can greatly increase the densities and flow velocities of O^+ ions and decrease their temperature. Centrifugal acceleration is therefore a possible explanation for the large outflow velocities observed in the mid-altitude polar magnetosphere with the Dynamics Explorer 1 and Akebono spacecraft [*Ho et al.*, 1994].

This paper is intended to continue the work of *Horwitz et al.* [1994] in studying how centrifugal acceleration affects the H^+ and O^+ outflows in the high latitude region. However, instead of using a collisionless semikinetic model and simply choosing parametric topside boundary conditions, we now specify these parameters through the low-altitude chemistry, heating and transport in our model. *Wilson* [1994] used a similar model to study the effect of $\mathbf{E} \times \mathbf{B}$ convection friction on ionospheric outflows, however, his model did not include H^+ ions and it consisted of only a short flux tube (300-1100 km). In another paper, *Wilson* [1992] studied the upward transport of H^+ ions at the polar latitudes in a static O^+ background. In that model, the O^+ dynamics was not considered. Nevertheless, these two papers [*Wilson*, 1992, 1994] represent the first serious attempts to model self-consistently the high latitude plasma outflows in the collisionless/collisional transition region by treating the ions kinetically. *Wilson* [1994] obtained transient streams of O^+ ions of large upward velocity, temperature, and heat flux. He also showed the evolution of the ion velocity distribution functions. These distribution functions are toroidal in shape in the F-region due to ion-neutral collisions (*Schunk and Walker*, 1972), but quickly approach a near Maxwellian through the effects of self collisions as they move from the F-region to higher but still ion-ion collision dominated altitudes. For H^+ ions flowing upward from a collisional to a collisionless region through a static O^+ background, *Wilson* [1992] found that the velocity distributions of H^+ ions deviated significantly from a Maxwellian, and the H^+ ions carry a large upward and then a large downward heatflux as they move up in altitude. At low altitudes where collisions are important, the major influence which shapes the H^+ distribution function are collisions with the dominant ion and neutral species. Even at altitudes far above the region where O^+ collisions dominate, an imprint of the effect of these collisions on the H^+ velocity distribution can still be seen.

In this paper, we seek the answers to the following questions: How significant is centrifugal acceleration at altitudes below the lower boundary of 4000 km used by *Horwitz et al.* [1994]? When the processes in the F-region such as ion-ion, ion-neutral collisions, chemical reactions and charge exchange, etc. are taken into account, how does the outflow of centrifugally accelerated ions in the collisionless region compare to the results of *Horwitz et al.* [1994] who used a collisionless model? How do H^+ ions behave in a dynamic O^+ background instead of the static one considered by *Wilson* [1992]?

A brief description of our model is given here, for further details, see *Wilson* [1994]. The model is semikinetic in that a PIC (particle-in-cell) scheme is applied to the ions while the electrons are treated as a fluid which obeys the Boltzmann relation. The motion of the ions is determined by several macroscopic forces (ambipolar electrical force, gravity, magnetic mirror force, and centrifugal force) and collisions and reactions which include O^+-O resonant charge exchange collisions, polarization collisions of O^+ with O and N_2, reactions of O^+ with O_2 and N_2, and self-collisions of O^+. The present model has been improved over that of *Wilson* [1994] by including the accidentally resonant charge exchange reaction, $H^+ + O \rightleftharpoons H + O^+$, Coulomb collisions of H^+ with O^+, and H^+ self-collisions. We also use a longer flux tube that now extends from 200 to 6000 km.

In order to maximize the centrifugal acceleration with reasonable physical conditions, we assume a convection electric field of 200 mV/m at the lower boundary (200 km) which corresponds to a drift speed of about 4 km/s. Convection of this magnitude may not be common but has been observed (e.g. *Anderson*, [1991]). We will treat four cases: (1) No convection, (2) Convection with centrifugal acceleration but no frictional heating, (3) Convection with frictional heating but no centrifugal acceleration, and (4) Convection with both centrifugal acceleration and frictional heating. Velocity distributions of H^+ and O^+ ions and their velocity moments will be shown for each case.

2. RESULTS

Figure 1 shows that the shape of the H^+ and O^+ ion velocity distributions (at steady-state, at 3000 km) are quite dependent on whether there is convection, and if there is convection, whether there is centrifugal acceleration and/or frictional heating. When there is no convection, the O^+ ions deviate slightly from a Maxwellian distribution and carry a downward heatflux. The H^+ ions form an inverted-bowl shaped distribution due to deflection by Coulomb collisions at lower altitudes. Under centrifugal acceleration, O^+ ions at 3000 km form a cold Maxwellian as a result of adiabatic cooling, while the H^+ distribution is positively skewed resulting in a large upward heat flow. Convection with heating but without centrifugal acceleration results in toroidal distributions of O^+ in the F-region. The toroids are still

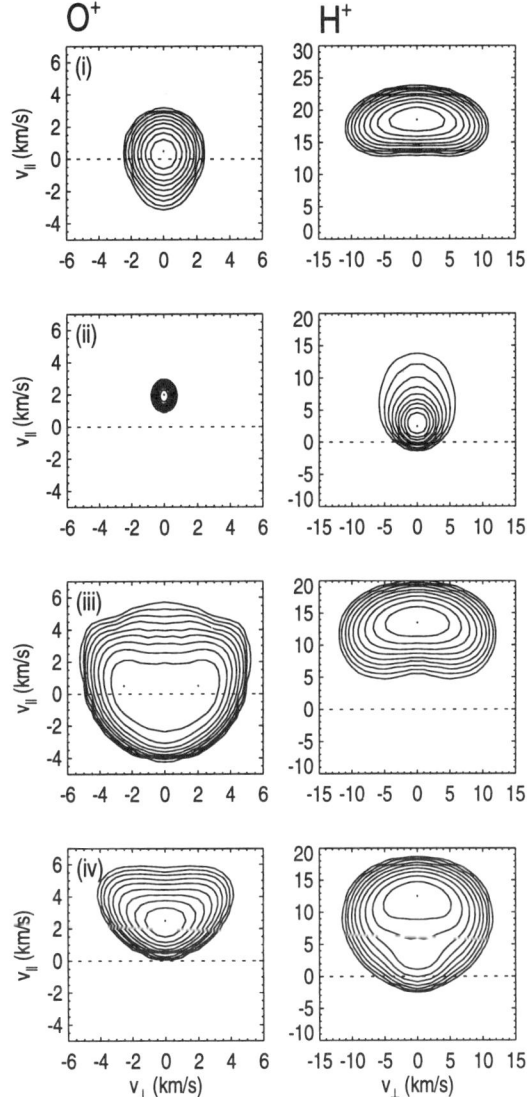

Figure 1. Velocity distribution functions of O^+ and H^+ ions at 3000 km at steady state for the cases of (i) no convection, (ii) convection with centrifugal acceleration but no frictional heating, (iii) convection with frictional heating but no centrifugal acceleration, and (iv) both. The distribution functions are normalized to 1 at the peak, and the contour lines are decreased by a factor of $e^{-1/2}$.

apparent at 3000 km, but are folded due to the mirror force. The H^+ distribution is very similar to the case when there is no convection; however, the H^+ ions are hotter due to energy exchange through collisions with the heated O^+ ions, and have a smaller drift speed due to a decrease in the ambipolar electric field. When centrifugal acceleration is also included, the O^+ velocity distribution indicates a cooler O^+ ion stream as a result of adiabatic cooling, and the H^+ ions exhibit a low speed tail indicating a downward heatflux.

Figure 2 shows the O^+ and H^+ velocity moment profiles at steady-state. The effects of convection on the O^+ profiles can readily be understood as follows: Removal of O^+ ions via chemical reactions with neutrals causes a large reduction in the O^+ density (the reduction of the H^+ density peak is due to its reduced production following the loss of O^+ ions). Frictional heating greatly enhances the F-region O^+ perpendicular temperature, and the O^+ parallel temperature is enhanced via O^+ self-collisions. Centrifugal acceleration increases the O^+ speed dramatically and reduces the parallel temperature by means of adiabatic cooling.

Because of their small mass, H^+ ions are much less affected by gravity and centrifugal acceleration in comparison to the ambipolar electric field (Figure 3). The overall acceleration is therefore determined by the electron density gradient. When there is frictional heating, density depletions in the F-region due to enhanced O^+ chemical loss cause large reductions in the ion density gradient, resulting in a large decrease in the ambipolar electric field. In the absence of convection, H^+ ions attain the highest speeds at most altitudes because of the relatively large ambipolar electric field. When there is only centrifugal acceleration, the H^+ ions have the smallest flow speeds compared to the other three cases, although the centrifugal acceleration has the second largest negative potential (Figure 3). This indicates that the elevated O^+ densities above 1000 km and the increased O^+ drift speed exert a significant collisional drag on H^+.

In studying the topside ionospheric outflows of H^+ ions from a collisional to collisionless transitional region, *Wilson* [1992] found that the H^+ velocity distribution functions are shaped mainly by collisions with O^+ ions at low altitudes and by macroscopic forces and self-collisions at high altitudes. Significant departures from Maxwellian occur, especially at the subsonic to supersonic transition point. The H^+ velocity distributions obtained sometimes have a downward extended tail, which is the result of the collisionless fast H^+ ions accelerating away from the collision-dominated near-zero-velocity H^+ ions. H^+ ions exchange energy with O^+ mostly efficiently if they have small velocities relative to O^+ (which in Wilson [1992] occurs at $v = 0$ for a static O^+ background). With O^+ dynamic, however, the H^+ behavior is quite different. For instance, in the two cases where centrifugal acceleration was included, the H^+ parallel temperature and heatflux profiles were

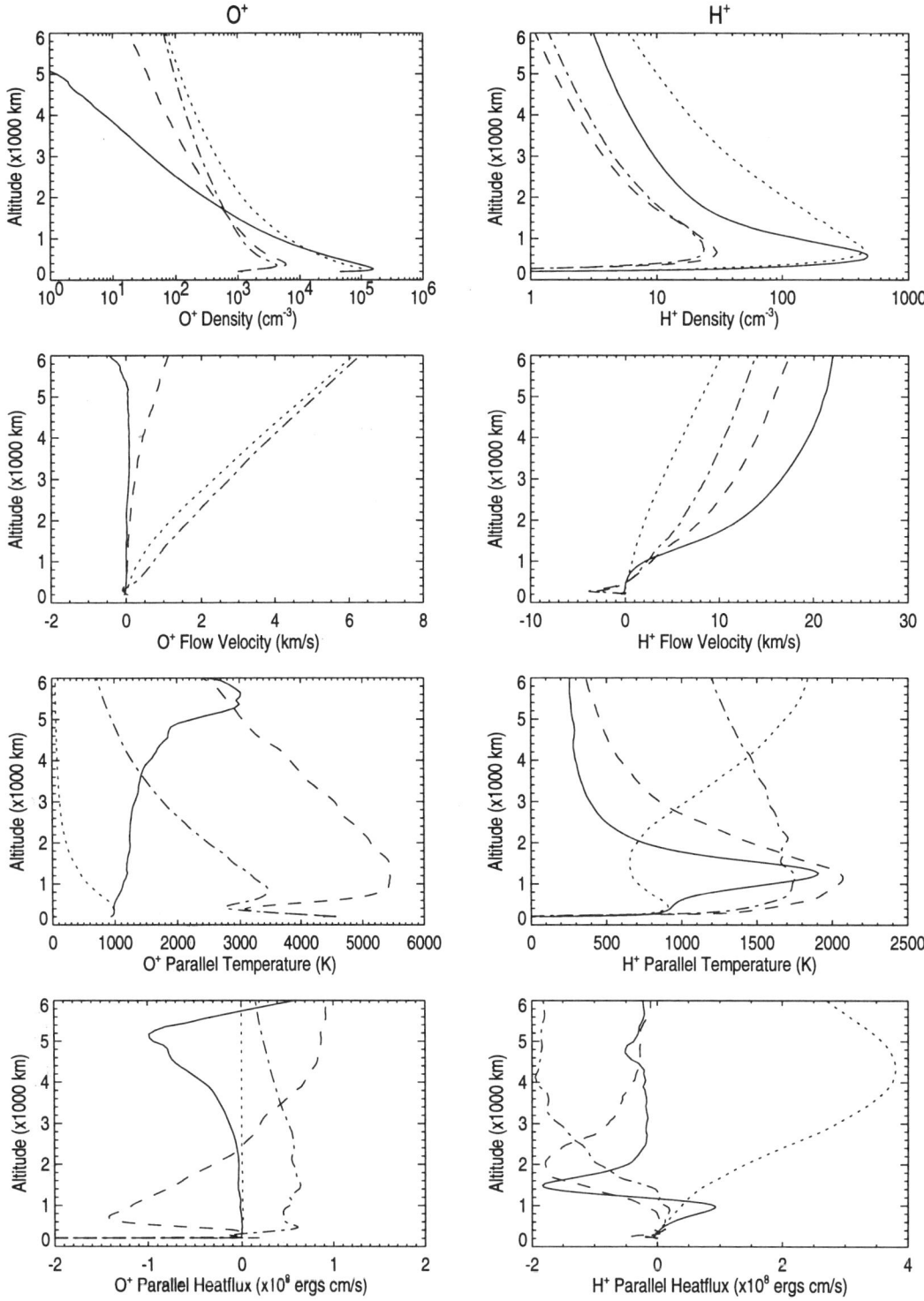

Figure 2. The H^+ and O^+ density, flow velocity, and parallel temperature and heatflux for the cases of no convection (solid), convection with centrifugal acceleration but no frictional heating (dotted), convection with frictional heating but no centrifugal acceleration (dashed), and convection with both centrifugal acceleration and frictional heating (dotted-dashed).

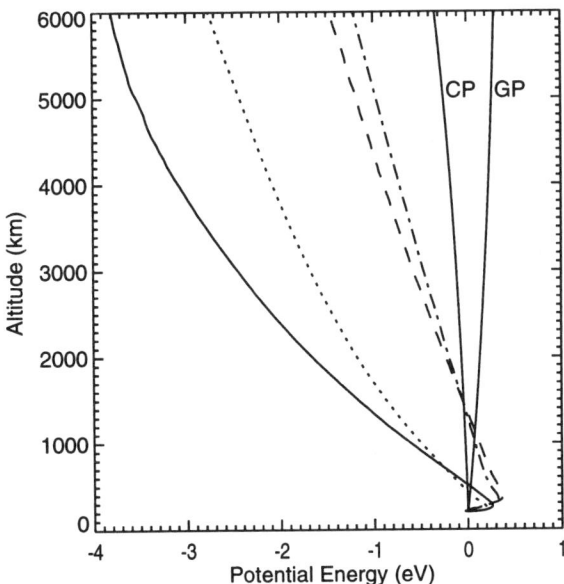

Figure 3. The total potential energy of H$^+$ ions for the same four cases as in Figure 2 (see Figure 2 for legend of lines). For the case of only centrifugal acceleration, the H$^+$ negative potential is the second largest, but the flow velocity is the smallest due to H$^+$-O$^+$ collisions (see texts). The centrifugal potential (marked CP) and gravitational potential (marked GP) are included for comparison, they are negligible compared to the ambipolar electric potential. The electrical potentials (not shown) are very close to the total potential of each case plotted here.

very different from the cases when there was no centrifugal acceleration. This is due to the dynamical response of the O$^+$ ions to centrifugal acceleration and the H$^+$-O$^+$ collisions in shaping the H$^+$ velocity distribution.

The shape of the H$^+$ velocity distribution is found to depend greatly on the collisional mean free path of the H$^+$ ions. The mean free path λ of an H$^+$ ion of mass m moving with speed u relative to the main ion stream of density n is given by (e.g., Chen, [1988])

$$\lambda = \frac{2\pi\epsilon_o^2 m^2 u^4}{e^4 n \ln\Lambda} \quad (1)$$

where Λ is the plasma parameter and $\ln\Lambda$ is taken to be 10. From (1), it is clear that, above a certain speed, H$^+$ ions become collisionless. In the case of only centrifugal acceleration, the large O$^+$ flow speed ensures that collisional energy exchange with H$^+$ ions is effective even for the fast H$^+$ ions. Therefore, the majority of H$^+$ ions are held back by collisions with the O$^+$ ions and few are free to stream away carrying an upward heatflux. In the case with both frictional heating and centrifugal acceleration (although the O$^+$ flow velocities are about the same as the case when there is only centrifugal acceleration), the smaller O$^+$ densities give rise to larger collision mean free paths for the H$^+$ ions, so more H$^+$ ions are able to flow upward without hindrance. The result is a downward heatflux.

We can therefore conclude that the large upward and then downward parallel H$^+$ heatflux as seen in *Wilson* [1992] is only strictly true in the absence of convection. The H$^+$ heatflux is generally downward when there is centrifugal acceleration and frictional heating, and it is upward when there is only centrifugal acceleration.

3. DISCUSSION

In studying the effects of centrifugal acceleration on the polar wind, *Horwitz et al.* [1994] assumed an upgoing Maxwellian distribution for both H$^+$ and O$^+$ at about 4000 km. In that study, the densities and flow speeds for both H$^+$ and O$^+$ at the lower boundary, although based on observations, remain the same for different values of the convection electric field. The results shown in the last section clearly demonstrate that the distribution functions of H$^+$ and O$^+$ ions and their bulk parameters at collisionless altitudes greatly depend on the collisional and chemical processes, as well as the centrifugal acceleration, occurring in the collisional region. For instance, *Horwitz et al.* [1994] found that the H$^+$ escape fluxes at 4000 km under their particular chosen boundary conditions is a constant of 6×10^6 cm^{-2}s^{-1} irrespective of the convection strength. We found that at the same altitude, the H$^+$ flux is somewhat less when there is centrifugal acceleration than when there is none. However, when the effects of frictional heating and enhanced chemical reaction with neutrals are included, the H$^+$ flux at the same altitude decreases by an order of magnitude from 10^7 cm^{-2}s^{-1} to 10^6 cm^{-2}s^{-1} primarily due to the drastic drop in H$^+$ density in the F-region.

In order to obtain O$^+$ velocity profiles in the 2 - 5 R_E range which are comparable with observations from Dynamics Explorer 1 and the Akebono, *Horwitz et al.* [1994] and *Ho et al.* [1994] assumed a relatively high ion temperature (5000K) at 4000 km in their collisionless GSK model. We found that the O$^+$ ions are cooled dramatically due to adiabatic cooling under a strong centrifugal acceleration, and their parallel and perpendicular temperatures decrease rapidly with altitude to about 1000-2000 K at 4000 km. Although weaker convection will result in a less steep temperature profile, less frictional heating will also yield a lower O$^+$ temperature at 4000 km altitude. It is almost certain that in order to get the high ion temperatures used by *Horwitz et al.* and *Ho et al.*, additional ion heating in and above the F-region may occur, for example, by means of wave-particle interactions [*Barakat and Barghouthi*, 1994] hot electrons effect [*Barakat and Schunk*, 1994] and kinetic ion heating [*Chen et al.*, 1992]. High O^+ ion temperatures (upper limit of $> 10^4$K) were observed by *Abe et al.* [1994].

In this study, we assume an ionospheric convection field of 200 mV/m. This is rather large for normal geophysical conditions although it has been observed [*Anderson et al.*, 1991]. Nevertheless, our study has shown the important fact that centrifugal acceleration can strongly influence the shape of the velocity distribution of the H^+ ions, and hence their moments, by decreasing the collisional mean free path of the H^+ ions. This occurs because the O^+ velocities are increased greatly by centrifugal acceleration. We can therefore conclude that centrifugal acceleration affects not only directly the transport of ions in the collisionless region as *Horwitz et al.* [1994] have demonstrated, but indirectly via the H^+-O^+ collisional processes. The results are often difficult to anticipate due to the complicated dynamic co-dependency of the H^+ and O^+ ions. More studies of this sort will certainly be necessary in order to understand more fully the complex processes of ionospheric plasma transport.

Acknowledgments. This work was supported by NASA grant NAGW-1554 and NSF grant ATM-9301024 to the University of Alabama in Huntsville.

REFERENCES

Anderson, P. C., R. A. Heelis, and W. B. Hanson, The ionospheric signatures of rapid subauroral ion drifts, *J. Geophys. Res.*, *96*, 5785, 1991.

Barakat, A. R., and R. W. Schunk, Effect of hot electrons on the polar wind, *J. Geophys. Res.*, *89*, 9771, 1984.

Barakat, A. R., and I. A. Barghouthi, The effect of wave-particle interactions on the polar wind O^+, *Geophys. Res. Lett.*, *21*, 2279, 1994.

Chen F. F., Introduction to plasma physics and controlled fusion, Second Edition, Vol. 1, *Plenum Press, New York*, 1988.

Chen M. W., M. Ashour-Abdalla, and T. E. Holzer, Dynamical polar wind and its response to kinetic ion heating, *J. Geophys. Res.*, *97*, 19,433, 1992.

Cladis, J. B., Parallel acceleration and transport of ions from polar ionosphere to plasma sheet, *Geophys. Res. Lett.*, *13*, 893, 1986.

Delcourt, D. C., C. R. Chappell, T. E. Moore, and J. H. Waite, Jr., A three-dimensional numerical model of ionospheric plasma in the magnetosphere, *J. Geophys. Res.*, *94*, 11,893, 1989.

Ho, C. W., J. L. Horwitz, and T. E. Moore, DE-1 observations of polar O^+ stream bulk parameters and comparison with a model of the centrifugally-accelerated polar wind, *Geophys. Res. Lett.*, *21*, 2459, 1994.

Horwitz, J. L., Model of the geomagnetic spectrometer in the magnetotail lobes, in *Magnetotail Physics*, Edited by A.T.Y. Lui, p. 291, The Johns Hopkins University Press, Baltimore, MD, 1987.

Horwitz J. L., C. W. Ho, H. D. Scarbro, G. R. Wilson, and T. E. Moore, Centrifugal acceleration of the polar wind *J. Geophys. Res.*, *99*, 15,051, 1994.

Miller, R. H., C. E. Rasmussen, T. I. Gombosi, G. V. Khazanov, and D. Winske, Kinetic simulation of plasma flows in the inner magnetosphere, *J. Geophys. Res.*, *98*, 19,301, 1993.

Schunk, R. W., and J. C. G. Walker, Ion velocity distributions in the auroral ionosphere, *Planet. Space Sci.*, *20*, 2175, 1972.

Swift, D. W., Simulation of the ejection of plasma from the polar ionosphere, *J. Geophys. Res.*, *95*, 12,103, 1990.

Wilson, G. R., Semikinetic modeling of the outflow of ionospheric plasma through the topside collisional to collisionless transition region, *J. Geophys. Res.*, *97*, 10,511, 1992.

Wilson, G. R., Kinetic modeling of O^+ outflows resulting from $\mathbf{E} \times \mathbf{B}$ convection heating in the high-latitude F region ionosphere *J. Geophys. Res.*, *99*, 17,453, 1994.

C. W. Ho, J. L. Horwitz, G. R. Wilson, and D. G. Brown, Department of Physics and Center for Space Plasma and Aeronomic Research, University of Alabama in Huntsville, Huntsville, Alabama 35899.

Observations of Lower-Hybrid Spikelet Phenomena: Topaz3 Particle Data

Kristina A. Lynch and Roger L. Arnoldy

Institute for the Study of Earth Oceans and Space, University of New Hampshire, Durham

Paul M. Kintner

School of Electrical Engineering, Cornell University, Ithaca, New York

Recent auroral sounding rocket flights have provided observations of ionospheric particle acceleration with spatial resolution for two dimensional distribution functions on the order of 250 m. The Topaz3 mission measured bursty ion acceleration events which are consistent with nonlinear collapse theories for VLF hiss, wherein packets of intense, short-wavelength VLF waves are Landau damped by ions in the direction perpendicular to B. In this brief review of the Topaz3 ion data analysis effort to date, we present particle data on microphysical scales (proton and oxygen distribution functions for oxygen and protons, with 0.25 km, 0.25 sec resolution), and outline plans for relating this data to the macrophysics of ion acceleration and outflow.

1. INTRODUCTION

High resolution sounding rocket and small satellite missions have allowed study of ion outflow processes in increasingly fine detail. These new data give rise to an interesting macrophysical/microphysical coupling study: how do the macrophysics of ionospheric ion outflow develop from the microphysics of ion acceleration in the lower ionosphere? Microphysical processes can serve as catalysts for, and/or diagnostics of, larger scale processes: we are investigating this general relationship in particular for the Topaz3 auroral sounding rocket data.

The Topaz3 auroral sounding rocket mission measured one form of microphysical ion acceleration, bursty transverse acceleration associated with bursts of lower hybrid wave activity. The measurement resolution was 250 m for 2-D distribution functions. These high resolution measurements allowed a new look at the microphysics of ion acceleration, and showed, at least in this case, the transverse acceleration to occur in localized bursts. Previous studies of the Topaz3 data [*Lynch et al.*(1994), *Vago et al.*(1992), *Kintner et al.*(1992), *Arnoldy et al.*(1992)] have examined individual events of ion acceleration and wave enhancement in great detail. In this paper, we provide a review of this previous work, and outline the beginnings of a study of how (and if) these localized events contribute to the macrophysics of auroral ion outflow.

The question we are investigating is the following: does the Topaz3 data show any relationship between the occurrence rate of the ion acceleration events, and the bulk parameters of the ambient ions (or the precipitating electrons)? We wish to discover the significance of these events for ionospheric outflow, and also to see if their presence is an indicator of some particular macroscopic process. To this end, we have begun a more thorough analysis of the Topaz3 ion data set. We present here the first results of this effort, microphysical-scale ion data which has been corrected for various spacecraft effects. We plan to include this corrected ion data in a macrophysical study incorporating the statistics of the bursty ion and wave events, the background VLF data, bulk parameter particle data from the Topaz3 Stics instrument, and Freja results.

The structure of the paper is as follows. Section 2 is a description of the Topaz3 mission, with a review of the microphysical events that were measured and their interpretation based on case studies of individual events. Section 3 presents some corrected detailed ion data, consisting of three examples of the ion acceleration events: one involving mostly oxygen, one involving mostly protons, and one involving both, with 250 m spatial resolution. Section 4 contains some discussion of the available macroscopic data, and of the analysis we plan to do.

2. THE TOPAZ3 SOUNDING ROCKET

Topaz3 was a sounding rocket mission to investigate the topside auroral zone. It had a useful flight time of 900 seconds, and reached an altitude of 1100 km. Topaz3 observed bursty transverse oxygen heating (TAI) in conjunction with bursts of lower hybrid wave activity and transient density depletions. Figure 1 shows an example of the bursty wave-ion events. Note that the figure covers only 2 seconds (roughly 2 km across B). The observations are consistent with the theory that VLF wave power in the auroral arc reaches nonlinear levels and collapses in short wavelength packets which are absorbed by Landau damping involving ambient ions. [*Musher and Sturman*(1975), *Sotnikov et al.*(1978), *Retterer et al.*(1986), *Retterer et al.*(1993), *Chang and Coppi*(1981), *Chang*(1993), *Pottelette et al.*(1992)] Several hundred of these events were seen throughout the flight, and the occurrence rate of the events as a function of different environmental parameters (such as electron precipitation and ion temperature) is important in evaluating the relationships between scale sizes.

Figure 2 shows various surveys of the different data as functions of flight time for the entire mission (200-1100 seconds), as an overview of the different relationships to be investigated. Two parameters which drive much of the data are altitude and electron precipitation. The occurrence rate of the wave bursts, in particular, is a strong function of B^2/n, an altitude-dependent function. This ratio is proportional to the ratio ω_{pe}^2/Ω_e^2, an important factor in the theoretical thresholds for lower hybrid wave nonlinear collapse. [*Sotnikov et al.*(1978)] Also, there appears to be a low-altitude cutoff for the ion events at about 800 km, but this may well be an instrument geometry factor limitation.

The electron precipitation is another important environmental factor, because of the VLF environment which it generates. Of all the available data, the occurrence rate of the ion events is most closely linked to the shape of the electron distribution function—itself a measure of the power of the VLF hiss. The ion events occur when the main electron precipitation is dominantly field aligned (indicating an active region in the arc), but does not contain a (temporally) bursty field aligned population. The spectrogram in the top panel of Figure 2 shows the electron precipitation. Topaz3 was within one of a series of multiple auroral arcs for most of its flight time. The lower energy, field-aligned, bursty population that can be seen in the spectrogram (for example, at time T+450-550) seems to be related to the ion events. On a 10-s or longer timescale, the two populations are correlated; on smaller timescales, they are anti-correlated. [*Lynch et al.*(1993)]

The existing analysis of the Topaz3 ion data has been based upon case studies of a few ion/wave events, and qualitative relationships between the occurrence of the events and some of the background environmental parameters. Before we can relate the microphysical events to the macrophysical environment in any quantifiable fashion, we need to generalize the ion event observations by examining events throughout the entire flight in a statistical fashion. Then we can use the observed VLF power levels to link the electron precipitation to the ion heating events. Finally, we can use the Freja results to put the Topaz3 results into perspective: the Topaz3 environment was fairly steady throughout the entire mission, and the Freja results can tell us how ion acceleration in that type of environment (inside an inverted-V type nightside auroral arc) fits into the ionospheric ion outflow overall.

3. MICROSCALE PHYSICS: THREE EXAMPLES OF WAVE-ION EVENTS

Before generating statistics of occurrence rates of ion events, it was necessary to clean the ion data of various spacecraft effects. Here are three different exam-

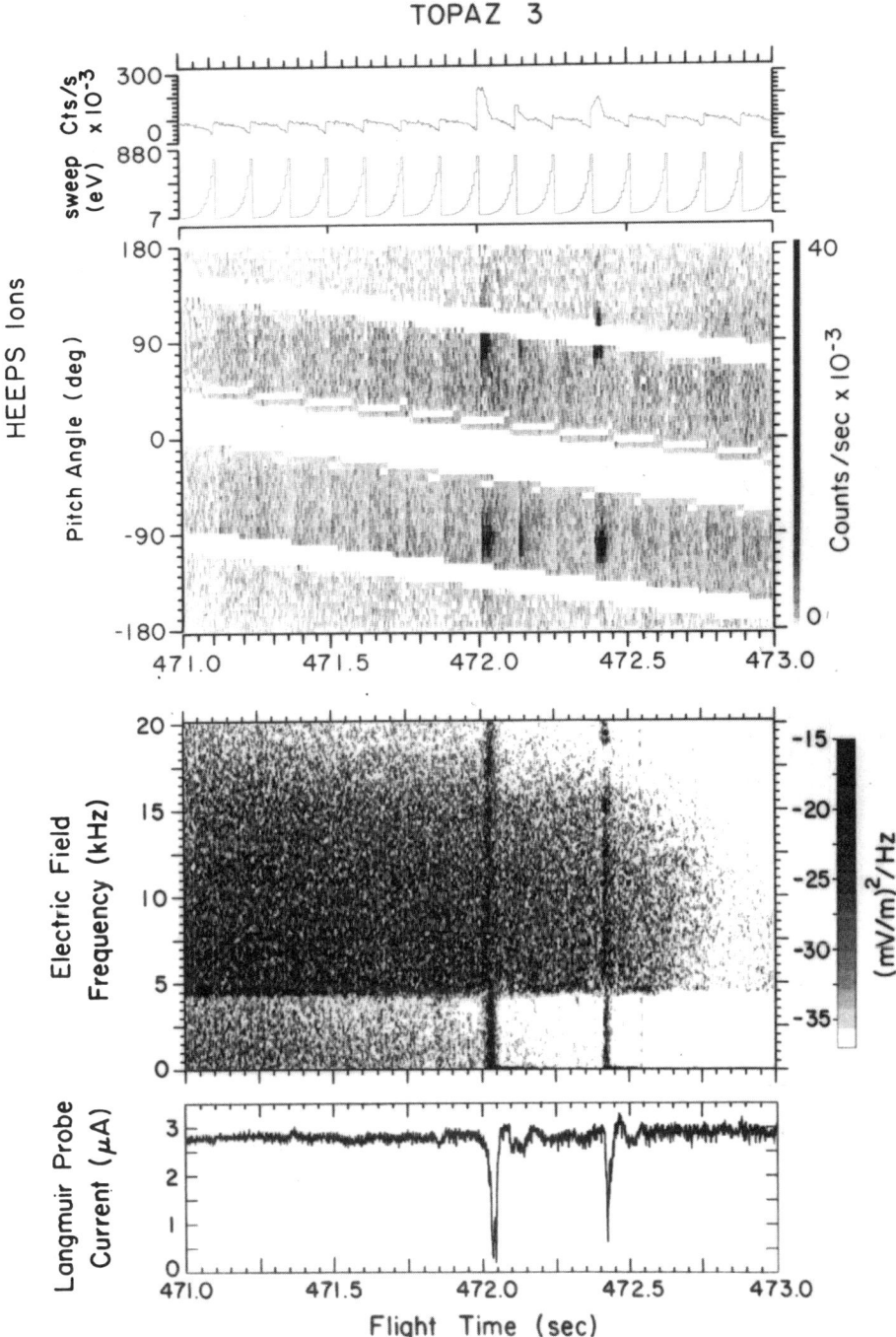

Figure 1. An example of ion/wave events on Topaz3. The top two panels show raw count rate summed over all pitch angles, and energy sweep, from the high-energy ion detector. The next panel shows the pitch angle image from this detector, indicating that the ion bursts at T+472.05 and T+472.4 have pitch angles near 90°. The white diagonal stripes through the image are instrument blind spots. The next panel is a VLF wave intensity spectrogram, with the grey scale indicating $10 \log ((mV/m)^2/Hz)$. This plot indicates that the ion events were accompanied by broadband bursts of wave activity. The lowest panel shows data from the Langmuir probe and indicates that the events were also accompanied by a strong decrease in density.

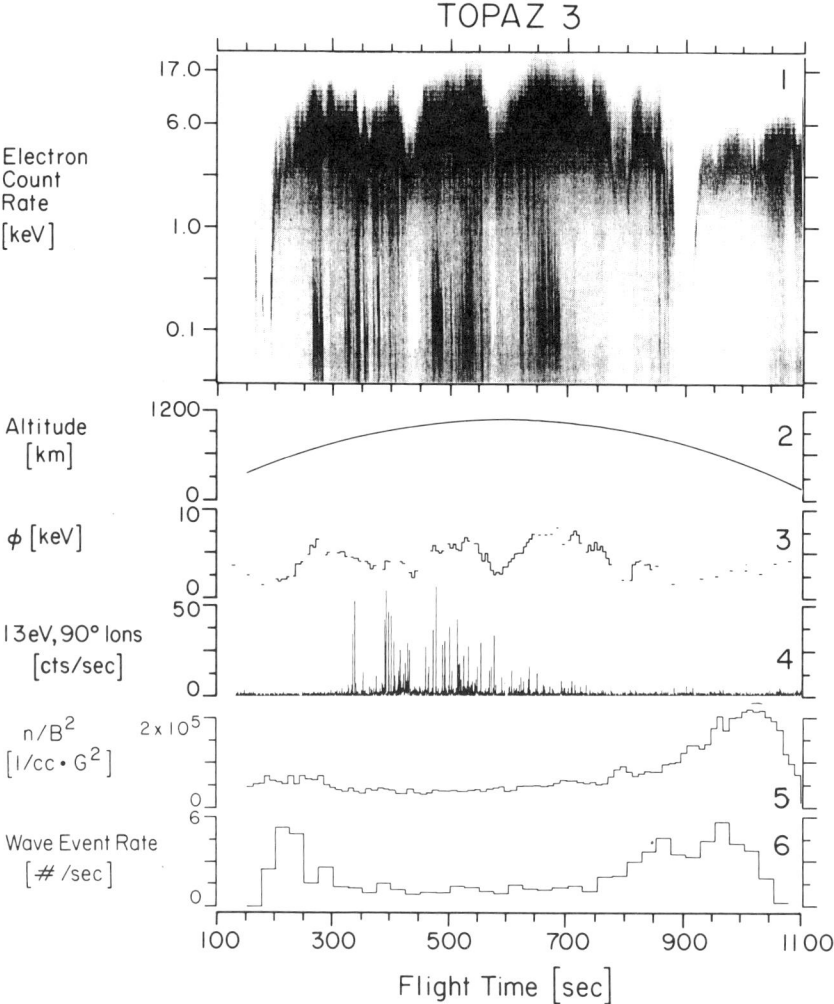

Figure 2. A summary plot of various data quantities as a function of time throughout the entire flight. (See text for discussion.) From top to bottom, the different panels indicate 1, a spectrogram of electron total count rate, summed over all pitch angles versus energy and time for the entire flight; 2, rocket altitude; 3, electron peak energy; 4, ion acceleration events at 13 eV and 90 degrees pitch angle; 5, n/B^2, which is the altitude dependence of the wave events; and 6, the observed rate of wave events.

ples of a wave-ion event (Figures 3,4, 5, and 6), after this processing. The event at T+577 is at an altitude of 1067 km; at T+477, 1018 km; at T+392, 922 km.

Figure 3 shows the different events as a function of time in a 10 sec window. The time of interest is marked with an arrow. Each panel contains data from the wave instrument and from the 2 kHz sampling, 13 eV, 90 degree pitch angle ion instrument. The line shows the count rate of the fast ion instrument, showing the timescale of the bursty ion events. The diamonds show the times of the wave bursts. There are many more wave events than ion events, but most of the large ion bursts are associated with a wave event. Note that the sampling of the wave instrument is much better than that of the ions: the ion detectors are often at the wrong energy or pitch angle at the critical times.

Figures 4, 5, and 6 show details of the distribution functions within and near the events, with data that have been corrected for spacecraft effects. Data are used only when significant counts remain after correcting for ram and spacecraft potential, ignoring any look directions which look into the wake, removing any energy steps with counts below the one count level, or above the saturation level, and finally, only using distribution functions which have some look directions within +/- 5 degrees of 90 degrees. As a result of all these restrictions, it becomes difficult to find events which were observed by all four ion channels, but those which

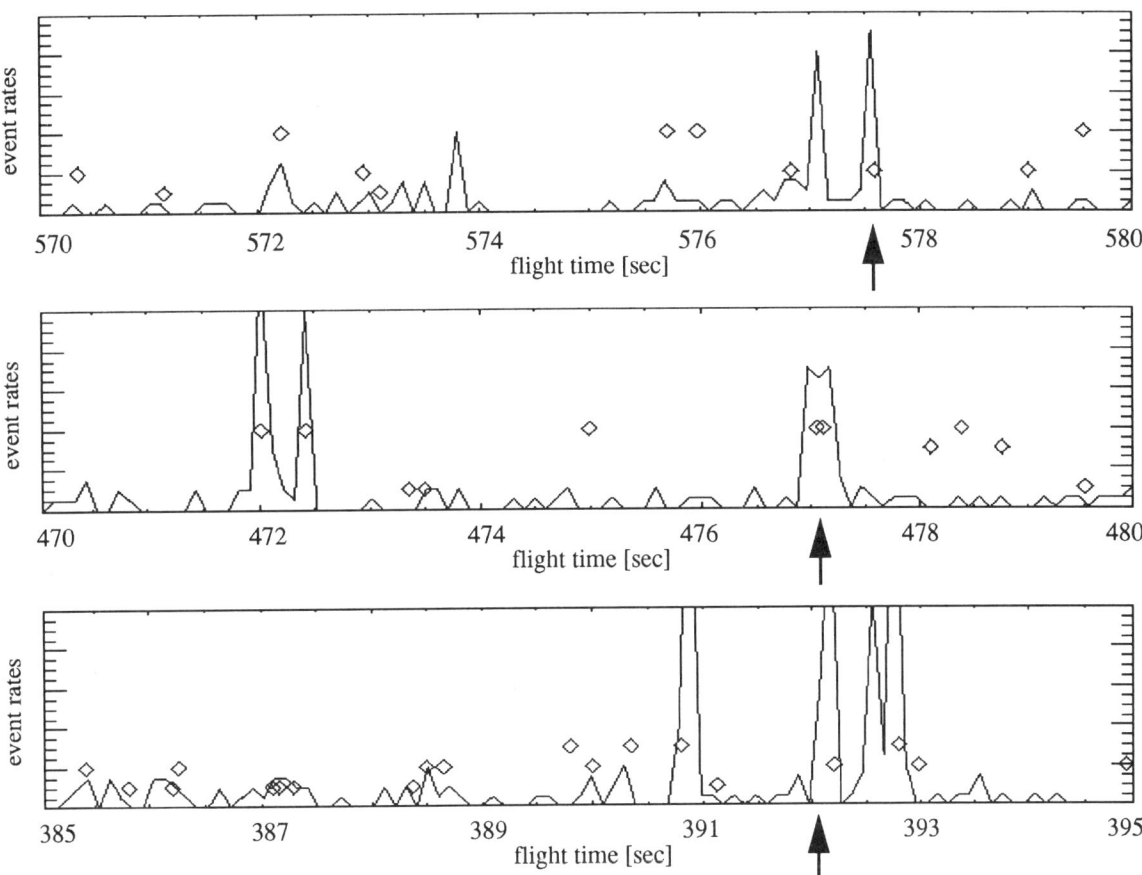

Figure 3. 10 second time history of the three separate events shown in the following figures. The diamonds indicate the times of the wave events, and the line indicates the relative count rate of the fast-sampling ion detector. In each panel, the arrow indicates the time of the events shown in the following figures.

remain are relatively uncontaminated by spacecraft effects. These figures show data from the high-geometry factor top-hat, and from the oxygen and proton channels of the mass-resolving top-hat. The high-geometry factor ion detector has better azimuthal coverage, but the mass-analyzing dectector extends to lower energies. In each of the three examples shown, the three top-hat channels and the 2 kHz ion channel all had valid data at the time of a wave burst.

Each event is shown, for each detector, both as a contour plot of the distribution function within and next to the event, and as a lineplot which is a cut through the two shown distribution functions at 90 degrees. The distribution function for the high-geometry factor top-hat is calculated assuming a dominant mass of oxygen (the oxygen:hydrogen ratio for this flight is nominally 10:1). The distribution functions f [sec^3/km^{-6}] are shown as functions of velocity perpendicular and (anti-)parallel to B. The lineplots showing log(f) vs. E are in units of [sec^3/km^{-6}] and eV; they can be used to evaluate the contour levels of the neighboring plots.

The event at T+577 (Figure 4) is the classic example of the entire flight. It has acceleration of both oxygen ions and protons, at 90 degrees pitch angle, in close conjunction with a strong, localized burst of wave activity just above the lower hybrid frequency. The oxygen acceleration is tail-like, and narrowly confined to 90 degrees pitch angle. The proton acceleration looks more like bulk heating; it is centered around 90 degrees, but is much wider in pitch angle. Note that the velocities of the oxygen tail range from 5 to 20 km/sec, and that the proton heating is over a similar velocity (not energy) range. The plots of f(E) show that the peak energy of the oxygen event is in the 5-20 eV range, while that

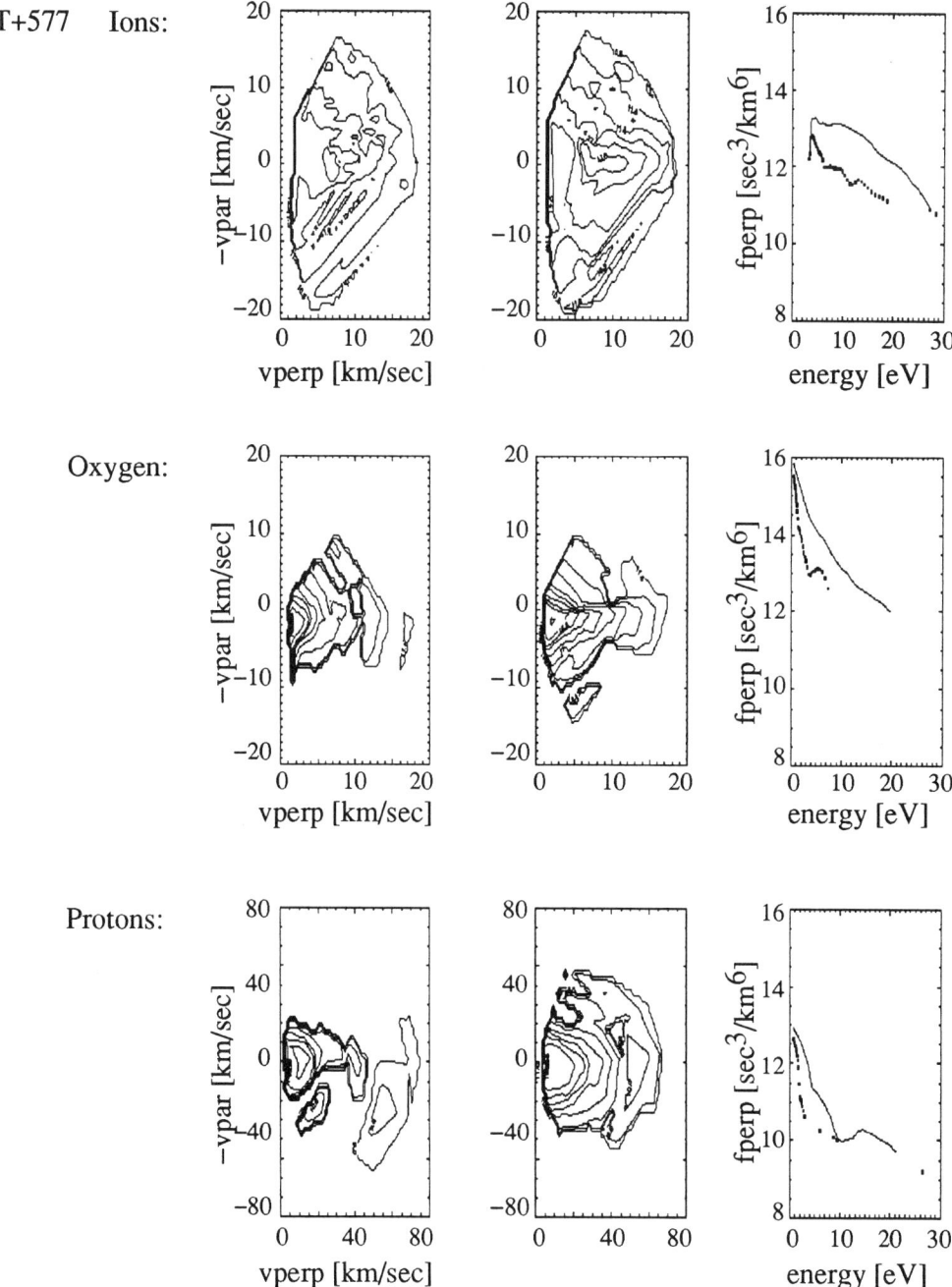

Figure 4. Particle and wave data for the ion event at T+577, at an altitude of 1067 km. Each row contains, from the high-geometry factor, oxygen, or proton channel, a contour plot of the distribution function next to (left) and within (right) the event, and a lineplot which is a cut through the two shown distribution functions at 90 degrees pitch angle. The distribution functions $f\,[\sec^3/\mathrm{km}^{-6}]$ are shown as functions of velocity [km/sec] perpendicular and (anti-)parallel to B. The lineplots showing log(f) vs. E are in units of $[\sec^3/\mathrm{km}^{-6}]$ and eV. The contours are unlabelled because of their size; the values of the contours can be read from the lineplot.

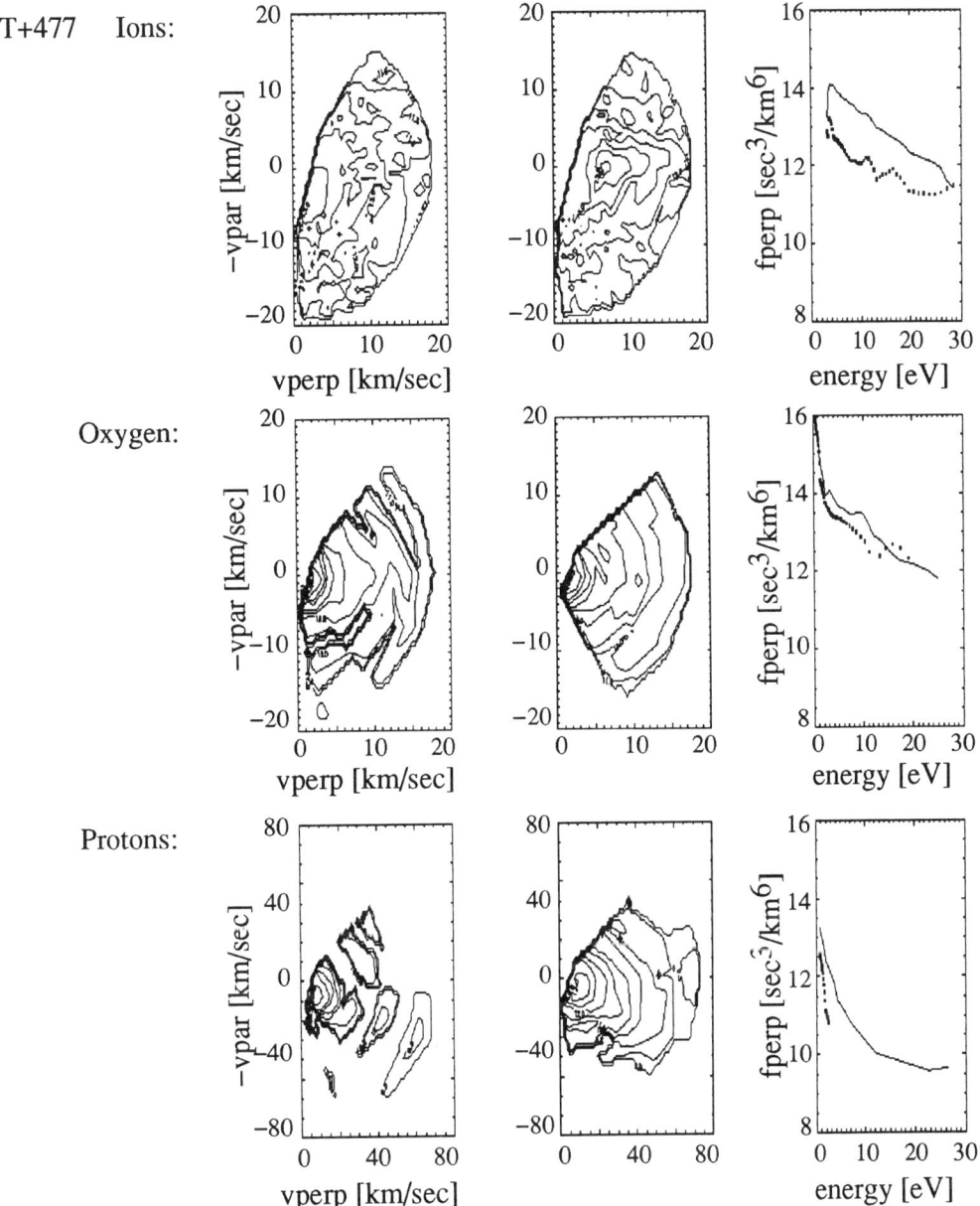

Figure 5. Particle and wave data for the ion event at T+477, at an altitude of 1018 km. The format is the same as in Figure 4.

of the protons is more like 1-5 eV. These features are consistent with a velocity-resonant effect.

The event at T+477 demonstrates that some of the events, with higher characteristic velocities, are seen more in the proton channel than the oxygen channel, despite the fact that both channels are looking in the right direction at the right time. The observed hydrogen tail starts at about 20 km/sec. Since the oxygen signature is seen by the (faster sampling) ion detector (0.125 s sooner), the oxygen acceleration region must be smaller than the corresponding proton region.

The event at T+392 demonstrates that some of the events, with lower characteristic velocities, are seen in the oxygen channel but not the proton channel, again

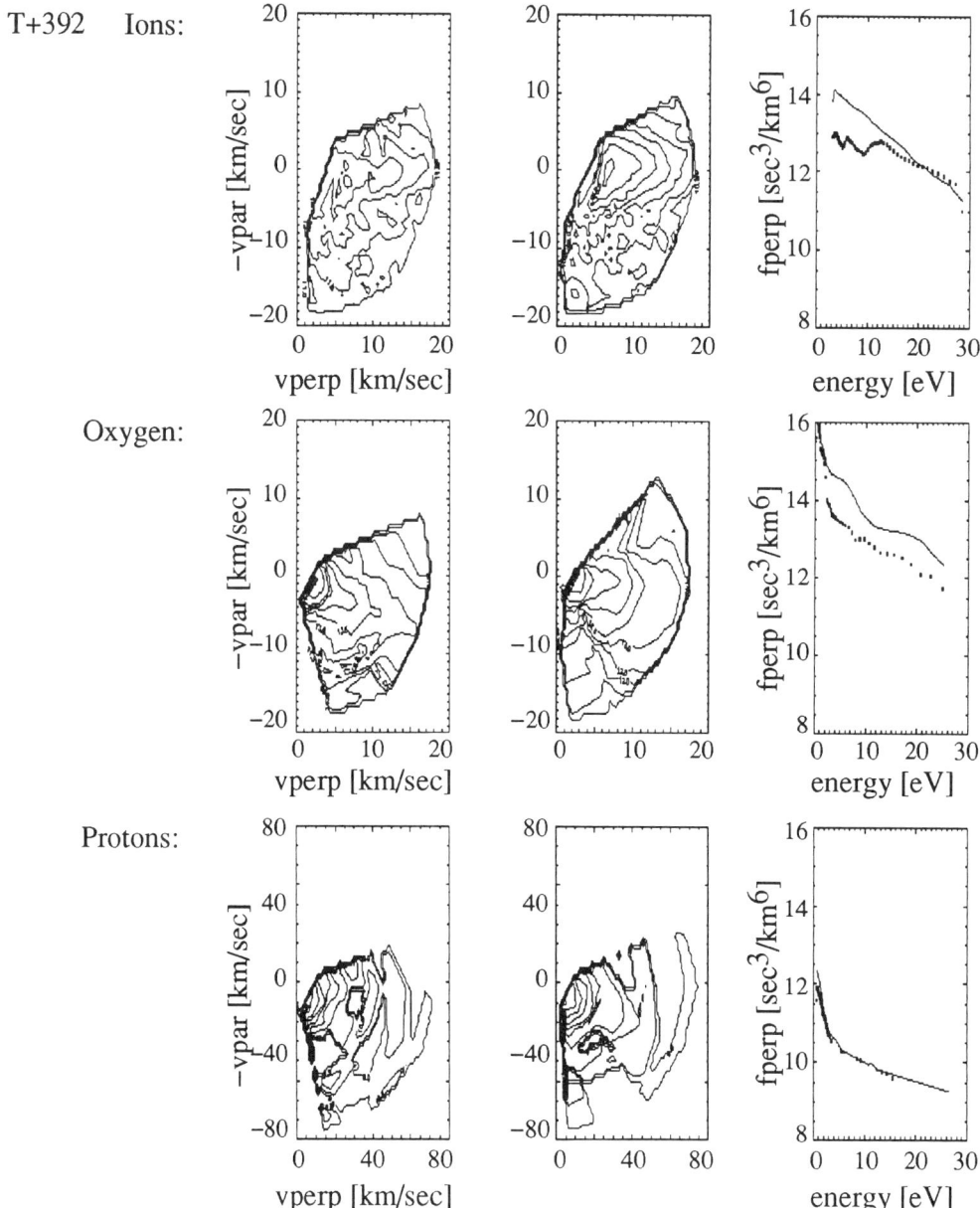

Figure 6. Particle and wave data for the ion event at T+392, at an altitude of 922 km. The format is the same as in Figure 4.

despite correct look directions. The interpretation here is that the dominant wavenumber within the lower hybrid wave packet is resonant with the dominant oxygen species. There may be a proton response as well, but the low phase velocity combined with the small proton mass makes the energy of the response below the resolution of this instrument. We hope in a future study to compare the measured dominant lower hybrid wavelengths in each event to the dominant velocity of the particle acceleration.

4. DISCUSSION AND ONGOING WORK: IS THERE A RELATIONSHIP BETWEEN THE MICROPHYSICAL AND THE MACROPHYSICAL SCALES?

We have restricted the ion data as described above. The data remaining is relatively unbiased by spacecraft effects; there are 46 oxygen events, 50 proton events, and 60 high-geometry-factor ion events remaining. We hope to use these events together with larger-scale data sets to understand the relationship to the macrophysics of ion outflow. Some of the questions we plan to pursue with this data set are as follows.

What is the altitude dependence of the ion events? The wave event occurrences are a strong function of B^2/n, or, ω_{pe}^2/Ω_e^2, indicating agreement with the theoretical thresholds for lower hybrid wave collapse. [*Sotnikov et al.*(1978)] The ion heating events appear to have a low altitude cutoff, which may be instrumental, and they become fewer and stronger with increasing altitude. Is this altitude profile a cause or an effect of decreasing density or of changes in other ambient ion parameters such as temperature? Preliminary results seem to indicate that the bursty events are isolated from the ambient background; they do not seem to heat the ambient population, at least over the limited altitude range of the Topaz3 mission. We intend to pursue this more thoroughly using the 3-D thermal ion data from the Topaz3 STICS instrument.

Are the observed ion events indicative of particular macrophysical processes, such as convective amplification of VLF leading to nonlinear levels of lower hybrid waves, or of strong shear? Typical parameters of the events, such as the velocity range of the accelerated particles, may restrict the available energy sources to particular processes. A comparison of dominant wavelengths within the wave events to dominant particle velocities in the ion events should help identify if the nonlinear wave collapse/Landau damping interpretation is correct. Preliminary analysis of a few case studies supports this interpretation. A comparison of the electron precipitation, the observed VLF power levels, and the ion heating events will quantify the relationship between the input energy source (the precipitating electrons), the intermediate link (the VLF) and the resulting ion heating, apparently through the collapse and damping of nonlinear wave packet structures.

Can the observed microphysical ion events, in some cumulative fashion, account for the observed bulk flow? Are the observed ion events a significant source of energy or number flux outflow from the ionosphere? How do the Topaz3 observations mesh with the Freja satellite observations of ion heating? Freja results do not indicate that these lower hybrid wave events occur within regions of strong ion heating. Instead, they are usually seen in regions without strong ion heating, perhaps together with small, isolated bursts of accelerated ions. [personal communication, D. Knudsen, 1994] Perhaps the Topaz3-type ion heating only occurs "when all else fails", that is, when no large-scale ion heating process is possible. Note that in the presence of significantly heated ion tails, the nonlinear evolution of the VLF wave activity would not proceed to the short wavelengths observed on Topaz3; the accelerated ion populations would Landau damp the waves at longer wavelengths. In a Topaz3-like environment, without a significant heated ion population, one would expect to see the most striking ion-wave events. The Topaz3-like events in this interpretation are not the dominant source of ion outflow for the ionosphere; but they are what happens inside inverted-V regions or in any region with strong VLF and a cold ion population.

REFERENCES

Arnoldy, R. L., K. A. Lynch, P. M. Kintner, J. Vago, S. Chesney, T. E. Moore, and C. J. Pollock, Bursts of transverse ion acceleration at rocket altitudes, *Geophys. Res. Lett.*, *19*, 413, 1992.

Chang, T., Lower hybrid collapse, caviton turbulence, and charged particle energization in the topside auroral ionosphere and magnetosphere, *Phys. Fluids B*, *5(7)*, 2646, 1993.

Chang, T., and B. Coppi, Lower hybrid acceleration and ion evolution in the suprauroral region, *Geophys. Res. Lett.*, *8*, 1253, 1981.

Kintner, P. M., J. Vago, S. Chesney, R. L. Arnoldy, K. A. Lynch, T. E. Moore, and C. J. Pollock, Localized lower hybrid acceleration of ionospheric plasma, *Phys. Rev. Lett.*, *68*, 2448, 1992.

Lynch, K. A., R. L. Arnoldy, P. M. Kintner, and J. L. Vago, Fine structure of auroral particle precipitation, *Physics of Space Plasmas (1993), SPI Conference Proceedings and Reprint Series*, 1993.

Lynch, K. A., R. L. Arnoldy, P. M. Kintner, and J. L. Vago, Electron distribution function behavior during localized transverse ion acceleration events in the topside auroral zone, *J. Geophys. Res.*, *99*, 2227, 1994.

Musher, S. L., and B. I. Sturman, On the collapse of plasma waves near the lower-hybrid resonance, *Sov. Phys. JETP Lett.*, Engl. Transl., *22*, 265, 1975.

Pottelette, R., R. A. Treumann, and N. Dubouloz, Generation of auroral kilometric radiation in upper hybrid wave-lower hybrid soliton interaction, *J. Geophys. Res.*, *97*, 12,029, 1992.

Retterer, J. M., T. Chang, and J. R. Jasperse, Ion acceleration by lower hybrid waves in the sprauroral region, *J. Geophys. Res.*, *91*, 1609, 1986.

Retterer, J. M., T. Chang, and J. R. Jasperse, Lower hybrid collapse and charged particle acceleration, in *Research Trends in Nonlinear Space Plasma Physics*, H. Alfvén, R. Bingham, R. Z. Sagdeev, and K. Quest, Eds., page 252, American Institute of Physics, 1993.

Sotnikov, V. I., V. D. Shapiro, and œ V. I. Shevchenko, Macroscopic consequences of collapse at the lower hybrid resonance, *Sov. J. Plasma Phys.*, *4*, 252, 1978.

Vago, J., P. M. Kintner, S. W. Chesney, R. L. Arnoldy, K. A. Lynch, T. E. Moore, and C. J. Pollock, Transverse ion acceleration by localized hybrid waves in the topside auroral ionosphere, *J. Geophys. Res.*, *97*, 16,935, 1992.

R. L. Arnoldy and K. A. Lynch, Institute for the Study of Earth, Oceans, and Space, University of New Hampshire, Durham, N. H. 03824. (e-mail: arnoldy@unhesp.unh.edu, lynch@unhesp.unh.edu)

P. M. Kintner, School of Electrical Engineering, Cornell University, Ithaca, N. Y. 14853. (e-mail: paul@magneto.ee.cornell.edu)

Effects of Auroral Electron Precipitation on Topside Ion Outflows

Phil G. Richards

Computer Science Department and Center for Space Plasma and Aeronomic Research, University of Alabama, Huntsville

This paper presents model calculations of the response of the ionosphere to soft auroral electron precipitation. An incident Maxwellian electron distribution with a characteristic energy of 300 eV can increase the F region electron density to 5×10^{-5} cm^{-3} and increase the electron temperature by 1500°K. These changes in ionospheric properties produce large upward ion flows in the major ion species followed by a large downward flows of O$^+$ and He$^+$ when the precipitation ends. In contrast, the H$^+$ flow remains large and upward even after the precipitation ceases.

1. INTRODUCTION

This paper presents model calculations of the response of the ionosphere to soft auroral electron precipitation. An incident Maxwellian electron distribution with a characteristic energy of 300 eV can increase the F region electron density to 5×10^5 cm^{-3} and increase the electron temperature by 1500°K. These changes in ionospheric properties produce large upward ion flows in the major ion species followed by a large downward flows of O$^+$ and He$^+$ when the precipitation ends. In contrast, the H$^+$ flow remains large and upward even after the precipitation ceases There are a large number of auroral precipitation models [e.g. *Strickland et al*., 1983; *Lummerzheim et al*., 1989; *Solomon et al*., 1988] and a similar number of ionospheric models [e.g. *Sojka and Schunk*, 1983 ; *Roble and Rees*, 1977; *Fuller-Rowell et al*., 1987; *Roble et al*., 1988; *Min and Watkins*, 1995]. Model studies can be local or global in nature depending on the parameters of interest. Most model studies of the aurora have concentrated on the emissions, which are the most obvious manifestation of the energetic particle precipitation.

Roble and Rees [1977] combined an auroral model with an ionospheric model to study the effect on the ionosphere of soft, medium, and hard electron precipitation. They found that the response of densities, temperatures and emissions depends on the spectral energy distribution and total energy flux of the auroral precipitation. They also found that soft aurorae affect mainly the F region while hard aurorae have little effect on the F region but greatly enhance the densities and temperatures in the $F1$ and E regions. Since the present paper is concerned with ionospheric ion outflows only soft precipitation is considered.

More recent studies, which included auroral data, have been conducted by *LaBelle et al*. [1989] *and Lilensten et al*. [1990]. In the former paper, rocket measurements of electron spectra and electron density showed high correlation between density spikes and electron fluxes. Model studies of the ionization rate confirmed that the electron fluxes could be the cause of the density enhancements. A set of measurements suitable for modeling of electron precipitation and ionospheric plasma has been published by *Lilensten et al*. [1990] who used it to examine the energy balance of the ionosphere during an aurora. The ionospheric properties were measured by the European incoherent scatter radar (EISCAT) and the coincident electron flux measurements were made by the Swedish spacecraft, Viking. *Stamnes et al*. [1986] studied the effects of electron precipitation on the daytime ionospheric using fluxes from the AUREOL-3 satellite and ionospheric densities and temperatures from the EISCAT radar. They found good agreement between the model densities and temperatures obtained from the observed electron flux and

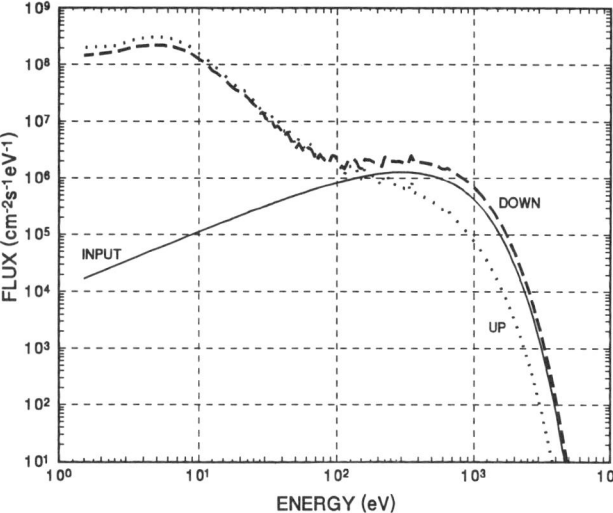

Fig. 1. Input electron flux at upper boundary (800 km) and calculated auroral electron fluxes at 300 km. The input flux is a Maxwellian with a characteristic energy of 300 eV and an energy flux of 2 erg cm^{-2} sec^{-1}.

the densities and temperatures measured by the radar.

There have been numerous model studies of the global effects of auroral activity most recently by *Parish et al.* [1994]. They studied the global impact of a magnetic storm on the emissions from atomic oxygen. A global simulation of the ionospheric response to a magnetic storm has been carried out by *Sojka et al.* [1994]. Their study showed that increased convection speed leads to an increase in ion temperatures, enhanced NO^+ densities, and decreases in both $NmF2$ and $hmF2$.

The present paper deals with the local effects of particle precipitation on the outflow of thermal ions from the topside ionosphere. Processes related to plasma convection are ignored.

2. MODELS

The field line interhemispheric plasma (FLIP) model has been developed over a period of more than ten years, and has been described previously by *Richards and Torr* [1988], by *Torr et al.* [1990], and by *Richards et al.* [1994a,b]. The main component of this one-dimensional model calculates the plasma densities and temperatures along entire magnetic flux tubes from 80 km in the northern hemisphere through the plasmasphere to 80 km in the southern hemisphere. The model uses a tilted dipole approximation to the Earth's magnetic field. The ion continuity and momentum equations for O^+, H^+, and He^+ and N^+ equations are solved to find their densities and velocities.

The electron and ion temperatures are obtained by solving the energy equations [*Schunk and Nagy*, 1978]. Electron heating due to photoelectrons is provided by a solution of the two-stream photoelectron flux equations using the method of *Nagy and Banks* [1970]. The solutions have been extended to encompass the entire field line on the same spatial grid as the ion continuity and momentum equations. The model photoelectron fluxes are in good agreement with the measured fluxes of *Lee et al.* [1980].

The three key inputs to the FLIP model are the neutral densities, the solar EUV fluxes, and the neutral winds. The mass spectrometer and incoherent scatter (MSIS-86) model [*Hedin*, 1987] was used for the neutral atmosphere, while the solar EUV fluxes are from the EUVAC model [*Richards et al.*, 1994a]. Both the EUVAC and the MSIS-86 models use the daily $F_{10.7}$ index as a measure of short term solar activity. The EUVAC and MSIS-86 models also have a dependence on long term solar energy deposition, which is monitored through the 81 day average $F_{10.7}$ index ($F_{10.7A}$), which changes little over a short period.

The neutral winds are particularly important for detailed comparisons between model and data because they affect the height of the F region peak density ($hmF2$) and consequently the magnitude of the peak density ($NmF2$) itself. The neutral winds for this study were selected to maintain $hmF2$ near 350 km at night in the absence of precipitation. This is a typical altitude for the F region peak density at night.

The auroral electron precipitation model, which is a two stream model similar to that used for the photoelectron flux, has been described by *Richards and Torr* [1990]. The model uses variable spatial and energy grids to extend the range of energies up to tens of keV. *Richards and Torr* [1990] showed that this model gives emission rates that agree well with the more sophisticated model of *Strickland et al.* [1983].

3. CALCULATIONS

The precipitating downward electron flux was assumed to be a Maxwellian with a characteristic energy of 300 eV. This flux was inserted as the upper boundary flux to our auroral electron model which then computed upward and downward electron fluxes. The energy deposition rate was 2 erg cm^{-2} sec^{-1}. The calculated fluxes were folded with the appropriate cross sections to obtain ion production rates and electron heating rates for input to the FLIP model. A soft precipitating flux was chosen in order to deposit significant amounts of energy in the F region. The soft precipitation may be accompanied by hard precipitation but this will have little effect on the F region because electrons with characteristic energies greater than about 1 keV deposit most of their energy below 200 km altitude. Soft electron fluxes normally occur on the poleward edge of the auroral oval on the night side and on the dayside near local noon [*Roble and Rees*, 1977]. Energy fluxes of the order of 1-5 erg cm^{-2} sec^{-1} are typical of hard auroral precipitation. In this paper a soft precipitation was chosen as a numerical experiment to illustrate the possible effects on the polar ion flows.

The calculated upward and downward electron fluxes at 300 km altitude are shown in Figure 1 along with the input downward flux (solid line). Above 100 eV, the calculated downward flux is similar to the input flux but has a slightly larger magnitude as a result of cascade. In this energy region the calculated upward flux is smaller than the calculated downward flux as well as the input flux. However, below 100 eV, cascade dominates and both the upward and downward fluxes become 3 orders of magnitude larger than the input flux. In the low energy region, the upward flux becomes slightly larger than the downward flux.

The FLIP model was run for 24 hours beginning about eight hours prior to the precipitation event which began at 2000 UT and ended at 2100 UT. The geographic location of the flux tube coincides with the EISCAT radar. The thermospheric conditions are appropriate for low solar ac-

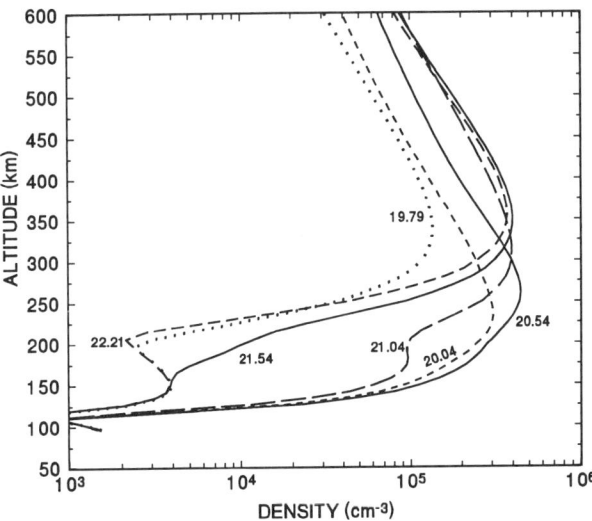

Fig. 3. The electron density profiles at different times before during and after the electron precipitation. The times indicated on the curves are fractional hours (20.50 ≡ 2030 UT). Model electron precipitation occurred between 2000 and 2100 UT.

tivity ($F_{10.7}$=110) and moderate magnetic activity (A_p=18). Figure 2 shows electron density contours for the entire simulation period. The precipitation event occurred several hours after local sunset and several hours prior to local sunrise at 3000 UT. The timing and the one hour duration of the event were chosen in order to illustrate the ionospheric effects both during and after the precipitation free of the additional complication of sunrise and sunset effects. This figure shows that the electron density is severely affected all the way down into the E region. The precipitation produces a long lived enhancement in the F region electron density.

Figure 3 shows FLIP model electron density altitude profiles at several times before, during, and after the precipitation event. The electron precipitation lowers $hmF2$ from 350 km to 250 km while increasing $NmF2$ from 1.5×10^5 to 5.0×10^5 cm^{-3}. Below 200 km, the low background electron density is maintained by production from starlight and by sunlight resonantly scattered from the geocorona. When the electron precipitation occurs, the electron density rapidly increases below 200 km and saturates after a few minutes as indicated by the curve labeled 20.04 (times on the graphs are in fractional hours). Thirty minutes after the precipitation has been turned off (21.54), the low altitude density has not yet decayed back to the background values. However, an hour after the cessation of precipitation, the low altitude density is back to normal. In the F region, the electron density increases for the first half hour of the precipitation but then stabilizes. After the

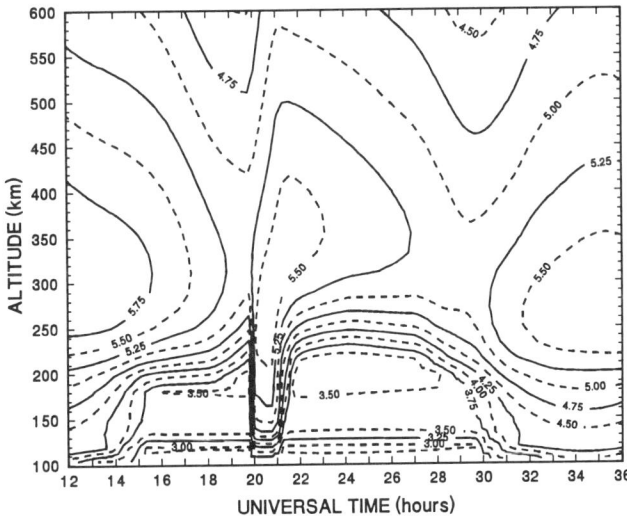

Fig. 2. Model contours of log_{10} electron density for a complete day starting from 1200 UT. The model electron precipitation occurred between 2000 and 2100 UT.

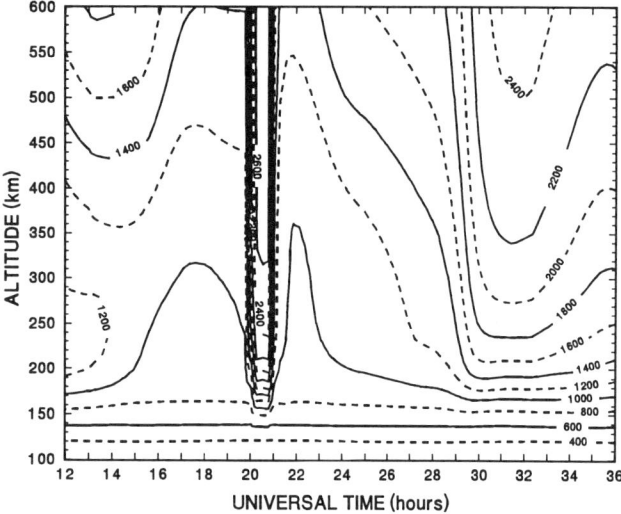

Fig. 4. Model contours of electron temperature for a complete day starting from 1200 UT. The model electron precipitation occurred between 2000 and 2100 UT.

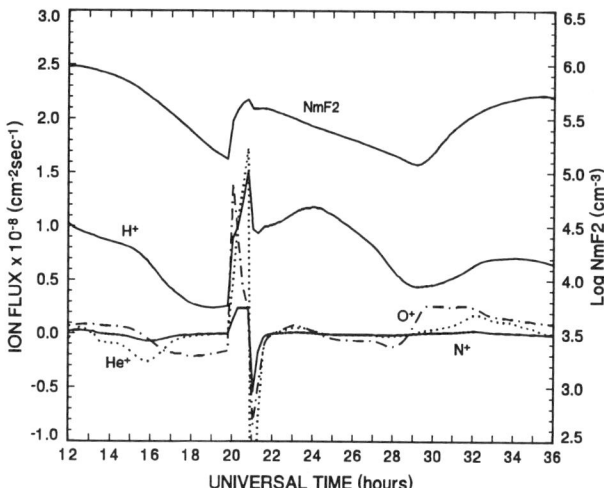

Fig. 6. Variation of the topside ion fluxes at 1500 km and the $F2$ peak electron density ($NmF2$) for a complete day starting from 1200 UT. Upward fluxes are positive. The model electron precipitation occurred between 2000 and 2100 UT.

precipitation has been turned off ionization flows back from the topside ionosphere and increases the electron density above 300 km. At the same time, the peak height moves back to its equilibrium value at 350 km as the bottom side of the F region decays away.

Figure 4 shows contours of the electron temperature. T_e decays after local sunset but increases dramatically during the precipitation event. Following the cessation of the precipitation, the temperature rapidly decays back to its

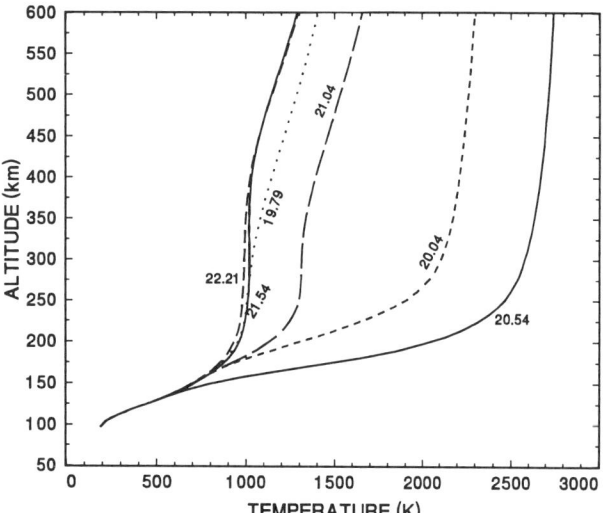

Fig. 5. Electron temperature profiles at different times before during and after the electron precipitation. Model electron precipitation occurred between 2000 and 2100 UT.

prior low values. An increase in heating occurs as a result of conjugate sunrise near 2300 UT. Of course, this heating effect from conjugate photoelectrons would only occur if the magnetic field lines were closed at this time. In fact, a rapid heating coinciding with conjugate sunrise might be used as a signature that the flux tubes are in fact closed. If the field lines were open, the conjugate heating source would be absent but other more variable heat sources may operate to complicate the temperature variation. There is a further dramatic increase in electron temperature at local sunrise (3000 UT). Altitude profiles of the electron temperature are shown in Figure 5. There is a rapid increase in T_e during the precipitation event with the F region electron temperature reaching 2700°K after half an hour of precipitation. One interesting aspect of these profiles is the very small temperature gradient above 300 km. The small temperature gradient results from the ionospheric heat source being comparable to the magnetospheric heat source.

Figure 6 shows the time variation of the F region peak density $NmF2$ and the ion fluxes through the 1500 km altitude level. The electron precipitation causes an immediate, large increase in the upward H^+, O^+, N^+ and He^+ fluxes. This large initial upward flow precedes the density build up and is a response to the sudden increase in temperature. The subsequent increase in H^+ and He^+ upward flows is generated by a large increase in ion production near the $F2$ peak. The O^+ flux peaks very quickly and then begins to decay even while the precipitation is in progress, whereas, the H^+ and He^+ fluxes continue to rise throughout the precipitation period and only decay when it ends. The very large downward O^+ and He^+ fluxes after precipitation ceases are caused by the decrease in T_e and in ion production rates. The H^+ flux shows a much smaller decrease because it is produced from O^+ whose density remains high. In contrast, when the precipitation ends, the sources of O^+, He^+, and N^+ essentially disappear.

Figure 7 shows the time variation of the height of the F region peak density ($hmF2$) and the electron temperature at $hmF2$. This figure clearly shows the large decrease in $hmF2$ and the large increase in T_e during the precipitation event. Both $hmF2$ and T_e respond rapidly to the initial precipitation and rapidly return to their ambient values once the precipitation ceases. Note that T_e stabilizes quickly after the initial response to the precipitation. This explains the initial rapid increase and later decrease in the O^+ flux shown in Figure 6. The fact that T_e rapidly stabilizes shows that the subsequent increases in H^+ and He^+ fluxes result from the increase in ion density.

The model ion densities at 20.54 UT are shown in Figure 8. O^+ is the dominant ion in the $F2$ region while NO^+ is the dominant ion in the $F1$ region a factor of 2 larger than the O_2^+ density. N^+, which can be important in the daytime ionosphere, is only a trace ion in this event even at high altitudes.

4. CONCLUSION

This paper shows that low energy auroral electron precipitation can lead to large outflows of thermal ions from the ionosphere. The F region densities, temperatures and flows respond to auroral precipitation in a matter of minutes. Ion fluxes continue to build up while the precipitation persists. However, there is a marked difference between the H^+ and other ions during this recovery phase. Whereas the H^+ flow continues strongly upward, the O^+, He^+ and N^+ flow quickly reverses. The downward O^+ flux contributes to an increase in the electron density above the $F2$ peak. The electron temperature shows a rapid increase during the event and decays abruptly as the heat source is turned off.

Acknowledgments. This work was supported by NSF grants ATM-9018165, ATM-9017201, and ATM-9202887; and NASA grant NAGW-996 at The University of Alabama in Huntsville.

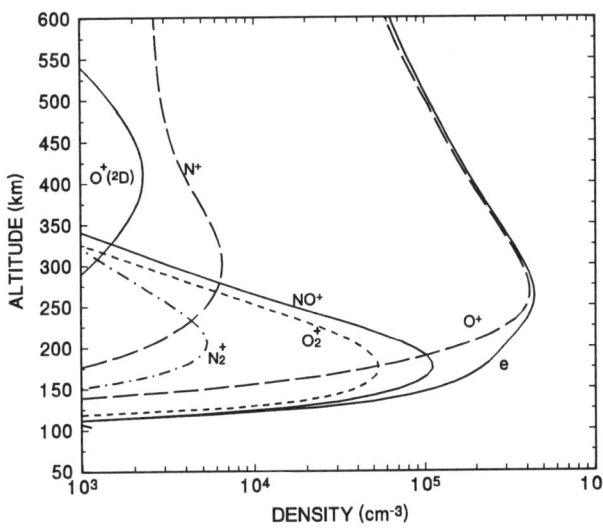

Fig. 8. Calculated ion density profiles at 20.54 hours (2032 UT).

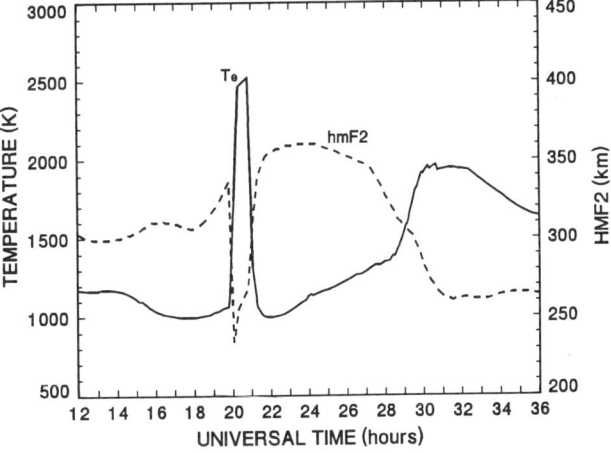

Fig. 7. Variation of the height of the peak electron density ($hmF2$) and electron temperature at $hmF2$ for a complete day starting from 1200 UT. The model electron precipitation occurred between 2000 and 2100 UT.

REFERENCES

Fuller-Rowell T. J., D. Rees, S. Quegan, R. J. Moffett, and G. J. Bailey, Interactions between neutral thermospheric composition and the polar ionosphere using a coupled ionosphere-thermosphere model, *J. Geophys. Res.*, *92*, 7744, 1987.

Hedin, A. E., MSIS-86 thermosphere model, *J. Geophys. Res.*, *92*, 4649, 1987.

LaBelle, J., R. J. Sica, C. Kletzing, G. D. Earle, M. C. Kelley, D. Lummerzheim, R. B. Torbert, K. D. Baker, and G. Berg, Ionization from soft electron precipitation in the auroral F region, *J. Geophys. Res.*, *94*, 3791, 1989.

Lee, J. S., J. P. Doering, T. A. Potemra, and L. H. Brace, Measurements of the ambient photoelectron spectrum from Atmosphere Explorer, I, AE-E measurements below 300 km during solar minimum conditions, *Planet. Space Sci.*, *28*, 947, 1980.

Lilensten, J., D. Fontaine, W. Kofman, L. Eliasson, C. Lathuillere, and E. S. Oran, Electron energy budget in the high-latitude ionosphere during Viking/EISCAT coordinated measurements, *J. Geophys. Res.*, *95*, 6081, 1990.

Lummerzheim, D., M. H. Rees, and H. R. Anderson, Angular dependent transport of auroral electrons in the upper atmosphere, *Planet. Space Sci.*, *37*, 109, 1989.

Min, Q.-L., and B. J. Watkins, Determination of auroral heat fluxes and thermal ion outflows using a numerical ionospheric model and incoherent-scatter radar data, *J. Geophys. Res.*, *100*, 251, 1995.

Nagy, A. F., and P. M. Banks, Photoelectron fluxes in the ionosphere, *J. Geophys. Res.*, *75*, 6260, 1970.

Parish, H. F., G. R. Gladstone, and S. Chakrabarti, Interpretation of satellite airglow observations during the March 22, 1979, magnetic storm, using the coupled ionosphere-thermosphere model developed at University College London, *J. Geophys. Res.*, *99*, 6155, 1994.

Richards, P. G., and D. G. Torr, Ratios of photoelectron to EUV ionization rates for aeronomic studies, *J. Geophys. Res.*, *93*, 4060, 1988.

Richards, P. G., and D. G. Torr, Theoretical modeling of the dependence of the N_2 second positive 3371 Å auroral emission on characteristic energy, *J. Geophys. Res.*, *95*, 10337-10344, 1990.

Richards, P. G., J. A. Fennelly, and D. G. Torr, EUVAC: A solar EUV flux model for aeronomic calculations, *J. Geophys. Res.*, *99*, 8981, 1994a.

Richards, P. G., D. G. Torr, B. W. Reinisch, R. R. Gamache, and P. J. Wilkinson, F2 peak electron density at Millstone Hill and Hobart: Comparison of theory and measurement at solar maximum, *J. Geophys. Res.*, *99*, 15,005, 1994b.

Roble, R. G., and M. H. Rees, Time-dependent studies of the aurora: Effects of particle precipitation on the dynamic morphology of ionospheric and atmospheric properties, *Planet. Space Sci.*, *25*, 991, 1977.

Roble, R. G., E. C. Ridley, A. D. Richmond, and R. E. Dickinson, A coupled thermosphere/ionosphere general circulation model, *Geophys. Res. Lett.*, *15*, 1325, 1988.

Schunk, R. W., and A. F. Nagy, Electron temperatures in the F region of the ionosphere: Theory and observations, *Rev. Geophys.*, *16*, 355, 1978.

Sojka, J. J., and R. W. Schunk, A theoretical study of the high latitude *F* region's response to magnetospheric storm inputs, *J. Geophys. Res.*, *88*, 2112, 1983.

Sojka, J. J., R. W. Schunk, and W. F. Denig, Ionospheric response to the sustained high geomagnetic activity during the March '89 great storm, *J. Geophys. Res.*, *99*, 21,341, 1994.

Solomon, S. C., P. B. Hays, and V. J. Abreu, The auroral 6300 Å emission: Observations and modeling. *J. Geophys. Res.*, *93*, 9867, 1988.

Stamnes, K., S. Perraut, J. M. Bosqued, M. H. Rees, and R. G. Roble, Ionospheric response to daytime auroral electron precipitation: Results and analysis of a coordinated experiment between the AUREOL-3 satellite and the EISCAT radar, *Ann. Geophys.*, *4*, 235, 1986.

Strickland, D. J., J. R. Jasperse, and J. A. Whalen, Dependence of auroral FUV emissions on the incident electron spectrum and neutral atmosphere, *J. Geophys. Res.*, *88*, 8051, 1983.

Torr, M. R., D. G. Torr, P. G. Richards, and S. P. Yung, Mid- and low-latitude model of thermospheric emissions, 1, $O^+(^2P)$ 7320 Å and N_2 (2P) 3371 Å, *J. Geophys. Res.*, *95*, 21,147, 1990.

P. G. Richards, Computer Science Department, The University of Alabama in Huntsville, Huntsville, AL 35899.

Fine Scale Auroral Beams and Conics

J. D. Perez, Chao Liu[1] and Lynne Lawson[2]

Physics Department, Auburn University, Auburn, Alabama

T. E. Moore

Space Science Laboratory, NASA Marshall Space Flight Center, Huntsville, Alabama

Low energy ion data from RIMS on-board DE 1 have been analyzed to reveal features in the phase space density, both conics, beams, and convecting plasma, on a spatial scale of approximately 30 km and/or a temporal scale of 6s. A particular pass through the nightside auroral region on October 24, 1981 is used to show how these features reflect larger scale structures and processes. In particular a convection reversal boundary that is tilted on the order of 5° with respect to the magnetic field lines is inferred, and the formation of an oblique double layer is modeled and shown to explain the observed electric potential. Two regions in which there are interspersed conics and beams are shown to have the conics correlated with downward current implying that the current is the free energy source for the perpendicular heating that produces the conics. Analysis of data from another region shows plasma convecting poleward at velocities of the order of 50 km/s. This is shown to be associated with a substorm and to be a signature of dipolarization.

1. INTRODUCTION

The Retarding Ion Mass Spectrograph (RIMS) [*Chappell et al.*, 1981] on-board the Dynamics Explorer 1 (DE 1) satellite provided a wealth of information regarding the composition and phase space density of low energy, 0 to 50 eV, ions in the Earth's magnetosphere. Applying a new method of deconvolving the RIMS data [*Perez et al.*, 1993], we have shown that structures in the phase space density on a spatial scale of approximately 30 km and/or a temporal scale of 6 s can be resolved. In this paper, we show how the phase space density can be used to infer information regarding mesoscale and macroscale structures in the magnetosphere.

---

[1]Current address is Center for Space Plasma and Aeronomic Research, University of Alabama in Huntsville, Huntsville, AL 35899

[2]Current address is Physics Department, Providence College, RI, 02918

Cross-Scale Coupling in Space Plasmas
Geophysical Monograph 93
Copyright 1995 by the American Geophysical Union

To illustrate how we determine large scale structural features from the phase space density, we examine a particular pass of DE 1 through the nightside auroral region on October 24, 1981. A spectrogram for 46 minutes of H^+ data from RIMS is shown in Figure 1. The horizontal axis is universal time (TIME) and altitude in Earth radii (RE), MacIlwain L value (L), magnetic local time (MLT) , and the magnetic latitude (MLT) are also shown. The vertical axis is detector look direction as a function of satellite spin, where 0° corresponds to looking into the RAM direction. The horizontal line in the upper part of the spectrogram shows the spin angle when the instrument is observing ions coming up the field line from the ionosphere, and the line in the lower portion shows when ions are going down the field line. When the peak flux is along one of these horizontal lines, we have a field-aligned beam. When there are symmetrical peaks at spin angle just off the field-aligned position, the ion velocity has a component both parallel and perpendicular to the field line forming a phase space density referred to as a conic.

Starting at 1326 UT as the satellite leaves the polar region and enters the nightside auroral region, we see H^+ conics that form progressively smaller angles with respect to the field line until the flow is field-aligned or beam-like. This is event E1 which occurs at approximately 70° invariant latitude (IL) and

128 BEAMS AND CONICS

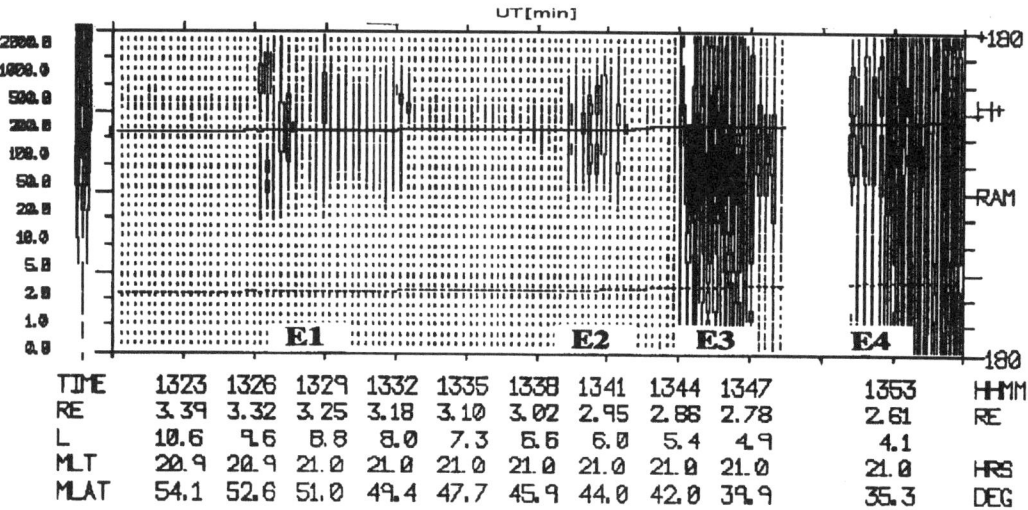

Fig. 1. Data from RIMS onboard DE 1 on day 297, October 24, 1981.

whose interpretation will be described in Section 2.

As the satellite moves equatorward reaching about 67° IL at 1340 UT, we see a pattern of interwoven conics and beams. This is the event marked E2 in Figure 1. The correlation between the conics and downward going current will be described in Section 3 along with a more detailed presentation of the data..

At 1344 UT, about 68° IL, we see superimposed upon a nearly isotropic background, an intense flux that is neither field-aligned nor in the RAM direction. Detailed analysis of this event, marked E3 in Figure 1, is presented in Section 4. It is

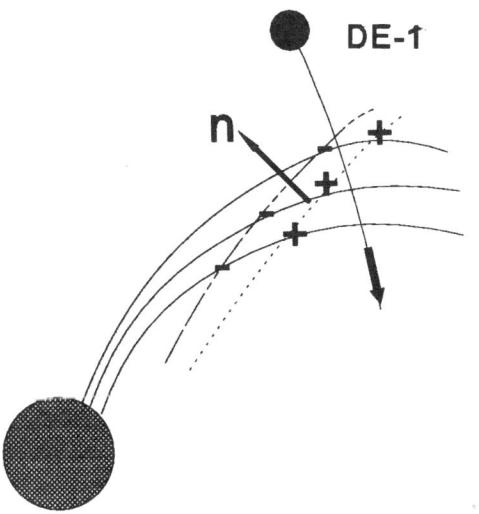

Fig. 2. Illustration of oblique double layer formed along convection reversal boundary. Symbols are defined in text.

inferred that these ions are $E \times B$ drifting poleward in response to dipolarization in the magnetotail.

Following a gap in the data, we see at 1351 UT, 61° IL, another pattern of intermixed conics and beams marked as event E4 in Figure 1. This time it is superimposed on an isotropic background, but again we find that the conics are correlated with intense downward going current as described in Section 3.

2. CONVECTION REVERSAL BOUNDARY AND THE FORMATION OF OBLIQUE DOUBLE LAYERS

A detailed analysis and modeling of four events similar to E1 was carried out by *Liu et al.* [1994c]. One physical mechanism that can be deduced from the phase space densities is illustrated in Figure 2. The satellite is depicted as the dark circle with its path moving downward across field lines connected to the Earth shown as the cross-hatched circle. The inferred convection reversal boundary makes an angle with respect to the magnetic field that is drawn larger than determined by the analysis for illustration purposes. The model assumes that weak shear at the convection reversal boundary produces electrostatic ion cyclotron waves perpendicularly heating the ions [*Ganguli* and *Lee*, 1985; *Ganguli et al.*, 1985, 1988; *Nishikawa et al.*, 1988, 1990]. The ions then respond to the mirror force and move away from the Earth along the field line where they are observed as conics by RIMS. Assuming conservation of energy and the first adiabatic invariant, the conic angle can be used to determine the distance of the perpendicular acceleration region from the spacecraft - the wider the conic angle, the closer to the satellite the conic was formed. Therefore the Y-shaped feature in the spin spectrogram in Figure 1 showing decreasing conic angle as the satellite moves equatorward indicates that the ions

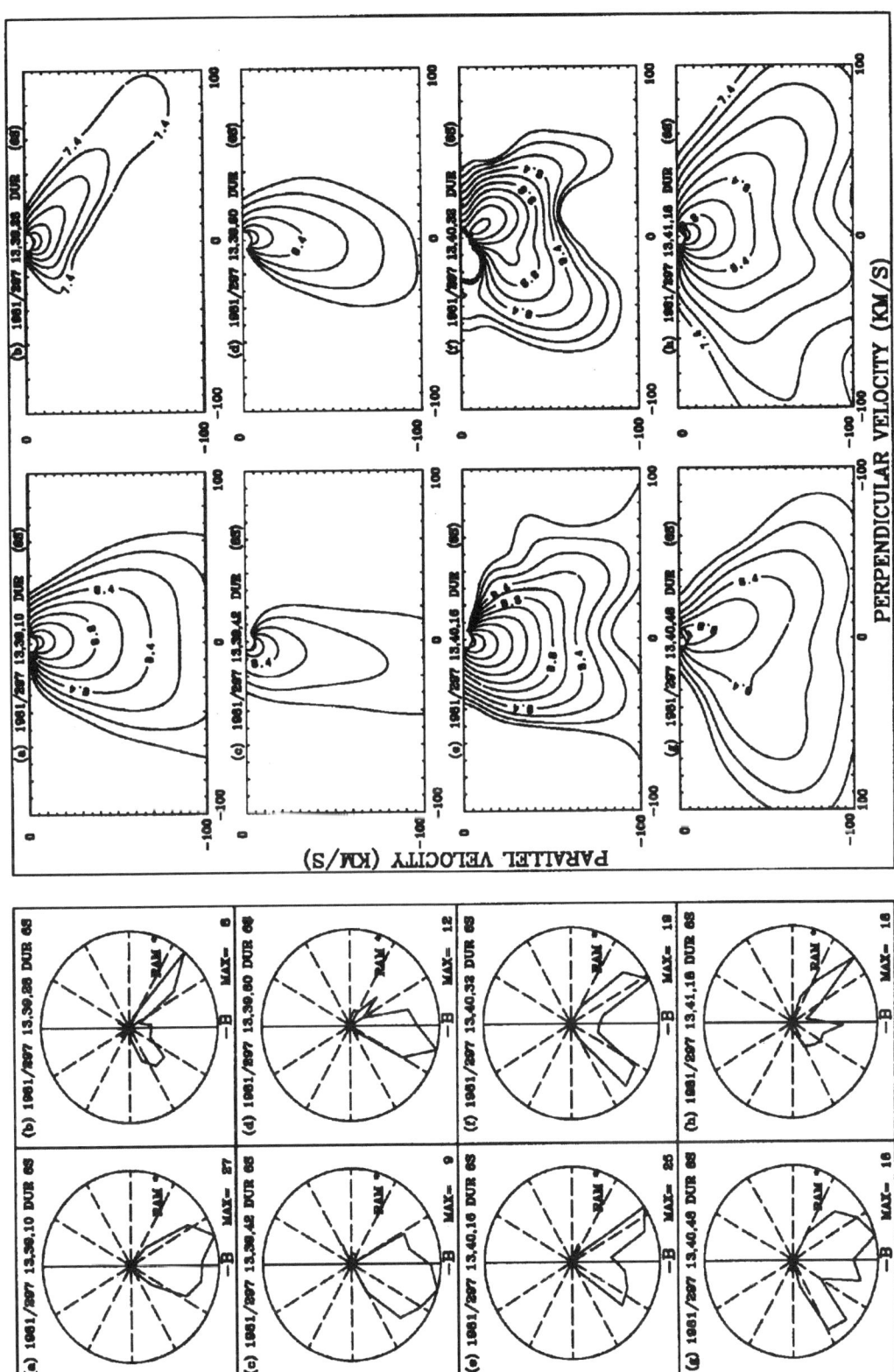

Fig. 3. On the left, each panel shows the normalized counts when the retarding ion potential is 0 vs. pitch angle for E2 of day 297. On the right are the corresponding contour plots of the logarithm of the phase space density.

130 BEAMS AND CONICS

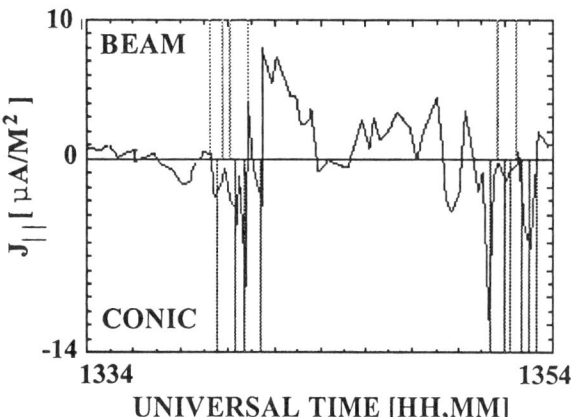

Fig. 4. Current and location of beams and conics for day 297, events E2 and E4.

were heated at locations that are increasingly far from the satellite. Plotting the altitude of the boundary where the conics were formed on field lines through the satellite position demonstrates that the convection reversal boundary forms an angle of approximately 5° with the magnetic field (See Plate 5 of *Liu et al.* [1994c]). This is shown as the dashed line with normal **n** in Figure 2 with an exaggerated angle for illustrative purpose. When the ions mirror up the field line, they leave electrons behind forming a double layer at an oblique angle to the magnetic field. Using electric field data from DE 1, and assuming that the electron and ion planes of the oblique double layer each have Gaussian charge distributions, parameters of the double layer can be inferred. In this manner, the distance of charge separation, *d* in Figure 2, is determined to be of the order 50 - 200 km, and the width of the Gaussian charge distributions, *w* in Figure 2, is found to be approximately 50 km.

3. CURRENT DRIVEN ION HEATING AND CONIC FORMATION

Events E2 and E4 in Figure 1 show a pattern of intertwined conics and beams. In order to display all the fine structure in such events, it is necessary to look on a finer temporal scale than the one shown in Figure 1. *Liu et al.* [1994b] examined each 6s spin of the satellite to obtain the best resolution possible and determined H^+ density, parallel and perpendicular velocity, and parallel and perpendicular temperature for each conic and beam for these events. They also showed a similar event on October 14, 1981. Typical results are shown in Figure 3 in two formats. On the left, a hodograph of the total flux as a function of spin angle is shown for eight different 6s intervals. The magnetic field direction points up, and we see that all the flux is in the lower half of the plane indicating that the particles are coming up the field line from the ionosphere. While this format may indicate whether the phase space density is conic-

or beam-like, only a deconvolution of the RIMS data can show what the H^+ distribution function really looks like. On the right, contour plots of the log of the deconvolved phase space density for each of the 6s intervals shown on the right are presented. Perpendicular velocity is plotted along the horizontal axis and parallel velocity along the vertical. Only negative parallel velocity, up the field line, is shown. Distributions with larger velocity spread in the parallel direction are classified as beams. Distributions with about equal velocity spread in the perpendicular and parallel directions and with a minimum at along the parallel direction are classified as conics. For some cases, e.g. (b) in Figure 3, only one wing of the conic is observed indicating that the fine structure is smaller than 30 km the distance traveled by the satellite during a 6s revolution. The bowl shape of the some of the conics, e.g. (f) in Figure 3, implies extended perpendicular heating along the field line [*Brown et al.*, 1991]. The distribution in (g), Figure 3, shows a wider spread in the perpendicular direction than is typical for a beam, but it does not show the minimum along the parallel axis typical of a conic. Its shape may be due to overlap of a beam and conic due to finer scale structure than can be resolved by the instrument, so its classification as a beam is somewhat problematic.

Figure 4 shows the field-aligned current obtained by *Marshall et al.* [1991] from the MAG-A [*Farthing et al.*, 1981] data as a function of universal time. The time of the conics and beams observed by RIMS in events E2 and E4 are marked by vertical lines above the time axis for the beams and below for the conics. It is clear that in almost all cases when the downward going current density exceeds approximately 1 $\mu A/m^2$, there are conics. This is the first time a correlation on a small scale between conics and field-aligned current has been observed.

Dusenbery and *Lyons* [1981] used quasi-linear theory to show that the resonant interaction of upgoing thermal ions with the thermal electrons can generate conics only in the downward current region and that ion conics should not occur simultaneously with the inverted-V keV electron precipitation. This is consistent with the correlation presented above and the fact that we do not see any low-energy ion conics in the region between events E2 and E3.

4. SUBSTORM DIPOLARIZATION

An intense, low-energy H^+ flux event (E3 in Figure 1), was seen as the satellite moved to lower latitude. It is neither field-aligned nor in the RAM direction and is superimposed upon an isotropic background that is indicative of an approach to the plasmasphere. Only examination of the specifics of the phase space density and its velocity moments can tell us what is the underlying physical mechanism responsible for the observations. *Liu et al.* [1994a] analyzed a similar event on

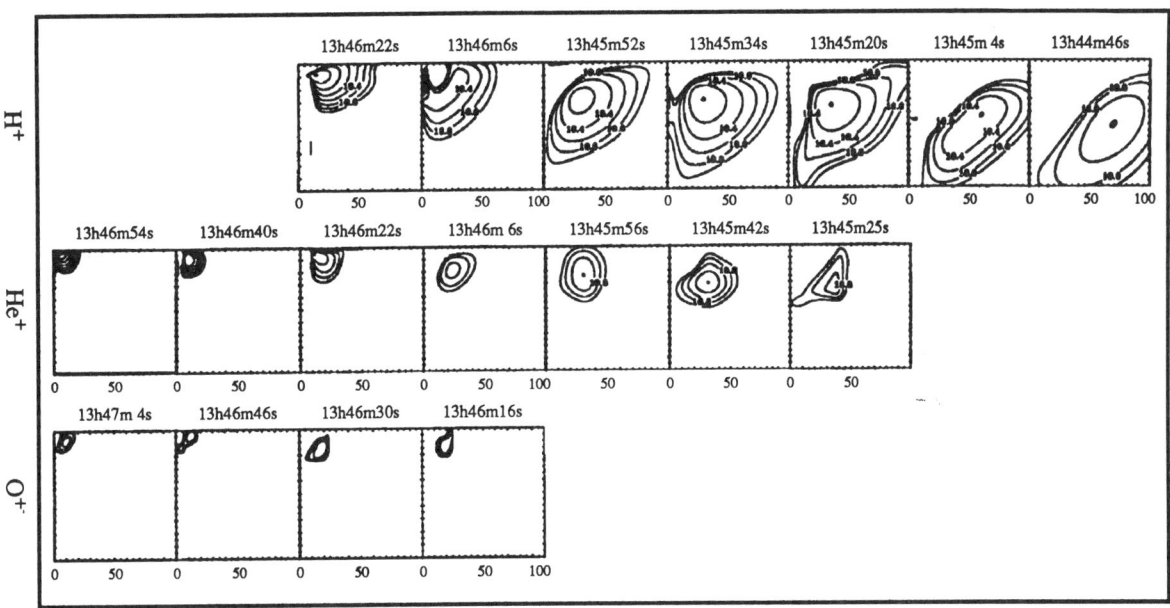

Fig. 5. Logarithm of phase space density contours for $v_{\parallel}$ (km/s) (horizontal axis) and $v_{\perp}$ (km/s) (vertical axis) for day 297 event E3. Positive $v_{\parallel}$ means toward the ionosphere along the magnetic field. Positive $v_{\perp}$ is perpendicular to the magnetic field in the poleward direction. Spacecraft velocity has been subtracted.

October 20, 1981 and determined that the plasma was $E \times B$ drifting poleward at velocities of the order 50 km/s. Data for event E3 is shown in Figure 5 where contour plots of the logarithm of the phase space density are plotted for H^+, He^+, and O^+. Each square is the deconvolution of the data from a 6s revolution of the satellite with the parallel velocity along the horizontal axis with negative values increasing to the right representing motion away from the ionosphere and with the perpendicular velocity along the vertical axis with positive values increasing upward representing velocity in the poleward direction. The latitude of the observations decreases from top to bottom of the figure. First we see that the peaks of the distribution functions show plasma flow velocities up the field line and poleward away from the plasmasphere. Second, the different ions all have the same flow velocities as would be expected if the cause is an electric field that is pointing toward dawn, i.e., in the eastward direction. The fact that the H^+ flow, then the He^+, and then the O^+ are seen as the latitude and the decreases is a reflection of the energy range of the RIMS instrument. At high latitude where the H^+ velocity is above 50 km/s, the energy of the He^+ and O^+ moving with the same velocity exceeds the range of sensitivity of the RIMS instrument. At low velocity and lower latitude, the energy of the H^+ is so low as to make deconvolution of the phase space density unreliable.

The consistent direction and the latitudinal dependence of the plasma flow velocity as determined from the phase space density is striking. It suggests the macroscopic process shown in Figure 6. When the current disruption associated with a substorm initiates dipolarization of the magnetic field lines that have been distorted into tail-like field lines, the snapping back of these field lines produces motion of the field lines and plasma that is poleward at intermediate altitudes where DE 1 made its observations. It is possible that such motion of the field lines could be associated with Pi 2 pulsations [*Lin et al.*, 1991].

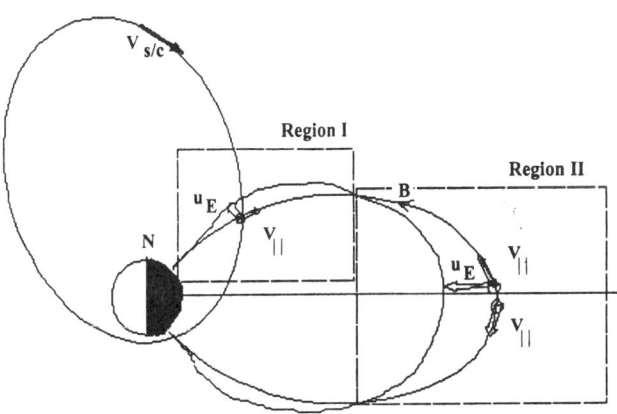

Fig. 6. Schematic drawing of particle motion associated with substorm dipolarization event E3 on day 297.

5. SUMMARY

Analysis of the phase space densities of low energy ions measured by the RIMS instrument aboard DE 1 as it passed through the nightside auroral region on October 24, 1981 illuminates several features of the mesoscale structure of the magnetosphere. Starting at approximately 70° IL, a convection reversal boundary that could produce electrostatic ion cyclotron waves that perpendicularly heat low energy ions is theorized to lie at angle to the magnetic field lines and to produce oblique double layers. At lower latitudes, regions, which have a fine structure of interlaced conics and field-aligned beams, are found to have the conics strongly correlated with the downward flowing parallel current density. Finally, near the plasmasphere, the low-energy signature of dipolarization in the magnetotail is discovered.

Acknowledgments. The work reported here was supported by the NASA JOVE program under contract NAG8-147.

REFERENCES

Brown, G. B., G. R. Wilson, J. L. Horwitz, and D. L. Gallagher, Self-consistent production of ion conics on return current region auroral field lines: a time-dependent, semi-kinetic model, *Geophys. Res. Lett., 18,* 1841, 1991.

Chappell, C. R., S. A. Fields, C. R. Baugher, J. H. Hoffman, W. B. Hanson, W. W. Wright, H. D. Hammock, G. R. Carignan, and A. F. Nagy, The retarding ion mass spectrometer on Dynamics Explorer-A, *Space Sci. Instrum., 5,* 477, 1981.

Dusenbery, P. B., and L. R. Lyons, Generation of ion-conic distribution by upgoing ionospheric electrons, *J. Geophys. Res., 86,* 7627, 1981.

Farthing, W. H., M. Sugiura, B. G. Ledley, and L. J. Cahill, Jr., Magnetic field observations on DE-A and -B, *Space Sci. Instrum. 5,* 551, 1981.

Ganguli, G., Y. C. Lee, Electrostatic ion-cyclotron instability caused by a nonuniform electric field perpendicular to the external magnetic field, *Phys. Fluids, 28,* 761, 1985.

Ganguli, G., P. Palmadesso, and Y. C. Lee, A new mechanism for excitation of electrostatic ion cyclotron waves and associated perpendicular ion heating, *Geophys. Res. Lett., 12,* 643, 1985.

Ganguli, G., Y. C. Lee, and P. J. Palmadesso, Kinetic theory for electrostatic waves due to transverse velocity shears, *Phys. Fluids, 31,* 823, 1988.

Lin, C. A., L. C. Lee, Y. J., and Y. J. Sun, *J. Geophys. Res., 96,* 21105, 1991.

Liu, Chao, J. D. Perez, T. E. Moore, and C. R. Chappell, Low energy particle signature of substorm dipolarization, *Geophys. Res. Lett., 21,* 229, 1994a.

Liu, Chao, J. D. Perez, T. E. Moore, C. R. Chappell, and J. A. Slavin, Fine structure of low-energy H^+ in the nightside auroral region, *J. Geophys. Res., 99,* 4131, 1994b

Liu, Chao, J. D. Perez, T. E. Moore, and C. R. Chappell, Boundary structure of low-energy ions associated with the nightside convection reversal, *J. Geophys. Res., 99,* 11401, 1994c.

Marshall, J. A., J. L. Burch, J. R. Kan, and J. A. Slavin, DE 1 observations of return current regions in the nightside auroral oval, *Geophys. Res. Lett., 18,* 45, 1991.

Nishikawa, K.-I., G. Ganguli, Y. C. Lee, and P. J. Palmadesso, Simulation of ion-cyclotron-like modes in a magnetoplasma with transverse inhomogeneous electric field, *Phys. Fluids, 31,* 1568, 1988.

Nishikawa. K.-I., G. Ganguli, Y. C. Lee, P. J. Palmadesso, Simulation of electrostatic turbulence due to sheared flows parallel and transverse to the magnetic field, *J. Geophys. Res., 95,* 1029, 1990.

Perez, J. D., Chao Liu, L. Lawson, T. E. Moore, and C. R. Chappell, A new technique for deconvolution of data from instruments that make integral measurements, e.g. RIMS on DE 1, *Annales Geophysicae 11,* 889, 1993.

J. D. Perez, Physics Department, Auburn University, Auburn, AL 36849.

Chao Liu, CSPAR, Huntsville, AL 35899.

Lynne Lawson, Physics Department, Providence College, River Avenue, Providence, RI 02918.

T. E. Moore, Space Science Laboratory/Code ES53, NASA Marshall Space Flight Center, Huntsville, AL 35812

Anisotropic Kinetic Effects of Photoelectrons on Polar Wind Transport

Sunny W. Y. Tam, Fareed Yasseen, Tom Chang

Center for Space Research, Massachusetts Institute of Technology, Cambridge

Supriya B. Ganguli

Science applications International Corporation, McLean, Virginia

John M. Retterer

Geophysics Directorate, Phillips Laboratory, Bedford, Massachusetts

There is increasing observational evidence that photoelectrons may affect polar wind dynamics. For example, suprathermal electron pitch-angle distributions in the photoelectron energy range have been observed in the high-altitude polar wind. These distributions contribute little to the polar wind density, but carry an appreciable outward heat flux. Evidence of such reflected photoelectron distributions at low altitudes have been attributed to field-aligned potential drop. More recently, measurements of day-night asymmetries in electron temperature and ion outflow provide further indications of the photoelectrons' impact on the polar wind. Such non-thermal fluxes can be explained by a mechanism relying on the earth's decreasing magnetic field, the field-aligned potential drop, and the energy dependence of the Coulomb collisional cross-sections. The description of this mechanism requires a kinetic approach. Such an approach was used in a test particle simulation of this mechanism, in agreement with the measured suprathermal fluxes. However, the effects of these fluxes on the polar wind itself require a self-consistent description. Unfortunately, a fully kinetic self-consistent description is at present not achievable. Instead, we suggest a hybrid approach, in which the background features of the polar wind are described by well-established fluid models, while the suprathermal features are described using a kinetic model. This approach retains the expediency of fluid theory while in effect extending its applicability. In this paper, we will review the physics underlying the mechanism mentioned earlier, discuss how the kinetic-fluid synthesis can best be achieved, and present our latest results. Our initial calculations show, for example, that the suprathermal electrons carry much of the polar wind heat flux, and may significantly increase the ambipolar electric field. This increase in the electric field can change the dynamics of the polar wind outflow.

1. INTRODUCTION

Why do fluid models apply much more generally to neutral gases than to plasmas? Fluid models usually assume systems close to thermodynamic equilibrium. In neutral gases, binary collision (which tend to establish equilibrium) are the microscopic process characterized by the fastest timescale and shortest scalelength, so that, for a macroscopic description, this assumption is generally true. On the other hand, plasmas are subject to Coulomb collisions, which also tend to establish equilibrium. When the self-consistent inhomogeneous electric and magnetic fields that occur in plasmas are characterized by much faster timescales and shorter scalelengths than those due to

Coulomb collisions, then the system will not necessarily be able to establish local thermodynamic equilibrium. Therefore, there are many field-related phenomena in plasmas that require a kinetic description. This argument may explain why kinetic models are considered necessary only in order to describe phenomena characterized by faster timescales and shorter scalelengths than Coulomb collisions.

In fact, the nature of Coulomb collisions itself sets a limit on the applicability of fluid models to plasmas. In neutral gases, the collisional cross-section has a very weak energy dependence, so that the mean free path of a test neutral particle will also have a very weak energy dependence. In this case it will be very easy to separate collisionless regimes requiring a kinetic description from collisional regimes adequately described by fluid models. The latter are characterized by a collisional mean free path much shorter than the macroscopic scale length. The macroscopic heat flux, for example, will result from a large number of short scale-length microscopic collisions, so that one can relate it to a local energy-averaged heat conductivity and a temperature gradient.

In plasmas, the situation is quite different. The cross-section for Coulomb collisions has a strong dependence on the inverse of the particle energy, so that the collisional path of a test particle will depend, roughly, on the square of its energy. While thermal particles in plasmas may have collisional paths shorter than the macroscopic scale lengths of the plasma configuration, the suprathermal particles may not. Thus, the macroscopic heat flux will be borne by thermal particles that have undergone a a large number of collisions, by energetic suprathermal particles that are collisionless, and particles in between. Because of these suprathermal particles, the heat flux acquires a non-local and anomalous character, in that it cannot be related to a local heat conductivity or temperature gradient.

Indeed, numerical simulations have shown that velocity-averaged local heat conductivities cannot describe the electron heat flux in inhomogeneous plasmas, and this inability has been attributed to the anomalous heat flux borne by the suprathermal electrons [Khan and Rognlien, 1981]. Such anomalous, field-aligned electron distributions have been observed in the solar wind [Scudder and Olbert, 1979]. These authors were the first to point out the non-local nature of these distributions, and to describe their formation using a global, kinetic collisional model. They also suggested that, due to their anomalous contribution to the energy flux, these suprathermal electrons may significantly increase the ambipolar electric field along the magnetic field lines, and thus "drive" the solar wind [Olbert, 1982].

2. MOTIVATION

Such anomalous field-aligned energy fluxes carried by suprathermal electrons were observed in the polar cap region by ISIS-1 [Winningham and Heikkila, 1974], and the DE-1 and -2 satellites [Winningham and Gurgiolo, 1982], and were shown by Yasseen et al. [1989] to be of ionospheric photoelectron origin. Applying Scudder and Olbert's arguments, they performed a kinetic test-particle simulation using an electron distribution that is consistent with the observed data at low altitude, and a background based on an empirical fit of the electron density profile produced by a fluid simulation and an assumed power-law electric field. Their results also indicated that the photoelectrons can give rise to the downstreaming suprathermal electron distribution observed in the low-altitude polar wind. Although photoelectrons represent only a small percentage of the total electron density, they may contribute significantly to the total heat flux in the polar wind.

The classical polar wind is an ambipolar outflow of plasma from the polar region of the ionosphere along open field lines. Energetic suprathermal electrons in the polar wind or in other ionospheric/magnetospheric settings have been considered by various authors. For example, kinetic collisional calculations by Khazanov et al. [1993] have examined the role of photoelectrons on plasmaspheric refilling. Collisionless kinetic calculations by Lemaire [1972] have shown that escaping photoelectrons may increase the ion outflow velocities in the polar wind. Collisionless kinetic calculations by Barakat and Schunk [1984] and semi-kinetic calculations by Ho et al. [1992] have examined the impact of hot magnetospheric electrons, and concluded that such particles may also increase the ion outflow velocities.

Enhanced ion outflow velocities [Abe et al., 1993a, b] and a marked day-night asymmetry [Abe et al., 1993; Yau et al., 1995] in several features of the polar wind have been observed recently. In this paper, we suggest that the photoelectrons should be further investigated as a likely source for the observed day-night asymmetry. Photoelectrons occur mainly in the sunlit ionosphere, so that they are a natural candidate to explain this asymmetry. By comparing their contribution to the heat flux (obtained from a test-particle collisional simulation) with the heat flux deduced from a moment-based polar wind simulation, we will demonstrate that the photoelectrons can make a significant contribution to the total electron heat flux. Further, using a heuristic argument based on quasineutrality considerations, we will suggest that the photoelectron contribution to the heat flux, due to the combined effects of Coulomb collisions, and magnetic mirror force can increase the ambipolar electric field considerably, thereby enhancing the ion outflow velocities. In this sense, the ambipolar electric field can be regarded as a polar wind control mechanism, which is influenced by the photoelectrons, and which can change the entire polar wind picture.

We should add that the several mechanisms that have been investigated theoretically may also be proposed as alternative explanations, in addition to the collisionless suprathermal effects mentioned earlier. Parallel ion acceleration driven by E×B convection was considered by *Cladis* [1986], and shown to significantly energize oxygen ions escaping to the polar magnetosphere. This force can also be seen as a centrifugal force in the convecting frame of reference, and was included in this form in the time-dependent, semi-kinetic model developed by *Horwitz et al.* [1994].

This paper is structured as follows. In Section 3, we introduce our hybrid model, in which the photoelectrons are modeled kinetically and the thermal components of the polar wind are described using a 16-moment generalized transport model. This model will enable us, in Section 4, to discuss the photoelectron contribution to the total electron heat flux. In Section 5, we will estimate the effect of the photoelectrons on the ambipolar electric field, using heuristic quasineutrality arguments. Finally, we will conclude with a discussion of the implications of this work, of its shortcomings and how to remedy them in our current and future work.

3. MODEL

With the assumption that only transport along geomagnetic lines is important in the classical polar wind, the electron distribution $f(s, v_\parallel, v_\perp)$ is governed by the following collisional kinetic equation:

$$\left[\frac{\partial}{\partial t} + v_\parallel \frac{\partial}{\partial s} - \frac{e}{m_e} E_\parallel \frac{\partial}{\partial v_\parallel} \right] f - v_\perp^2 \frac{B'}{2B} \left[\frac{\partial}{\partial v_\parallel} - \frac{v_\parallel}{v_\perp} \frac{\partial}{\partial v_\perp} \right] f = L_{FP} f, \quad (1)$$

where s is the distance along the field line B, $E_\parallel$ represents the field-aligned electric field, $B' \equiv dB/ds$, and L_{FP} is the Fokker-Planck collision operator for Coulomb interactions, which is the dominant type of collision above 300 km. Equation (1) includes the major forces an electron experiences as it travels along the field line: field-aligned electric force, mirror force, and forces that are due to Coulomb collisions. Because photoelectrons only constitute a minor portion of the total electron distribution, we treat them as test particles in our study. To simulate the forces represented in Equation (1), we apply the Monte Carlo procedure developed for such test particles [*Retterer et al.*, 1983] to follow the evolution of the suprathermal electrons, initially taken to be distributed as an upper-half Maxwellian, with energy ranging from 2 to 62 eV, as suggested by AE-E measurements [*Lee et al.*, 1995].

In order to establish the significance of the suprathermal electrons, we should first compare their contributions to the polar transport with those due to thermal electrons. Thus, the density and temperature profiles of the thermal background, together with the electric field consistent with these profiles, are required. An ideal choice of background would be actual *in situ* experimental data collected by polar-orbiting satellites, e.g., Akebono [*Abe et al.*, 1993a, b], DE-2 [*Winningham and Gurgiolo*, 1982], and ISIS-1,2 [*Winningham and Heikkila*, 1974; *Johnstone and Winningham*, 1982]. However, the available data are unfortunately not detailed enough for our purpose. In this work, we shall therefore use a polar wind background obtained by solving a steady-state transport model based on the sixteen-moment approximation, which provides us with values for the field-aligned heat fluxes [*Ganguli et al.*, 1992]. Generalized transport models which include the sixteen-moment approximation have been discussed in detail in the literature [*Schunk*, 1977; *Barakat and Schunk*, 1982]. This moment-based background, generated by a code developed by S. Ganguli, will enable the comparison of the suprathermal with the thermal contributions to the various transport quantities, the heat fluxes in particular. (For consistency, we apply the boundary conditions used in *Ganguli et al.* [1987]). The code calculates transport of thermal proton and electron species in a stationary background of oxygen ions, which are assumed to be bound by the terrestrial gravitational field because of their heavier mass. Figure 1 shows the polar wind profiles of the electric field, together with the thermal proton and electron temperature and density profiles, obtained from the moment calculations. These profiles are used as the background in the kinetic simulations.

We note that the sixteen-moment equations form a set of stiff equations that are highly sensitive to the boundary conditions of higher moments, such as heat fluxes. A slight variation in the heat flux boundary conditions will change the solutions quantitatively, perhaps even qualitatively. We are, therefore, particularly interested in the heat flux of the photoelectrons, and should compare it with the flux carried by the thermal electrons.

4. CALCULATED HEAT FLUXES

Because our kinetic simulation uses the test-particle approximation, we need to determine the photoelectron

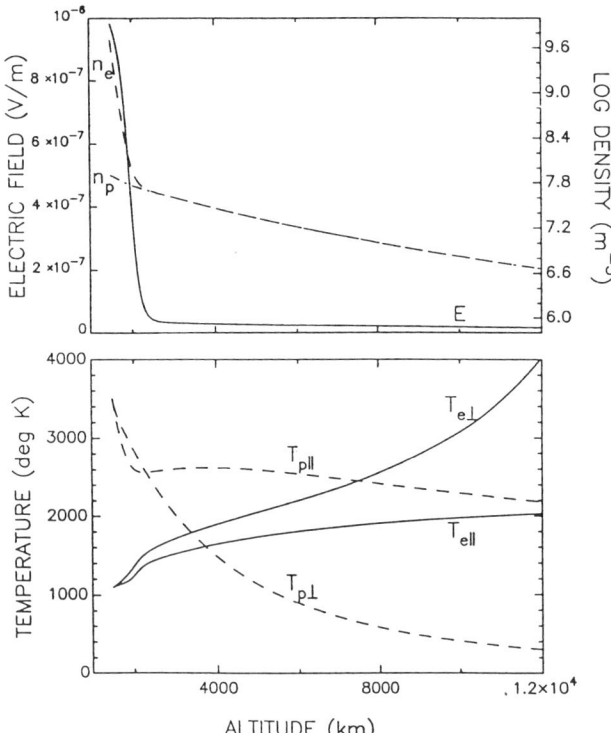

Fig. 1. Profiles obtained from 16-moment solutions. Top panel: electric field E (solid), proton density n_p (dashed), and electron density n_e (dashed). Bottom panel: electron parallel and perpendicular temperatures (solid), and proton parallel and perpendicular temperatures (dashed).

density at some reference altitude in order to assess the suprathermal contribution to the total heat flux. Recent measurements by Akebono provide an estimate for the suprathermal to thermal electron density ratio [*Yau et al.*, 1995]. Because of the limitations inherent to the instruments, there are uncertainties in the determination of the electron density, and in comparing the observational data with our simulations. For example, *Yau et al.* use 10 eV (the low-energy cutoff of their high-energy instrument) as the threshold to distinguish between thermal and suprathermal electrons. In contrast, our calculations consider 2 eV (still a large multiple of the electron temperature) to be within the suprathermal limit.

Let us calculate this density ratio at higher altitudes. Assuming a density ratio of 10^{-3} at 1500 km, we can determine the suprathermal and thermal electron densities throughout our simulation range (1500 – 12000 km) from the kinetic simulation, and the moment-based background calculation, respectively. Figure 2 shows that the density ratio increases by an order of magnitude at higher altitude, *i.e.* $n_{supra}/n_{therm} \sim 10^{-2}$. The sharp increase in the density ratio up to 2000 km is due to the rapid decrease in the background electron density (see Figure 1), the latter being due to a sharp transition from O^+-dominated to H^+-dominated regime, as calculated from the sixteen-moment equations.

Figure 3(a) shows the average parallel and perpendicular heat fluxes per particle carried by the thermal electrons, obtained using the sixteen-moment equations. Notice that the heat fluxes are directed downward (by convention, negative sign), consistent with the idea that the heat flux is in the direction opposite to the temperature gradient (see Figure 1). Figure 3(b) shows the contribution, per suprathermal particle, to the *total* electron heat fluxes, *i.e.* to the heat fluxes that would be obtained if the thermal and suprathermal components were combined a single population. The increase in $q_\parallel$ and the decrease in $q_\perp$ with altitude reflect the transfer of the perpendicular energy into parallel energy by the mirror force on the ascending suprathermal electrons. But more importantly, note that the total heat flux contribution carried by the suprathermal electrons is outwardly directed, and is also on average three orders of magnitude larger than the moment-generated heat fluxes per thermal particle (*i.e.* $|q_{supra}/q_{therm}| \sim 10^3$).

Knowing the density and the average heat flux contribution of the suprathermal electrons, we can now compare the total heat fluxes modified by the suprathermal electrons with the original moment-generated heat fluxes. The average heat fluxes per particle and the density ratio as calculated above suggest that the heat fluxes, $Q_{\perp,\parallel} = n \; q_{\perp,\parallel}$, carried by the suprathermal electrons are larger than their

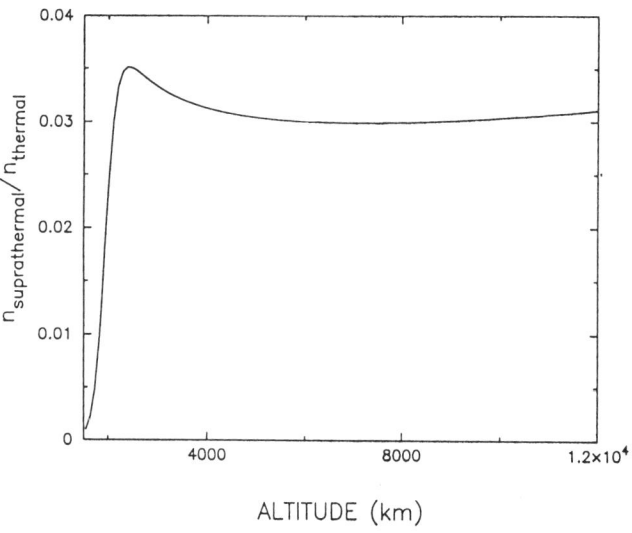

Fig. 2. Plot of the suprathermal to thermal electron density ratio. The density ratio is taken to be 10^{-3} at 1500 km. At higher altitudes, this ratio is to the order of 10^{-2}.

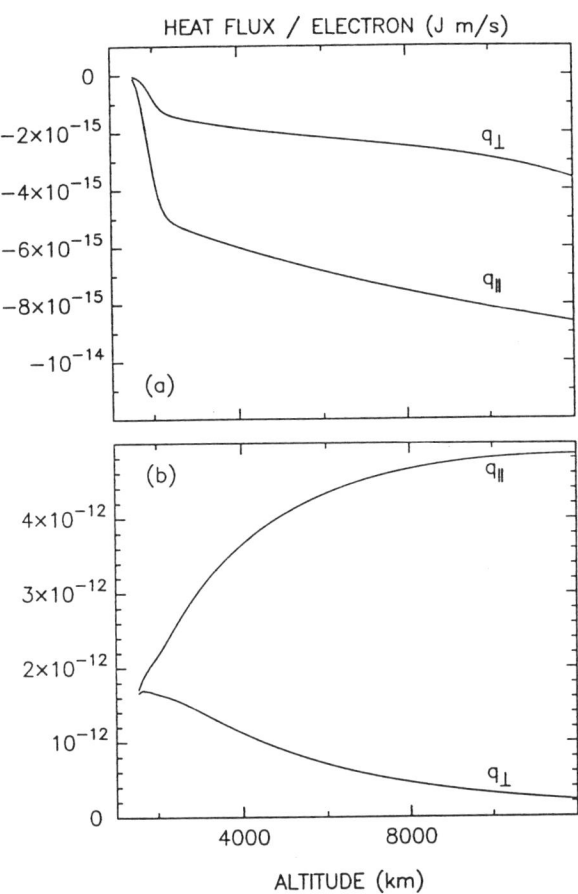

Fig. 3. Comparison of average parallel and perpendicular heat fluxes per particle, $q_\parallel$ and $q_\perp$, carried by (a) thermal electrons in 16-moment calculations, and (b) photoelectrons in Monte Carlo simulations. Upwardly directed heat flux is positive. The average heat fluxes carried by a photoelectron is upward, and is three order of magnitude larger than those carried by a thermal electron.

thermal counterparts by one order of magnitude. In other words, the suprathermal contribution will dominate in the heat fluxes of the total electron population. The total electron heat fluxes ($Q_{\perp,\parallel supra} + Q_{\perp,\parallel therm}$) are therefore upward, consistent with *Yau et al.* [1994].

5. AMBIPOLAR ELECTRIC FIELD

Because of the large amount of energy flux associated with the photoelectrons, one can expect them to affect the polar wind electric field. Only a full, self-consistent calculation including photoelectrons and all other thermal polar wind components can provide a qualitative description of the expected change in the electric field. However, an indication of this change may be obtained much more simply using the following heuristic argument. The inclusion of suprathermal electrons into our model will violate two major assumptions of the classical polar wind: quasi-neutrality and currentless flow. To restore these conditions, a stronger ambipolar field is required. We can calculate the potential that corresponds to this field by modeling the proton dynamics kinetically. More specifically, we shall use a proton distribution suggested by the moment-generated background at the bottom of our simulation range; we let it fold along the flux tube, under a potential profile, ϕ, that is to be determined. By calculating the proton density and mean velocity, which now depend on ϕ, we can find the potential drop that is required to sustain the conditions of quasi-neutrality and currentless flow. This potential drop is calculated to be 13 V across our simulation range, a value comparable to that suggested by *Winningham and Gurgiolo* [1982], based on measurements of electron return fluxes by DE-2, and that used in *Yasseen et al.* [1989]. In Figure 4, the corresponding electric field is shown and compared with the background field obtained from the sixteen-moment code. These results indicate that photoelectrons may increase the polar wind electric field by one order of magnitude. Note that earlier exospheric calculations have also indicated that photoelectrons will increase the polar wind electric field [*Lemaire*, 1972].

6. DISCUSSION

We have shown that the ionospheric photoelectrons (described using a global kinetic collisional treatment) carry the bulk of the heat flux in the polar wind (described using a generalized transport 16-moment model). A heuristic quasineutrality argument indicates that this flux may induce an order-of-magnitude increase in the polar wind electric field. As noted earlier, such a large increase in the polar wind electric field may account for the day-night asymmetry in polar wind outflow, as well as the enhanced ion outflow velocities, observed recently. If we neglect collisions, the photoelectron contribution to the heat flux would, of course, be even more dominant, because photoelectrons would be more likely to maintain their high outflow velocities, thereby more likely to escape. As a result, one would expect an even larger enhancement in the ambipolar electric field.

To extend the work presented here further, it will be necessary to remedy its two shortcomings. Specifically, we need to include oxygen ion dynamics, and to deduce the photoelectron effect on the electric field self-consistently, taking into account all polar wind components, including the bulk thermal populations.

In the model presented here, oxygen ions are in a

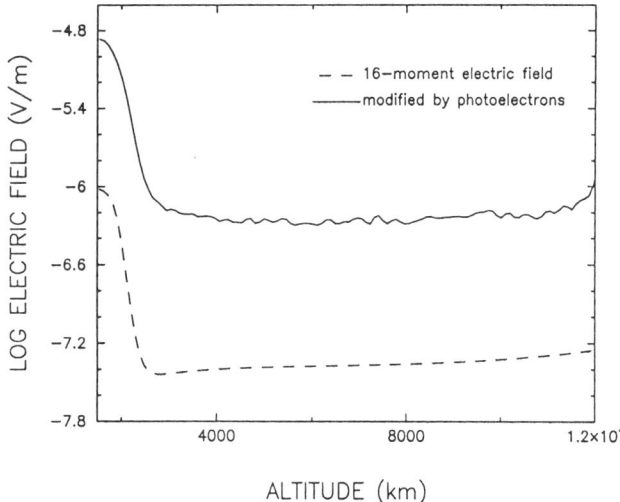

Fig. 4. Ambipolar electric field. The electric field in the moment description of the polar wind (dashed), when modified by photoelectron kinetic effects, becomes one order of magnitude larger (solid).

stationary barometric equilibrium, as they are assumed to be bound by the terrestrial gravitational force because of their heavier mass. However, this assumption may no longer be valid if we want to include the possibility of an order-of-magnitude enhancement in polar wind electric field. Indeed, data from Akebono [Abe et al., 1993a, b] indicate that polar oxygen ions have larger outflow velocities on the dayside, the sector where photoelectrons are created. Inclusion of oxygen dynamics may not be a straightforward matter. It probably requires the use of a description other than the 16-moment approximation, because its stiffness may increase with the inclusion of the dynamics of a heavier ion.

Also, the calculation of the electric field enhancement itself needs to be improved, as it neglects the bulk electron dynamics. While the increase in the electric field enhances the ion outflow, it can also suppress the outward flow of the bulk electrons. Such modifications of the thermal particle transport will have a feedback effect on the polar wind electric field itself. In fact, all the polar wind components (ion species, thermal electrons, photoelectrons) and the ambipolar electric field all influence each other. To determine the electric field, we need to take these self-consistent effects into account.

The results presented here clearly warrant a more detailed investigation of the role of photoelectrons in the polar wind. For this purpose, we are currently developing a model that should remedy the shortcomings discussed above, namely to include heavy-ion dynamics and to determine the ambipolar electric field self-consistently. In this model, both ions and suprathermal photoelectrons are treated kinetically, while the bulk electrons are described using a robust low-order fluid model. The model is still a kinetic-fluid hybrid, but it now has an expanded collisional kinetic part. The self-consistent ambipolar field is determined by using an iterative scheme. Our preliminary results indicate that this model is quite tractable, and that the iterative scheme converges well. These results will be discussed in a forthcoming publication [Tam et al., 1995].

Acknowledgements. The authors would like to thank Andrew W. Yau for discussions, and Christian T. Dum for his assistance in helping us transport the sixteen-moment code to the YMP at NCSA. This research is partially supported by NASA Grant Numbers NAG5-225 and NAGW-1532, AFOSR Grant Number F49620-93-1-0287, and Phillips Laboratory Contract Number F19628-91-K-0043.

REFERENCES

Abe, T., B. A. Whalen, A. W. Yau, R. E. Horita, S. Watanabe, and E. Sagawa, EXOS D (Akebon1) suprathermal mass spectrometer observations of the polar wind, *J. Geophys. Res.*, 98, 11191, 1993a.

Abe, T., B. A. Whalen, A. W. Yau, S. Watanabe, E. Sagawa, and K. I. Oyama, Altitude profile of the polar wind velocity and its relationship to ionospheric conditions, *Geophys. Res. Lett.*, 20, 2825, 1993b.

Barakat, A. R. and R. W. Schunk, Transport equations for multicomponent anisotropic space plasmas: A review, *Plasma Phys.*, 24, 389, 1982.

Barakat, A. R. and R. W. Schunk, Effect of hot electrons on the polar wind, *J. Geophys. Res.*, 89, 9771, 1984.

Cladis, J. B., Parallel acceleration and transport of ions from polar ionosphere to plasma sheet, *Geophys. Res. Lett.*, 13, 893, 1986.

Ganguli, S. B., H. G. Mitchell, Jr., and P. J. Palmadesso, Behavior of ionized plasma in the high latitude topside ionosphere: the polar wind, *Planet. Space Sci.*, 35, 703, 1987.

Ganguli, S. B., T. Chang, F. Yasseen, and J. M. Retterer, Plasma transport modeling using a combined kinetic and fluid approach, in *Physics of Space Plasmas (1992), SPI Conference Proceedings and Reprint Series, no. 12*, edited by T. Chang and J. R. Jasperse, p. 393, Scientific Publishers, Inc., Cambridge, MA, 1992.

Ho, C. W., J. L. Horwitz, N. Singh, G. R. Wilson, and T. E. Moore, Effects of magnetospheric electrons on polar plasma outflow: a semikinetic model, *J. Geophys. Res.*, 97, 8425, 1992.

Horwitz, J. L., C. W. Ho, H. D. Scarbro, G. R. Wilson, and T. E. Moore, Centrifugal acceleration of the polar wind, *J. Geophys. Res.*, 99, 15051, 1994.

Johnstone, A. D. and J. D. Winningham, Satellite observations of suprathermal electron bursts, *J. Geophys. Res.*, 87, 2321, 1982.

Khan, S. A. and T. D. Rognlien, Thermal heat flux in a plasma for arbitrary collisionality, *Phys. Fluids*, 24, 1442, 1981.

Khazanov, G. V., M. W. Liemohn, T. I. Gombosi, and A. F. Nagy, Non-steady-state transport of superthermal electrons in the plasmasphere, *Geophys. Res. Lett.*, 20, 2821, 1993.

Lee, J. S., J. P. Doering, T. A. Potemra, S. B. Ganguli, and L. H. Brace, Measurements of the ambient photoelectron spectrum from Atmosphere Explorer: II. AE-E measurements from 300 to 1000 km during solar minimum conditions, *Planet. Space Sci., 28*, 973, 1980.

Lemaire, J., Effect of escaping photoelectrons in a polar exospheric model, *Space Res., 12*, 1413, 1972.

Olbert, S., Role of thermal conduction in the acceleration of the solar wind, *NASA Conf. Publ.*, 149, 1982.

Retterer, J. M., T. Chang, and J. R. Jasperse, Ion acceleration in the suprauroral region: a Monte Carlo model, *Geophys. Res. Lett., 10*, 583, 1983.

Schunk, R. W., Mathematical structure of transport equations for multispecies flow, *Rev. Geophys. Space Phys., 15*, 429, 1977.

Scudder, J. D. and S. Olbert, A theory of local and global processes which affect solar wind electrons: 1. The origin of typical 1 AU velocity distribution functions – steady state theory, *J. Geophys. Res., 84*, 2755, 1979.

Tam, S. W. Y., F. Yasseen, T. Chang, and S. B. Ganguli, Kinetic photoelectron effects on the polar wind, *CSR-95-02*, MIT, 1995, To appear in Geophys. Res. Lett..

Winningham, J. D. and W. J. Heikkila, Polar cap auroral electron fluxes observed with ISIS-1, *J. Geophys. Res., 79*, 949, 1974.

Winningham, J. D. and C. Gurgiolo, DE-2 Photoelectron measurements consistent with a large scale parallel electric field over the polar cap, *Geophys. Res. Lett., 9*, 977, 1982.

Yasseen, F., J. M. Retterer, T. Chang, and J. D. Winningham, Monte-Carlo modeling of polar wind photoelectron distributions with anomalous heat flux, *Geophys. Res. Lett., 16*, 1023, 1989.

Yau, A. W., T. Abe, T. Chang, T. Mukai, K. I. Oyama, and B. A. Whalen, Akebono observations of electron temperature anisotropy in the polar wind, to appear in *J. Geophys. Res.*

S. W. Y. Tam, F. Yasseen, T. S. Chang, Center for Space Research, Massachusetts Institute of Technology, Cambridge, MA 02139.

S. B. Ganguli, Science Applications International Corporation, McLean, VA 22101.

J. M. Retterer, Geophysics Directorate, Phillips Laboratory, Bedford, MA 01731.

Coupling of Micro- and Mesoscale Processes in the Polar Wind Plasma Transport: A Generalized Fluid Model with Microprocesses

Supriya B. Ganguli

Plasma Physics Division, Science applications International Corporation, McLean, Virginia

The classical polar wind is an ambipolar outflow of thermal plasma from the terrestrial ionosphere at high latitudes along convecting flux tubes. Theory and observations have proved that wave-particle interactions play a major role in polar wind plasma outflow. Time-dependent one-dimensional generalized fluid simulations show that when the effects of plasma microscopic processes are included in the mesoscale field-aligned plasma outflow modeling there is a major impact on the results. The polar wind ion gas experiences adiabatic cooling as it expands along the diverging geomagnetic flux tubes with temperature anisotropy $T_{p\parallel} > T_{p\perp}$. However, when the effects of plasma microprocesses were considered, plasma energization in the direction perpendicular to the geomagnetic field lines occurs. The polar wind H^+ ion anisotropy is reversed to $T_{p\perp} > T_{p\parallel}$, consistent with ion energization observations. O^+ ion energizations due to microprocesses explain the presence of O^{+} ions in the polar magnetosphere. We have also performed three-dimensional time-dependent simulations of the polar wind. The results indicate the presence of significant cross-field transport that modifies the field-aligned flow. How this cross-field transport will modify the results of our one-dimensional simulations, that couple the micro and mesoscale processes in the plasma outflow modeling, remains to be investigated.

INTRODUCTION

The "classical" polar wind is an ambipolar outflow of thermal plasma from the high latitude terrestrial ionosphere to the magnetosphere. Measurements [*Waite et al.*, 1985, *Lockwood et al.*, 1985, *Yau et al.*, 1985 and *Moore et al.*, 1986] have indicated that the polar wind contains suprathermal components of both light and heavy ions. Although it was initially believed that O^+ ions play a major role only at low altitudes, it is now clear from observations that relatively large amounts of suprathermal and energetic O^+ ions are present in the polar magnetosphere.

It has been demonstrated by both observations and theory that the effects of kinetic wave particle interactions play an important role on the macroscopic polar plasma outflow [*Ganguli and Palmadesso*, 1987; *Barakat et al.*, 1995]. Plasma energization [*Lundin and Eliasson*, 1991; *Hultqvist*, 1991; *Moore et al.*, 1986; *Reiff et al.*, 1988] is one of the important mechanisms by which ionospheric plasma escapes into the magnetosphere, and this energization is a consequence of transport processes. Consequently, many mesoscale models were developed to incorporate these micro-processes. Several mechanisms have been proposed theoretically to account for O^+ acceleration to high altitudes based on coupling between micro and mesoscale processes [*Ganguli et al.*, 1991, *Brown et al.*, 1991, *Chen and Ashour-Abdalla*, 1992] and *Barakat et al.* [1994a, b].

A three-dimensional kinetic transport model is in principle most desirable for this task. However, given the state-of-the-art in computer technology this is not currently feasible. Therefore, it is generally necessary to use fluid theory to construct numerical models of large-scale phenomena even when kinetic effects play an important role in the dynamics. The temporal and spatial scales of kinetic theory are tied to plasma frequencies, gyro radii, etc., which are orders of magnitude smaller than the characteristic scale times and sizes associated with large or mesoscale phenomena. In general, existing computers are not yet capable of directly handling this broad range of scale sizes. For this reason the most optimal approach is to use a generalized fluid system of equations, where higher moments can deal with temperature and heat flow anisotropies of the magnetosphere and macroscopic effects of microscopic instabilities present in the magnetosphere can be included via anomalous transport coefficients.

In principle, the major macroscopic effects of most, if not all, of these processes can be modeled within the framework of a multimoment, multispecies simulation code with anomalous transport co-efficients, but each process would need to be studied separately and incorporated in such a way that the essential physics is preserved. Clearly, the development of generalized models of this type must proceed in a careful and systematic way, beginning with a relatively simple and tractable subset of microscopic phenomena and increasing complexity in steps. Ganguli and Palmadesso [1987] were the first to introduce a scheme, whereby the macroscopic effects of micro-processes on large-scale plasma transport, based on a generalized 16- moment system of transport equations, can be modeled.

In this paper we will demonstrate how the inclusion of micro-processes can change the mesopheric field-aligned plasma transport processes. Higher dimensional effects, such as cross-field transport, may modify these results. In order to investigate such effects we have generalized our model to three-dimensions.

MODEL AND RESULTS

Ganguli et al. [1985, 1987] have simulated the time-dependent polar wind outflow. The model used the one-dimensional version of the 16-moment set of equations of Barakat and Schunk [1982] (see Appendix I) and solved the coupled system of equations for supersonic H^+, O^+, and e^- plasma outflow extending from 1500 km to 10 R_E. The 16-moment set of equations consider continuity, momentum, both parallel and perpendicular heat flows in conjunction with parallel and perpendicular temperatures. The collision terms used are Burger's [1969] for the case of Coulomb collisions with corrections for finite species velocity differences (see Appendix II). The collision-dominated region was below 2500 and similar ion and electron anisotropies were obtained. The effect of the mirror force was studied. For classical polar wind H^+ ions, $T_{p\parallel} > T_{p\perp}$ and the direction of heat flow is positive (upward flow from the ionosphere to the magnetosphere).

DE-1 satellite observations of Biddle et al. [1985] demonstrated a capability to observe transport effects in plasmas as higher moments of the distribution function, making possible comparisons with theories. The observed H^+ ion heat flux was positive as predicted by Ganguli et al. [1985, 1987]. The Mach number calculated from this model compared well with the observations of Nagai et al. [1984].

Ganguli and Palmadesso [1987] and Ganguli et al. [1988; 1991] introduced coupling of micro and mesoscale processes in space plasma transport. They have studied the important consequences of plasma collective effects, such as the high frequency ($\omega \sim \Omega_i$) current driven ion cyclotron instability (EIC) on transport phenomena and vice-versa. The EIC instability is well known to produce strong ion heating and play a role in ion conic formation [Crew et al., 1990; André and Chang, 1992]. H^+ energization due to light ion cyclotron waves, O^+ energization due to heavy ion cyclotron waves, and an EIC-related dc anomalous resistivity process for electrons, are studied. It is demonstrated that when anomalous resistivity and ion heating due to this instability are included via anomalous transport coefficients, there is a major impact on the results. For example, for the polar wind outflow the ion gas will experience adiabatic cooling as it expands along the diverging geomagnetic flux tubes. However, when the effects of the observed ion cyclotron waves in this region [Cattell et al., 1991] were considered, plasma energization in the direction perpendicular to the geomagnetic field lines occurs. The polar wind H^+ ion anisotropy is reversed to $T_{p\perp} > T_{p\parallel}$. Satellite observations have indicated similar ion energization [Moore et al., 1986].

Ganguli and Palmadesso [1987] have shown that the EIC instability turns on progressively at lower altitudes with the increase in the magnitude of the field-aligned current. Low-altitude EIC-induced transverse bulk ion heating decreases the magnitude of the ambipolar electric field, by increasing the H^+ scale height, and increases the mirror force experienced by the ions. As a result the H^+ in the polar wind are denser and gain speed more slowly at low altitudes but achieve higher final velocities than in the polar wind.

The critical drift velocity for exciting the EIC instability is an increasing function of the ratio $T_{i\perp}/T_{e\parallel}$ (here, i represents H^+ or O^+ depending on which instability is excited) [Ganguli et al., 1988; 1991]. When the electrons are heated simultaneously by the onset of anomalous resistivity (a microprocess) (Figure 1a), the critical velocity for exciting the EIC instability is lowered and the relative drift velocity between the ions and the electrons is increased (Figure 1b). This simultaneous electron heating due to a kinetic instability produces much higher H$^+$ and O$^+$ temperatures (macroprocess) (Figure 1c). This process demonstrates a positive feedback loop and the complex nature of interaction between microprocesses and large-scale macroscopic parallel dynamics in plasma outflow processes. Higher dimensional effects, capable of moderating these results, remain to be investigated.

Ganguli et al. [1991] show that heavy ion cyclotron waves are active below the O$^+$ - H$^+$ crossover point. This is due to the fact that the growth rate of the instability is proportional to the gyrofrequency (eB/mC), which means that lower altitudes in the presence of higher B field heavier oxygen ions are more unstable. O$^+$ perpendicular heating and subsequent acceleration of oxygen ions by oxygen cyclotron waves explains another possible mechanism by which the oxygen ions can be accelerated to higher altitudes.

These theoretical models are, however, restricted to only one-dimensional aspects of the transport phenomena. Cross-field transport, which can significantly impact the field-aligned dynamics, was not considered by these studies. Particle flows along the magnetic field are present almost everywhere in the magnetosphere and these flows are often inhomogeneous. Observations of sheared parallel flows (i.e., $dv_\parallel/dx \neq 0$, where $v_\parallel$ is the flow along the geomagnetic field) in the magnetosphere have been reported by D'Angelo, 1973; D'Angelo et al., 1974; Potemra et al., 1978; Heelis et al., 1984; Loranc, 1988; and Loranc et al., 1991. These sheared parallel flows can excite a low frequency ($\omega \ll \Omega_i$, the ion cyclotron frequency) instability [*D'Angelo, 1965*]. This was first observed in a laboratory experiment by D'Angelo and VonGoeler [1966] and in space by D'Angelo et al., 1974, and Potemra et al., 1978. D'Angelo [1965] termed this instability the Kelvin-Helmholtz mode. However, in order to distinguish this mode from the classical Kelvin-Helmholtz mode [*Raleigh, 1896; Drazin and Howard, 1966*] which is driven by the gradient of shear in transverse flows (i.e., $d^2v_\perp/dx^2$), and to classify a number of other high frequency shear driven modes, G. Ganguli et al. [1991, 1994] identified the $dv_\parallel/dx$ driven low frequency mode as the D'Angelo mode.

These low frequency instabilities can generate significant cross-field transport that can modify the field-aligned dynamics. It is evident from three-dimensional

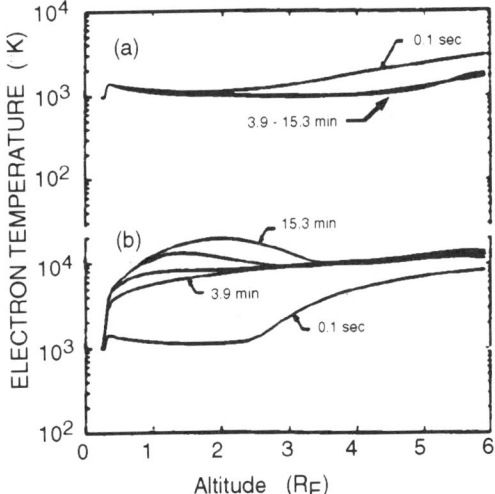

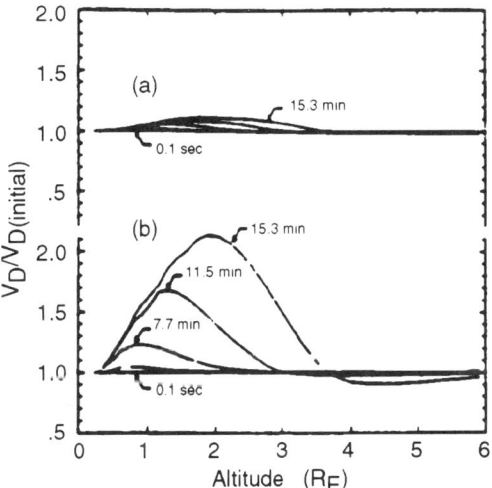

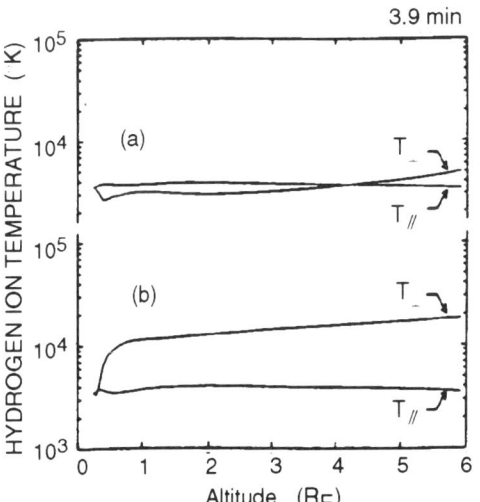

Figure 1: (a) Electron temperaure at different times with and without the effects of electron heating (b) relative drift velocity between the electrons and the ions with and without the effects of electron heating and (c) H$^+$ ion temperatures at different times with and without the effects of electron heating.

144 COUPLING OF MICRO- AND MESOSCALE PROCESSES

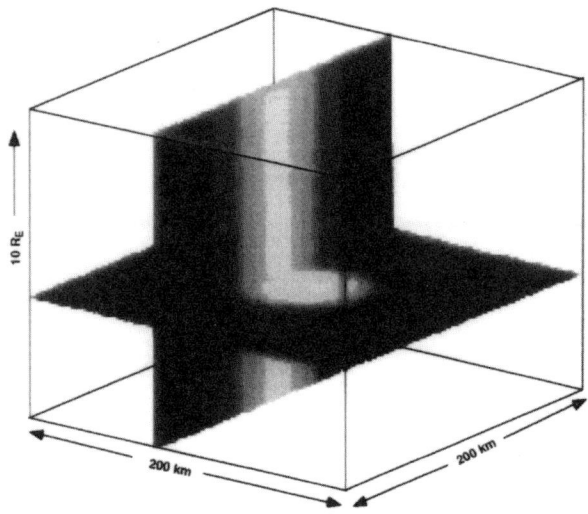

Figure 2: The initial velocity configuration at t = 0. Here, black indicates the highest and white the lowest velocity.

simulations of Dimitis et al. [1991] that the cross-field dynamics play an important role in plasma transport processes. They have also shown that cross-field transport generated by two-dimensional simulations is smaller than that generated by more realistic 3D simulations.

We have developed a 3D multimoment, multifluid model to study the polar wind [*Ganguli and Guzdar*, 1992; 1993]. Our 3D model allows for the self-consistent generation of a perpendicular electric field that initiates cross-field transport while at the same time, it is capable of preserving the extensive field-aligned dynamics of our 1D treatment [*Ganguli and Palmadesso*, 1987]. In this paper we solve the continuity and momentum equations in full 3D. The change from one to three dimensions, even at this level, introduces a number of interesting physical features which are of importance to transport processes. Here we will discuss the preliminary results of our study.

Our transport model is based on the set of 3D multimoment transport equations for multicomponent space plasmas given by Barakat and Schunk [1982]. The one-dimensional version of these equations are used in our 1D transport model [*Ganguli and Palmadesso*, 1987]. The 3D model is an explicit time-dependent solution of the coupled nonlinear set of the continuity equation and the momentum equation. These equations are as follows:

Continuity

$$\frac{\partial \rho_s}{\partial t} + \nabla \cdot (\rho_s \mathbf{u}_s) = 0 \qquad (1)$$

Momentum

$$\rho_s \frac{D_s \mathbf{u}_s}{Dt} + \nabla_\perp p_{s\perp} + \nabla_\parallel p_{s\parallel}$$
$$+ \nabla \cdot \tau_s - n_s e_s \left(\mathbf{E} + \frac{1}{c} \mathbf{u}_s \times \mathbf{B} \right) +$$
$$(p_{s\parallel} - p_{s\perp}) \nabla \cdot (\mathbf{e}_3 \mathbf{e}_3) = \delta M_s / \delta t \qquad (2)$$

where, **E** is the electric field, $(D_s/Dt) = (\partial/\partial t) + \mathbf{u}_s \cdot \nabla$ is the convective derivative, $r_s = n_s m_s$ is the mass density, $p_\parallel = nkT_\parallel$, $p_\perp = nkT_\perp$, $\mathbf{e}_3$ = unit vector along **B**, and t = stress tensor. The RHS of equation (A2) represent collision terms.

For simulations presented here, we have assumed that the plasma is collisionless and the temperature is isotropic. The interaction between electrons and ions is quasi neutral. The dynamics transverse to the ambient magnetic field is the $\mathbf{E} \times \mathbf{B}$ motion. The electric field will be calculated from the electron momentum equation (2). For this case, the electron response is assumed adiabatic with $e\phi/T_e = \ln(n)$, where $\mathbf{E} = -\nabla\phi$.

The computational geometry is such that z is the direction along the geomagnetic field lines and x, y are the two transverse directions. The region modeled extends from 1500 km to 10 R_E in the field-aligned direction (z) and 200 km in each of the two transverse directions (x and y). High resolution grid points are used with $N_x = N_y = 75$ and $N_z = 61$. The magnetic field is uniform in this simulation. The magnetic field inhomogeneity due to the dipolar geometry of the Earth's magnetic field is not expected to affect this investigation in a major way. The dipolar nature will introduce an inhomogeneity along the magnetic field (e.g., mirror force) which is likely to affect the location and the axial extent of the potential structures. Work, however, is in progress to incorporate the dipolar magnetic field geometry and will be reported later.

The leap-frog trapezoidal time stepping and spatial finite differencing schemes used in this simulation were developed by Zalesak [1981]. The boundary conditions are the same as that used by Ganguli and Palmadesso [1987]. The initial velocity configuration at t = 0 consists of a field-aligned flow that is uniform along the magnetic field and is cylindrical (Figure 2); it has maximum intensity at the center and decreases radially. This can be represented as $v_\parallel = v_{\parallel 0} \exp(-((x-0.5)^2 + (y-0.5)^2)/L^2_v)$ with $v_{\parallel 0} = 10$ km/sec and $L_v = 100$ km. The electron temperature, $T_e = 1000°K$ and the ion temperature, $T_i = 3000°K$.

The results of this current-free polar wind simulation are shown in Figures 3 - 7. Figure 3 shows contour plots of

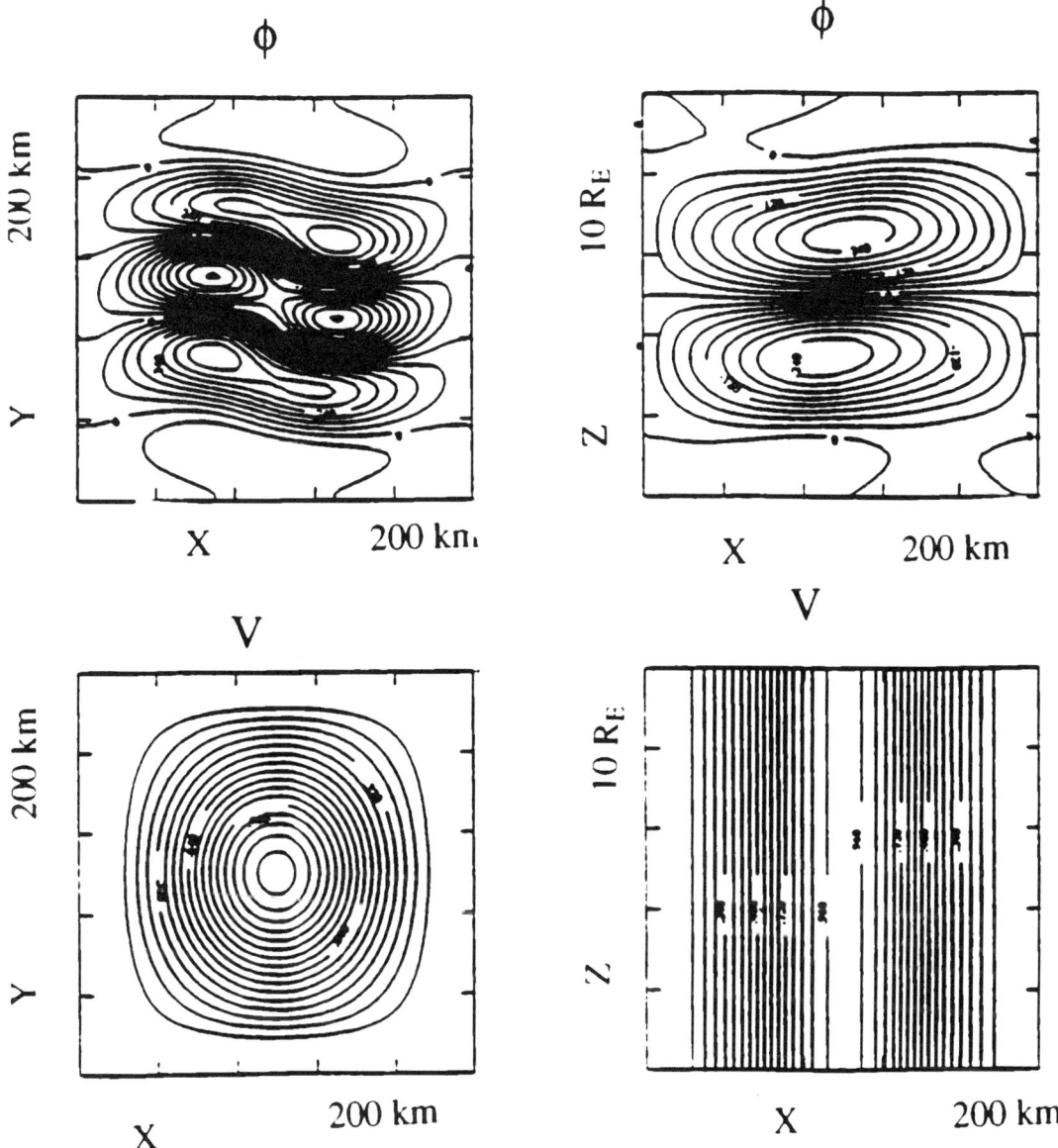

Figure 3: Contour plots of (a) the potential ϕ in the transverse plane (x, y) at 5 R_E, (b) the potential in the field-aligned plane (x, z), (c) the velocity in the transverse plane (x, y) at 5 R_E, (d) the field-aligned velocity in the (x, z) plane at t = 10 s.

(a) the potential ϕ in the transverse plane (x, y) at 5 R_E, (b) the potential in the field-aligned plane (x, z), (c) the velocity in the transverse plane (x, y) at 5 R_E, (d) the field-aligned velocity in the (x, z) plane at t = 10 s. The maximum potential at this time is 0.1 eV. The velocity shear-driven D'Angelo instability develops with dominant mode m = 3. In this early linear phase, contours of $V_\parallel$ (Figures 3 c, d) are not large enough to affect the equilibrium flow. The average flow velocity at this time is still the initial flow introduced to the system by the source.

The average momentum flux is also small at this time since fluctuations are not yet large enough to produce significant transport.

Figure 4 shows similar 2D contour plots of the nonlinear stage of the instability at t = 60 seconds. The nonlinear convection of $V_\parallel$ excites the D'Angelo instability with dominant m = 4 mode (Figure 4a). The instability generates significant cross-field transport which alters the field-aligned flow. The magnitude of the field-aligned flow decreases by 25% of its initial value.

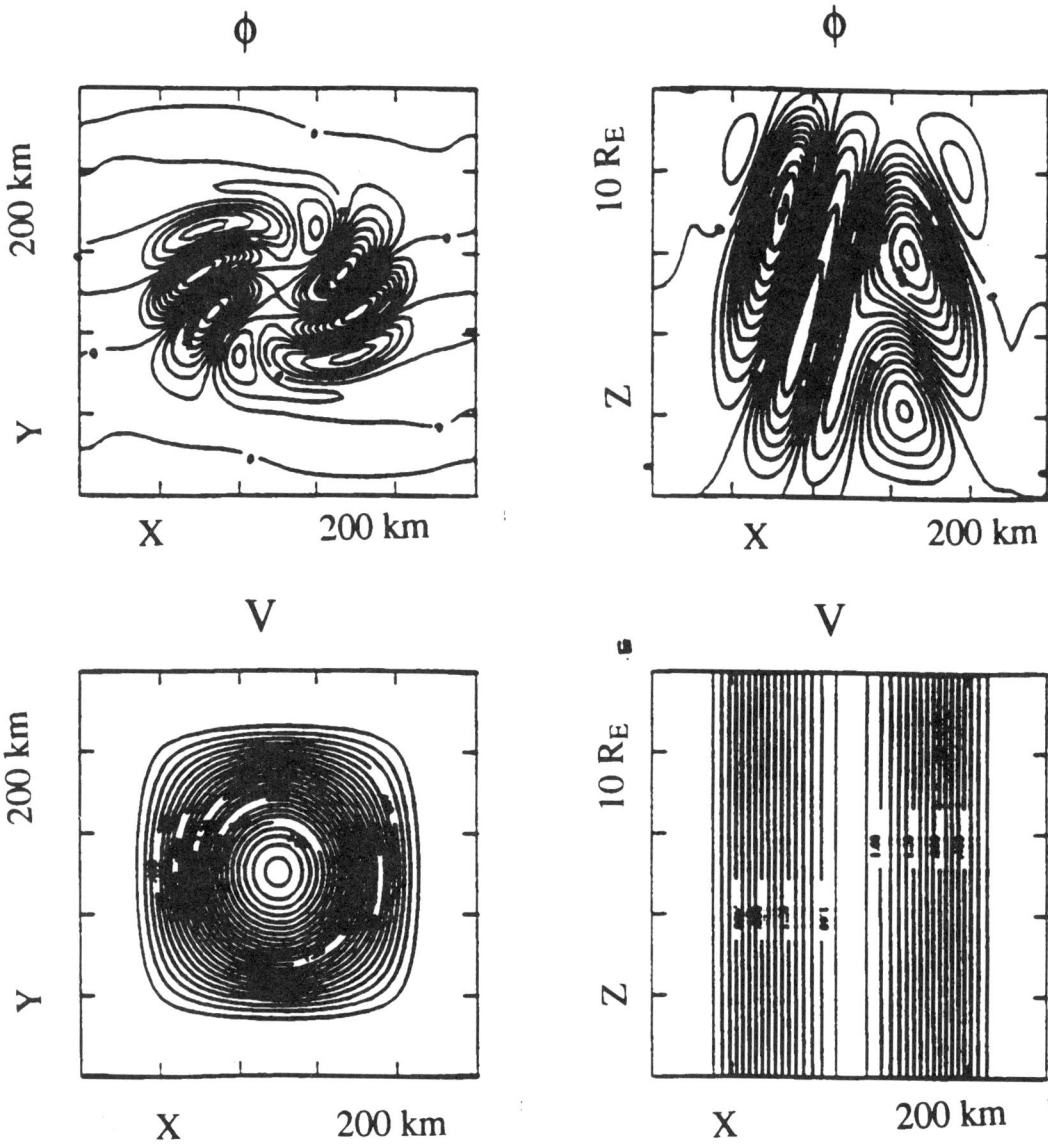

Figure 4: The same contour plots as in Figure 2 for t = 60 s.

The 3D morphology and its evolution can be seen in gray scale in Figures 5 - 7. Here, the light gray color indicates the highest and dark gray the lowest intensity. These 3D representations can be directly compared with experimental observations. Figure 5 shows the density distribution on a velocity surface of 7 km/sec. Figure 6 shows potential distribution at 40 second, with maximum potential = 1.0 eV. Figure 7 shows isosurface of potential distribution. Currently, we are working with experimentalists to further develop our 3D representations to facilitate comparison of direct experimental data.

In the future, we will add detailed physics, namely temperature anisotropy, heat flow anisotropy and kinetic instabilities to our model, as we have done in our 1D studies [*Ganguli et al.*, 1988]. Coupling between micro and mesoscale phenomena will be studied using the three-dimensional model. The three-dimensional treatments are considerably more complex and computationally intensive. The simplicity of the one-dimensional formalism is lost. The one-dimensional models are well suited for investigations of field aligned aspects of the transport problem and develop initial pictures which are

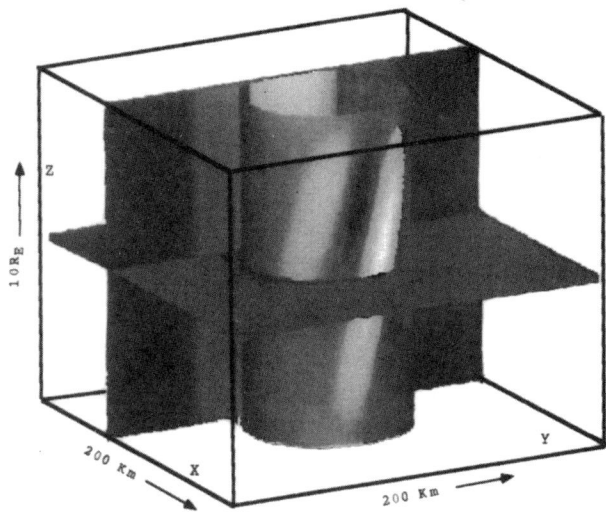

Figure 5: The density distribution on a velocity surface of 7 km/sec.

useful in guiding a full three-dimensional treatment. The three-dimensional studies on the other hand are necessary for other aspects such as cross-field dynamics, its contributions to the overall transport problem, and its role in influencing the field aligned dynamics. Here, we provide a formalism in which the important effects of the low frequency instabilities on space plasma transport processes can be studied self-consistently.

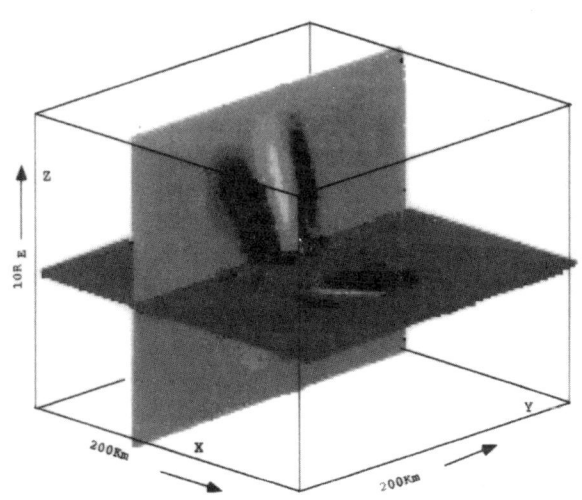

Figure 6: The potential distribution with maximum potential = 1.0 eV.

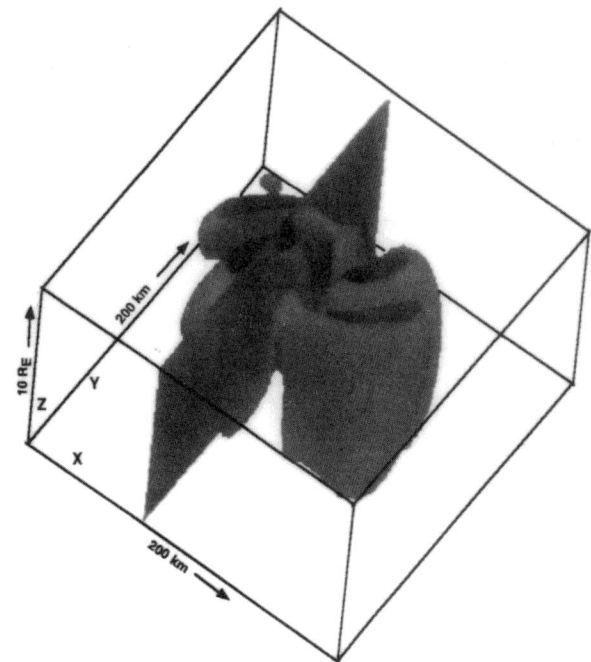

Figure 7: Isosurface of potential distribution.

APPENDIX I

Continuity

$$\frac{\partial n}{\partial t} + \frac{\partial nv}{\partial r} + \dot{A}nv = \frac{\delta n}{\delta t} \quad (1)$$

Momentum

$$\frac{\partial v}{\partial t} + v\frac{\partial v}{\partial r} + \frac{k}{mn}\frac{\partial nT_{\parallel}}{\partial r} + \frac{GM}{r^2} - \frac{eE}{m}$$
$$+ \frac{\dot{A}k}{m}(T_{\parallel} - T_{\perp}) = \frac{\delta v}{\delta t} \quad (2)$$

Parallel energy

$$k\frac{\partial T_{\parallel}}{\partial t} + kv\frac{\partial T_{\parallel}}{\partial r} + \frac{2}{n}\frac{\partial nh_{\parallel}}{\partial r} + 2\dot{A}(h_{\parallel} - h_{\perp})$$
$$+ 2kT_{\parallel}\frac{\partial v}{\partial r} = k\frac{\delta T_{\parallel}}{\delta t} \quad (3)$$

Perpendicular energy

$$k\frac{\partial T_{\perp}}{\partial t} + kv\frac{\partial T_{\perp}}{\partial r} + \frac{1}{n}\frac{\partial nh_{\perp}}{\partial r}$$
$$+ \dot{A}(2h_{\perp} + kT_{\perp}v) = k\frac{\delta T_{\perp}}{\partial t} \quad (4)$$

Heat flow per particle for parallel energy

$$\frac{\partial h_\parallel}{\partial t} + v\frac{\partial h_\parallel}{\partial r} + 3h_\parallel\frac{\partial v}{\partial r} + \frac{3}{2}\frac{k^2 T_\parallel}{m}\frac{\partial T_\parallel}{\partial r} = \frac{\delta h_\parallel}{\delta t} \quad (5)$$

Heat flow per particle for perpendicular energy

$$\frac{\partial h_\perp}{\partial t} + v\frac{\partial h_\perp}{\partial r} + h_\perp\frac{\partial v}{\partial r} + \frac{k^2 T_\parallel}{m}\frac{\partial T_\perp}{\partial r}$$

$$+ \dot{A}\left(vh_\perp - \frac{T_\perp^2 k}{m}(T_\parallel - T_\perp)\right) = \frac{\delta h_\perp}{\delta t} \quad (6)$$

The scale of this model is large compared to the electron Debye length, so the transport equation (1) for electron number density may be replaced by an expression for charge neutrality,

$$n_e = n_p + n_o \quad (7)$$

where

n	number density;
v	species velocity;
A	cross-sectional area of a flux tube (proportional to 1/B)
$\dot{A}$	$=\frac{1}{A}\frac{\partial A}{\partial r} = -\frac{1}{B}\frac{\partial B}{\partial r} = \frac{3}{r}$;
B	magnetic field of the earth;
E	electric field parallel to the field line;
G	gravitational constant;
M	mass of the earth;
m	mass of the particular species;
$T_\parallel$	temperature of the species parallel to the field line;
$T_\perp$	temperature of the species perpendicular to the field line;
$h_\parallel$	flow of parallel thermal energy along B, per particle;
$h_\perp$	flow of perpendicular thermal energy along B, per particle;
k	Boltzmann constant;
e	electrons;
p	Hydrogen ions;
o	oxygen ions.

For a given moment F of the distribution function, δF represents the change in F due to the effects of collisions anomalous transport effects associated with plasma turbulence. The collision terms used in this model are Burgers [1969] collision terms for the case of Coulomb collisions with Burgers' corrections for finite species velocity differences.

Using equations (1) and (2), the oxygen ion density is calculated. We have assumed that the total flux tube current remains constant along the tube

$$I = eA(n_p V_p - n_e V_e)$$

which implies

$$V_e = \frac{1}{n_e}\left(n_p V_p - \frac{I}{eA}\right)$$

Using equations (2), (7), and (8), the electric field E parallel to the field line is calculated.

$$E = \frac{m_e}{en_e A}\frac{\partial}{\partial r}\left(n_p v_p^2 A - n_e v_e^2 A\right)$$

$$-\frac{k}{e}\left[\frac{\partial T_{e\parallel}}{\partial r} + \frac{T_{e\parallel}}{n_e}\frac{\partial n_e}{\partial r} + \frac{(T_{e\parallel} - T_{e\perp})}{A}\frac{\partial A}{\partial r}\right]$$

$$-\frac{n_o m_e GM}{n_e e r^2} + \frac{m_e}{e}\left[\frac{\delta v_e}{\delta t} - \frac{n_p}{n_e}\frac{\delta v_p}{\delta t}\right]$$

APPENDIX II

The collision terms used in this paper for the 16 moment system of transport equations are shown here. As explained before, these are Burgers' [1969[collision terms for the case of Coulomb collisions with corrections for finite species velocity differences. The subscripts b and a represent the species, in this case either the electrons or H+ or O+.

$$\frac{\delta n_b}{\delta t} = 0 \quad (1)$$

$$\frac{\delta v_b}{\delta t} = \sum_a v_{ba}(v_a - v_b)(1 + \Phi_{ba}) \quad (2)$$

$$k\frac{\delta T_{b\parallel}}{\delta t} = \sum_a \frac{m_b v_{ba}}{(m_b + m_a)} \times \left\{\frac{6}{5}kT_{a\perp} - \left[2 + \frac{4m_a}{5m_b}\right]kT_b\right.$$

$$+ \frac{4}{5}kT_{a\perp} + \frac{4m_a}{5m_b}kT_{b\perp}$$

$$\left. + \left[2kT_a + \left(4 + 6\frac{m_a}{m_b}\right)kT_b\right]\Phi_{ba}\right\} \quad (3)$$

$$k\frac{\delta T_{b\perp}}{\delta t} = \sum_a \frac{m_b v_{ba}}{(m_b + m_a)}$$

$$\left[3kT_a - 3kT_b + m_a(v_a - v_b)^2(1 + \Phi_{ba})\right]$$

$$-\frac{k}{2}\frac{\delta T_{b\parallel}}{\delta t} \quad (4)$$

$$\frac{\delta h_{b\parallel}}{\delta t} = -v_b h_{b\parallel} \quad (5)$$

$$\frac{\delta h_{b\perp}}{\delta t} = -v_b h_{b\perp} \quad (6)$$

Each sum includes all charged particle species in the simulation. The velocity-corrected Coulomb collision frequency v_{ba} is given by:

$$v_{ba} = \frac{\alpha n_a \sqrt{\Pi}}{3 m_b^2 m_a \alpha_{ba}^3} \, e_b^2 e_a^2 (m_b + m_a) \ln A \exp(-x_{ba}^2) \quad (7)$$

ln A is the Coulomb logarithm and

$$T_b = \frac{1}{3} T_{b\parallel} + \frac{2}{3} T_{b\perp} \quad (8)$$

$$\alpha_{ba}^2 = \frac{2kT_b}{m_b} + \frac{2kT_a}{m_a} \quad (9)$$

$$x_{ba}^2 = \frac{(v_b - v_a)^2}{\alpha_{ba}^2} \quad (10)$$

$$\Phi_{ba} = \frac{2}{5} x_{ba}^2 + \frac{4}{35} x_{ba}^4 + \frac{8}{315} x_{ba}^6 \quad (11)$$

APPENDIX III

In order to simulate ion heating and resistivity it is necessary to add an anomalous friction to our model. The anomalous collision frequency for the EIC instability is assumed to have a simple form,

$$v_{epH}^* = \alpha_H \Omega_p \left(\frac{n_p}{n_e}\right)\left(\frac{V_D}{V_{cH}} - 1\right) \quad \text{for} \quad V_D \geq V_{cH}$$
$$= 0 \quad \text{for} \quad V_D < V_{cH} \quad (1)$$

where V_D is the relative drift velocity between the electrons and the ions, Ω_p is the hydrogen ion cyclotron frequency, and α_H is an adjustable parameter which is chosen for numerical stability and to express qualitatively the effects of the anomalous collision processes.

The critical velocity for exciting the EIC instability (V_{cH}) is given by [Lee, 1972]

$$\frac{V_{cH}}{V_{therm_p}} \cong \frac{1 + \Delta(\mu_p^*)}{\Delta(\mu_p^*)} \left(\ln\left[2C\Gamma_1(\mu_p^*)\right]\right)^{1/2} \quad (2)$$

where

$$C \equiv \left(\frac{T_{p\parallel}}{T_{p\perp}}\right)\left(\frac{T_{e\parallel}}{T_{p\parallel}}\right)^{3/2}\left(\frac{m_p}{m_e}\right)^{1/2}$$

$$\Delta \equiv \frac{\Gamma_1(\mu_p)}{1 - G + \left(\frac{T_{p\perp}}{T_{p\parallel}}\right)\left(\frac{T_{p\parallel}}{T_{e\parallel}}\right)}$$

$$G \equiv \Gamma_1(\mu_p) + \frac{1 - \Gamma_o(\mu_p)}{\mu_p}$$

$$\mu_p = \frac{k_\perp^2 V_{th\perp p}^2}{2\Omega_p^2}$$

$$\Gamma_n(\mu_p) = e^{-\mu_p} I_n(\mu_p)$$

I_n is the modified Bessel function of order n; μ_p^* is the value of μ_1 which minimizes $(1+\Delta)/\Delta$, namely, $\mu_p^* \cong 1$, for this value of $\mu_p^* \cong 1$, $(1-G) = 0.258$, and $\Gamma_1(1) \cong 0.21$.

The perpendicular ion heating rate due to the EIC instability is given in terms of the effective collision frequency by

$$k \frac{\delta T_{p\perp}}{\delta t} = v_{peH}^* \, m_p V_D^2 \quad (3)$$

where

$$n_e m_e v_{epH}^* = n_p m_p v_{peH}^*$$

The anomalous collision frequency for the resistivity is given by

$$v_{epR}^* = \alpha_R \Omega_e \left(\frac{n_p}{n_e}\right)\left(\frac{V_D}{V_{cR}} - 1\right)^2 \quad V_D > V_{cH} \quad V_D > V_{cR}$$

where α_R is an adjustable parameter and depends on the details of the microphysics involved.

$$v_{epR}^* = 0 \quad V_D < V_{cR} \quad (4)$$

The critical velocity for anomalous resistivity (V_{cR}) is assumed to scale as follows.

$$V_{cR} = \frac{1}{4}\left(\frac{\Omega_e}{\omega_{pe}}\right) V_{ther\parallel e}$$

where ω_{pe} is the plasma frequency, Ω_e is the electron cyclotron frequency, and $V_{ther\parallel e}$ is the electron thermal velocity.

The flux is conserved, and therefore B is calculated from the total area A of the flux tube, such that AB = const. The electron heating is isotropic, and the heating rate due to anomalous resistivity is also expressed in terms of an effective collision frequency given in equation (4):

$$k \frac{\delta T_e}{\delta t} = v^*_{epR} m_e V_D^2 \ .$$

Acknowledgments. Stimulating discussions with Drs. G. Ganguli, T. E. Moore, R. W. Schunk, and J. L. Burch, are acknowledged. This work is supported by NASA, NSF and PSC Institute.

REFERENCES

André, M., and T. Chang, Ion heating perpendicular to the magnetic field, *Physics of Space Plasmas* (1992), Scientific Publishers, Inc., p. 35, 1992.

Barakat, A.R., and R.W. Schunk, *Plasma Phys.*, *24*, 389, 1982

Barakat, A.R. and R.W. Schunk, Stability of the polar wind, *J. Geophys. Res.*, *92*, 3409-3415, 1987.

Barakat, A. R. and I. A. Barghouthi, The effect of wave-particle interaction on the polar wind: Preliminary results, *Planet. Space Sci.*, in press, 1995.

Brown, D. G., G. R. Wilson, J. L. Horwitz, and D. L. Gallagher, *Geophys. Res. Lett.*, 18, 1841, 1991.

Biddle, A.P., T.E. Moore, and C.R. Chappell, Evidence for ion heat flux in the light ion polar wind, *J. Geophys. Res. 90*, 8552, 1985.

Bugers, J.M., Flow equations for composite gases, *Academic Press*, New York, 1969.

Cattell, C. A., et al., *J. Geophys. Res., 96,* 11421, 1991.

Chen, M.W., M. Ashour-Abdalla, and T.E. Holzer, *J. Geophys. Res., 92,* 19433, 1992.

Crew, G, et al., *J. Geophys. Res.,* 89, 2185, 1984.

D'Angelo, N., *Phys. Fluids, 8*, 1748, 1965.

D'Angelo, N., and S. VonGoeler, *Phys. Fluids, 9*, 309, 1966.

D'Angelo, N., *J. Geophys. Res., 78*, 1206, 1973.

D'Angelo, N., et al., *J. Geophys. Res., 79*, 3129, 1974.

Dimitis, et. al., *Phys. Fluids B, 3*, 620, 1991.

Drazin, P.G., and L.N. Howard, *Adv. Appl. Mech.*, Vol 9, Ch 1.

Ganguli, G., et al., Modeling magnetospheric plasma processes, *AGU Mono #62,* AGU, 17, 1991.

Ganguli, S. B., H.G. Mitchell, and P.J. Palmadesso, *Planet. Space Sci.*, *35*, 703, 1987.

Ganguli, S. B., and P.J. Palmadesso, *J. Geophys. Res.*, *92*, 8673, 1987.

Ganguli, S. B., P.J. Palmadesso, and H.G. Mitchell, *Geophys. Res. Lett., 15*, 1291, 1988.

Ganguli, S. B., H.G. Mitchell, and P.J. Palmadesso, p. 197, in *Physics of Space Plasmas,* T. Chang and G. B. Crew editors, 1991.

Ganguli, S.B., and P.N. Guzdar, Three-dimensional plasma transport in planetary magneto-spheres, 3rd Huntsville workshop on M/I plasma models, MS1, 1992.

Ganguli, S.B., and P.N. Guzdar, 3D polar plasma outflow, *EOS, AGU*, Fall 1993.

Heelis, R.A., et al., *J. Geophys. Res., 89*, 3893, 1984.

Hultqvist, B., *J. Atmos. and Terr. Phys., 53*, 3, 1991.

Lockwood, M., et al., *J. Geophys. Res., 90,* 1985.

Loranc, M., PhD. thesis, UTD, 1988.

Loranc, M., et al., *J. Geophys. Res., 96*, 3627, 1991.

Lundin, R., and L. Eliasson, *Ann. Geophysicae, 9,* 202, 1991.

Moore, T. E., C. J. Pollock, R. L. Arnoldy and P. M. Kintner, Preferential O^+ heating in the topside ionosphere, *J. Geophys. Res.*, 13, 901, 1986.

Nagai, T., J.H. Waite, J.L. Green, C.R. Chappell, R.C. Olsen, and R.H. Comfort, First measurements of supersonic polar wind in the polar magnetosphere, *Geophys. Res. Lett. ,11*, 669, 1984.

Potemra, T., et al., *J. Geophys. Res., 83*, 3877, 1978.

Reiff, P. H., et al., *J. Geophys. Res., 93*, 7441, 1988.

Lord Rayleigh, Theory of Sound, MacMillan, London, 1896 (reprinted 1940), Vol. II, Ch 21.

Waite, J.H., et al., *J. Geophys. Res., 90*, 1619, 1985.

Yau, A.W., et al., Physics of Space Plasmas, *Scientific Pub., 77,* 1985.

Zalesak, S.T., *J. Comput. Phys., 40,* 497, 1981.

S.B. Ganguli, Applied Physics Operation, Science Applications International Corporation, 1710 Goodridge Drive, McLean, VA 22102.

Single Ion Dynamics and Multiscale Phenomena

P. L. Rothwell

Geophysics Research Directorate, Phillips Laboratory, Hanscom AFB, Bedford, Massachusetts

M. B. Silevitch

Center for Electromagnetics Research, Northeastern University, Boston, Massachusetts

Lars P. Block and Carl-Gunne Fälthammer

Division of Plasma Physics, Alfvén Laboratory, Royal Institute of Technology, Stockholm, Sweden

The magnetosphere is populated by hot, tenuous plasma. Therefore, it is expected that at times electric fields will dominate the single ion dynamics which invalidates the usual fluid MHD description. We have found two such examples which we review in this paper. (1) The effect of a large scale electric field gradient on the single ion dynamics which leads to density striations and possible auroral arc formation. (2) Large spatial variations of the electric field on the scale of the ion gyro radius which causes chaotic untrapping of O^+ ions to occur.

1. INTRODUCTION

In this paper we review the interplay between micro and meso-scale phenomena that we have found and highlight the conditions underwhich an MHD approach is not valid. This is done by examining single ion dynamics in a spatially varying electric field. For example, we examined [Rothwell et al. 1994] the effect of the electric field variation near the equatorial Harang discontinuity on single ions as they drift earthward from the magnetotail. We found that under substorm growth phase conditions single ion trajectories were modified and caused macroscopic density striations if the electric field gradient is sufficiently strong. Conservation of the associated inertial current implied a connection between the striations and auroral arcs. Similarly if the electric field has a sufficiently large second derivative in the electric field then the ion gyro orbits become very distorted with the gyro velocity being highly variable over a gyro orbit. The problem is analogous to that of a finite pendulum. Just as a finite pendulum if driven sufficiently hard will pass from an oscillating mode to a rotating one a gyrating ion will become unmagnetized if the second derivative of the electric field is sufficiently large. This can cause heavy ions to become chaotically untrapped [Rothwell et al., 1995]. In this paper we briefly review this work with emphasis on the physical concepts.

2. TWO EXAMPLES OF MULTISCALE PHENOMENA

Constant First Derivative in E_x. We begin by looking at the simple case of a constant electric field gradient. The equations of motion [Cole 1976] are given by

$$\frac{dV_x}{dt} = \frac{e}{M}(E_x(x(t)) + V_y B)$$

$$\frac{dV_y}{dt} = -\frac{e}{M} V_x B \tag{1}$$

which can be combined into a single equation

$$\frac{d^2 V_x}{dt^2} + (\omega^2 - \frac{e}{M}\frac{dE_x}{dx})V_x = 0 \tag{2}$$

152 SINGLE ION DYNAMICS

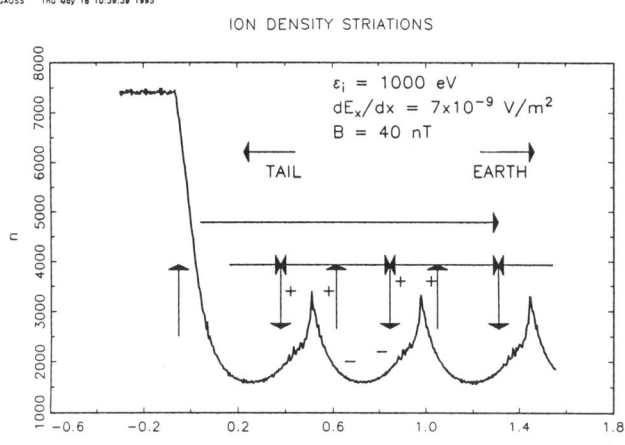

Fig. 1. A computer simulation of O^+ ions ExB drifting towards the earth and encountering an electric field gradient at x=0. Phase bunching causes density striations to form in the gradient region which could be a source of multiple auroral arcs.

where we define the gyro-frequency to be

$$\Omega^2 = \omega^2 - \frac{e}{M}\frac{dE_x}{dx} \qquad (3)$$

The main effect of a spatial gradient in E_x is to modify the gyrofrequency. The symbol ω denotes the gyrofrequency eB/M. It is immediately seen that if $\Omega^2 < 0$ then V_x has an exponential rather than an oscillatory solution. That is, if the electric field gradient is too steep the ions become locally untrapped. This effect becomes significant in regions of the magnetosphere where the magnetic field is weak and the electric field gradient is strong. One such region is the equatorial Harang discontinuity. For example, (3) predicts that O^+ ions will become untrapped in a 40γ magnetic field if $dE_x/dx > 9.6 \times 10^{-9}$ V/m².

The next questions to ask relates to what happens when an ensemble of O^+ ions EXB drift from the magnetotail into a region of significant earthward (positive) electric field gradient. It has been noted by Daglis et al. [1991] and others that during active periods there is an efficient transport of ionospheric O^+ to the plasma sheet. What effect does this global transport have interfacing with the mesoscale electric field structure of the Harang discontinuity? How does this interfacing create microscale structure and do the different processes acting at different scales reach some form of equilibrium?

Density Striations. Oxygen ions injected into the plasma sheet drift earthward due to a cross-tail electric field E_y. The solution to (1) with an E_y term was reported by Rothwell et al. [1994]. The earthward drift velocity V_{xd} in the region of finite dE_x/dx becomes

$$V_{xd} = \frac{\omega^2}{\Omega^2}\frac{E_y}{B} \qquad (4)$$

so that if dE_x/dx is positive then $V_{xd} > E_y/B$. Earthward drifting ions that encounter a region of $dE_x/dx > 0$ acquires a higher drift velocity than they had outside the region. Before encountering the E-field gradient the ions are uniformly distributed in phase angle in the E_y/B drift frame. Upon encountering the gradient region the ions acquire a faster drift velocity. The acceleration of the drift frame imparts a negative velocity component in that frame to each ion. This causes the ions to bunch in phase in the tailward direction. They then gyrate as a group. Where the ions have their turning points in the gradient region, density enhancements or striations form [Rothwell et al. 1994].

Figure 1 illustrates the effect using a computer simulation. Monoenergetic 1 keV ions are injected tailward of the gradient region which begins at x=0. The increase in the ion drift velocity in the gradient region causes the density n to drop from its previous value which requires an upward field aligned current near x=0 to maintain charge quasi-neutrality. The long horizontal arrow denotes an earthward inertial current J_x which is the continuation of this upward current in the magnetosphere. In this picture the current J_x closes on the earthward side of the Harang discontinuity to the ionosphere, creating a macroscopic radial current system between the ionosphere and the equatorial plane.

The presence of density striations modulates the earthward inertial current J_x and is denoted by the shorter arrows in Figure 1. The modulation creates a series of smaller current wedges that are related to the mutiple arcs which are symbolized by the upward pointing arrows. Physical insight can by gained by deriving an expression for J_x. We assume a coordinate system such that positive x is earthward, y points westward and z is parallel to B, the magnetic field. We also assume that the number flux of ions is conserved as they drift earthward. This means that the earthward ion flux F_i in the electric field gradient region is the same as outside (i.e. $F_i = N_o E_y/B$ where N_o is the ambient ion number density in the plasma sheet). However, in the gradient region the ions are drifting faster than the electrons according to (4). Therefore, the average ion density in the gradient region is $N_o/(\omega^2/\Omega^2)$. Charge neutrality requires that the electron number density is the same as the ion number density. This can only be achieved by a magnetic-field-aligned electron flux at the onset of the gradient region. The electron flux in the gradient region is $F_e = N_o(\Omega^2/\omega^2)(E_y/B)$. The net electric current is equal to $e(F_i - F_e)$ which can by using (3) be written as

$$J_x = (\frac{\rho_o E_y}{B})\frac{1}{B^2}\frac{dE_x}{dx}$$

$$J_x = \frac{\omega^2}{\Omega^2}\frac{\rho E_y}{B^3}\frac{dE_x}{dx} \qquad (5)$$

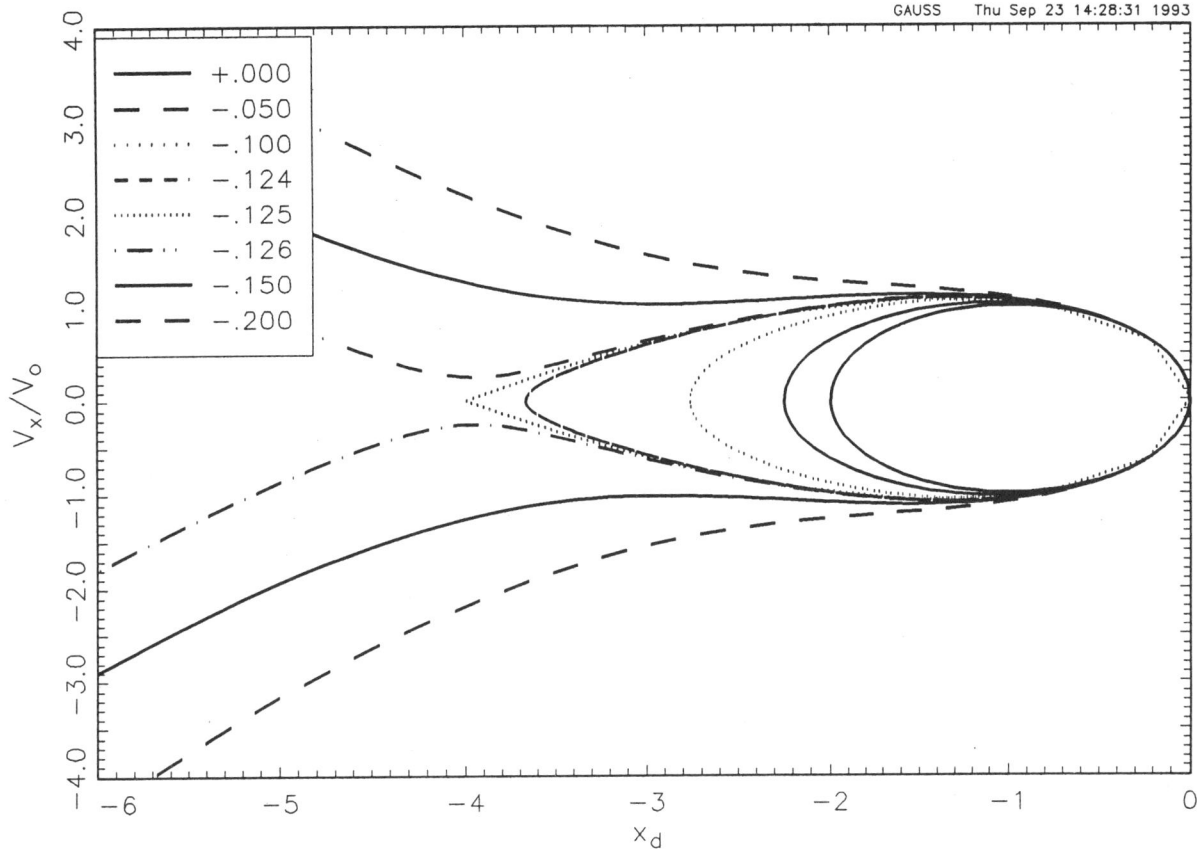

Fig. 2. Orbit shapes for negative d^2E/dx^2 and for $\alpha_o = -\pi/2$. Note that the transition to open (untrapped) orbits for $A_d \leq -1/8$ where A_d is a dimensionless representation of d^2E/dx^2. $x_d = -4$ is a critical point and is located at twice the ion gyro-diameter corresponding to the $d^2E/dx^2 = 0$ orbit. Similarly untrapping occurs for $\alpha_o = +\pi/2$, $x_d = +4$ when $A_d \geq +1/8$.

The symbol ρ denotes mass density in the gradient region and ρ_o refers to the mass density in the gradient-free region. The mass flux inside the bracket will be constant if the ions are conserved as they drift earthward. In this case J_x varies as B^{-2} rather than B^3.

The principle of quasi-neutrality requires that locally $N_e(x) \approx N_i(x)$. The electrons have an earthward flux $F_e = N_e(x)E_y/B$ where $N_e(x)$ approximates the local number density of the ions which is spatially dependent due to phase bunching. Satisfaction of the quasi-neutrality principle requires that electrons are free to move along magnetic field lines between the ionosphere and magnetosphere. In other words, the presence of density striations implies magnetospheric-ionospheric coupling and the formation of periodic auroral structures. Density striations and the principle of charge quasi-neutrality lead to perturbations in the earthward inertial current J_x (Figure 1) associated with the gradient region. Note that in regions of density enhancements the perturbation in J_x is tailward and in regions of density depletion it is earthward. This requires a downward field-aligned current where the density is increasing and an upward field-aligned current where the density is decreasing. Since upward field-aligned currents are carried in part by precipitating electrons this is where we locate the auroral arcs in this model. Reference is made to Rothwell et al. [1994] for more details. This is an example of how the meso-scale properties of the electric field modifies the micro-scale orbital characteristics of the single ions so as to produce unexpected micro-scale structure that may have geophysical significance.

For example, periodic arc structures spaced 35 km apart in the ionosphere correspond to magnetospheric density striations spaced approximately 2×10^6 apart at the equator. A simple calculation shows that the required electric field gradient in the equatorial plane is 5×10^{-9} V/m^2. This assumes $E_y = 1 \times 10^{-3}$ V/m and B = 40 nT.

Constant Second Derivative in E_x. We now look at how electric field structure on the scale size of an ion gyroradius can modify the gyro motion. The effect of a second spatial derivative of E_x will now be considered. The presence of a constant second derivative in E_x can be examined by expanding the first derivative about the initial position x_o of the ion.

$$\frac{dE_x}{dx} = \frac{dE_x}{dx}\Big|_{x=x_o} + \frac{d^2E_x}{dx^2}(x-x_o) \quad (6)$$

Then equation (1) can be rewritten as

$$\frac{d^2V_x}{dt^2} + \Omega_o^2 V_x - \frac{e}{M}\frac{d^2E_x}{dx^2}(x-x_o)V_x = 0 \quad (7)$$

where Ω_o is the gyro-frequency as defined in equation (3) with $dE_x/dx = dE_x/dx|_{x=xo}$. This is to be distinguished from the gyro-frequency Ω which reflects a constant second derivative in E_x. Equation (7) can be easily integrated by noting that $V_x = dx/dt$ in the second and third terms. A subsequent integration is also trivial after the previous result is multiplied by V_x. The final result is cast into the following form.

$$(\frac{dx_d}{d\tau})^2 = A_d(x_d - a_d)(x_d - b_d)(x_d - c_d) \quad (8)$$

where the subscript 'd' refers to dimensionless quantities.

Equation (8) is solved in terms of Jacobian Elliptic functions [Byrd and Friedman 1971]. The key point is that the usual concept of uniform circular gyration about the magnetic field line is not true in this case. The orbits are highly distorted by the electric field structure and the orbital speed is highly variable. Figure 2 shows the solutions to (8) for large, negative d^2E_x/dx^2. If the second derivative in E_x is sufficiently large the ions can become locally untrapped. There is a critical point at $A_d = -1/8$, where A_d is proportional to d_x^2E/dx^2. It is found that the untrapping criterion is dependent on the initial azimuthal phase angle of the ion and the sign of the second derivative. This untrapping criterion will be satisfied for 5 keV O^+ ions in a 40 nT magnetic field if $|d^2E/dx^2| \geq >40$ (mV/m)/R_E^2.

3. DISCUSSION AND CONCLUSIONS

Multi-scale phenomena has been shown for the two cases considered. In the first case a large scale electric field gradient was shown to produce a set of nested current systems between the ionosphere and the magnetosphere. Presently, we are investigating the self-consistency of the structure shown in Figure 1. That is, the upward current regions are associated with field-aligned potential drops. The question is whether the equatorial electric fields associated with these currents are sufficiently strong as to scatter the ions and, therebye, destroy the striations. The result depends on the auroral arc model used. In the second case small scale electric field structure strongly affected the orbital dynamics of trapped ions. In both examples that were considered the unifying idea is that when (3) becomes small or negative then finite orbit effects become important. Another important example that depends on this concept is the stochastic heating of ions [Rothwell et al. 1992].

A negative value of (3) implies that in the x-direction the ion's increase in momentum due to the electric field gradient is larger than the ion's decrease in momentum due to the magnetic field. This is the physical basis of untrapping.

4. REFERENCES

Byrd, P. F., and M. D. Friedman, *Handbook of Elliptic Integrals for Engineers and Scientists, 2nd edition, revised,* Springer-Verlag Publ., New York, 1971.

Cole, K. D., Effects of crossed magnetic and (spatially depen dent) electric fields on charged particle motion, *Planet. Space Sci.,* 24, 515-518, 1976.

Daglis, I. A., E. T. Sarris, and G. Kremser, Ionospheric contribution to the cross-tail current during the substorm growth phase, *J. Atmos. Terr. Phys.,* 53, 1091-1098, 1991.

Rothwell, P. L., M. B. Silevitch, L. P. Block, and C.-G. Fälthammar, Acceleration and stochastic heating of ions drifting through an auroral arc, *J. Geophys. Res.,* 97, 19,133-19,339, 1992.

Rothwell, Paul L., Michael B. Silevitch, Lars P. Block, and Carl-Gunne Fälthammar, O^+ phase bunching and auroral arc structure, *J. Geophys. Res.,* 99, 2461-2470, 1994.

Rothwell, Paul L., Michael B. Silevitch, Lars P. Block, and Carl-Gunne Fälthammar, Particle dynamics in a spatially varying electric field, accepted for publication in *J. Geophys. Res.,* 1995.

Problems in Simulating Ion Temperatures in Low Density Flux Tubes

R. H. Comfort and P. G. Richards

CSPAR, University of Alabama in Huntsville

P. D. Craven and M. O. Chandler

Space Science Laboratory, NASA Marshall Space Flight Center, Huntsville, Alabama

Observed ion temperatures in the outer plasmasphere, where densities are on the order of a few hundred per cubic centimeter, frequently exceed 10,000 K. Hydrodynamic models, such as the Field Line Interhemispheric Plasma (FLIP) model, have difficulty producing high ion temperatures for these low density flux tubes by the usual means of heating the ions via the thermal electrons. We present Dynamics Explorer observations of ion temperatures and illustrate these difficulties parametrically. We find that there are practical limits, related to Coulomb collisional energy transfer, for heating thermal ions through the electrons, so that observed ion temperatures cannot be produced. We then demonstrate that direct heating of the thermal ions does not suffer from this constraint, so that observed ion temperatures are produced. However, other effects seen in the FLIP model results and associated with achieving the high ion temperatures at high altitudes are in conflict with observations, including topside ionosphere temperatures which are too high and heavy ion densities at high altitudes which are too large. We suggest that these mesoscale effects are consequences of excessive thermal energy being transported to the ionosphere through microscale thermal conduction in the simulation.

INTRODUCTION

Observed ion temperatures in the outer plasmasphere (refilling flux tubes) are frequently high (> 10,000K) [*Comfort et al.*, 1988]. Even in the inner plasmasphere, numerical simulations made by the Field Line Interhemispheric Plasma (FLIP) code have required an additional source of heating for the thermal electrons, other than that provided by the conventional photoelectron heating, in order to produce observed the ion temperatures [*Newberry et al.*, 1989]. In this inner plasmasphere case, *Newberry et al.* [1989] found that if approximately 55% of the remaining photoelectron energy (after the conventional theory had been applied) were somehow trapped in the flux tube and used to heat the thermal electrons, the resulting thermal ion modeled temperatures would correspond to the levels observed, ~ 2000 K to 5000 K, depending on the local time. In the outer plasmasphere, the observed ion temperatures are much higher, typically by a factor of two or more and densities are lower, frequently by an order of magnitude or more, depending on the level of refilling. In addition, other physical mechanisms for ion heating are plausible in this region, such as Coulomb collisions with ring current ions [*Kozyra et al.*, 1987; *Fok et al.*, 1993] and wave particle interactions [*Khazanov*, 1995].

In this study, we investigate the response of the thermal ions to both indirect and direct ion heating. In the indirect case, the ions are heated through collisions with thermal electrons, which have themselves been heated by, for example, collisions with photoelectrons. In both cases, the heating mechanism is unspecified. The purpose is to examine the efficiency of these two paths in raising the ion temperature, rather than to study particular heating mechanisms. As will be seen below, these results can place restrictions on the types of heating mechanisms

156 ION TEMPERATURES AT LOW DENSITIES

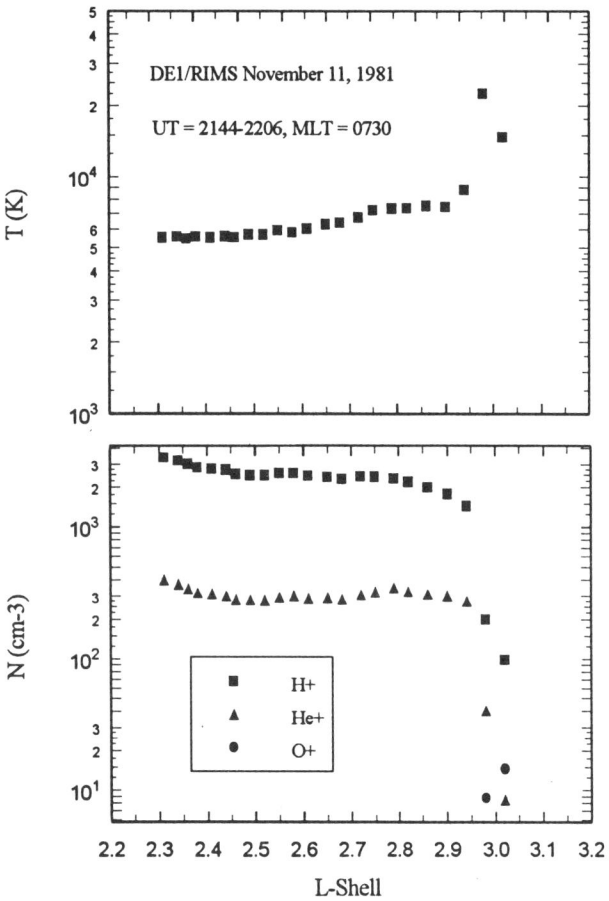

Figure 1. DE1/RIMS observations as a function of L-shell on an outbound plasmasphere transit for November 11, 1981. Upper Panel: H^+ temperatures. Lower panel: Ion densities for H^+, He^+, and O^+.

which need be considered in the outer plasmasphere.

We use satellite observations of ion temperature and densities of the major ions (H^+, He^+, and O^+), to provide realistic benchmarks for the results. Our approach is to study the temperature response of the ions parametrically, first through electron heating and then through direct ion heating. This is done by adding specific increments of energy in a simulation code and determining the ion temperature response. We also look briefly at the temporal response of the ion temperature to the heating and to subsequent cooling when the heat source is removed.

DATA AND MODEL

To assure that conditions are realistic, we use a case study approach. The observations we are trying to simulate were obtained at high altitudes by the Retarding Ion Mass Spectrometer (RIMS) on the Dynamics Explorer 1 spacecraft.

At low altitude, we also use a temperature measurement made by the RPA on DE2 within an hour of the high altitude measurements and approximately on the same field line. The RIMS system is described in detail by *Chappell et al.* [1981] and the data analysis methods by *Comfort et al.* [1982, 1985]. The RIMS observations are shown in Figure 1. The flux tube to be simulated is for L = 3, which is clearly in the refilling region of the outer plasmasphere. It is seen that the temperature in this region is near 15,000 K and the H^+ density is near 100 cm^{-3} at almost 13,000 km altitude. The magnetic local time of the observations is about 0730 hours, which is the reference local time for the numerical simulations. The RPA on DE2 is described in detail by *Hanson et al.* [1981]. The RPA temperature value, displayed below, is 2100 K at about 740 km altitude (from Figure 7 of *Horwitz et al.*, 1990]).

The simulation code employed is the FLIP model. It is a one-dimensional, hydrodynamic code which solves the electron and ion continuity and momentum equations along a magnetic flux tube from 120 km in the northern hemisphere to 120 km in the southern hemisphere to obtain (in this study) H^+, O^+ and He^+ densities. It also solves the two-stream photoelectron flux equations and the electron and ion energy equations, using a *Spitzer-Harm* [1953] thermal conductivity coefficient, along the entire flux tube to determine electron heating and T_e and T_i. It contains all known ionospheric chemistry for the ion species included, with H^+ and O^+ treated as major ions. The neutral atmosphere is provided by MSIS [*Hedin*, 1987]. Three-hour geomagnetic activity Ap indices are input as are the average F10.7 cm solar flux values. A more complete description is provided by *Richards et al.* [1994] and references therein.

Using the FLIP code, we first initialize the state of plasma in an L = 3 flux tube by running the code, with only the standard heating terms, for 74 hours to allow a diurnal cycle to be established and reduce sensitivity to any initial inconsistencies. Then we provide additional heating to the plasma in two ways, each one starting from the same initialized state: (1) we increase the trapping of photoelectrons which provides additional heat to thermal electrons, which in turn heat the thermal ions; (2) we heat the thermal ions directly. In both cases, the heating rate is increased parametrically until a clear pattern of the resulting ion temperature response is established. While we are attempting to match the observed ion temperature, we are at the same time trying to obtain the observed H^+ density, which represents most of the plasma number density. This was a primary factor in the selection of the initial density, and is affected by the length of time we allow the flux tube to fill. We found that it was necessary to begin the heating period (at 1200 LT) with an H+ density of 500 cm^{-3} in order to arrive at the local time of observation (0730) with the observed density of about 100 cm^{-3}. The imposed heating caused the H+ density to decrease, so that starting with lower densities initially (closer to that observed) resulted in densities that were too small at the local time when the comparison with observations was to be made. This is shown in simulations below.

RESULTS

Thermal Electron Heating

First, we examine ion heating through the intermediate process of heating thermal electrons, as typically occurs in the inner plasmasphere through photoelectron heating of thermal electrons. We do this by progressively increasing the trapping factor, α, of the photoelectrons. A trapping factor of 1 would indicate that all photoelectrons that escape from the ionosphere are trapped in the flux tube in making the transit to the conjugate ionosphere and eventually give up their energy to the thermal electrons in the flux tube. We begin with the initialized L = 3 flux tube plasma and continue the calculations in the diurnal cycle, but include additional energy input to the thermal electrons by means of the trapping factor, as described by *Newberry et al.* [1989]. In Figure 2 we show the ion temperature profiles for the initial state and for α = 0.7. Figure 3 shows how the modeled ion temperature at the equator, typically the maximum ion temperature in the flux tube,

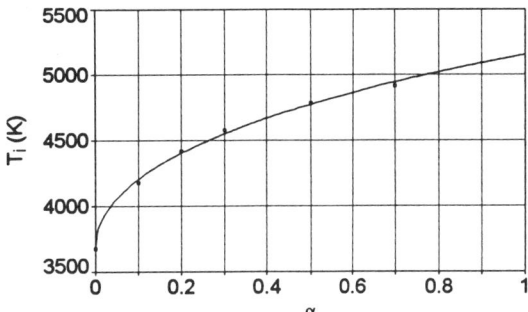

Figure 3. Simulated ion temperature at the top of the L = 3 flux tube for increasing trapping factors. The solid curve is an empirical fit, as described in the text. The fitting function is $T_i^2 = a + b\alpha^{0.5}$, where $a = 1.351 \times 10^7$ and $b = 1.3025 \times 10^7$, with correlation coefficient $r^2 = 0.997$.

increases with increasing trapping factor. The solid line is an empirical function of the trapping factor, fitted to the modeled equatorial ion temperatures. This function had the highest correlation coefficient from results of fitting the data with TableCurve 2D curve-fitting software for more than 55 two-parameter trial functions. Relevant fitting statistics are included in the figure caption. As is evident in the figure, although the ion temperature increases with the trapping factor, it is well below the observed temperature for a trapping factor of 1 and therefore will not achieve the observed temperature for any physically reasonable values of α.

We can understand the physical reasons for this by considering the processes involved. As seen in Figure 2, the electron temperature is increased much more than the ion temperature when photoelectron trapping is the heating mechanism. When we consider the collisional coupling term which transfers energy from electrons to the ions, we find that the maximum energy transfer rate occurs for $T_e = 3 T_i$ and decreases for higher T_e. From Figure 2, we see that this maximum rate is not achieved. But, although the electron temperature can be further increased by increasing α, Figure 3 shows that T_i will not approach observed magnitudes for physically reasonable values of α by this mechanism because it increases much more slowly with α than T_e.

Direct Heating of Thermal Ions

Having found indirect heating of the ions through the thermal electrons to be inadequate to produce observed values, we now examine whether direct heating of the ions can produce them. Such heating could come from wave particle interactions [*Khazanov*, 1995] or from Coulomb

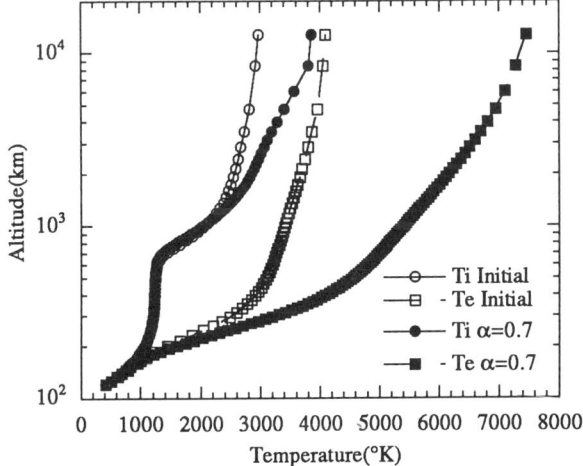

Figure 2. FLIP simulations of altitude profiles for T_i and T_e for initial flux tube conditions and for conditions after a photoelectron trapping factor (α) of 0.7 had been imposed for more than 24 hours.

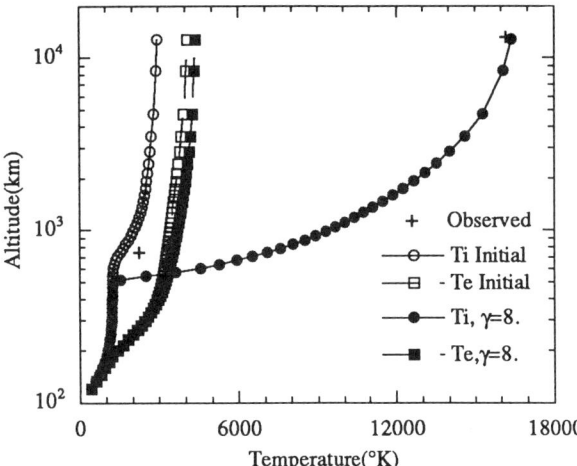

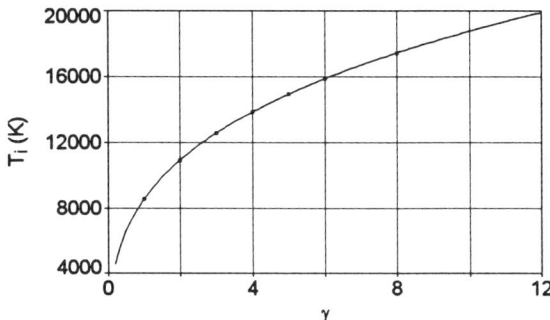

Figure 4. FLIP simulated ion and electron temperature profiles for initial conditions and those resulting from direct ion heating for $\gamma = 8$. For comparison, ion temperatures observed at high altitude (from DE1/RIMS, see Figure 1 above) and at low altitude (DE2/RPA from *Horwitz et al.* [1990], Figure 7) are also shown.

Figure 5. FLIP simulated equatorial (top of flux tube) ion temperatures as a function of heating factor γ. The solid curve represents an empirical fitting function given by $T_i = a + b\gamma^c$, where $a = 1906.6$, $b = 10440$, and $c = 0.29607$, with correlation coefficient $r^2 = .99997$.

collisions with ring current ions [*Kozyra et al.*, 1987; *Fok et al.*, 1993]. From the initialized flux tube plasma we continue the usual diurnal computations, including refilling through ionization production at appropriate local times, but we add energy, in the form of a source term in the energy equation, in increments of 10^9 eVcm^{-2}s^{-1} distributed over the flux tube above 3000 km altitude. These energy increments are denoted in subsequent figures by the parameter γ. This amount of energy is small compared to the energy made available by a trapping factor of one for the photoelectrons, so it is not unreasonable to use several units of energy input to produce the observed temperatures.

We find that for $\gamma = 8$, we achieve close agreement between the simulations and the observed values of T_i and H$^+$ density. This agreement is displayed Figure 4, which shows both initial and final temperature profiles for electrons and ions. Note that with direct ion heating, T_i is considerably larger than T_e. In Figure 5, we show the empirical relationship between T_i and γ, obtained in the same way as with the trapping factor in Figure 3 above. From this it appears that ion temperatures of several eV can be attained if relatively small amounts of energy are supplied directly to the ions. These and the above results for electrons suggest that when high ion temperatures (i.e. $\geq \sim 10^4$ K) are observed in the outer plasmasphere, some mechanism for direct heating of the ions must be operating; ion heating through the thermal electrons alone is inadequate.

Matching the observed ion and electron temperatures is but one test that must be satisfied in plasmaspheric numerical simulations (established by *Newberry et al.* [1989]). Since the full flux tube is simulated in the model, and production, loss, and transport (low speed) are included, a holistic approach must be taken [*Craven et al.*, 1995]. It must be determined if the simulated ion composition, related to the temperature, is also consistent with observations. Figure 6 shows ion density profiles for H$^+$, He$^+$ and O$^+$, both the initial profiles and the final profiles for $\gamma = 8$. The ion heating has clearly had three effects on the composition: H$^+$ densities have decreased to observed levels at DE1 altitudes (initial densities were chosen to insure this); He$^+$ densities have increased to near observed levels at DE1 altitudes (see Figure 1); and O$^+$ has become

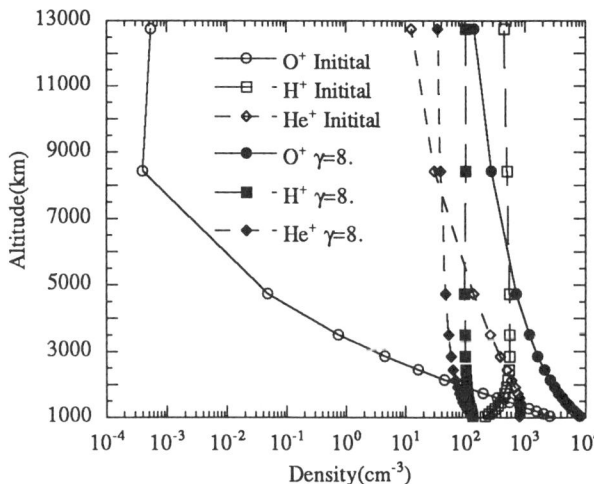

Figure 6. FLIP simulated ion density altitude profiles for H$^+$, He$^+$ and O$^+$ for initial conditions and for conditions following direct heating of ions for $\gamma = 8$, which produced the ion temperature observed by DE1/RIMS (see Figure 5).

the dominant ion species at all altitudes, which is not observed (Figure 1). Comparing initial and final profiles for He$^+$ and O$^+$ in Figure 6, we see that these result from the significantly increased effective scale heights of these two heavier species. That, in turn is a result of the dramatically larger ion temperatures at all altitudes above about 600 km, well into the low altitude region where O$^+$ is normally the dominant ion species. Bulk flow velocities in all these simulations remain quite subsonic.

DISCUSSION

With regard to ion temperatures at low altitudes, we have plotted the ion temperature observed by the DE2 RPA (from *Horwitz et al.*[1990]) at 740 km with the model profiles in Figure 4. In this figure it is clear that at low altitudes, ion temperatures from the simulations far exceed those observed by the DE2 RPA. This appears to be responsible for the high O$^+$ densities at topside ionosphere and higher altitudes in the simulations. These high temperatures at low altitudes result from thermal conduction from high altitudes down to the topside ionosphere. One reason the thermal conduction is so high is because of its temperature dependence. The Spitzer-Harm thermal conductivity for ions varies as $T_i^{5/2}$, which is a consequence of the dependence of the Coulomb collision cross section on velocity.

A final aspect of our study is to examine the temporal response of T_i to the heat source turning on and off. Figure 7 shows the evolution of T_i at the top of the flux tube

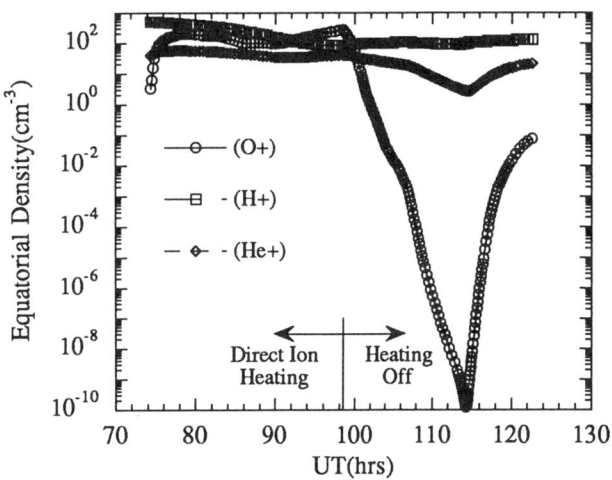

Figure 8. Response of equatorial ion densities for H$^+$, He$^+$, and O$^+$ to the thermal variability in Figure 8.

for the case of direct ion heating. While there is some small diurnal variation associated with the relatively small change of the (conventional) photoelectron heat source, it is clear that the low energy ions come into thermal equilibrium with the heat source and transport processes within about two hours after both onset and removal of the heat source. This is for an L = 3 flux tube; we can expect that higher L-shell flux tubes will take somewhat longer to respond, while lower L flux tubes will respond somewhat more rapidly, but in any event the time scales are on the order of a few hours.

It is instructive to view the response of the ion densities to these same events; this is shown in Figure 8. The most obvious feature of this figure is the dramatic way O$^+$ virtually disappears from the top of the flux tube with removal of the direct heat source for the ions. While it is tempting to look at the overall drop in O$^+$ density in estimating the time scale, for observational purposes we need consider only how long it takes for the density to drop below observational levels of ~ 0.1 particle per cubic centimeter, and this again is on the order of about two hours. Higher densities in the flux tube would tend to increase the response time to both heating and cooling because mean free paths would be reduced, thus increasing the time required for transport processes to reach steady-state conditions. What appears to be a bounce off the bottom in the O$^+$ temporal variation is associated with heating at local sunrise. Note that the H$^+$ density shows little response except the gradual density decrease noted above, while the response of He$^+$ is much smaller than that of O$^+$, but nevertheless significant. This result suggests that the presence of thermal O$^+$ at high altitudes is a signature of local ion heating in some form and that these effects are not long lasting. A

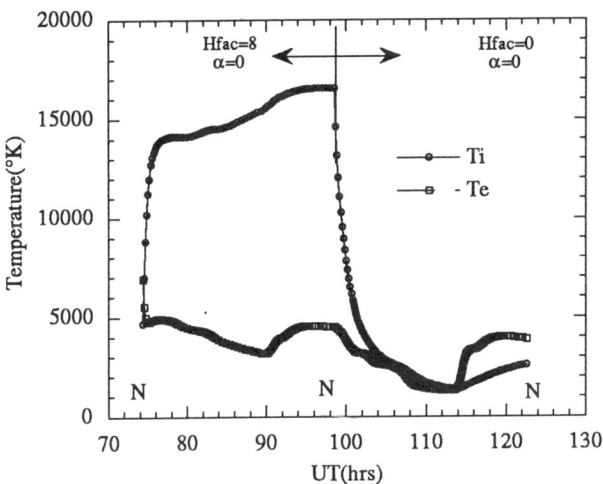

Figure 7. FLIP simulation showing the temporal response of the electron and ion temperature at the equator for turning on direct ion heating for γ = 8 for 24 hours, then removing it, with conventional photoelectron heating giving a diurnal variation in the background.

relative abundance of O^+ at high altitudes appears to be associated with relatively high temperatures in low density flux tubes.

CONCLUSIONS

We have simulated the response of ion temperatures to heating of the plasma in outer plasmasphere low-density flux tubes in two ways and compared the results with satellite observations of ion temperatures. Heating of ions indirectly through heating thermal electrons was found to be inadequate to produce observed ion temperatures. Direct heating of ions was able to produce ion temperatures of the observed magnitude at high altitudes; however, it also had undesirable side effects. Temperatures in the topside ionosphere were much higher than observed and O^+ became the dominant ion at all altitudes. These side effects appear to be the result of large heat transport through conduction. From these results we suggest that for outer plasmasphere flux tubes with low densities and high ion temperatures, direct heating of the ions must be taking place. If this is the case, the ion temperature will likely exceed the electron temperature. In addition, it appears likely that the Spitzer-Harm thermal conductivity microscale process, as it stands, is too large for low density conditions, and that it must be reduced in some way to produce mesoscale ion temperature and composition profiles which are consistent with observations. This could be due to the development of non-Maxwellian features of the plasma distribution associated with small collision frequencies in the low density plasma.

Acknowledgments. The research of RHC was partially supported by NASA grants NAG8-239 and NAGW-1630, while that of PGR was partially supported by NASA grants NAGW-1630 and NAGW-996.

REFERENCES

Chappell, C. R., S. A. Fields, C. R. Baugher, J. H. Hoffman, W. B. Hanson, W. W. Wright, H. D. Hammack, G. R. Carignan, and A. F. Nagy, The retarding ion mass spectrometer on Dynamics Explorer-A, *Space Sci. Instrum.*, 5, 477, 1981.

Comfort, R. H., C. R. Baugher, and C. R. Chappell, Use of the thin sheath approximation for obtaining ion temperatures from the ISEE-1 limited aperture RPA, *J. Geophys. Res., 87,* 5709, 1982.

Comfort, R. H., I. T. Newberry, and C. R. Chappell, Preliminary statistical survey of plasmaspheric ion properties from observations by DE-1/RIMS, *Modeling Magnetospheric Plasma,* ed. T. E. Moore and J. H. Waite, Jr., American Geophysical Union, p. 107-114, 1988.

Comfort, R. H., J. H. Waite, Jr., and C. R. Chappell, Thermal ion temperatures from the retarding ion mass spectrometer on DE-1, *J. Geophys. Res., 90,* 3475, 1985.

Craven, P. D., R. H. Comfort, P. G. Richards, and J. Grebowsky, Comparisons of modeled N^+, O^+, H^+, and He^+ in the mid-latitude ionosphere with mean densities and temperatures from Atmospheric Explorer, *J. Geophys. Res., 100,* 257, 1995.

Fok, M.-C., J. U. Kozyra, A. F. Nagy, C. E. Rasmussen, and G. V. Khazanov, Decay of equatorial ring current and associated aeronomical consequences, *J. Geophys. Res., 98,* 19381, 1993.

Hanson, W. B., R. H. Heelis, R. A. Power, C. R. Lippincott, D. R. Zuccaro, B. J. Holt, L. H. L. Harmon, and S. Sanatani, The retarding potential analyzer for Dynamics Explorer-B, *Space Sci. Instrum., 5,* 503, 1981.

Hedin, A. E., MSIS-86 thermosphere model, *J. Geophys. Res., 92,* 4649, 1987.

Horwitz, J. L., R. H. Comfort, P. G. Richards, M. O. Chandler, C. R. Chappell, P. Anderson, W. B. Hanson, and L. H. Brace, Plasmasphere-ionosphere coupling II: ion composition measurements at plasmaspheric and ionospheric altitudes and comparison with modeling results, *J. Geophys. Res., 95,* 7949, 1990.

Khazanov, G. V., J. U. Kozyra, and A. F. Nagy, Modeling of the thermal structure of the outer plasmasphere, *Adv. Space Res.,* in press, 1995.

Kozyra, J. U., E. G. Shelley, R. H. Comfort, L. H. Brace, T. E. Cravens and A. F. Nagy, The role of ring current O^+ in the formation of Stable Auroral Red arcs, *J. Geophys. Res., 92,* 7487-7502, 1987.

Newberry, I. T., R. H. Comfort, P. G. Richards, and C. R. Chappell, Thermal He^+ in the plasmasphere: comparison of observations with numerical calculations, *J. Geophys. Res., 94,* 15, 265, 1989.

Richards, P. G., D. G. Torr, B. E. Reinisch, R. R. Gamache, and P. J. Wilkinson, F2 peak electron density at Millstone Hill and Hobart: Comparison of theory and measurement at solar maximum, *J. Geophys. Res., 99,* 15,005, 1994.

Spitzer, L., and R. Harm, Transport Phenomena in a completely ionized gas, *Phys. Rev., 89,* 977, 1953.

R. H. Comfort and P. G. Richards, CSPAR, University of Alabama in Huntsville, Huntsville, AL 35899

M. O. Chandler and P. D. Craven, Space Sciences Laboratory, NASA/Marshall Space Flight Center, Huntsville, AL 35812

Ring Current-Plasmasphere Coupling Through Coulomb Collisions

Mei-Ching Fok, Paul D. Craven, and Thomas E. Moore

Space Sciences Laboratory, NASA Marshall Space Flight Center, Huntsville, Alabama

Philip G. Richards

CSPAR, The University of Alabama in Huntsville, Huntsville, Alabama

In the ring current-plasmasphere region, two plasma populations interact with each other via multiple processes, such as Coulomb collisions and wave-particle interactions. In this paper, the consequences of coupling between ring current ions and the plasmasphere through Coulomb collisions are examined. A kinetic Ring current-Atmosphere interaction Model (RAM) has been used to solve the temporal evolution of the ring current ion distribution and obtain the instantaneous ring current heating to the plasmasphere through Coulomb collisions. A buildup of low-energy (< 1 keV) ion population is found as a result of energy degradation of ring current ions in a background of thermal plasma. The "drift-holes" in the ring current ion energy spectra are also somewhat smoothed out by Coulomb interactions. Energy transferred from ring current ions to the plasmasphere is a source of plasma heating and results in enhanced plasma temperatures at high altitudes. The ion temperatures calculated from the Field Line Interhemispheric Plasma (FLIP) model, taking into account the additional heat source from the ring current, appear to be consistent with the enhanced ion temperatures observed in the high altitude regime.

INTRODUCTION

In the region of overlap between the ring current and the plasmasphere, two plasma populations interact with each other via multiple processes, such as wave-particle interactions and Coulomb collisions. Energy of the ring current ions is a source of free energy to excite plasma waves. This energy, in turn, will be redistributed amongst the thermal and energetic populations as the plasma waves undergo damping. Energy contained in the ring current can also be transferred to the plasmasphere through Coulomb collisions. Previous studies on the interactions between the ring current and the plasmasphere and the effects on both plasma populations are summarized in the review of *Kozyra and Nagy* [1991].

Cross-Scale Coupling in Space Plasmas
Geophysical Monograph 93
Copyright 1995 by the American Geophysical Union

In this paper, results on the coupling between the ring current ions and the plasmasphere through Coulomb collisions are presented. Hereafter, ring current ions are designated as energetic plasma and the plasma (ions and electrons) in the plasmasphere as thermal plasma. When energetic ions move through a background of thermal plasma in the plasmasphere, they experience energy loss but very small angular deflection. Since the energy transfer rate is maximum when the velocity of the energetic ions and that of the thermal species are comparable (function G in Eq. (21) of *Fok et al.*, 1993), most of the energy received by the thermal plasma goes to the electrons. However the conductivity of thermal ions is about one-fortieth that of thermal electrons. A small amount of heating from the ring current may produce enhanced ion temperatures comparable to, or higher than, the electron temperatures in the plasmasphere. Interactions with the plasmasphere also have a significant effect on the energetic population as it undergoes energy degradation.

A previously established Ring current-Atmosphere

interaction Model (RAM) [*Fok et al.*, 1993, 1995] is used to study the coupling between ring current ions and the plasmasphere. RAM is a kinetic model solving the temporal evolution of ring current ion phase space density, considering drift motion, charge exchange with the hydrogen geocorona, and Coulomb interactions with the plasmasphere. The instantaneous heating rate to the plasmasphere is also calculated. The response of the plasmasphere to the additional heating by the ring current is investigated using the Field Line Interhemispheric Plasma (FLIP) model [*Richards and Torr*, 1985; *Torr et al.*, 1990]. For comparison, plasmaspheric ion temperatures are calculated from FLIP, with and without heating from the ring current.

THE MODEL: RAM

The temporal and spatial variations of the phase space density, f_s, of ring current ion species, s, can be obtained by solving the following kinetic equation, considering drift motion, charge exchange with the neutral hydrogen geocorona, and Coulomb collisions with the plasmasphere

$$\frac{\partial \bar{f}_s}{\partial t} + \langle \dot{R}_0 \rangle \frac{\partial \bar{f}_s}{\partial R_0} + \langle \dot{\phi} \rangle \frac{\partial \bar{f}_s}{\partial \phi} = -v\sigma_s \langle n_H \rangle \bar{f}_s + \frac{1}{M^{1/2}} \frac{\partial}{\partial M}\left(\langle \dot{M} \rangle M^{1/2} \bar{f}_s\right) \quad (1)$$

where R_0 is the radial distance at the equator, ϕ is the magnetic local time, M is the magnetic moment, and $\bar{f}_s$ is the average f_s along the field line between mirror points. Since the bounce period of ions in the ring current energy range is much shorter than the decay lifetimes, f_s is assumed to be constant along the field line and thus $\bar{f}_s$ can be replaced by the distribution function at the equator. $\langle x \rangle$ is the bounce-averaged value of quantity x.

For the collisional terms in (1), σ_s is the cross section for charge exchange of species s with neutral hydrogen, and n_H is the hydrogen density. $\dot{M}$ is the rate of change of magnetic moment due to the Coulomb drag from the plasmasphere and it is proportional to the background plasma density. A time-dependent, two-dimensional plasmasphere model of *Rasmussen et al.* [1993] is used to calculate $\dot{M}$. This is the unique feature of RAM, which incorporates a time-dependent plasmasphere model to calculate effects of Coulomb collisions between the energetic and the thermal population. The instantaneous heating rate from the ring current to the plasmaspheric species p is given by

$$Q_p = \frac{1}{\Delta t} \sum_s \left(\int \Delta f_s \frac{1}{2} m v^2 d^3 v \right) \quad (2)$$

where Δf_s is the change of f_s in Δt due to Coulomb collisions with species p. More details of RAM can be found in *Fok et al.* [1993, 1995].

COULOMB COLLISION EFFECTS ON THE ENERGETICS

As a result of frictional forces exerted on the ring current ions by the background plasma in the plasmasphere, the energy spectra of the energetic ions shift toward low energies. This effect on the energetic ion distribution is clearly illustrated in Figure 1. Given an initial H$^+$ (Figure 1a) and O$^+$ (Figure 1b) ion distribution, peaking at 10 keV (top panel) and 40 keV (middle panel) in a background thermal (1 eV) plasma with density of 2000 cm^{-3}, the temporal evolution of these distributions is calculated taking into account only Coulomb interactions. The variation in rate of energy loss as a function of ion energy (bottom panel) is provided as a convenient reference. Energy loss rate in energy space peaks at 4 keV for H$^+$ (bottom panel, Figure 1a) and at 50 keV for O$^+$ (bottom panel, Figure 1b). Ions will be removed rapidly from regions of peak energy loss rate and will build up in regions of lower energy loss rate. The effect is dramatically illustrated in the top panels of Figures 1a and 1b. A low-energy (< 1 keV) flux of H$^+$ builds up after 12 hours as a result of energy loss by the "10-keV" H$^+$. However, this low-energy population diminishes after 2 days of interaction. Low-energy O$^+$ fluxes build up more slowly. They only reach values comparable to the peak low-energy H$^+$ fluxes after 2 days have elapsed. In contrast, the peak low-energy H$^+$ fluxes appear after 1 day of decay. This buildup of low-energy ions is also predicted by the model of *Jordanova et al.* [1994].

The buildup of low-energy ions from a 40-keV initial ion distribution is much less dramatic (middle panels, Figures 1a and 1b). The 40-keV peak in the H$^+$ flux occurs in a region of increasing energy loss rate as ions degrade in energy. Ions move from a region of slower to faster energy loss (toward the left in the middle panel, Figure 1a). As a consequence, the H$^+$ distribution is eroded, with only a weak buildup at the lowest energies. The 40-keV peak in the O$^+$ flux (middle panel, Figure 1b), occurs in a region of nearly constant energy loss rate. For this case, the O$^+$ peak is convected in energy space, almost without change in amplitude or shape, to lower energies. Eventually, as time increases beyond 2 days, a flux buildup at the low energies is expected to occur.

The buildup of the low-energy ion population predicted above is reproduced when calculating the whole phase space densities of ring current ions during the recovery of a model storm of moderate intensity. The initial conditions (when the recovery starts) of the model storm are taken

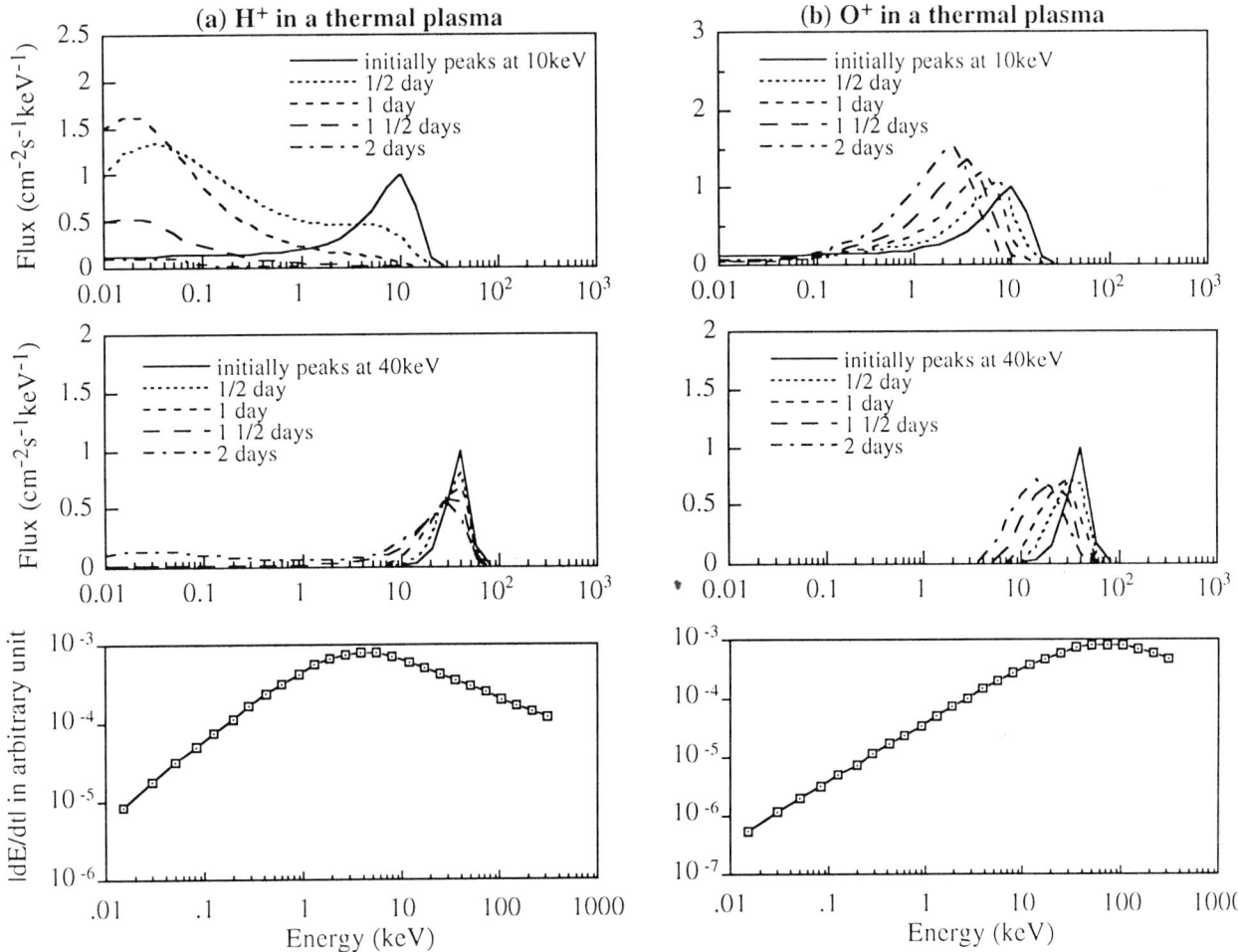

Fig. 1. Time variations of (a) H$^+$ and (b) O$^+$ fluxes with initial energy peaking at 10 keV (top panel) and 40 keV (middle panel) in a background of thermal plasma (1 eV, density = 2000 cm^{-3}). The rate of energy loss as a function of ion energy is plotted in the bottom panel [taken from *Fok et al.*, 1993].

from the average stormtime spectra reported by *Kistler et al.* [1989]. The Kp value is assumed to be 6 initially and decreases at a constant rate in 15 hours to the post recovery value of 1 (top panel, Plate 1). The 10–500 eV fluxes derived from the ring current H$^+$ and O$^+$ as a function of elapsed time are calculated. Plate 1 shows the initial fluxes, and the fluxes at 12 hours and 2 days after the main phase of the storm. For both ions, the maximum fluxes are initially located in the region of L between 3–4, at midnight local time (Plate 1a). The initial fluxes on the dayside are assumed to be 0. Twelve hours later (Plate 1b), fluxes of the same order of magnitude or higher than the initial peak are found at a lower L shell (L ~ 2.5). Particles which have open drift paths move toward the upper spatial boundary of the model and are lost. Therefore, the longest-lived fluxes are confined to a range of L values from about 2.5 to 3.5, where drift paths are closed.

After 2 days of recovery (Plate 1c), there is an order of magnitude decrease in the maximum H$^+$ flux. The maximum flux at this time has moved outward to L ~ 4. The temporal history of the low-energy O$^+$ flux is much different than that of the low-energy H$^+$ ions. The maximum O$^+$ flux in the energy range 10–500 eV is located at L ~ 2.75, but does not appear until 2 days into the recovery phase. The different times for buildup and disappearance of each ion species in this energy range (10–500 eV) are consistent with the simulations shown in Figure 1 and can be explained by their different rates of energy loss to the plasmasphere. After the low-energy H$^+$ is built up, it will be lost by energy degradation in a few hours. It takes about 1 day for low-energy O$^+$ to be formed from high-energy particles, as a result of its comparatively long Coulomb lifetime; therefore, it lasts for a few days.

The other effect on the energetic ions due to Coulomb

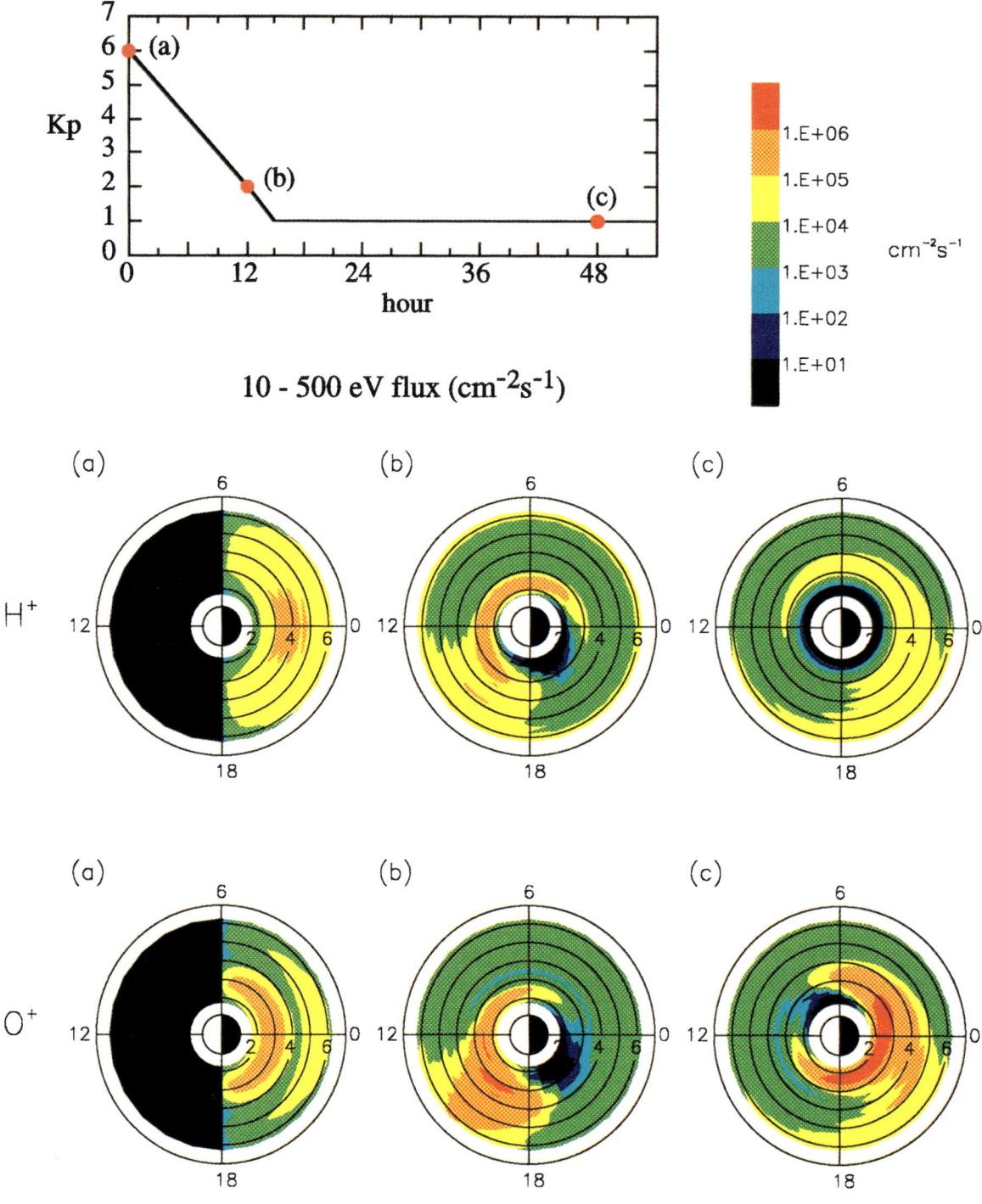

Plate 1. Upper panel: Model decay of *Kp* after the main phase of a moderate storm. Lower panels: 10-500 eV H$^+$ and O$^+$ fluxes in cm^{-2} s^{-1} at different times during the recovery phase of the storm: (a) 0 hour, (b) 12 hours, and (c) 2 days [taken from *Fok et al.*, 1993].

collisions with the plasmasphere is the shallower minimum in the energy distribution. Dips or sharp drop-offs in the ion energy spectra for $1 \leq E \leq 10$ keV are consistent features in observations [*McIlwain*, 1972; *Lennartsson et al.*, 1981; *Kistler et al.*, 1989]. Corotation dominates the motion of low-energy ions, whereas high-energy ions drift westward due to the gradient and curvature in the magnetic field. Ions of intermediate energies experience eastward and westward drifts that are similar in magnitude, resulting in very slow drift velocity. The hole in the energy spectra corresponds to particles at these energies which have either not yet reached the observation point or which have experienced significant losses during their slow drift to the observation point. *Cornwall* [1972] and *Spjeldvik* [1977] found that Coulomb collisions with the plasmasphere made the dips caused by drift motion shallower when they calculated the equilibrium density structure of the magnetosphere ions.

In order to see the effect of Coulomb drag on the shape of the "drift hole," ion energy spectra are calculated for different cases: (1) considering only drift motion, (2) drifts plus charge exchange loss, and (3) same as (2) except Coulomb collisions are also included. Figure 2 shows these test results of H^+ at two locations. At $L = 2.5$ (Figure 2a), the inclusion of Coulomb collisions results in a slight reduction of the distribution function at high energies (> 50 keV). However, a low-energy (< 1 keV) ion population is formed by the energy degradation of high-energy ions. Coulomb collisions also make the dip caused by drift motion shallower. At L values outside the plasmapause ($L = 4$, Figure 2b) the Coulomb collision effects are much weaker. However, the partial filling-in of the drift hole is still observed. *Kistler et al.* [1989] only considered drifts and charge exchange loss of ring current ions. They found that at low energies, the observed fluxes were higher than predicted and that the dips in some of their modeled spectra were deeper than the measurements. The inclusion of Coulomb collisions in our model leads to a better agreement with observations, especially at low energies. Coulomb collisions also have pronounced effects on the low-energy distribution of ring current He^+ and O^+ ions.

RING CURRENT HEATING OF THE PLASMASPHERE

Under the assumptions of uniform plasmasphere density along field lines and constant phase space density between mirror points for the ring current ions, RAM is able to calculate the instantaneous volume heating rate (at a given local time, latitude, and altitude) to the plasmasphere that is a consequence of collisions between thermal ions and ring current ions. Only heating to the thermal ions is dis-

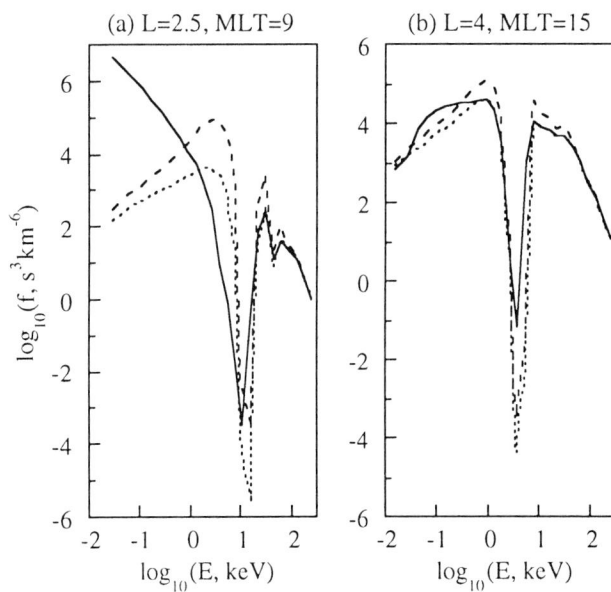

Fig. 2. Phase space distribution functions of ring current H^+ at 12 hours after the model moderate storm main phase. Dashed lines represent results when only drift is considered. Dotted lines contain effects of drift and charge exchange, while solid lines represent results under the same conditions except Coulomb collisions are also included.

cussed in this work. In order to see what ring current species and energies contribute to the heating of thermal ions, the rate of change of the energy of energetic ions moving through a background thermal ion gas is calculated and plotted in Figure 3 as a function of the energy of the incoming ions. As shown in the figure, low-energy (< 10 keV) O^+ is the main source of heating to the thermal ions via Coulomb interactions, unless the ring current is overwhelmingly dominated by H^+. The energy transfer rates are fairly constant over a wide range of thermal ion temperature observed in the plasmasphere (0.2–2 eV) [*Comfort et al.*, 1988]. In contrast, the efficiency of energy transfer to the thermal electrons is sensitive to the thermal electron temperature.

We calculate the volume heating rate to the thermal ions in the plasmasphere during the recovery phase of a modeled major storm similar to that which occurred in early February 1986. The initial distribution of ring current ions (at the minimum Dst) is taken from AMPTE/CCE spacecraft observations spanning the L value range $L = 2.25$ to 6.75. Ion fluxes at energies of 10 eV to 1.52 keV, which is the lowest energy bin of CCE measurement, are estimated by linear extrapolation. The simulated temporal evolution of the ring current ions during the recovery of this storm is presented in *Fok et al.* [1995]. In general, the calculated

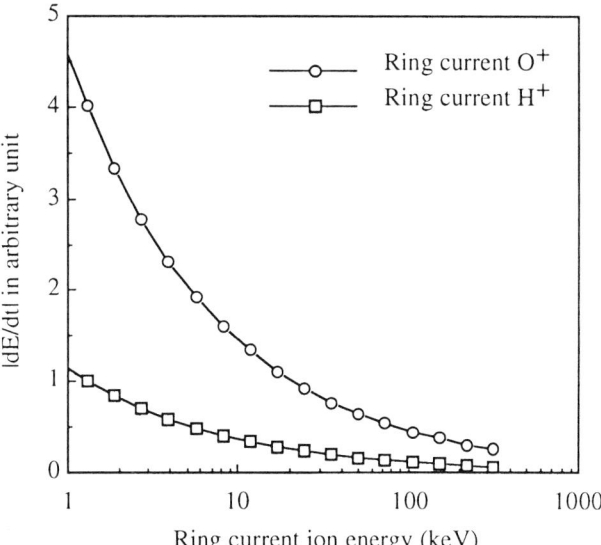

Fig. 3. Rate of energy change of ring current ions moving through a background thermal ion (1 eV H$^+$) gas as a function of the energy of the incoming ions.

ion fluxes are consistent with observations, except for H$^+$ fluxes at tens of keV, which are always over-estimated. However this discrepancy has only a small effect on the plasmaspheric heating rate because the main contribution of heating comes from the ring current O$^+$ ions. Plate 2 shows the volume heating rate to the thermal ions at the equator (left panels) and the noon-midnight meridian (right panels) at (a) 2 and (b) 52 hours after the start of the recovery phase, together with the corresponding Kp values. At 2 hours, the maximum heating rate is on the order of 1 eV cm^{-3} s^{-1} and is located at L shell of 2 to 2.5, with local time extending from midnight to dawn. The region of high heating rate corresponds to the location of peak density of low-energy (< 10 keV) ring current ions during the early recovery. This region of high heating rate drifts eastward and reaches noon in the next few hours as the low-energy ring current ions corotate. The isocontours of heating rate at L shells larger than 3 roughly follow the thermal density calculated from the model of *Rasmussen et al.* [1993]. The plasmasphere bulge located between noon and dusk during this active period can be inferred in the equatorial view of Plate 2a. The meridian view shows that the heating rate peaks at the equator near the inner edge of the ring current and is fairly uniform along field lines at high L shells. The localized heating near the Earth is a consequence of ring current ions which have an anisotropic pitch-angle distribution (peaks at 90°) caused by strong charge exchange loss at low L shells. The distribution of the heating rate is a result of the combined effect of the densities of the source

(ring current ions with energy less than 10 keV) and the sink (thermal ions).

In late recovery (Plate 2b), isocontours of heating rate expand and fall off with L smoothly as a result of the refilling of the plasmasphere during the storm recovery. The high heating rate (on the order of 1 eV cm^{-3} s^{-1}), which is seen at $L < 2.5$ at 2 hours, is diminished due to the charge exchange losses of low-energy (< 10 keV) ions at that location. In contrast, the heating rates at high L's are higher at late recovery than at early recovery. In the meridian view, the peak heating rate at the equator is more pronounced and extends to higher L shells as a consequence of strong ring current pitch-angle anisotropy during late recovery of the storm [*Fok et al.*, 1995].

PLASMASPHERIC RESPONSE TO THE HEATING FROM RING CURRENT

The energy loss of the ring current from Coulomb decay is small compared with that of the charge exchange loss. However, this small amount of energy may be a significant heat source to the thermal plasma in the plasmasphere. The energy received by the thermal plasma in the plasmasphere is, in turn, conducted down along field lines to the ionosphere and produces observable signatures. A number of works have shown that the energy transferred to the plasmasphere through Coulomb collisions with the ring current ions is responsible for the enhanced plasmaspheric temperature and resulting ionospheric electron temperature peaks and associated stable auroral red (SAR) arc emissions [*Cole*, 1965; *Kozyra et al.*, 1987; *Chandler et al.*, 1988; *Fok et al.*, 1993].

In order to see the plasmaspheric response to the heating from the energetic ions, the thermal ion temperatures are calculated using the RAM-generated heat source for the February 1986 storm as input to the FLIP model. The volume heating rate calculated from RAM is scaled by the ratio of plasmaspheric densities obtained from the model of *Rasmussen et al.* [1993] and FLIP. We modified the standard FLIP model slightly in order to accommodate direct ion heating from the ring current source. Otherwise the model is the same as described by *Richards and Torr* [1985], and *Torr et al.* [1990]. FLIP, as used here, solves the continuity, momentum, and energy equations along the flux tube from 120 km in one hemisphere to 120 km in the conjugate hemisphere. All ion species are assumed to have the same temperature, but are different from those of thermal electrons. Since the ring current heating to the thermal electrons depends on the thermal electron temperature, RAM and FLIP have to be run interactively in order to have consistent results. Because of this, heating to the

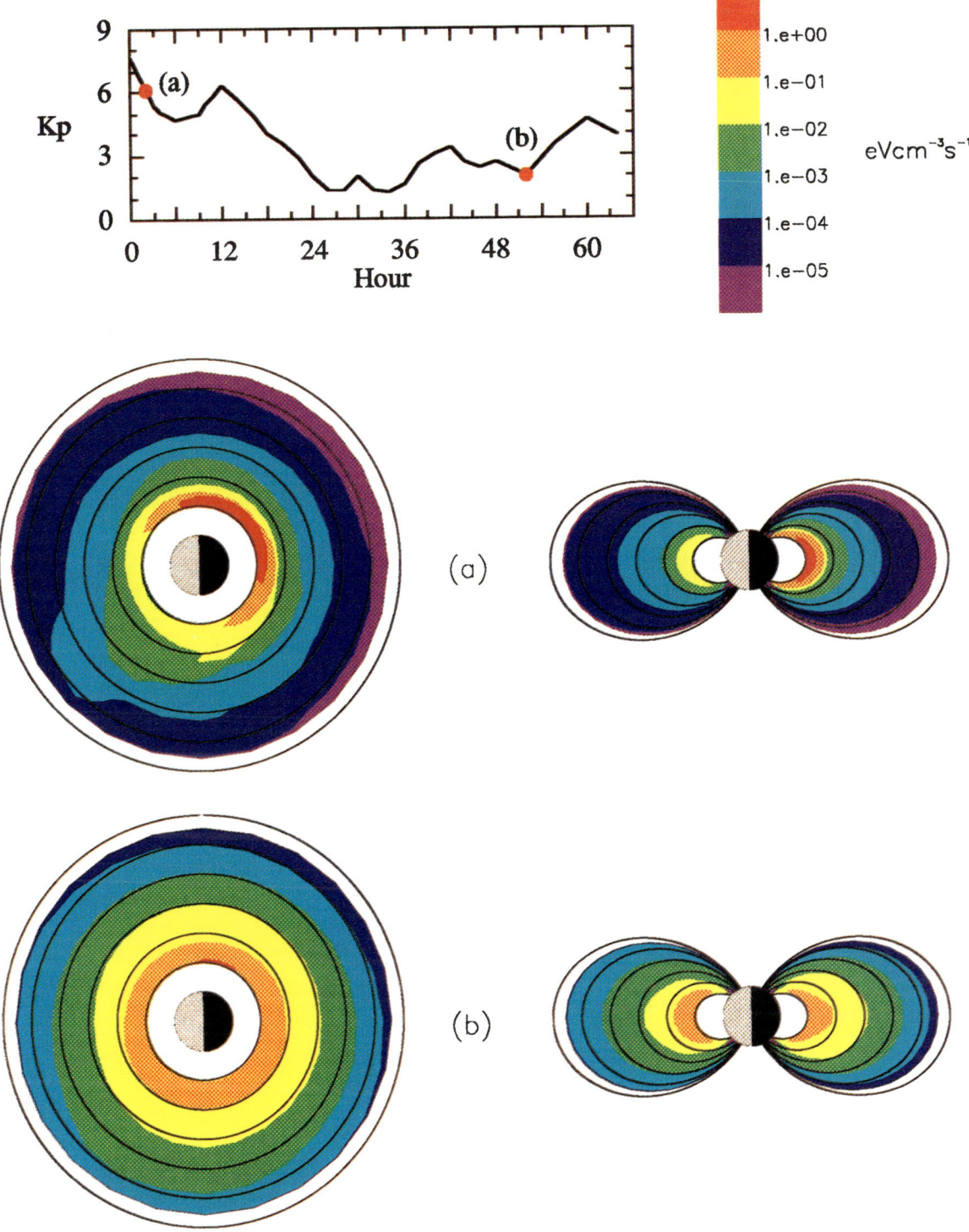

Plate 2. Plasmaspheric ion volume heating rate at the equator (left) and the noon-midnight meridian (right) at (a) 2 hours and (b) 52 hours after the main phase of the model storm. The *Kp* values, for which the heating rates are shown, are indicated by red dots.

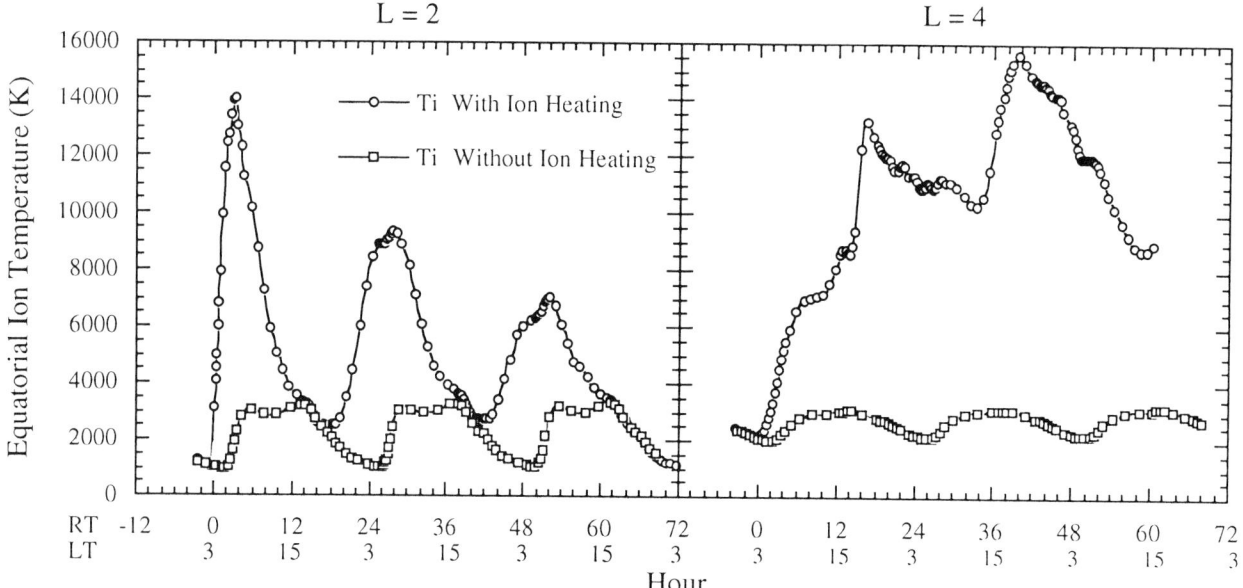

Fig. 4. Simulated equatorial ion temperature for $L = 2$ (left) and $L = 4$ (right) as a function of time from the start of the storm recovery (RT = 0) and local time (LT). Results with and without heating from the ring current are shown in both panels.

thermal electrons and the resulting electron temperature enhancements are excluded in the present study. This problem will be addressed in our future work.

The equatorial ion temperatures calculated from FLIP for $L = 2$ and 4 as a function of elapsed time from the beginning of the recovery phase are plotted in Figure 4. Because the flux tube corotates, the local time changes with the elapsed time. Local time is also shown in Figure 4. The storm starts at 0 hour recovery time (RT) and decays to quiet time rates at 60 hours RT. The insert in Plate 2 shows the Kp history. Results with and without heating from ring current are displayed. As shown in Figure 4, at $L = 2$, including the ring current heat source greatly increases the ion temperature from the nightside to dawn but has almost no effect from noon to dusk. The local time asymmetry of the magnetospheric heat source enhances the diurnal variation in the ion temperature and causes the daily minimum temperature to shift from the nightside to dusk. The decrease in the ring current heating at dusk, at $L = 2$, is a result of low-energy (< 10 keV) ring current ions being significantly removed by charge exchange before reaching the dusk side. The high heating rate at $L = 2$ fades away as the storm recovery proceeds (Plate 2). The maximum ion temperature thus decreases in the same manner. At $L = 4$, the effects of the ring current heating of the thermal ions are obvious at all local times because the ring current is fairly constant with local time at this L value (Plate 2). Also the ion temperature enhancement persists during the long recovery since the source experiences few losses at $L = 4$ and low-energy ring current ions are continuously replenished from the tail.

The altitude profiles of ion temperature at $L = 2$ and 4, on the morning and evening side, at different elapsed times during the storm recovery as calculated from FLIP are plotted in Figure 5. In all cases, heating from the ring current increases the ion temperature at altitudes above the heat sink due to the neutral atmosphere, about 500 km. Once again the temperature at $L = 2$, 0900 LT, increases rapidly during the early recovery phase and then gradually returns to the quiet time level. On the other hand, T_i at $L = 4$ is increasing throughout the recovery until $t \sim 50$ hours after which it decreases. There is a two- to five-fold increase in temperature compared with no heating from ring current except at $L = 2$, 2200 LT, where the heating rate is low for this particular storm (Plate 2). For comparison, the mean DE 1/RIMS H^+ temperatures during October and November of 1981 [Comfort et al., 1988] are also shown. In general, heating from the ring current can more than account for the ion temperature observed. In particular, the high temperature ($T_i \sim 14000$ K) near dawn at $L = 2$ predicted by this study has not been observed. The overestimation of the low-energy (< 1.52 keV) ion fluxes by linear extrapolation may be responsible for producing this high ion temperature. If the initial low-energy (< 1.52 keV)

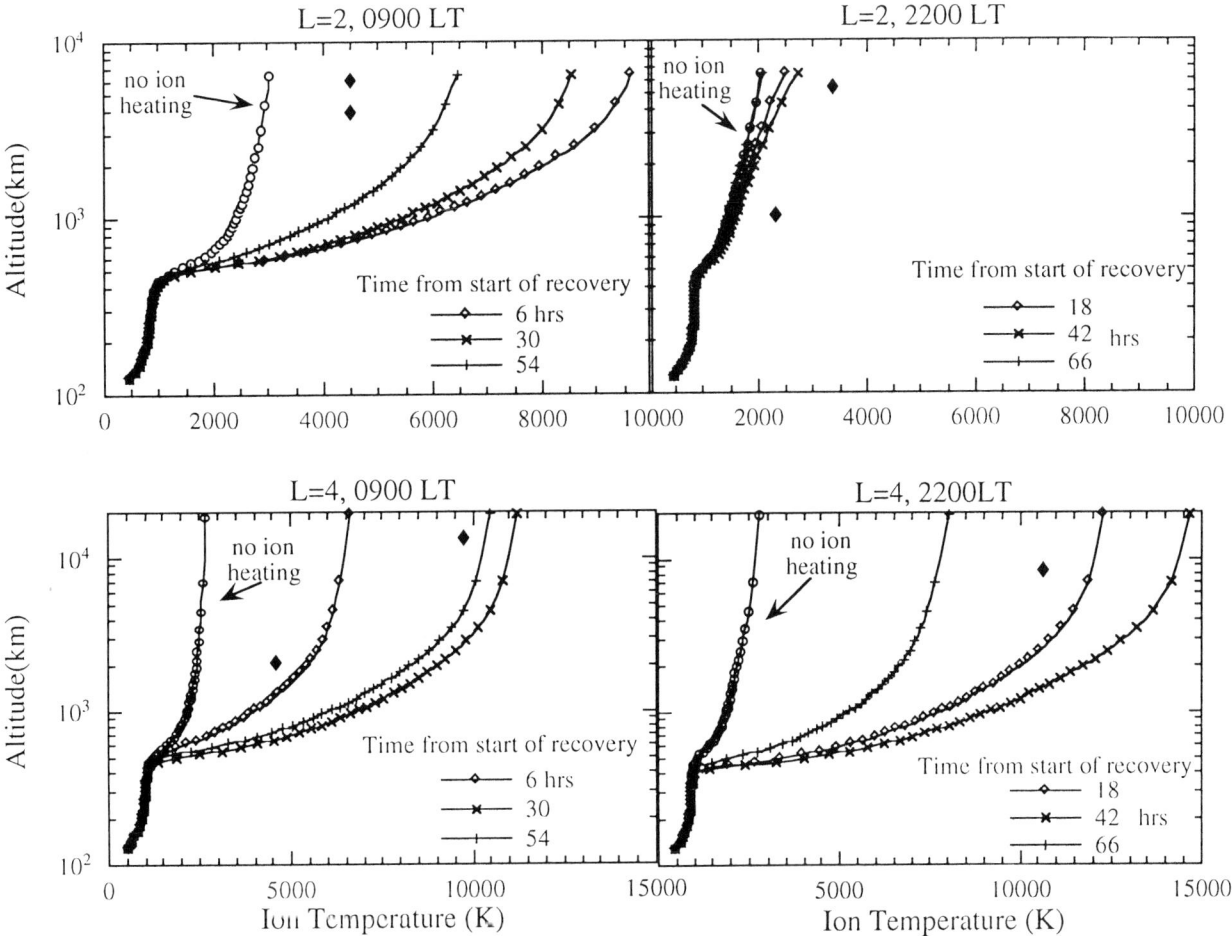

Fig. 5. Altitude profiles of ion temperatures for $L = 2$ (upper panel) and $L = 4$ (lower panel), morning (left) and evening(right) at different times relative to the start of the recovery. The temperature profile at $L = 2$, 2200 LT for 66 hours RT falls on top of that with no ion heating. Mean DE 1/RIMS data in 1981 are shown with the ♦.

ion fluxes are set equal to the 1.52 keV flux measured by AMPTE/CCE, the resulting heating rates to the thermal ions at $L < 2.5$ are approximately an order of magnitude lower than those presented in Plate 2. The corresponding equatorial ion temperature near dawn at $L = 2$ only reaches a peak value of 4700 K during the early recovery. Since the ring current ion energy spectra become flat in shape at low energies for $L > 3.5$ (Figure 2, Fok et al., 1995), the way of extrapolating initial ion fluxes down to 10 eV does not much affect the plasmaspheric heating and the resulting ion temperature for $L > 3.5$. Accurate modeling of the energy range below 1.52 keV will require measurements in that energy range as initial conditions.

In contrast to the predicted high temperature on the dawn side at $L = 2$, the ion temperature at $L = 2$, 2200 LT increases by less than 1000 K with the heating from the energetic ions (upper right panel, Figure 5). However, if the heating of plasmaspheric electrons from the ring current is considered, the enhanced electron temperature is expected to raise the ion temperature to a value closer to measurements. Although we are comparing the simulated results with the observed temperature averaged over the period of October to November 1981, in which geomagnetic activity did not reach a level as high as the February 1986 storm, we have clearly shown that the enhanced ion temperature in the plasmasphere can be explained with the heating from a magnetospheric source. Moreover, the relationship between the ion temperature and geomagnetic activity is surely not trivial. It may depend on the instantaneous activity level, perhaps even the past history of geomagnetic activity, also on the phase of the storm at which the measurements are made.

DISCUSSION AND SUMMARY

The observed temperature enhancements at high altitudes indicate the existence of a magnetospheric heat source [*Horwitz and Chappell*, 1979; *Farrugia et al.*, 1989; *Craven et al.*, 1991]. Previous studies [*Chandler et al.*, 1988; *Fok et al.*, 1993] and this study suggest that energy transferred from the ring current ions to the plasmasphere via Coulomb interactions is a significant heat source to power the plasmaspheric temperatures to the observed levels and can produce significant diurnal variations in the thermal plasma temperature. In the future, heating of thermal electrons in the plasmasphere from ring current ions will also be considered in order to have a better picture of plasmaspheric heating. Other mechanisms, such as wave-particle interactions, may also contribute to the plasmaspheric heating. Various kinds of plasma waves have been observed [cf., *Korth et al.*, 1984; *LaBelle et al.*, 1988] and theoretical studies suggest that the unstable plasma waves heat the thermal population [*Roth et al.*, 1990; *Gorbachev et al.*, 1992]. Although wave-particle interactions are presently not included in RAM, RAM can serve as a baseline model to distinguish or separate the wave effects on the plasmaspheric heating.

The role of Coulomb collisions on the ring current dynamics sometimes does not receive much attention. However, we found that ring current Coulomb interactions with the plasmasphere are important processes for both energetic and thermal populations. The buildup of a low-energy ion population, predicted by RAM, may be a source of the low-energy ions observed in the inner magnetosphere [*Lennartsson and Sharp*, 1982; *Shelley et al.*, 1985].

In summary, we found the following consequences of the ring current-plasmasphere coupling through Coulomb collisions:

(1) A low-energy (< 1 keV) ion population is formed as ring current ions degrade in energy.

(2) Dips in ring current ion energy spectra caused by drift motion are somewhat smoothed out by Coulomb drag.

(3) Energy transferred from ring current ions to the plasmasphere is a significant heat source for the thermal plasma at high altitudes.

Acknowledgments. The authors thank R. H. Comfort for several helpful discussions. This work was performed while one of us, M.-C. Fok, held a National Research Council-Marshall Space Flight Center Research Associateship. Other work at MSFC was supported by the NASA Space Physics Division under UPN 432-20-00. Support at the University of Alabama was from NASA grant NAGW 996.

REFERENCES

Chandler, M. O., J. U. Kozyra, J. L. Horwitz, R. H. Comfort, and L. H. Brace, Modeling of the thermal plasma in the outer plasmasphere - A magnetospheric heat source, in *Modeling Magnetospheric Plasma, Geophys. Monogr. Ser., vol. 44*, edited by T. E. Moore and J. H. Waite Jr., pp. 101-105, AGU, Washington, DC, 1988.

Cole, K. D., Stable auroral red arc, sinks for energy of Dst main phase, *J. Geophys. Res., 70*, 1689-1706, 1965.

Comfort, R. H., I. T. Newberry, and C. R. Chappell, Preliminary statistical survey of plasmaspheric ion properties from observations by DE 1/RIMS, in *Modeling Magnetospheric Plasma, Geophys. Monogr. Ser., vol. 44*, edited by T. E. Moore and J. H. Waite Jr., pp. 107-114, AGU, Washington, DC, 1988.

Cornwall, J. M., Radial diffusion of ionized helium and protons: A probe for magnetospheric dynamics, *J. Geophys. Res., 77*, 1756-1770, 1972.

Craven, P. D., R. H. Comfort, D. L. Gallagher, and R. West, A study of the statistical behavior of ion temperatures from DE1/RIMS, in *Modeling Magnetospheric Plasma Processes, Geophys. Monogr. Ser., vol. 62*, edited by G. R. Wilson, pp. 173-182, AGU, Washington, DC, 1991.

Farrugia, C. J., D. J. Young, J. Geiss, and H. Balsiger, The composition, temperature, and density structure of cold ions in the quiet terrestrial plasmasphere: GEOS 1 results, *J. Geophys. Res., 94*, 11,865-11,891, 1989.

Fok, M.-C., J. U. Kozyra, A. F. Nagy, C. E. Rasmussen, and G. V. Khazanov, Decay of equatorial ring current ions and associated aeronomical consequences, *J. Geophys. Res., 98*, 19,381-19,393, 1993.

Fok, M.-C., T. E. Moore, J. U. Kozyra, G. C. Ho, and D. C. Hamilton, Three-dimensional ring current decay model, *J. Geophys. Res., 100*, 9619-9632, 1995.

Gorbachev, O. A., G. V. Khazanov, K. V. Gamayunov, and E. N. Krivorutsky, A theoretical model for the ring current interaction with the Earth's plasmasphere, *Planet. Space Sci., 40*, 859-872, 1992.

Horwitz, J. L., and C. R. Chappell, Observations of warm plasma in the dayside plasma trough at geosynchronous, *J. Geophys. Res., 84*, 7075-7090, 1979.

Jordanova, V. K., J. U. Kozyra, G. V. Khazanov, A. F. Nagy, C. E. Rasmussen, and M.-C. Fok, A bounce-averaged kinetic model of the ring current ion population, *Geophys. Res. Lett., 21*, 2785-2788, 1994.

Kistler, L. M., F. M. Ipavich, D. C. Hamilton, G. Gloeckler, B. Wilken, G. Kremser, and W. Stüdemann, Energy spectra of the major ion species in the ring current during geomagnetic storms, *J. Geophys. Res., 94*, 3579-3599, 1989.

Korth, A., G. Kremser, S. Perraut, and A. Roux, Interaction of particles with ion cyclotron waves and magnetosonic waves. Observations from GEOS 1 and GEOS 2, *Planet. Space Sci., 32*, 1393-1406, 1984.

Kozyra, J. U., and A. F. Nagy, Ring current decay-coupling of ring current energy into the thermosphere/ionosphere system,

J. Geomag. Geoelectr., *43*, 285-297, 1991.

Kozyra, J. U., E. G. Shelley, R. H. Comfort, L. H. Brace, T. E. Cravens, and A. F. Nagy, The role of ring current O^+ in the formation of stable auroral red arcs, *J. Geophys. Res.*, *92*, 7487-7502, 1987.

LaBelle, J., R. A. Treumann, W. Baumjohann, G. Haerendel, N. Sckopke, G. Paschmann, and H. Luhr, The duskside plasmapause/ring current interface: Convection and plasma wave observations, *J. Geophys. Res.*, *93*, 2573-2590, 1988.

Lennartsson, W., and R. D. Sharp, A comparison of the 0.1–17 keV/e ion composition in the near equatorial magnetosphere between quiet and disturbed conditions, *J. Geophys. Res.*, *87*, 6109-6120, 1982.

Lennartsson, W., R. D. Sharp, E. G. Shelley, R. G. Johnson, and H. Balsiger, Ion composition and energy distribution during 10 magnetic storms, *J. Geophys. Res.*, *86*, 4628-4638, 1981.

McIlwain, C. E., Plasma Convection in the vicinity of the geosynchronous orbit, in *Earth's Magnetospheric Processes*, edited by B. M. McCormac, pp. 268-279, D. Reidel, Hingham, Mass., 1972.

Rasmussen, C. E., S. M. Guiter, and S. G. Thomas, Two-dimensional model of the plasmasphere: Refilling time constants, *Planet. Space Sci.*, *41*, 35-43, 1993.

Richards, P. G., and D. G. Torr, Seasonal, diurnal, and solar cyclical variations of the limiting H^+ flux in the Earth's topside ionosphere, *J. Geophys. Res.*, *91*, 5261-5268, 1985.

Roth, I., B. I. Cohen, and M. K. Hudson, Lower hybrid drift instability at the inner edge of the ring current, *J. Geophys. Res.*, *95*, 2325-2332, 1990.

Shelley, E. G., D. M. Klumpar, W. K. Peterson, A. Ghielmetti, H. Balsiger, J. Geiss, and H. Rosenbauer, AMPTE/CCE observations of the plasma composition below 17 keV during the September 4, 1984 magnetic storm, *Geophys. Res. Lett.*, *12*, 321-324, 1985.

Spjeldvik, W. N., Equilibrium structure of equatorially mirroring radiation belt protons, *J. Geophys. Res.*, *82*, 2801-2808, 1977.

Torr, M. R., D. G. Torr, P. G. Richards, and S. P. Yung, Mid- and low-latitude model of thermospheric emissions 1: $O^+(^2P)$ 7320 Å and N_2 (2P) 3371 Å, *J. Geophys. Res.*, *95*, 21,147-21,168, 1990.

M.-C. Fok, P. D. Craven, and T. E. Moore, ES83, Space Sciences Laboratory, NASA Marshall Space Flight Center, Huntsville, AL 35812.

P. G. Richards, CSPAR, The University of Alabama in Huntsville, Huntsville, AL 35899.

Plasmasphere Modeling with Ring Current Heating

S. M. Guiter[1], M.-C. Fok[1], and T. E. Moore

Space Sciences Laboratory, NASA Marshall Space Flight Center, Huntsville, Alabama

Coulomb collisions between ring current ions and the thermal plasma in the plasmasphere will heat the plasmaspheric electrons and ions. During a storm such heating would lead to significant changes in the temperature and density of the thermal plasma. This was modeled using a time-dependent, one-stream hydrodynamic model for plasmaspheric flows, in which the model flux tube is connected to the ionosphere. The model simultaneously solves the coupled continuity, momentum, and energy equations of a two-ion (H^+ and O^+) quasineutral, currentless plasma. Heating rates due to collisions with ring current ions were calculated along the field line using a kinetic ring current model. First, diurnally reproducible results were found assuming only photoelectron heating of the thermal electrons. Then results were found with heating of the H^+ ions by the ring current during the recovery phase of a magnetic storm.

1. INTRODUCTION

Significant heating of the thermal plasma in the plasmasphere can result from interactions with ring current ions. This was studied by *Chandler et al.* [1988] using the FLIP model; for this work the thermal plasma was assumed to be heated by Coulomb collisions with the ring current ions, with equatorial heating rates calculated using the methods of *Kozyra et al.* [1987] and with a Gaussian profile in latitude. More recently, *Gorbachev et al.* [1992] investigated the heating of the thermal plasma by MHD waves which would be generated by the ring current; they used a hydrodynamic model which included the heat flow equations.

For this work the heating of H^+ ions due to Coulomb collisions with ring current ions during the recovery phase of a magnetic storm was studied. Heating rates per particle were derived from a three-dimensional kinetic ring current model [*Fok et al.*, 1995a]. A time-dependent hydrodynamic plasmasphere model was used; the magnetic field was assumed to be a corotating dipole. It is fully interhemispheric and no diffusive equilibrium assumptions are made.

2. MODEL DESCRIPTION

For this work the model is essentially the same as described in Guiter et al. (Modeling of O^+ ions in the plasmasphere, submitted to *J. Geophys. Res.*, 1995), with some modifications. It includes the time-dependent continuity, momentum, and energy equations for O^+ and H^+ ions, and the energy equation for electrons; the plasma is assumed to be quasineutral and currentless [cf. *Guiter and Gombosi*, 1990]. Ionization, charge exchange, recombination, collisions, and heat conduction are included; external heat sources are allowed. Photoelectron heating of thermal electrons is described using analytic formulas which include the effect of trapped photoelectrons. Neutral parameters were found using the MSIS-86 [*Hedin*, 1987] (densities and temperatures) and HWM-90 [*Hedin et al.*, 1991] (winds) models, for February 9, 1986. Also, the energy loss of H^+ ions due to charge exchange with H atoms was included. This was calculated using the cross section for this process as a function of energy, as given in *Barnett* [1990].

The decay of the ring current during the February, 1986 storm was simulated using a kinetic ring current model [*Fok et al.*, 1995a]. In this model, the distribution function is found as a function of pitch angle and energy; drifts, and losses due to charge exchange and Coulomb collisions are

[1] NAS/NRC Research Associate.

included. The initial distributions of the ring current species are found by extrapolating from AMPTE/CCE observations on the dawn and dusk sides of the inner magnetosphere; boundary values on the nightside are taken from CCE ion flux data at $L = 6.75$ during the storm. The ring current fluxes along the field line are inferred from the equatorial distribution function [cf. *Fok et al.*, 1995a]. The simulation began at 0300 UT on February 9, 1986. At various times in the simulation average heating rates per particle were calculated, as a function of local time, L shell, and magnetic latitude. It was assumed that only H^+ ions were heated by the ring current interactions, as the heating rate per particle for O^+ is much smaller than that for H^+ (personal communication, M.-C. Fok, 1995). For $L = 2$ and $L = 3$, these heating rates were set to zero for magnetic latitudes greater than 40 degrees, which is the boundary of the ring current model for these L shells.

For the plasmasphere modeling, a longitude was chosen so that 0 LT would be approximately equal to 0300 UT; the magnetic longitude was set to 25 degrees. For the $L = 2$ flux tube, the model was run for several days assuming only photoelectron heating, until the results were approximately diurnally reproducible. Then the model was allowed to run for one day with the ring current heating of the H^+ ions included. For the $L = 3$ flux tube, results with ring current heating were found starting from a solution which corresponds to a partially filled flux tube.

3. RESULTS

Figure 1 shows results found when the model was run on an $L = 2$ flux tube. The solid curve shows results found assuming photoelectron heating only, the dotted line gives results when the ring current heating started at 0 LT, and the dashed line gives results when the ring current heating started at 1200 LT. For the case when the heating starts at 0 LT the equatorial H^+ temperature rises to a value of about 35,000 K after 4 hours and then decreases more gradually; the equatorial H^+ density drops as the temperature increases. For this case the maximum ion heating rate per particle is about 2×10^{-3} eV/s. The equatorial temperature increase causes downward flows due to the increased pressure gradients, with H^+ velocities reaching 10 km/s near 500 km altitude (velocity profiles not shown). In the case when the ring current heating starts at 1200 LT the temperature does not change much until 12 hours into the simulation, when the temperature starts to rise. At this time the flux tube is on the nightside, which is where the ring current heating rates are highest. The equatorial H^+ temperature only rises to about 22,000 K in this case.

Figure 2 shows results found on an $L = 3$ flux tube; the solid line gives results found assuming photoelectron heating only, while the dotted line gives results when the ring current heating was included and assumed to start at 0 LT. In the first four hours T_{eq} (H^+) rises to about 12,000 K, and rises again in the afternoon, reaching a maximum of about 20,000 K. In addition, the plasmasphere density is lower when the ring current heating is included.

4. DISCUSSION AND SUMMARY

We have given results found using a time-dependent plasmasphere model when ring current heating of the H^+ ions is included. It was found that this would lead to significant enhancements of the equatorial H^+ temperature and that the equatorial H^+ density decreases when the temperature rises. A similar study [*Fok et al.*, 1995b] was done using the FLIP model; they found a similar time profile for the ion temperature, although the maximum temperature in their results is only about 14,000 K on an $L = 2$ field line. This difference is probably a result of the assumption in the

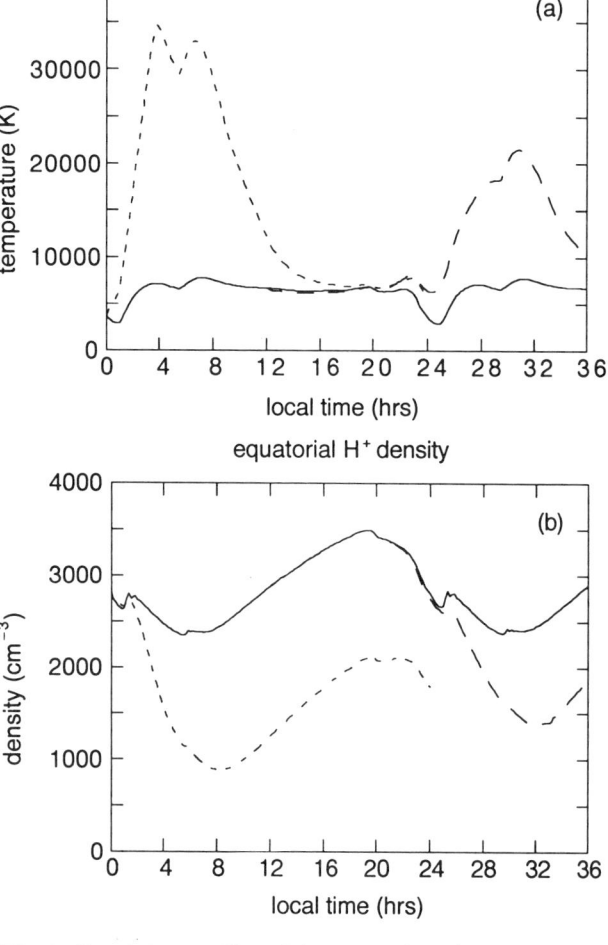

Fig. 1. Local time profiles of the equatorial H^+ temperature (a) and density (b) for $L = 2$. The solid line shows results when only photoelectron heating of electrons is included; the dotted line gives results when the ring current heating of H^+ ions started at 0 LT; the dashed line gives results when the ring current heating of H^+ ions started at 1200 LT.

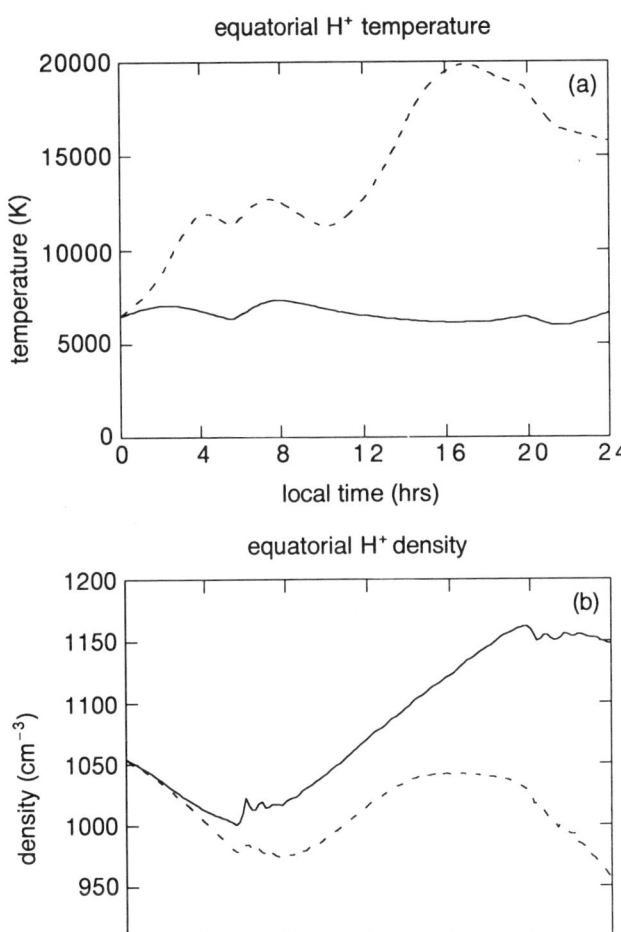

Fig. 2. Local time profiles of the equatorial H⁺ temperature (a) and density (b) for $L = 3$. The solid line shows results when only photoelectron heating of electrons is included; the dotted line gives results when the ring current heating of H⁺ ions started at 0 LT.

FLIP model that there is only one ion temperature, whereas in our plasmasphere model an energy equation is solved for each species. Assuming that the different ion species have the same temperature implies that collisions are strong enough to keep them in thermal equilibrium, and that the per particle heating rates are the same for H⁺ and O⁺. However, the per particle heating rates due to ring current interactions are roughly twenty times larger for H⁺ than for O⁺ (personal communication, M.-C. Fok, 1995), and the momentum transfer collision frequency of H⁺ with O⁺ is inversely proportional to the H⁺ temperature raised to the power 1.5, which means that the two ion species should rapidly be thermally decoupled. Furthermore, to force the ion temperatures to be the same in our model would require an enhancement of the thermal coupling between H⁺ and O⁺, which would be equivalent to extra energy loss for the H⁺ ions.

The temperature rise is probably unrealistically high in our $L = 2$ results. This could be a result of the ion heating rates being too high; changing the method for obtaining the initial ion fluxes at energies lower than the lowest energy measured by AMPTE can reduce the heating rates by an order of magnitude at lower L shells [cf. *Fok et al.*, 1995b]. With the ion heating rates reduced by a factor of ten, the maximum equatorial H⁺ temperature is about 10,000 K for $L = 2$. These results show that after the main phase of a magnetic storm ring current heating would have a significant effect on the thermal plasma in the plasmasphere.

Acknowledgments. This work was performed while two of the authors (S. M. Guiter and M.-C. Fok) held National Research Council-Marshall Space Flight Center Research Associateships. Acknowledgment is also made to Dr. Paul Craven for helpful discussions.

REFERENCES

Barnett, C. F., *Atomic Data for Fusion, vol. 1: Collisions of H, H2, He, and Li Atoms and Ions with Atoms and Molecules*, Technical Report ORNL-6086/V1, Oak Ridge National Laboratory, Oak Ridge, TN, 1990.

Chandler, M. O., J. U. Kozyra, J. L. Horwitz, R. H. Comfort, and L. H. Brace, Modeling of the thermal plasma in the outer plasmasphere -- a magnetospheric heat source, in *Modeling Magnetospheric Plasma, Geophys. Monogr. Sr.*, vol. 44, ed. by T. E. Moore and J. H. Waite, Jr., pp. 101-105, AGU, Washington, D. C., 1988.

Fok, M.-C., T. E. Moore, J. U. Kozyra, G. C. Ho, and D. C. Hamilton, Three-dimensional ring current decay model, *J. Geophys. Res., 100,* 9619, 1995a.

Fok, M.-C., P. D. Craven, T. E. Moore, and P. G. Richards, Ring current-plasmasphere coupling through Coulomb collisions, this monograph, 1995b.

Gorbachev, O. A., G. V. Khazanov, K. V. Gamayunov, and E. N. Krivorutsky, A theoretical model for the ring current interaction with the earth's plasmasphere, *Planet. Space Sci., 40,* 859, 1992.

Guiter, S. M., and T. I. Gombosi, The role of high speed plasma flows in plasmaspheric refilling, *J. Geophys. Res., 95,* 10,427, 1990.

Hedin, A. E., MSIS-86 thermospheric model, *J. Geophys. Res., 92,* 4649, 1987.

Hedin, A. E., M. A. Biondi, R. G. Burnside, G. Hernandez, R. M. Johnson, T. L. Killeen, C. Mazaudier, J. W. Meriwether, J. E. Salah, R. J. Sica, R. W. Smith, N. W. Spencer, V. B. Wickwar, and T. S. Virdi, Revised global model of thermosphere winds using satellite and ground-based observations, *J. Geophys. Res., 96,* 7657, 1991.

Kozyra, J. U., E. G. Shelley, R. H. Comfort, L. H. Brace, T. E. Cravens, and A. F. Nagy, The role of ring current O⁺ in the formation of stable auroral red arcs, *J. Geophys. Res., 92,* 7487, 1987.

S. M. Guiter, M.-C. Fok, and T. E. Moore, Space Sciences Laboratory, NASA Marshall Space Flight Center, Huntsville, AL 35812.

Equatorial Warm Ion Thermalization by Coulomb Collisions with Cool Outer Plasmaspheric Ions

Jinsoo Lee, J. L. Horwitz, G. R. Wilson, J. Lin, and D. G. Brown

*Department of Physics and Center for Space Plasma and Aeronomic Research
The University of Alabama in Huntsville, Huntsville, Alabama*

Coulomb collisions between dense cool plasmaspheric plasmas and the warm equatorially-trapped ions (frequently observed in the outer plasmasphere) will cool and isotropize these warm trapped ions to a thermodynamic equilibrium with the plasmasphere. In this paper, we examine the thermalization of anisotropic warm ions ($n(H_h^+)_{eq}=10$ ions cm^{-3}, $T_{\perp eq}=10$eV, $T_{\parallel eq}=5$eV) by denser ($n(H_c^+)=100$ ions cm^{-3}), cool isotropic (T=0.25eV) background ions through Coulomb collisions at L=4. For the present, we incorporate only ion-ion self collisions in our GSK (Generalized Semi-Kinetic) simulation. Collisions contribute to both energy degradation and pitch angle scattering of the warm ion population. It is found that the lowest energy portion of the perpendicular equatorial warm ion population (≤ 3 eV) cools within 10 hours to nearly the temperature of the cool ion population, whereas the more energetic (≥ 3 eV) tail of the warm ions tends to retain its spectral shape in energy while losing particles. These effects are tentatively explained in terms of the v^{-4} dependence of the Coulomb collision cross-section: the lowest energy ions experience frequent quasi-local thermalizing collisions with the background plasmasphere, whereas the more energetic ions may experience more infrequent pitch angle scattering collisions which cause them to free-stream down the field line where they are either thermalized or lost from the magnetic flux tube. The collisions also heat the original cool plasmasphere population; for these simulation conditions, the original plasmasphere heats from 0.25 eV to 0.4 eV by 5 hours and to 0.5 eV by 10 hours after the simulation is initiated.

INTRODUCTION

Equatorially-trapped ions are frequently observed within the outer plasmasphere (L=4-7) [e.g., *Horwitz and Chappell*, 1979], either as the apparently-dominant population [e.g., *Olsen et al.*, 1987] or mixed with a denser, cold plasmasphere background [*Horwitz et al.*, 1981]. These equatorially-trapped warm ions are probably generated by wave-particle interactions with hot ring current ions, though Coulomb collisions with such ring current ions may also play a role [*Kozyra*, 1992].

However these equatorially-trapped warm ions are initially created, we may expect that the subsequent evolution of such a population will involve energy degradation of the warm trapped population and heating of the cold dense plasmasphere population through Coulomb collisional energy transfer as the two populations come to a thermodynamic equilibrium. In this paper, we simulate this evolution with our Generalized SemiKinetic model for closed field lines.

THEORETICAL BACKGROUND

Using the Fokker-Planck expression [*Krall and Trivelpiece*, p303, 1986] for the slowing-down time of a test ion (in hours), and including only the effects of the ion-ion collision terms, we obtain:

$$\tau \sim 81 \times \frac{(kT)^{\frac{3}{2}}}{n_e}$$

for a test ion of characteristic energy kT(eV) colliding with a cold ion background of density n_e (cm^{-3}). We find that the thermalization time for a 10 eV test ion in a local background cool (T= 0.25 eV) plasma of 100 ions cm^{-3} is approximately 25 hours. If ion-electron collisions are included with the same background parameters, the estimated thermalization time is about 5 hours. We intend to incorporate such ion-electron collisions into investigations of the warm ion thermalization in the near future.

SIMULATION METHODS

Generation of Initial Warm and Cool Plasma Distributions

The closed field line GSK simulation of the motion of the ionospheric ion gyrocenters and anisotropic electron fluid is similar to that of Lin et al. [1994]. Here we impose additional warm and cool plasmas as an initial condition. The warm plasma distribution along the magnetic field line at L=4 is obtained by mapping equatorially-specified bi-Maxwellian ion and electron populations along the magnetic field line down to the ionosphere, with a self consistent potential as given by the condition of quasi-neutrality [*Olsen et al.*, 1994]. In this paper, the initial warm ion population at the equator is taken to be parameterized by n(H_h^+)$_{eq}$ =10 ions cm^{-3}, $T_{\perp eq}$ =10eV, $T_{\parallel eq}$ =5eV. Isotropic, cool background ions are initially distributed along the field line with equal density of 100 ions cm^{-3} and temperature of 0.25 eV. The accompanying electron fluids are initialized with the same density as that of the cool and warm populations. The temperature of cool electrons distributed along the field line is taken to be 0.25 eV, while warm electron temperatures are mapped from an isotropic thermal energy of 5 eV at the equator.

Evolution of the Plasma System

The cool plasmasphere plasma is maintained with continued refilling by injecting upgoing portions of cool isotropic (T=0.25eV) ionospheric ion distributions from the northern and southern topside ionospheric boundaries at 1900 km altitude, with the same density parameter of 100 ions cm^{-3} [e.g., *Lin et al.*, 1994]. Both injected and pre-existing ions are moved along the magnetic field line by gravitational, magnetic mirror and ambipolar electric forces. Coulomb collisions among all protons (warm and cool as initialized) are conducted with the Takizuka and Abe [1977] procedure, as discussed in Lin et al. [1994] and Wilson et al. [1992].

RESULTS

Figure 1 displays warm and cool ion characteristic parameter profiles at t= 0, 5, 10, 20 and 30 hours following the start of the simulation. In the parameters for the cool population it is evident that the density remains essentially at 100 ions cm^{-3} near the equator for the duration, and approaches 200 ions cm^{-3} toward the topside ionosphere. During the first 10 hours we can clearly see the heating of the cool plasmaspheric ions from 0.25 eV to about 0.4 eV after 5 hours and about 0.5 eV after 10 hours. Hence, such warm ions, however they are established, are a possible heat source for explaining the relatively high plasmaspheric ion temperatures frequently observed in the outer plasmasphere [e.g., *Horwitz et al.*, 1984, 1990; *Comfort et al.*, 1985] by collisional energy transfer.

Regarding the evolution of the warm ion population, Figure 1-(a) shows that both ion densities and temperatures (perpendicular is displayed) diminish strongly all along the field line over the 30 hours of the simulation as a result of the warm collisional energy and particle loss. The decline of the warm ion density/content may be interpreted primarily (see also discussion of Figure 2 below) as a result of the energetic portion of the equatorially-trapped ions being pitch angle scattered to smaller pitch angles. Such ions stream to lower altitudes and are eventually lost through the topside ionospheric boundary, possibly after one or more further pitch-angle scattering collisions. The warm ion population (about 5 ions cm^{-3}) remaining after 30 hours is interpreted as the lower-energy portion of the warm ion population which has thermalized quasi-locally under more frequent collisions with the cool ion background and has remained near the equator.

With regard to the warm ion temperature decrease, we note that the equatorial perpendicular temperature declines from 10 eV to about 3 eV, i.e., about a factor of 1/e, over the 30 hours of the simulation. This is reasonably consistent with the earlier estimate of 25 hours for the characteristic slowing down time for a thermal warm ion. However, as Figure 2 indicates, the equatorial perpendicular proton distribution function evolution is more complex and depends strongly on ion energy. It is evident from Figure 2 that the lower energy portion (say $\leq$ 3 eV) has come to approximate

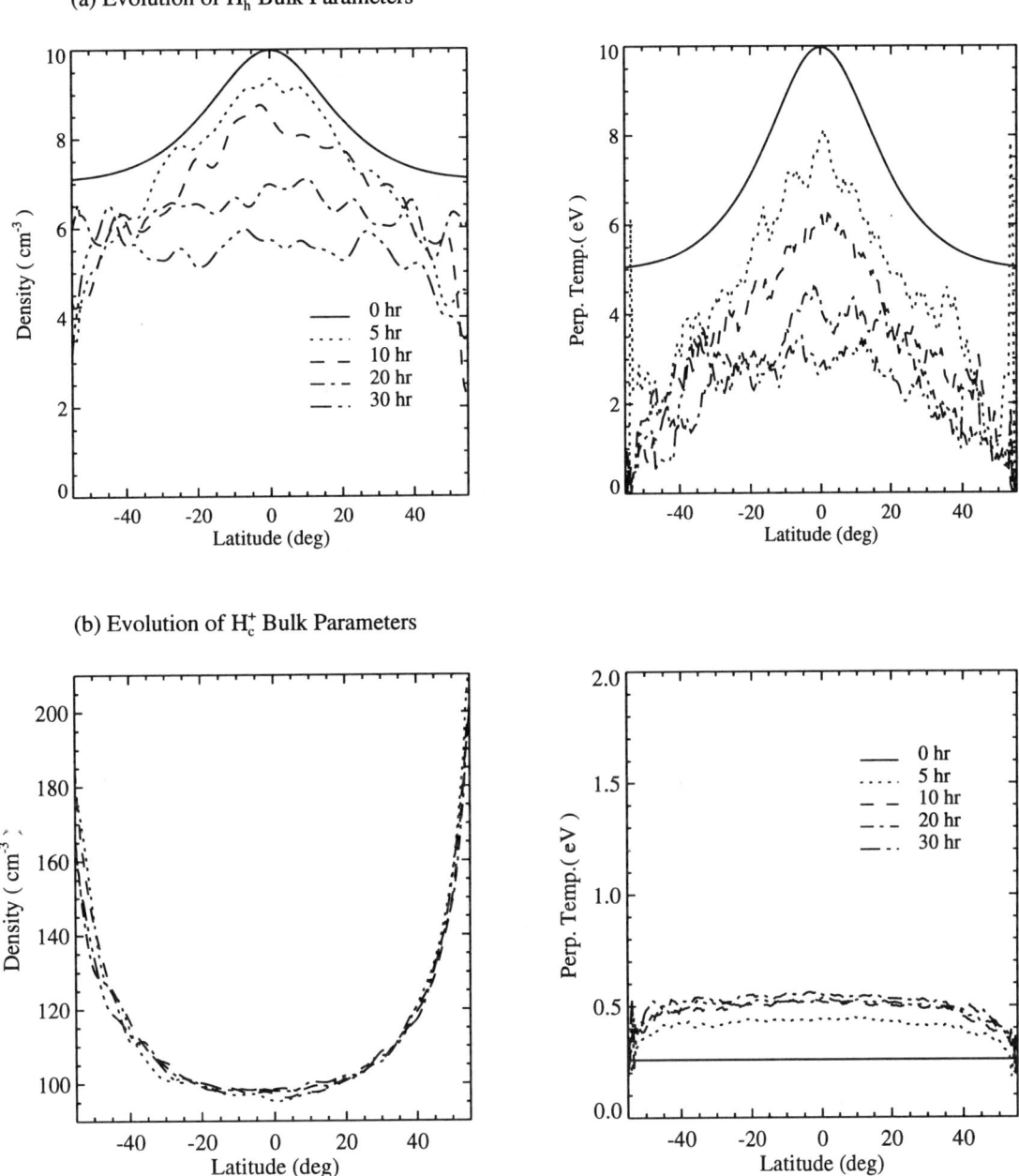

Fig. 1. Evolution of selected warm(a) and cool(b) proton bulk parameters.

thermodynamic equilibrium with the cool ion population by about 5 hours, while a superthermal tail (3-20 eV) persists, with relatively constant effective temperature but a decline in the overall phase space density level.

The separate behavior of these two (less than and greater than about 3 eV) portions of the warm ion population is tentatively interpreted as follows. Noting the v^{-4} dependence of the Coulomb collision cross-section, it can be surmised that the lowest energy ions experience frequent quasi-local thermalizing collisions with the background plasmasphere, whereas the more energetic ions may experience more infrequent pitch angle scattering collisions which cause them to free-stream down the field line where they are either thermalized or lost from the magnetic flux tube. Hence, the lowest energy portion rapidly cools to near the background cool ion temperature and has an energy spectrum (slope

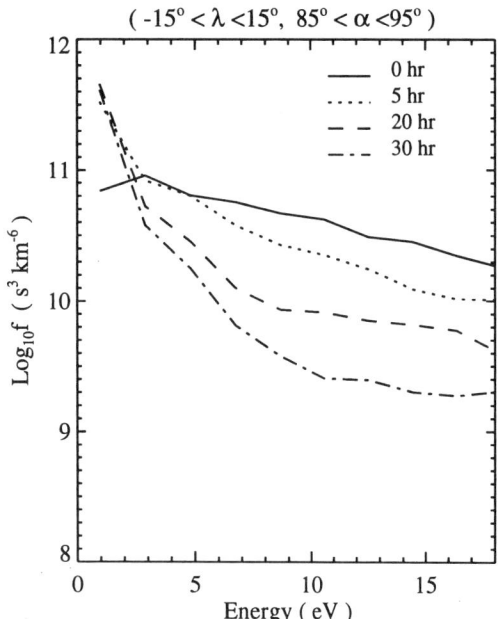

Fig. 2. Evolution of perpendicular part of the equatorial warm distribution function.

below 3 eV) consistent with this, whereas the higher energy spectrum portion is largely associated with loss of particles without a substantial change in the spectrum over time. Such behavior might be consistent with the mixtures of cold plus low density warm pancake-shaped ion distributions observed by Horwitz et al. [1981] in the plasmasphere.

Acknowledgments. This work was supported by NASA grant NAG8-134 and NSF grant ATM-9301024 to the University of Alabama in Huntsville.

REFERENCES

Comfort, R. H., J. H. Waite, Jr., and C. R. Chappell, Thermal ion temperatures from the retarding ion mass spectrometer on DE 1, *J. Geophys. Res., 90,* 3475, 1985.

Horwitz, J. L., and C. R. Chappell, Observations of warm plasma in the dayside plasma trough at geosynchronous orbit, *J. Geophys. Res., 84,* 7075, 1979.

Horwitz, J. L., C. R. Baugher, C. R. Chappell, E. G. Shelley, and D. T. Young, Pancake distributions of warm ions observed by ISEE-1, *J. Geophys. Res., 86,* 3311, 1981.

Horwitz, J. L., R. H. Comfort, and C. R. Chappell, Thermal ion composition measurements of the new outer plasmasphere and double plasmapause during the storm recovery phase, *Geophys. Res. Lett., 11,* 701, 1984.

Horwitz, J. L., R. H. Comfort, P. G. Richards, M. O. Chandler, C. R. Chappell, P. C. Anderson, W. B. Hanson, and L. H. Brace, Plasmasphere-ionosphere coupling, 2, Ion composition measurements at plasmaspheric and ionospheric altitudes and comparison with modeling results, *J. Geophys. Res., 95,* 7949, 1990.

Kozyra, J. U., Magnetic storm effects in the inner magnetosphere-The decay of the Earth's ring current, In T. Chang and J. R. Jasperse, editors, *Physics of Space Plasmas (1992),* volume 12, page 185. Scientific Publishers, Inc., Cambridge, Mass., 1992.

Krall, N. A., and A. W. Trivelpiece, *Principles of Plasma Physics,* San Francisco Press, San Francisco, 1986.

Lin, J., J. L. Horwitz, G. R. Wilson, and D. G. Brown, Equatorial heating and hemispheric decoupling effects on early inner magnetosphere core plasma evolution, *J. Geophys. Res., 99,* 5727, 1994.

Olsen, R. C., S. D. Shawhan, D. L. Gallagher, J. L. Green, C. R. Chappell, and R. R. Anderson, Plasma observations at the earth's magnetic equator, *J. Geophys. Res., 92,* 2385, 1987.

Olsen, R. C. L. J. Scott, and S. Boardsen, Comparison between Liouville's theorem and observed latitudinal distribution of trapped ions in the plasmapause region, *J. Geophys. Res., 99,* 2191, 1994.

Takizuka, T., and H. Abe, A binary collision model for plasma simulation with a particle code, *J. Comput. Phys., 25,* 205, 1977.

Wilson, G. R., J. L. Horwitz, and J. Lin, A semikinetic model for early stage plasmasphere refilling, 1, Effects of Coulomb collisions, *J. Geophys. Res., 97,* 1109, 1992.

D. G. Brown, J. L. Horwitz, J. Lee, J. Lin, and G. R. Wilson, Center for Space Plasma and Aeronomic Research, The University of Alabama in Huntsville, Huntsville, Alabama, 35899.
(Internet: brownd@cspar.uah.edu, horwitzj@cspar.uah.edu, leej@cspar.uah.edu, linj@cspar.uah.edu, wilsong@cspar.uah.edu)

Nonsteady State Coupling Processes in Superthermal Electron Transport

M. W. Liemohn and G. V. Khazanov

Space Physics Research Laboratory, Department of Atmospheric, Oceanic, and Space Sciences, University of Michigan, Ann Arbor, Michigan

Numerical solutions to the nonsteady state kinetic equation which describes the transport of superthermal electrons in the ionosphere and plasmasphere between the magnetically conjugate regions of the ionosphere are presented. The distribution function in time, distance along arbitrary geomagnetic field lines, energy, and pitch angle are among the parameters calculated by the model. This model represents a unified approach by self-consistently coupling the interaction of the two hemispheres and the trapping of superthermal electrons in the plasmasphere. Our calculations take into account the various ionization and excitation processes and the effect of an inhomogeneous magnetic field (i.e. magnetic mirroring of precipitating electrons and focusing of escaping electrons along magnetic field lines). Omnidirectional flux spectra and pitch angle distributions are shown for various L shells and situations, and the features are described in detail. Nonsteady state calculations predict that a depleted flux tube can take several hours to reach steady state levels again. Heating rates along the entire geomagnetic field line are presented as well as total energy deposition rates into the plasmasphere.

1. INTRODUCTION

A characteristic feature of near-earth space plasma is the presence of superthermal electrons in the energy range of 1 to 500 eV. The superthermal electron population is formed due to ionization of neutral atmospheric atoms and molecules by photoionization (photoelectrons) or by impact ionization from energetic particles of magnetospheric origin (secondary electrons). Superthermal electrons play a very important role in a large number of ionospheric and plasmaspheric processes. A fairly detailed knowledge of the superthermal electron distribution function is required when modeling upper atmospheric phenomena, such as the heating of thermal plasma, optical emissions, ionization processes, plasma instabilities, or plasma wave generation.

Several methods have been used to calculate superthermal electron fluxes in the ionosphere (the differential flux is easily related to the distribution function). The simplest approach is a local equilibrium approximation [e.g. *Victor et al.*, 1976], and over the last few decades many models have been developed to describe the transport and energy degradation (nonelastic collisions) of photoelectrons [cf. *Banks and Nagy*, 1970; *Mantas*, 1975; *Oran and Strickland*, 1978; *Prather et al.*, 1978; *Lejeune*, 1979; *Torr et al.*, 1990; *Link* 1992; *Khazanov et al.*, 1992] as well as auroral (secondary) electrons [*Banks et al.*, 1974; *Mantas and Walker*, 1976; *Strickland et al.*, 1976; *Khazanov* 1979; *Gefan et al.*, 1985; *Porter et al.*, 1987; *Lummerzheim et al.*, 1989]. Except for *Khazanov* [1979] and *Torr et al.* [1990], these calculations were based on separate treatments of the ionosphere and plasmasphere. *Khazanov and Gefan* [1982] and *Khazanov et al.* [1994] developed a comprehensive model which is equally valid in the ionosphere and in the plasmasphere and self-consistently couples the two hemispheres. *Khazanov et al.* [1994] have also shown that this self-consistent approach produces significant changes in the superthermal electron distributions compared to "pure" ionospheric or plasmaspheric calculations.

Time-dependent models may be needed when the local source of superthermal electrons increases or decreases rather sharply, such as during sunrise and sunset, auroral precipitation events, or plasmaspheric refilling. *Gefan and*

Khazanov [1990] have demonstrated the ability to model the time-dependent behavior of superthermal electrons. They simplified the problem by bounce-averaging the trapped population, and the time constants for filling the trapped region were found. It was also shown that for the case when the flux tubes are depleted of thermal plasma, as after a magnetic storm, the refilling time constants may be considerably longer then the storm time constants, thus meriting a time-dependent description of the problem. Recently, *Khazanov et al.* [1993a] have made the next step and considered the nonsteady state superthermal electron transport in the plasmasphere (at altitudes greater than about 1000 km) based on the kinetic equation in the guiding center approximation.

In this paper, results of a numerical study of nonsteady state ionosphere-plasmasphere coupling are presented. This model is based on the generalization of the numerical codes in *Khazanov et al.* [1993a, 1994] and *Khazanov and Liemohn* [1995].

2. KINETIC EQUATION

The kinetic equation includes the effects of elastic and inelastic collisions with both neutral and charged particles, transport in the inhomogeneous geomagnetic field, and particle sources can be written as [*Khazanov et al.*, 1993a, b]:

$$\frac{\beta}{\sqrt{E}}\frac{\partial \phi}{\partial t} + \mu \frac{\partial \phi}{\partial s} - \frac{1-\mu^2}{2}\frac{1}{B}\frac{\partial B}{\partial s}\frac{\partial \phi}{\partial \mu} = Q + S_{ee} \\ + \sum_i S_{ei}^0 + \sum_\alpha S_{e\alpha}^0 + \sum_\alpha S_{e\alpha}^* + \sum_\alpha S_{e\alpha}^+ \quad (1)$$

where $\beta = 1.7 \times 10^{-8}$ eV$^{1/2}$ cm^{-1}s, $\phi = \phi(t, E, \mu, s)$ is the differential flux of electrons, μ is the pitch angle cosine, E is the energy, s is the coordinate parallel to the geomagnetic field B, and Q is the electron production rate due to photoionization of neutral particles. The collision terms, given by S_{ee}, S_{ei}^0, $S_{e\alpha}^0$, $S_{e\alpha}^*$, and $S_{e\alpha}^+$, represent the collision integrals of superthermal electrons with thermal electrons, thermal ions, elastic scattering with neutral particles, inelastic excitation scattering with neutral particles, and inelastic ionization scattering with neutral particles, respectively. The details of this formulation are given in *Khazanov et al.* [1993a, 1994], including a transformation from (μ, s) to (μ_0, s), by

$$\mu_0 = \frac{\mu}{|\mu|}\sqrt{1 - \frac{B_0}{B(s)}(1-\mu^2)} \quad (2)$$

with B_o and μ_o denoting the magnetic field and the cosine of the pitch angle at the magnetic equator of the flux-tube, to decrease undesirable computational effects associated with approximation errors of the derivatives $\partial/\partial s$ and $\partial/\partial \mu$, and superthermal electron initial and boundary conditions.

To perform the calculations in this paper we used the following input for our superthermal electron model. Solar EUV and X-ray radiation spectra were obtained using the *Hinteregger* [1981] model, while neutral thermospheric densities and temperatures were given by MSIS-90 [*Hedin*, 1991]. The electron profile in the ionosphere was calculated based on the IRI model [*Bilitza*, 1990] and extended in the plasmasphere region using the assumption that the electron thermal density distribution in the plasmasphere is proportional to the geomagnetic field. Photoabsorption and photoionization cross sections for O, O_2, and N_2 were taken from *Fennelly and Torr* [1992]. Partial photoionization cross sections for O_2, and N_2 were obtained from *Conway* [1988], while partial photoionization cross sections of *Bell and Stafford* [1992] were adopted for atomic oxygen. Cross sections for elastic collisions, state-specific excitation and ionization were taken from *Solomon et al.* [1988]. All of the calculations were performed for a local time of noon on March 21, 1986, with $F_{10.7}$ and $<F_{10.7}>$ values of 150, chosen so the atmospheric conditions are symmetric and the solar radiation is at an average intensity level.

3. MODELING RESULTS

The main objective of this paper is to demonstrate the capabilities of the nonsteady state ionosphere-plasmasphere superthermal electron transport model and to indicate the possibilities for future research now available. A nonsteady state calculation is needed to understand the superthermal electron behavior when the source changes rapidly, such as sunrise or sunset conditions, auroral precipitation events, artificial injection of energetic electrons from the spacecraft, or plasmaspheric refilling. During the refilling process, for instance, the streaming of electrons into the plasmasphere creates an electric field and couples the superthermal electrons to the ions. As the plasma flows up and out of the topside ionosphere, the flow conditions change from subsonic to supersonic, from collision-dominated to collisionless, and from O^+ dominance to H^+ dominance. In the collisionless regime, the ion velocity distributions become highly non-Maxwellian and the coupling between various plasma species occurs through the development of a self-consistent potential. For the purposes of this study, we excluded the electric field from (1) and assume that the thermal plasma density is quasineutral and constant with time. Even without the electric field, the kinetic equation given by (1) allows for time-dependent calculations for the entire geomagnetic field line.

To simulate the refilling process correctly, the proper initial and boundary conditions must be used. This includes not only superthermal electron conditions, but also the thermal plasma distribution. Unfortunately, these conditions are not yet definitively known. For the nonsteady state results presented here, the calculations begin with no superthermal electrons in the flux tube. There is experimental [*Carpenter and Park*, 1973; *Park et al.*, 1978; *Corcuff et al.*, 1972] and theoretical [*Khazanov et al.*, 1984] evidence that during geomagnetic disturbances, the equatorial thermal plasma can decrease to levels as low as 10^{-1} to 1 cm^{-3}. A nonsteady state thermal plasma transport model in the ionosphere and plasmasphere should be used to simulate this density decrease and the subsequent refilling, but the coupling of a thermal plasma model with our superthermal electron model is beyond the scope of this presentation and will be discussed in an upcoming publication. Nevertheless, there are several ways of modeling this plasmaspheric density decrease. One method is to use a different density distribution in the plasmasphere, such as $n_e \sim B^2$ rather than $n_e \sim B$. Another method would be to use the IRI correction method proposed by *Buonsanto* [1989] to scale down the density in the topside ionosphere. *Buonsanto* [1989] showed that incoherent scatter radar observations of the thermal density in the ionosphere are lower than IRI densities at altitudes above the F$_2$ peak, and he provides a simple correction to the IRI model. *Kozyra et al.* [1990] uses this thinner F$_2$ layer model when calculating the solar cycle variations of stable auroral red arcs. By choosing the correct scaling factors, this method can be used to maintain the IRI model F$_2$ peak but decrease the density above this altitude to a target topside ionosphere density. Both of these methods will be demonstrated.

The first results presented use the IRI densities in the ionosphere with the plasmaspheric densities proportional to the square of the magnetic field. Upward flowing fluxes are shown for 5 eV and 30 eV in Figure 1 for four altitudes ((a) 300 km, (b) 500 km, (c) 800 km, and (d) 10,000 km) along an $L=2$ field line. The first altitude is near the thermal electron density peak, the next altitude is in the transition region of the ionosphere, the third altitude is the topside ionosphere, and the last altitude is at the equatorial plane in the plasmasphere. Five times are shown for each altitude and energy. The times presented are chosen because of their relation to the elastic collision timescale, the fastest collisional timescale in the calculations. Specifically, the ratio of the real time to the elastic collision timescales are 0.1, 0.5, 1, and 2, with the final result being the steady state solution.

The 5 eV flux calculations, Figures 1(a)-(d), show the greatest interaction with the thermal plasma and neutral species of the two energies presented. In Figure 1(a), each time presented shows a uniform distribution because the elastic scattering with the thermal plasma and the neutral particles is very efficient. Notice how quickly the solution reaches the steady state solution. This is because the collision timescale is a tenth of a second. The fluxes in Figure 1(b) show a decrease at large pitch angles, especially in the steady state results. This is because of the path length to traverse a given field-aligned distance is longer for a large pitch angle electron compared to an electron with a smaller pitch angle, and so these longer-path particles undergo more collisions. This decrease at larger pitch angles is also seen at 800 km (Figure 1(c)). Once in the plasmasphere, however, the focusing of the electrons due to the changing magnetic field strength creates a maximum in the pitch angle distribution in the loss cone, which is 20° at the equator for $L=2$. Coulomb collisions with the thermal plasma scatters electrons into the trapped zone, though, and Figure 1(d) shows a gradual decrease with pitch angle of two orders of magnitude from the loss cone to 90°. Notice that the loss cone fluxes are continually decreasing, dropping by a factor of 30 from the F$_2$ peak to the equatorial plane.

The 30 eV upward flowing fluxes presented in Figures 1(e)-(h) show many of the same features, but the differences are worth noting. First, the decrease for large pitch angles in Figures 1(f)-(g) are less pronounced than the decrease seen in Figures 1(b)-(c). This is because the Coulomb cross section is proportional to E^{-2}, so these electrons are affected less by collisions with the thermal plasma than the 5 eV electrons. Second, the trapped zone in Figure 1(h) is less populated than the trapped zone in Figure 1(d), dropping five orders of magnitude between 20° and 50°. Also, the fluxes in the loss cone do not drop as much as the 5 eV loss cone fluxes, decreasing by only a factor of two from 300 km to 10,000 km. This is explained by the decreased effect of Coulomb collisions on the higher energy electrons.

It should be noted that for this thermal density distribution, none of the results at two elastic collision timescales have reached the steady state results in Figure 1, indicating that another elastic scattering timescale or more is needed for convergence.

With a nonsteady state model, the early stages of the refilling process can be examined in detail. This might be important in sunrise or sunset conditions, where a relatively empty flux tube with no source in either ionosphere rotates onto the sunlit side of the earth, and one or both ionospheres experience photoionization and a front of superthermal electrons traverses the plasmasphere. This situa-

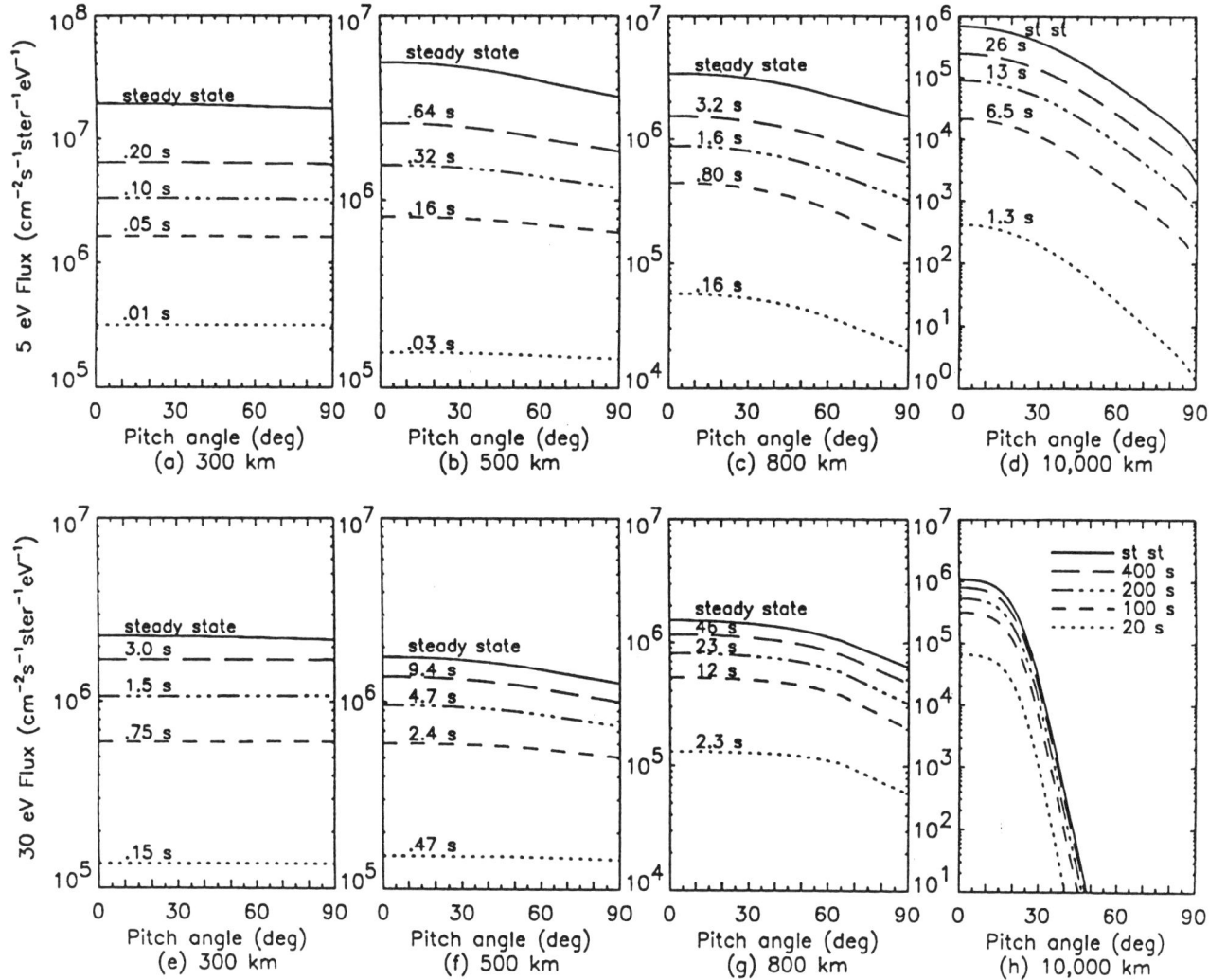

Fig. 1. Upward flowing pitch angle distributions for $L=2$ with $n_e \sim B^2$ at the indicated altitudes for (a)-(d) 5 eV and (e)-(h) 30 eV electrons.

tion was modeled for an initially empty $L=4$ flux tube with $n_e \sim B$, then one ionosphere is illuminated while the conjugate ionosphere is kept dark.

Figure 2 shows omnidirectional flux spectra, given in $eV^{-1}cm^{-2}s^{-1}ster^{-1}$, for this case at (a) 5 seconds, (b) 10 seconds, (c) 15 seconds, and (d) 20 seconds after the illumination begins. The sunlit ionosphere is to the left of each plot, and the dark ionosphere is on the right. First, notice some of the features of the omnidirectional flux spectra. There is a maximum near 20 eV close to the sunlit ionosphere. This is due to photoionization by the HeII-304 Å resonance line. The differences in energy thresholds for the ionization states of the neutral species creates a range of structure in the photoelectron spectrum, which moves out of the ionosphere. The decrease in flux at low energies is due to depletion by Coulomb collisions. These particles not only have a larger cross section with the thermal plasma, but they also move slower than the higher energy electrons. At higher energies, the effect of Coulomb collisions is greatly reduced, so the particles stream through the plasmasphere much easier. The sharp drop in the flux around 70 eV is due to a corresponding decrease in the primary photoelectron spectrum. Also, in the ionospheres, the loss cones cover the entire 180° of pitch angle (definition of the loss cone), but at the equatorial plane, the loss cones account for only 12.8° of pitch angle (for $L=4$), 6.4° in each direction along the magnetic field. Therefore, at the equatorial plane, the omnidirectional flux value is

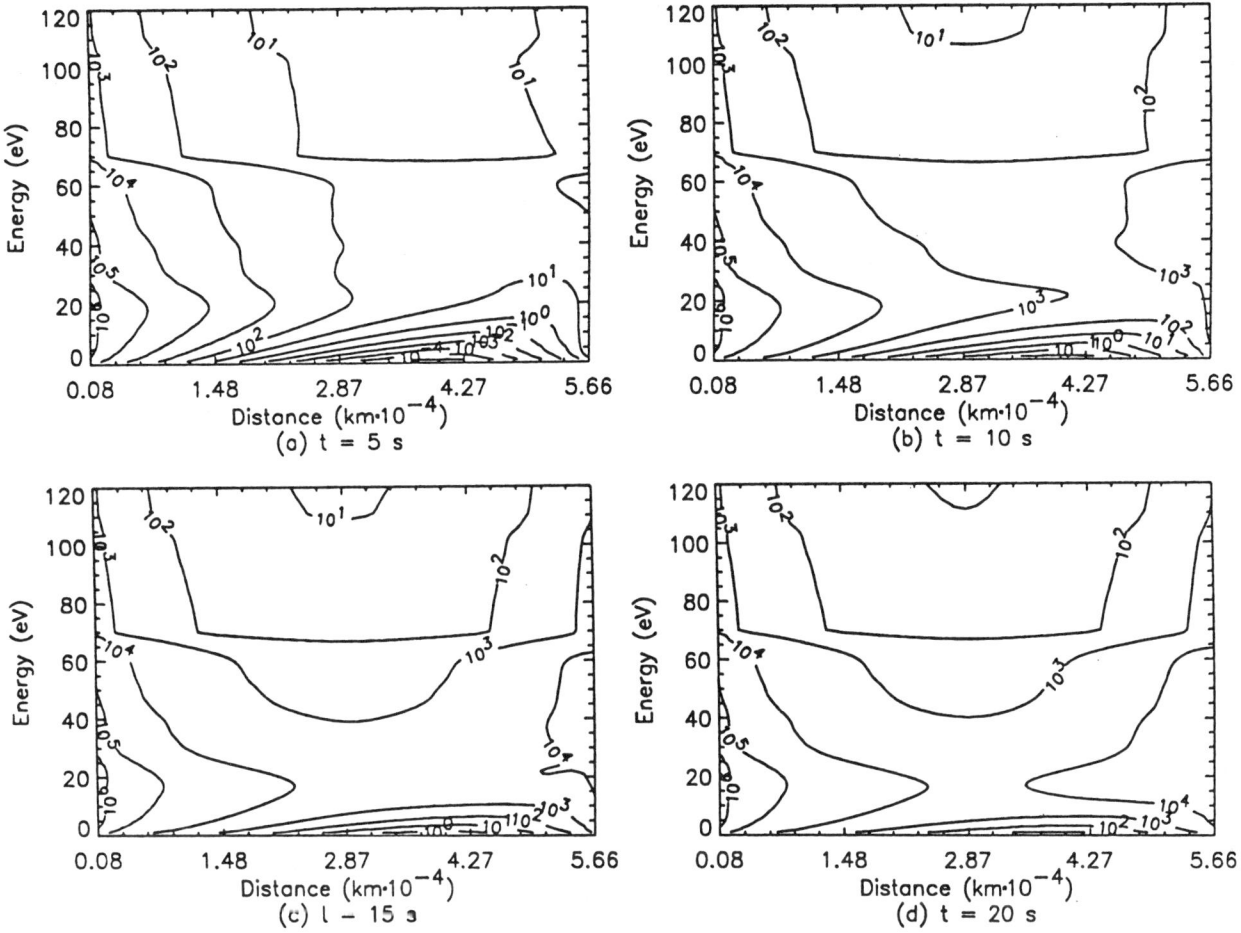

Fig. 2. Omnidirectional fluxes in the plasmasphere for $L=4$ with $n_e \sim B$ showing the early stages of the refilling process, at (a) 5 s, (b) 10 s, (c) 15 s, and (d) 20 s after illumination of one ionosphere began.

heavily influenced by the number of particles with pitch angles in the trapped zone, which is lower than the number of particles with pitch angles in the loss cone. So, omnidirectional fluxes decrease towards the equatorial plane due to this effect.

At $L=4$, a 100 eV electron takes 9.4 seconds to traverse the plasmasphere, while a 5 eV electron takes 42 seconds to reach the conjugate ionosphere. So in Figure 2(a), the plasmasphere is still mostly depleted because this is at 5 seconds of refilling. The high energy electrons have reached the equatorial plane, while the lower energy electrons are just entering the plasmasphere. By 10 seconds (Figure 2(b)), the high energy particles have reached the conjugate ionosphere. This is evident by the near symmetric appearance of the high energy omnidirectional flux. The high energy fluxes are still slightly higher near the sunlit ionosphere because there has only been time for particles to make one pass through the plasmasphere, so no electrons have backscattered out of the dark ionosphere and scattering to the trapped zone is minimal. The maximum near 20 eV is evident in Figure 2(b), also. The omnidirectional fluxes after 15 seconds (figure 2(c)) shows a backscatter contribution at high energies. The low energies are continuing to refill, with fluxes of at least 10^3 cm^{-2}eV^{-1}s^{-1}ster^{-1} for 20 eV electrons. The final frame, Figure 2(d), shows that the high energy omnidirectional fluxes are close to their steady state values since the upward flowing and downward flowing loss cones are filled (the loss cone fluxes are the main contributors to the omnidirectional fluxes). The low energies are still refilling, not only because of transport from the sunlit ionosphere, but also due to high energy electrons Coulomb energy decaying.

The development of the pitch angle distribution in the trapped zone of the plasmasphere is another distinguishing feature of the model. This is demonstrated for the case of a depleted flux tube using the IRI correction method of

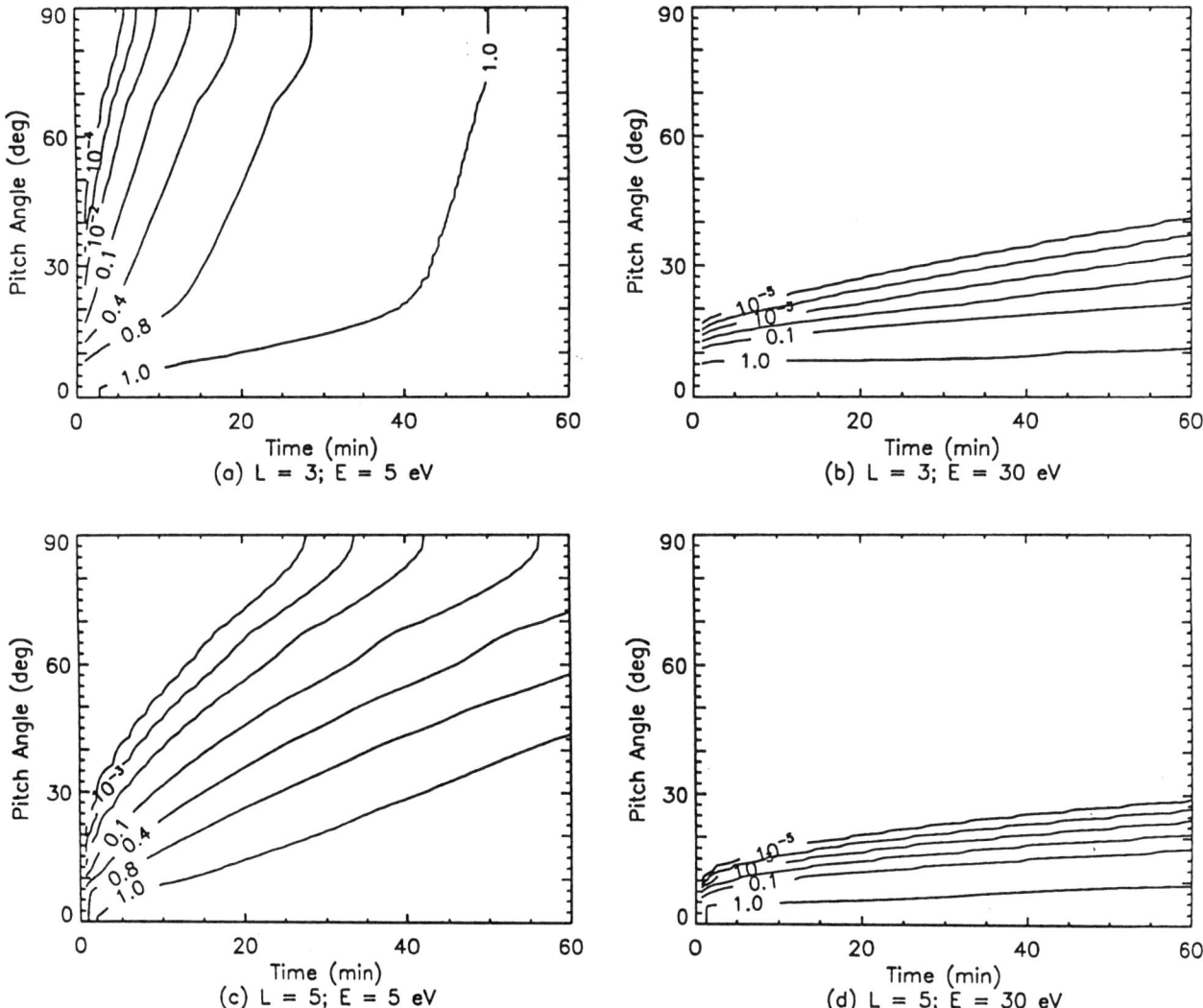

Fig. 3. Time-development of equatorial pitch angle distributions at (a)-(b) $L=3$ and (c)-(d) $L=5$ with $n_e \sim B$ and a thinned topside F_2 layer. Distributions at 5 eV are shown in (a) and (c), and 30 eV fluxes in (b) and (d). The fluxes are normalized to the steady state distribution at each pitch angle.

Buonsanto [1989]. By choosing the correct scaling factors, the ionospheric thermal plasma is thinned above the F_2 peak to make the topside ionosphere density 550 cm^{-3}, and then $n_e \sim B$ is used in the plasmasphere. For $L=3$, this results in an equatorial plasma density of 17 cm^{-3}, and for $L=5$, the density reaches 4 cm^{-3} in the equatorial plane. These depleted densities, although lower than the equatorial densities of 780 cm^{-3} for Figure 1 and 1700 cm^{-3} for Figure 2, is still within the depleted density range for geomagnetically disturbed times [*Carpenter and Park*, 1973; *Park et al.*, 1978; *Corcuff et al.*, 1972; *Khazanov et al.*, 1984].

Figure 3 shows equatorial pitch angle distributions for (a) 5 eV electrons at $L=3$, (b) 30 eV electrons at $L=3$, (c) 5 eV electrons at $L=5$, and (d) 30 eV electrons at $L=5$. The results have been normalized by dividing the fluxes by the steady state solution at each pitch angle. Thus, the contour line "1.0" represents convergence to the steady state solution. These plots show the development of the distribution function in the trapped zone during a refilling process. Figures 3(a)-(b) show results at $L=3$ for 5 eV and 30 eV, respectively. The loss cone, located at pitch angles of less than 10°, fills very quickly because these pitch angles have a direct source in the ionosphere. The trapped zone, however, takes much longer to fill because its source is Coulomb scattering from the loss cone. For instance, for 5 eV, the 20° pitch angle fluxes reach 80% of the steady state value in 12 minutes, and the steady state flux levels are reaches after 47 minutes of refilling. The 5 eV 90° pitch angle fluxes, however, take 28 minutes to reach the 80% level and 54 minutes to reach 100% of the steady state flux. In Figure 2(b), the 30 eV fluxes in the trapped zone do not reach steady state after an hour of refilling. To illustrate, the 20° fluxes need over 50 minutes of refilling to reach 10% of the steady state flux value, and there are essentially no 30 eV electrons with 90° pitch angles after an hour of refilling.

For comparison, the refilling of an $L=5$ flux tube is shown in Figures 3(c)-(d), for 5 eV and 30 eV, respectively. Due to the lower plasma density at the equator, scattering into the trapped zone is slower for this case than for the $L=3$ flux tube. This is evident in Figure 3(c) by noticing that only the first 42° of pitch angle have reached steady state after one hour of refilling. The 90° fluxes take 56 minutes to reach 10% of the steady state level. Figure 3(d) is also not close to full convergence, and is slower than the convergence of the $L=3$ 30 eV fluxes in Figure 3(b). At $L=5$, the loss cone is the first 4° of equatorial pitch angle, so part of the trapped zone is converging to steady state levels, but the majority of the trapped zone is still devoid of particles after an hour of refilling. Therefore, the superthermal electron plasmaspheric refilling process could take hours and a time-dependent model must be used. Also, the low energy flux levels are dependent on the high energy fluxes because of the energy cascading due to Coulomb collisions, so the 5 eV fluxes will continue to adjust closer to the convergent values until the higher energy fluxes reach steady state with the high energy refilling timescale.

As for the depletion of a trapped population in the plasmasphere, we refer the reader to our earlier work [*Khazanov et al.*, 1993a, Fig. 6]. It was shown that the low energy superthermal electrons are depleted rapidly, with the trapped zone population decreasing to 10^{-3} of its steady state level in an hour. Higher energy electron distributions take much longer to decay, and it was shown that 30 eV fluxes were still close to the steady state levels in the trapped zone after an hour of depletion. The higher energy particles will decay to lower energies and will give some contribution to the heating of the plasmasphere. This process is very sensitive to the integral content of the thermal plasma in the flux tube, which depends on its temperature. Thus, in order to give a quantitative description of the nighttime energy deposition from the trapped superthermal population, it is necessary to perform a self-consistent calculation with the thermal plasma and superthermal electrons, which we plan to do in the near future.

4. IONOSPHERIC-PLASMASPHERIC ENERGY DEPOSITION

Knowledge of the distribution of superthermal electrons is required when solving many geophysical problems: heating of the thermal ionospheric and plasmaspheric plasma; optical emissions and ionization of the upper atmosphere; and wave generation in and stability of the ionospheric and plasmaspheric plasma. The results of this model can be used to examine any of these situations, and in this section the deposition of energy along the field line will be considered. The superthermal electrons scatter with the thermal electrons via Coulomb collisions, which in turn transfer energy to the thermal ions. The importance of these results in the present study is that the heating rates have been calculated in the plasmasphere and the two conjugate ionospheres with one spatially-coupled model.

Figure 4 shows steady state heating rates, Q_e, as a function of distance along the field line and L shell for (a) symmetric conditions of illumination, and (b) nonsymmetric conditions where the conjugate ionosphere has no photoelectron source. The field-aligned distance has been normalized to the length of the flux tube at each L shell, which is necessary because the plasmaspheric distance increases with L shell.

Notice that Q_e in the sunlit hemisphere is the same for the two conditions, with a maximum at 200 km for all L shells. This maximum of Q_e changes from 7.8×10^3 eV cm^{-3}s^{-1} at $L=2$ down to 3.8×10^3 eV cm^{-3}s^{-1} at $L=6$. This is because the solar zenith angle increases with L shell, so the total photoelectron production rate decreases, and thus the heating rate also decreases. For Figure 4(a), these are also the maximum values in the conjugate ionosphere, but in Figure 4(b), the dark conjugate hemisphere has a maximum near 400 km which varies from 21 eV cm^{-3}s^{-1} for $L=2$ to 2.6 eV cm^{-3}s^{-1} for $L=6$. The maximum is at a higher altitude in this ionosphere because the source of superthermal electrons is from the plasmasphere, and the decrease with L shell is due to the decreasing production rate in the source ionosphere which is due to the increasing solar zenith angle. Another reason for this decrease is because the volume of the flux tube increases with L shell, so there is more absorption of energy in the plasmasphere. The heating rates then decrease to a minimum at the equatorial plane, which also decreases with increasing L shell.

Spatially integrating these heating rates in the plasmasphere yields the total energy deposition rate to the plasmasphere, P_E. P_E can also be thought of as a column heating rate, where the column is the plasmaspheric flux tube.

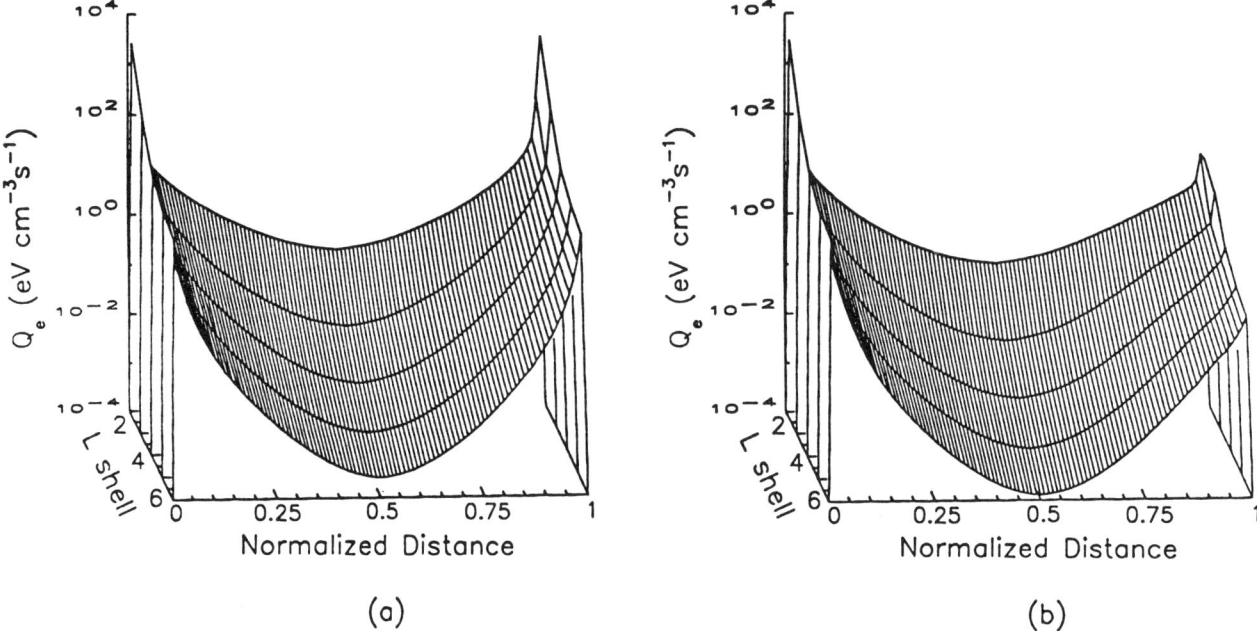

Fig. 4. Heating rates along the field line for several L shells for (a) symmetric illumination conditions and (b) nonsymmetric illumination conditions. The field-aligned distance is normalized to the length of the flux tube for each L shell.

These results are shown in Table 1 as a function of L shell for symmetric and nonsymmetric illumination conditions. Although the heating rates decrease with increasing L shell, this quantity increases because the volume of the flux tube is increasing with L shell faster than Q_e is decreasing. The total energy deposition rate for the symmetric case is close to twice the rate for the nonsymmetric case since one of the ionospheres is without a photoelectron source. The reason for this is because the kinetic equation is linear, so removing half of the source should decrease the total energy deposition by half as well. The linearity of this decrease is also an indication of the energy conservation of our numerical scheme.

Using these results, it is possible to determine the portion of energy deposited to the plasmasphere and to check the energy conservation in our model at plasmaspheric heights. Since P_E is an explicit calculation of the plasmaspheric energy deposition, dividing it by the energy entering the plasmasphere gives an explicit calculation of the portion of energy deposited in the plasmasphere, ε_1. An implicit method of finding this quantity is through the portion of energy leaving the plasmasphere. There is a plasmas-

Table 1. Plasmaspheric energy deposition rates

	Symmetric				Nonsymmetric			
L	ε_1	ε_2	% Diff.[a]	P_E[b]	ε_1	ε_2	% Diff.[a]	P_E[b]
2	.595	.684	13.0	8.06	.601	.688	12.6	4.20
3	.720	.755	4.6	9.26	.722	.755	4.4	4.66
4	.782	.799	2.1	9.81	.782	.799	2.1	4.91
5	.807	.816	1.1	9.99	.808	.816	1.0	5.00
6	.817	.824	0.8	10.03	.818	.824	0.7	5.02

[a] % Diff. = $|\varepsilon_2-\varepsilon_1|/\varepsilon_2 \times 100\%$
[b] P_E in eV cm^{-2} s^{-1} × 10^{-9}

pheric energy balance equating the energy flowing into the plasmasphere to the two possible losses: absorption in the plasmasphere or transport back to the ionospheres. Therefore, dividing the energy flowing out of the plasmasphere into the ionospheres by the energy flowing out of the ionospheres into the plasmasphere, and subtracting this from unity yields the second calculation of the portion of energy deposited in the plasmasphere, ε_2. The percent difference between ε_1 and ε_2 indicates the energy conservation of the numerical model. These results are also shown in Table 1. The percent difference decreases with increasing L shell, and is consistent with the numerical error of the calculations. Also, it can be seen that ε is independent of the ionospheric conditions. This is because the ionospheric environments change the amount of energy being deposited, but not the portion of energy that is absorbed, because in our calculations the thermal population in the plasmasphere is unchanged between the two ionospheric cases.

5. SUMMARY AND FUTURE DEVELOPMENT OF THE MODEL

Numerical solutions to the nonsteady state superthermal electron kinetic equation have been presented. The calculations were performed in the plasmasphere and the two conjugate ionospheres, self-consistently coupling the three regions. The model also includes various neutral particle ionization and excitation processes, scattering, and the effect of the inhomogeneous magnetic field. The distribution function in time, distance along arbitrary geomagnetic field lines, energy, and pitch angle are among the parameters calculated by the model. The results presented here show the complicated structure of the superthermal electron distribution, and this model is another step towards understanding the exchange of energy between spatial regions and particle populations in the ionosphere-plasmasphere system. The model can handle any magnetic field configuration, neutral atmosphere and thermal plasma description, and phase space grid.

The superthermal electron flux distribution was shown in several ways and the features of these distributions were described in detail. Pitch angle distributions at specific altitudes and plasmaspheric omnidirectional flux spectra were used to display the nonsteady state results. The time needed to refill a depleted flux tube was found to be several elastic scattering timescales, which means several hours is required to reach the steady state flux levels for high energy superthermal electrons in the geomagnetic trap.

One application of these results that was shown is the energy deposition along the geomagnetic field line. Heating rates for the entire ionosphere-plasmasphere system can be calculated with this model. Total energy deposition into the plasmasphere for several L shells and conditions of illumination were given, and from this an energy conservation check of the model was conducted.

There are several processes which will be incorporated into this model and discussed in upcoming publications. One development is coupling this model with a thermal plasma transport model. This will permit the calculation of the self-consistent electric field, which is generated when the highly mobile superthermal electrons stream into the plasmasphere ahead of the heavy ions and charge neutrality is violated. The inclusion of the self-consistent electric field allows for the proper coupling between the thermal and the superthermal plasmas. External electric fields, such as the electric field of magnetospheric convection, will also be incorporated.

Another effect which must be included at some time is the interaction with waves generated by an external source, such as ion-cyclotron waves from the ring current particles, plasmaspheric hiss, Alfven and magnetosonic waves, and electrostatic electron-cyclotron waves. There is some evidence that these waves can interact with the superthermal electrons; *Erlandson et al.* [1993] observed an enhancement in the superthermal electron fluxes in regions of increased wave activity by DE-2. If the energy density of the wave is large compared to the energy density of the superthermal electrons, the wave fields can be taken as constant functions in the term which describes the interaction with these waves. The inclusion of diffusion coefficients describing these waves would allow this model to describe these processes.

Plasma waves can also be amplified by instabilities in the superthermal electron distribution, and so a self-consistent feedback is needed to properly describe this process. As the distribution function becomes unstable, linear theory predicts that plasma waves will be amplified. These waves can then interact via diffusion terms and alter the superthermal electron distribution, and thus the amplification of the plasma waves will be changed. Incorporation of this self-consistent plasma wave interaction will also be conducted in the near future.

Acknowledgments. This work was supported at the University of Michigan by the National Science Foundation under contract ATM-9412409.

REFERENCES

Banks, P.M., and Nagy, A.F., Concerning the influence of elastic scattering upon photoelectron transport and escape, *J. Geophys. Res., 75*, 1902, 1970.

Banks, P. M., C. R. Chappell, and A. F. Nagy, A new model for the interaction of auroral electrons with the atmosphere: spectral degradation, back scatter, optical emission and ionization., *J. Geophys. Res., 79*, 1459, 1974.

Bell, K.L., and Stafford, R.P., Photoionization cross sections for atomic oxygen, *Planet. Space Sci., 40*, 1419, 1992.

Bilitza, D., Progress report on IRI status, *Adv. Space Res., 10*, (11) 3, 1990.

Buonsanto, M. J., Comparison of incoherent scatter observations of electron density, and electron and ion temperature at Millstone Hill with the International Reference Ionosphere, *J. Atmos. Terr. Phys., 51*, 441, 1989.

Carpenter, D. L., and C. G. Park, On what ionospheric workers should know about the plasmapause-plasmasphere, *Rev. Geophys. Space Phys., 11*, 133, 1973.

Conway, R.R., *Photoabsorption and photoionization cross sections: A compilation of recent measurements*, NRL Memo Rep. 6155, Naval Research Laboratory, 1988.

Corcuff, P., J. Corcuff, D. L. Carpenter, C. R. Chappell, J. Vigneron, and N. Kleimenova, La plasmasphère en période de recouvrement magnétique. Etude combinée des données satellites OGO 4, OGO 5 et des sekfflements reçus au sol, *Ann. Geophys., 28*, 679, 1972.

Erlandson, R. E., T. L. Aggson, W. R. Hogey, and J. A. Slavin, Simultaneous observations of subauroral electron temperature enhancements and electromagnetic cyclotron waves, *Geophys. Res. Lett., 20*, 1723, 1993.

Gefan, G. D., A. A. Trukhan, and G. V. Khazanov, A method of calculating auroral electron fluxes, *Annales Geophys., 3*, 135, 1985.

Gefan, G. D., and G. V. Khazanov, Non-steady-state conditions of filling up the geomagnetic trap with superthermal electrons, *Ann. Geophys. 8*, 519, 1990.

Fennelly, J. A., and D. G. Torr, Photoionization and photoabsorption cross sections of O, N_2, O_2, and N for aeronomic calculations, *At. Data and Nuc. Data Tab., 51*, 321, 1992.

Hedin, A.E., Extension of the MSIS thermospheric model into the middle and lower atmosphere, *J. Geophys. Res., 96*, 1159, 1991.

Hinteregger, H. E., Representations of solar EUV fluxes for aeronomical applications, *Adv. Space Res., 1*, 39, 1981.

Khazanov, G. V., *The kinetics of the electron plasma component of the upper atmosphere* (in Russian), Moscow, Nauka, 1979 [English translation: Washington, D.C., National Translation Center, #80-50707, 1980].

Khazanov, G. V., and G. D. Gefan, The kinetics of ionosphere-plasmasphere transport of superthermal electrons, *Phys. Solariterr., 19*, 65, 1982.

Khazanov, G. V., M. A. Koen, Yu. V. Konikov, and I. M. Sidorov, Simulation of ionosphere-plasmasphere coupling taking into account ion inertia and temperature anisotropy, *Planet. Space Sci., 32*, 585, 1984.

Khazanov, G. V., T. I. Gombosi, A. F. Nagy, and M. A. Koen, Analysis of the ionosphere-plasmasphere transport of superthermal electrons: 1. Transport in the plasmasphere, *J. Geophys. Res., 97*, 16,887, 1992.

Khazanov, G. V., M. W. Liemohn, T. I. Gombosi, and A. F. Nagy, Non-steady-state transport of superthermal electrons in the plasmasphere, *Geophys.Res. Lett., 20*, 2821, 1993a.

Khazanov, G. V., T. Neubert, G. D. Gefan, and A. A. Trukhan, E. V. Mishin, A kinetic description of electron beam ejection from spacecraft, *Geophys.Res. Lett., 20*, 1999, 1993b.

Khazanov, G. V., T. Neubert, and G. D. Gefan, Kinetic theory of ionosphere-plasmasphere transport of suprathermal electrons, *IEEE Transactions on Plasma Science, 22*, 187, 1994.

Khazanov, G. V., and M. W. Liemohn, Non-steady-state ionosphere-plasmasphere coupling of superthermal electrons, *J. Geophys. Res., 100*, in press, 1995.

Kozyra, J. U., C. E. Valladares, H. C. Carlson, M. J. Buonsanto, and D. W. Slater, A theoretical study of seasonal and solar cycle variations of stable aurora red arcs, *J. Geophys. Res., 95*, 12219, 1990.

Lejeune, G., "Two stream" photoelectron distributions with interhemispheric coupling: A mixing of analytical and numerical methods, *Planet. Space Sci., 27*, 561, 1979.

Link, R., Feautrier solution of the electron transport equation, *J. Geophys. Res., 97*, 159, 1992.

Lummerzheim, D., M. N. Rees, and H. R. Anderson, Angular dependent transport of auroral electrons in the upper atmosphere, *Planet. Space Sci., 37*, 109, 1989.

Mantas, G. P., The theory of photoelectron thermalization and transport in the ionosphere, *Planet. Space Sci., 23*, 337, 1975.

Mantas, G. P., and J. C. G. Walker, The penetration of soft electrons into the ionosphere, *Planet. Space Sci., 24*, 409, 1976.

Oran, E. S., and D. J. Strickland, Photoelectron flux in the Earth's ionosphere, *Planet. Space Sci., 26*, 1161, 1978.

Park, C. G., D. L. Carpenter, and D. B. Wiggin, Electron density in the plasmasphere: Whistler data on solar cycle, annual, and diurnal variations, *J. Geophys. Res., 83*, 3137, 1978.

Porter, H. S., F. Varosi, and H. G. Mayr, Iterative solution of the multistream electron transport equation, 1, Comparison with laboratory beam injection experiments, *J. Geophys. Res., 92*, 5933, 1987.

Prather, M. J., M. B. McElroy, and J. Rodriguez, Photoelectrons in the upper atmosphere: A formulation incorporating effects of transport, *Planet. Space Sci., 26*, 131, 1978.

Solomon, S. C., P. B. Hays, and V. J. Abreu, The auroral 6300 Å emission: Observations and modeling, *J. Geophys. Res.,*

93, 9867, 1988.

Strickland, D. J., D. L. Book, T. P. Coffey, and J. A. Fedder, Transport equation techniques for the deposition of auroral electrons, *J. Geophys. Res., 81,* 2755, 1976.

Torr, M. R., D. G. Torr, P. G. Richards, and S. P. Yung, Mid- and Low-latitude model of thermospheric emissions 1. $O^+(2P)$ 7320 A and $N_2(2P)$ 3371 A, *J. Geophys. Res., 95,* 21,147, 1990.

Victor, G. A., K. Kirby-Docken, and A. Dalgarno, Calculations of the equilibrium photoelectron flux in the thermosphere, *Planet. Space Sci.,* 24, 679, 1976.

M.W. Liemohn and G.V. Khazanov, Space Physics Research Laboratory, Department of Atmospheric Oceanic, and Space Sciences, University of Michigan, Ann Arbor, MI 48109.

Proton Cyclotron Wave-Ion Interactions Observed by AMPTE/CCE

Brian J. Anderson

The Johns Hopkins University Applied Physics Laboratory, Laurel, Maryland

Electromagnetic ion cyclotron (EMIC) waves occur routinely in the Earth's equatorial magnetosphere and interact both with the energetic protons which drive the waves as well as low energy ions. Compressions promote EMIC wave activity confirming indirectly that the waves are driven by the hot proton temperature anisotropy. Study of magnetosheath EMIC waves show that linear theory successfully predicts both the occurrence and the spectral structure of the waves. The magnetosheath studies also show that the temperature anisotropy of the unstable ions is regulated by the waves and the same mechanism should apply to magnetospheric populations of hot protons. The waves therefore influence macro-scale dynamics because pressure balance and pressure gradient driven currents are both affected by this regulation. Thermal proton and He^+ responses to EMIC waves are qualitatively different. Low energy protons experience a few eV heating at 90° pitch angles but He^+ heats resonantly where the wave frequency approaches the He^+ gyrofrequency.

1. INTRODUCTION

Magnetic pulsations in the 0.1 to 5 Hz range (Pc 1-2) are prominent in ground observations [*Saito*, 1969] and the proton cyclotron instability is generally regarded as responsible for these signals [*Cornwall*, 1965]. The instability is driven by a positive proton temperature anisotropy, $A_p = T_{\perp p}/T_{\parallel p} - 1$, and the waves are thought to be amplified within a few degrees of the magnetic equator and propagate toward the ionosphere along the magnetic field lines [*Horne and Thorne*, 1993]. Spacecraft observations have established that the waves occur in the magnetosphere [*Bossen et al.*, 1976] and that they are electromagnetic ion cyclotron (EMIC) waves [*Mauk and McPherron*, 1980]. Near the equator the observed wave frequencies have a pronounced gap at the He^+ gyrofrequency (F_{He^+}) [*Fraser and McPherron*, 1982] indicating that low energy heavy ions affect EMIC wave growth and propagation [*Roux et al.*, 1982; *Mauk*, 1982, 1983]. Studies using data from the Active Magnetosphere Particle Explorers/Charge Composition Explorer (AMPTE/CCE), hereinafter CCE, have addressed the interaction of the waves with the hot protons which drive the waves as well as the low energy H^+ and He^+. This paper discusses how EMIC waves modify the ion populations.

2. OCCURRENCE DISTRIBUTIONS

Statistical analysis of Pc 1-2 recorded by CCE shows that EMIC waves are commonplace in the outer magnetosphere [*Anderson et al.*, 1992]. Figure 1 shows the distribution and normalized occurrence rate of EMIC wave events observed by CCE. Most of the events occur on the dayside with the greatest concentration near noon and in the afternoon. A significant population of events occurs in the morning as early as 0300 MLT. The occurrence rate of EMIC waves, defined as the duration of waves divided by the residence time of the CCE in each bin, increases monotonically with L and maximizes in the early afternoon at about 20% for L = 8-9. Few events occur during the nighttime. For $L > 8$ near midnight, EMIC wave occurrence is potentially underrepresented in this analysis however [cf. *Anderson et al.*, 1992]. The occurrence rates of Figure 1 are the chance of observing waves at a given place and time and are useful for showing the relative spatial distribution of wave activity. The probability that EMIC waves occur

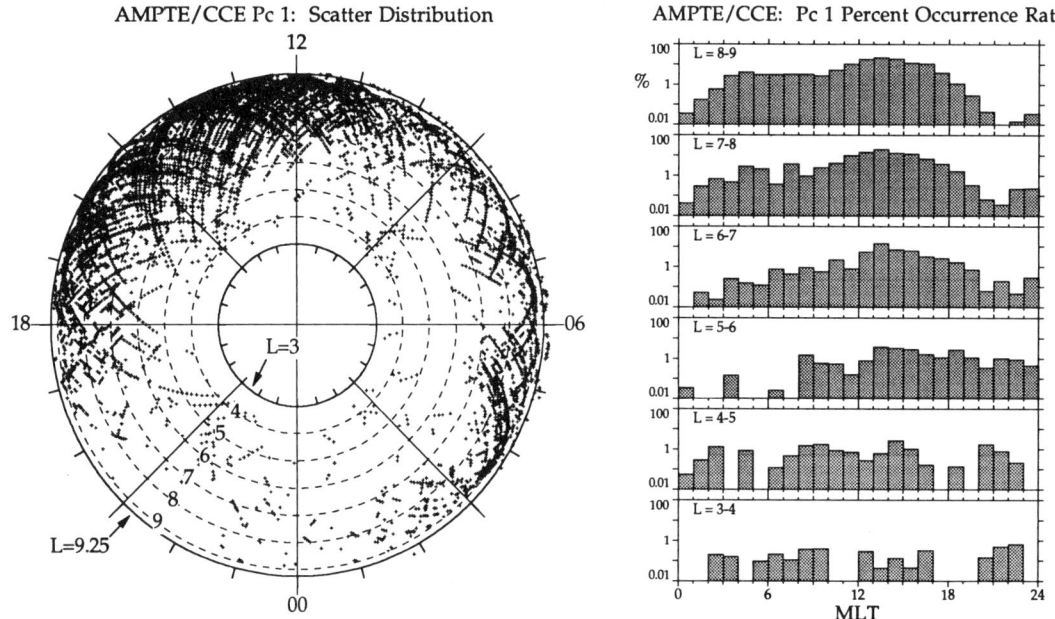

Figure 1. Scatter plot (left) and normalized occurrence rate (right) of EMIC wave events observed by CCE [from *Anderson et al.*, 1992a].

somewhere in the magnetosphere at any time, or at a given place over a longer time, is larger than the rates of Figure 1. In fact, EMIC waves occurred during nearly all passes of the CCE spacecraft in the afternoon sector [*Anderson et al.*, 1992].

3. ENERGETIC PROTONS

It is generally accepted that hot protons drive EMIC waves by the proton cyclotron instability but there is little direct observational confirmation of this. Because the instability reduces the temperature anisotropy, the proton distributions during EMIC waves should not be strongly unstable. *Roux et al.* [1982] and *Mauk and McPherron* [1980] have evaluated growth rates and net amplifications for a number of cases and found that positive wave growth generally coincided with the frequencies of observed waves but that the results were sensitive to the assumed He+ concentration. Because of this sensitivity to plasma composition [*Gendrin et al.*, 1984], quantitative comparison to instability theory is subject to considerable uncertainty.

3.1. *Compressions and EMIC Waves*

The temporal development of EMIC wave activity provides indirect evidence that the waves are driven by the hot proton anisotropy. Ground Pc 1 are initiated or enhanced at times of sudden impulse (SI) events [*Kangas et al.*, 1986]. It has been suggested that magnetospheric compressions associated with SIs trigger EMIC wave growth by convecting the protons inward, heating them adiabatically and producing enhanced temperature anisotropy [*Olson and Lee*, 1983].

In space, many dayside Pc 1 are correlated with increases in local field strength. Figure 2 from *Anderson and Hamilton* [1993] shows an EMIC wave event recorded by CCE and its correlation with increases in magnetic field strength. Panels show the transverse power from 0 to 2 Hz and local He+ gyrofrequency, F_{He^+}, (white trace); the magnetic field magnitude (B) and adjusted model field (B_{T87}^*), [cf. *Anderson and Hamilton*, 1993]; $B/B_{T87}^* - 1$; and the spectral power integrated from F_{He^+} to F_p.

The correlation between enhancements in field strength and wave power is apparent and was found to be strongest for $L > 7$ on the dayside. This correlation was studied statistically by examining magnetic field plots for sudden jumps (up and down) to identify compression/expansion events [*Anderson and Hamilton*, 1993]. Table 1 gives the percentage of compressions (B-up)/expansions (B-down) associated with: wave onset, %ON; wave turn-off, %OFF; continuous waves, %CONT; and no waves before or after, %NONE. About half of the B-up events were associated with EMIC wave onset. The B-down events correspond to EMIC wave onset in one event and cessations occurred in

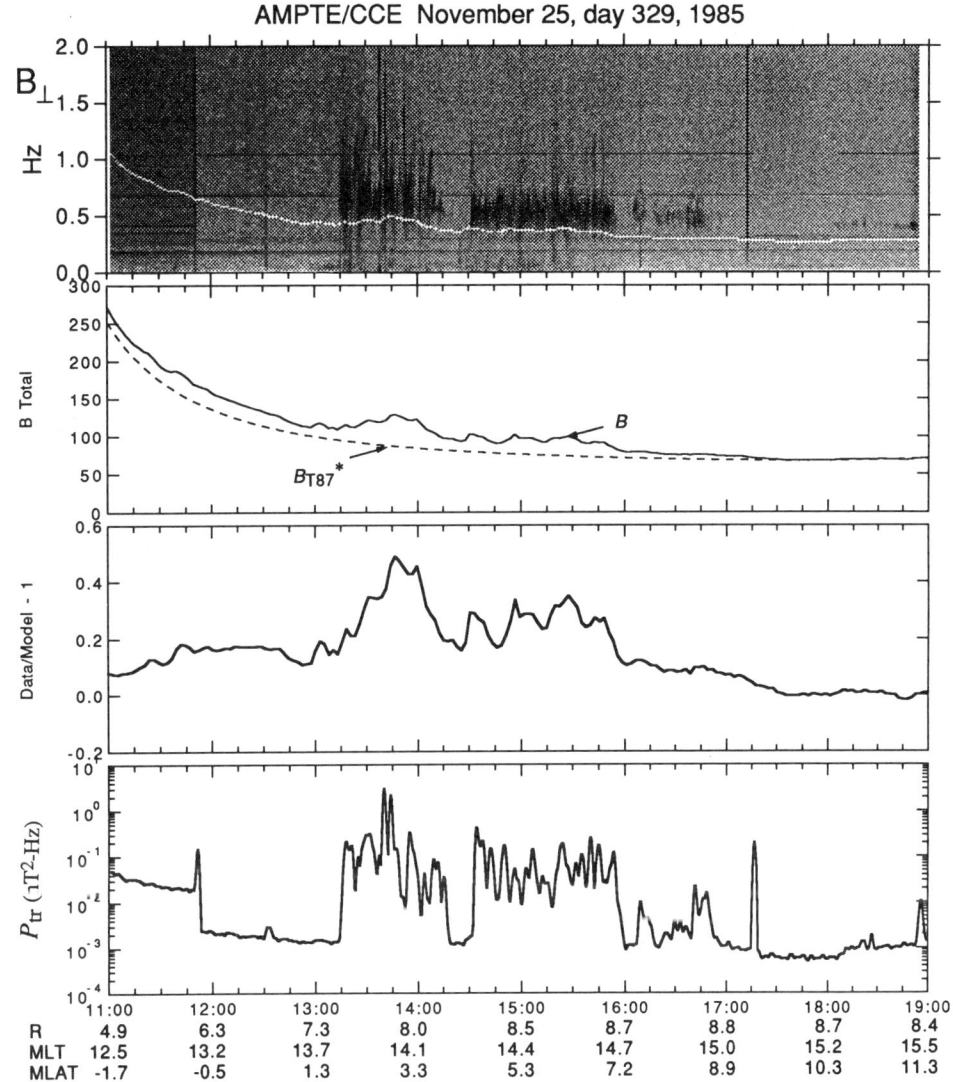

Figure 2. CCE magnetic field data for 1100 - 1900 UT, November 25, day 329, 1985 showing correlation between transient enhancements in field strength and EMIC wave emissions. See text for details. From *Anderson and Hamilton* [1993].

nearly half. Of the 23 B-down events preceded by waves (sum of OFF and CONT events) waves persisted through the expansion in only 4 cases.

Anderson and Hamilton [1993] examined the hot proton distributions for three events and found that the compressions caused dramatic enhancement of EMIC wave growth rates. The CCE results confirm the conclusions of *Olson and Lee* [1983] that compressions produce enhanced wave growth via adiabatic heating. The correlation of magnetic compressions with EMIC waves therefore provides indirect confirmation that the waves are driven by the hot proton temperature anisotropy.

3.2. EMIC Waves in the Magnetosheath

The magnetosheath is an ideal environment to compare the waves with instability theory because all of the ion populations have temperatures well above spacecraft potential and are therefore accurately measured. Ion cyclotron waves in the magnetosheath were first reported by *Fairfield* [1976]. *Anderson and Fuselier* [1993] identified numerous intervals of cyclotron-like and mirror-mode like fluctuations occurring during CCE magnetosheath encounters. *Gary et al.* [1993] used the measured H^+ and He^{2+} distributions to predict whether the mirror or ion-

Table 1. Statistics of EMIC waves and compression (B-up) and expansion (B-down) events [from *Anderson and Hamilton*, 1993].

Type	No.	%ON	%OFF	%CONT	%NONE
B-up:	115	47.0	0.9	15.6	36.5
B-down:	41	2.4	48.8	9.8	39.0

cyclotron instability had the larger growth rate. (In the magnetosheath He^{2+} is the dominant heavy ion.) During intervals with cyclotron-like (mirror-like) fluctuations the plasma was most unstable to the proton cyclotron (mirror) mode. *Anderson et al.* [1994] showed that the spectral signatures of the fluctuations evolved from mirror-like to proton-cyclotron like. Figure 3 illustrates the evolution of spectra they found, from predominantly compressional (MIR), to compressional with an admixture of low frequency (< F_{He2+}) transverse fluctuations (MRL), to low frequency transverse fluctuations (LOW), to transverse fluctuations extending continuously across F_{He2+} (CON), and finally to transverse fluctuations bifurcated near F_{He2+} (BIF). Solid bars show the frequency range for which the ion-cyclotron mode linear growth rate, γ, satisfied $\gamma/\omega_p > 0.01$. The evolution of transverse fluctuation spectra agrees with the theoretical predictions. *Denton et al.* [1994] showed that this agreement held for the ensemble of CCE magnetosheath observations: linear theory correctly predicted which mode was dominant and the growth rate spectral structures agreed with the observations.

The magnetosheath observations have also shown that the proton cyclotron instability regulates the proton temperature anisotropy. Figure 4 shows A_p versus $\beta_{\|p} = 2\mu_0 n_p T_{\|p}/B^2$ for the *Anderson et al.* [1994] events; solid triangles (circles) for the bifurcated (mirror) events. The remaining categories are not distinguished in this plot. There is an inverse A_p-$\beta_{\|p}$ correlation described by the least squares fit

$$A_p^* = 0.85 \beta_{\|p}^{-0.48}$$

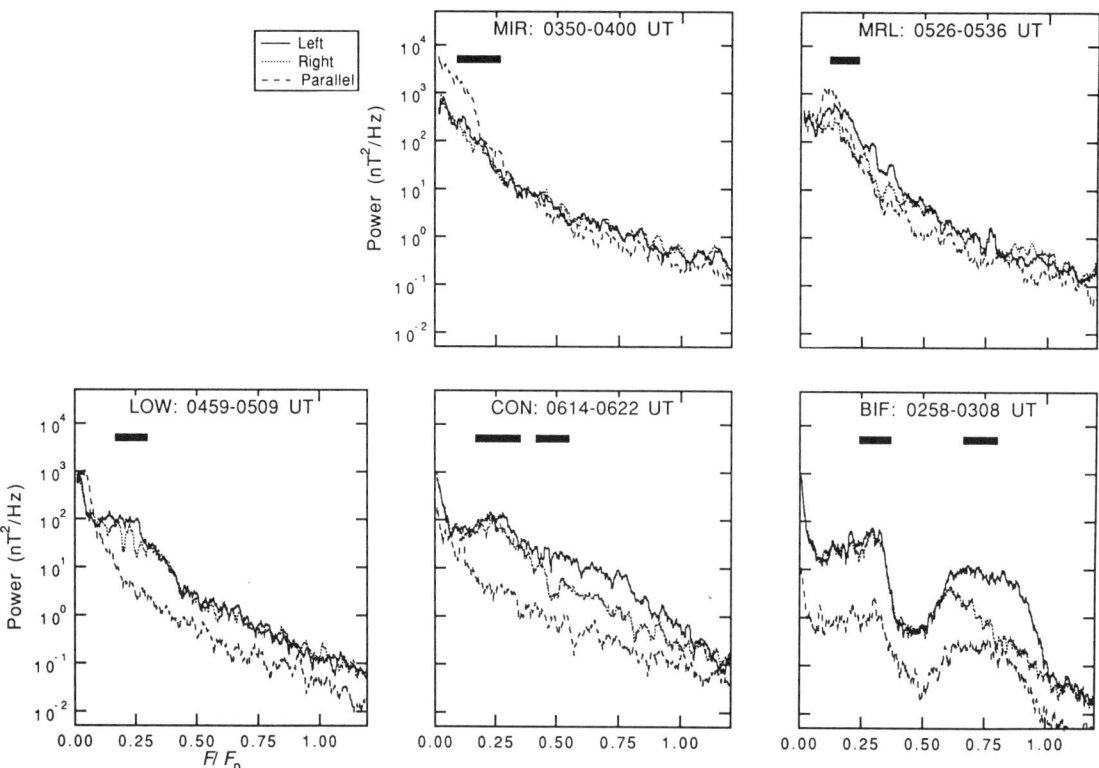

Figure 3. Example spectral characteristics observed by CCE during a magnetosheath encounter on day 280, 1984. Traces show left and right handed transverse and parallel power of magnetic field fluctuations. Frequency scale is normalized to the local proton gyrofrequency. Solid bars indicate the frequency range over which the ion cyclotron instability had $\gamma/\omega_p > 0.01$. [*Anderson et al.*, 1994].

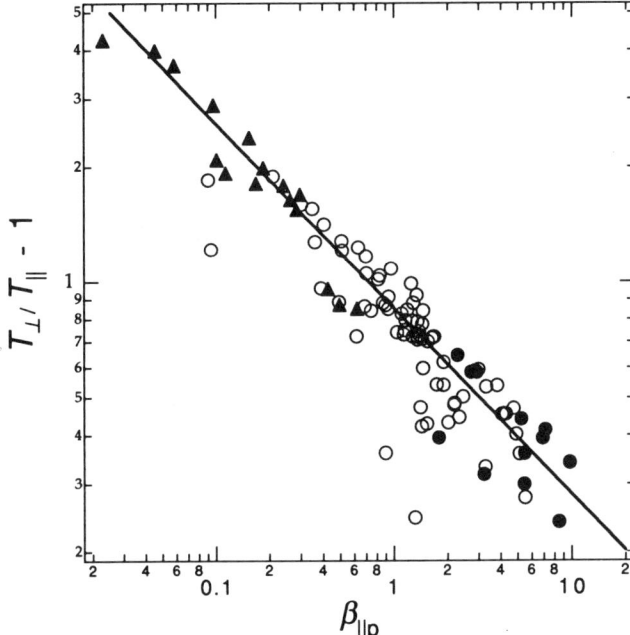

Figure 4. Proton temperature anisotropy versus parallel beta for quasi-perpendicular upstream conditions [Anderson et al., 1994].

This relationship is a necessary consequence of the proton cyclotron instability since when A_p exceeds A_p^* strong wave growth occurs which reduces A_p toward marginal stability [Gary et al, 1994].

An A_p-$\beta_{\parallel p}$ relation of the form observed and can be derived in several ways [Gary et al., 1994; and Gary and Lee, 1994]. Protons in resonance with the waves satisfy the condition

$$\omega = \Omega_p + \mathbf{k} \cdot \mathbf{v}$$

For strongest wave growth, the bulk of the distribution must resonate with the waves, so $v_\parallel \sim v_{th}$. The cold plasma dispersion relation and the approximation, $\omega/k = V_A$, give

$$\beta_{\parallel p} \sim (\Omega_p/\omega - 1)^2$$

Since only frequencies with

$$\omega/\Omega_p < A_p/(A_p+1)$$

are unstable [Kennel and Petschek, 1966] one may write

$$A_p > 1/(\Omega_p/\omega - 1)$$

thus giving the condition for strong wave growth:

$$A_p > \beta_{\parallel p}^{-0.5}$$

The success of cyclotron wave dynamics in explaining the inverse A_p-$\beta_{\parallel p}$ relation is strong evidence indicating that the cyclotron mode regulates the ion temperature anisotropies.

Regulation of the temperature anisotropy by EMIC waves has important large scale dynamical consequences. Because the ions carry nearly all of the plasma pressure, the exchange of perpendicular and parallel pressures affects large scale force balance. Moreover, the current densities associated with perpendicular and parallel pressure gradients are also affected.

4. LOW ENERGY IONS

Interaction of EMIC waves with low energy ions is important for wave propagation and ion heating [Roux et al., 1982]. Cold ions heavier than protons, mainly He^+ in the magnetosphere, produce a dispersion stop band [Roux et al., 1982] and introduce slots in the growth rate spectrum above the heavy ion gyrofrequencies [Gendrin, 1984] which widen with cold heavy ion concentration. He^+ also introduces a bi-ion resonance that may reflect waves propagating from the equator to lower altitudes [Rauch and Roux, 1982]. Mauk [1982] and Fraser and McPherron [1982] showed that the observed spectral slot is centered on F_{He^+} rather than lying above F_{He^+}. This result suggests that ion-wave interactions occur which absorb wave energy at frequencies in the vicinity of F_{He^+} [Mauk, 1983] and signatures of resonant He^+ heating signatures have been observed [Roux et al., 1982; Mauk, 1982].

The GEOS 1-2 observations [e.g. Roux et al., 1982] indicate a high correlation between He^+ energy density and EMIC waves. The result was taken to imply that increased He^+ concentration caused increased wave amplification [Roux et al., 1982; Rauch and Roux, 1982], but it has also been suggested that preferential heating of He^+ by the waves might account for the He^+-wave correlation [Mauk, 1992]. The ion spectrometers used in the GEOS and CCE measurements did not sample the ions below spacecraft potential, that is, the coldest part of the distribution was not measured [Olsen et al., 1985]. Hence, the measurements only reflect the composition of the ions that have energies higher than the spacecraft potential. Preferential heating of He^+ therefore produces an apparent increase in He^+ concentration.

Anderson and Fuselier [1994] examined He^+ and H^+ distribution functions during sustained EMIC wave activity (Active) and prolonged absence of waves (Quiet). Figure 5 shows distribution functions for H^+ (7a) and He^+ (7b) for two events, October 15, 1984, Active, and December 18, 1984, Quiet. During the Quiet period, the H^+ and He^+ distributions have the

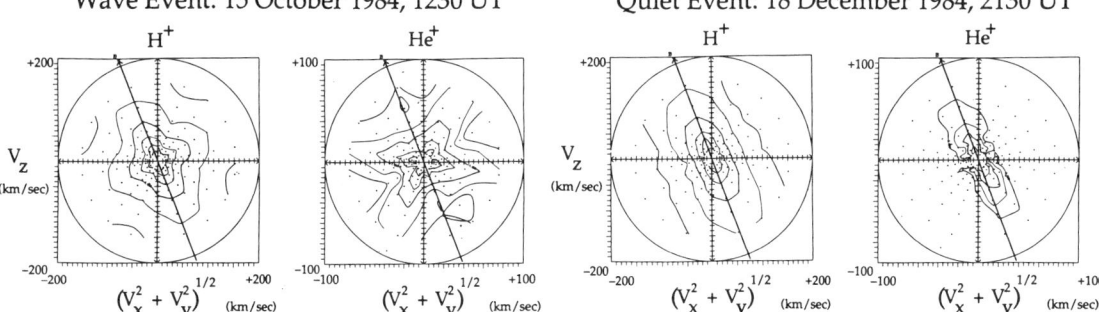

Figure 5. Contour plots and pitch angle distributions (at 3.1 eV) of H$^+$ (a) and He$^+$ (b) from the Hot Plasma Composition Experiment on CCE for an EMIC wave event, October 15, 1994; and during a period of prolonged absence of EMIC waves, December 18, 1994. [*Anderson and Fuselier*, 1994]

same "cigar" shape with $T_\parallel \gg T_\perp$. During the Active event the distributions are qualitatively different: H$^+$ displays a small, $T_\perp \sim 4$ eV, enhancement at 90° pitch angle giving the appearance of a "cross" whereas He$^+$ has maximum fluxes at pitch angles near 55° giving the appearance of an "X". Moreover, the He$^+$ temperature at 55° pitch angle is 130 eV. Table 2 shows the statistical results from many Active and Quiet intervals [*Anderson and Fuselier*, 1994]. The He$^+$ consistently displayed "X"-type distributions during Active events whereas H$^+$ displayed occasional enhancements at 90° pitch angles. Quiet events consistently displayed "cigar" type distributions in both species. On average the $T_\perp$ enhancement was about 30 eV for He$^+$ but 5 eV for protons.

The "X"-type He$^+$ distributions indicate that He$^+$ heating occurs at lower altitudes. For the October 15 event, wave power peaked sharply near $1.45 F_{He^+}$ and the "X" feature was observed at 55° pitch angle. Particles at 55° pitch angle mirror where the field is 1.49 times greater than at the observation point, that is, where $F \sim F_{He^+}$ suggesting that the feature is due to a resonance interaction.

Figure 6 is plot of apparent He$^+$ concentration, $\eta = n_{He^+}/(n_p + n_{He^+})$, versus T_{He^+}/T_p [*Anderson and Fuselier*, 1994]. T_{He^+}/T_p and η are well correlated whether or not EMIC waves are present: the regression coefficients between $\log(\eta)$ and $\log(T_{He^+}/T_p)$ are 0.86 and 0.77 for the Active and Quiet periods, respectively. Since T_{He^+}/T_p is so closely correlated with η, the EMIC wave-η association could well be due to preferential wave heating of He$^+$.

5. SUMMARY AND CONCLUSIONS

Recent observations have shown that ion cyclotron waves occur commonly in the magnetosphere, particularly on the dayside, so it is important to understand their effects on magnetospheric ion distributions. The correlation between modest compressions of the magnetosphere and onset of EMIC waves provides indirect evidence that the waves are driven by enhanced hot proton temperature anisotropy. Magnetosheath observations have been used to test the linear instability theory and show that the theory predicts both the occurrence and spectral structure of the waves. The magnetosheath observations also show that EMIC waves regulate the ion temperature anisotropies. Magnetospheric EMIC waves should limit the proton temperature anisotropy in much the same way. Since both pressure balance and pressure gradient driven currents are affected by this

Table 2. Number of proton and He$^+$ distributions of various general types during Active and Quiet EMIC wave intervals. [from *Anderson and Fuselier*, 1994].

Species	EMIC	No.	$T_\parallel \gg T_\perp$	Cross	X-type	$T_\perp \gg T_\parallel$	$T_\perp \approx T_\parallel$
H$^+$	Active	24	7	8	0	5	4
	Quiet	24	19	1	0	0	4
He$^+$	Active	18	3	1	11	2	1
	Quiet	19	18	0	0	0	1

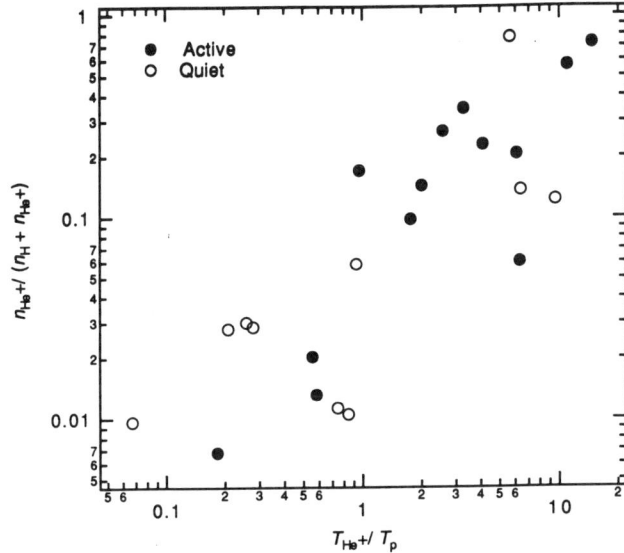

Figure 6. Plot of apparent He$^+$ concentration, $\eta \equiv n_{He^+}/(n_p + n_{He^+})$, versus the ratio T_{He^+}/T_p. Solid symbols indicate EMIC wave events and open symbols indicate quiet EMIC wave events. [*Anderson and Fuselier*, 1994]

regulation, the waves may have a significant effect on large scale dynamics. The effects of magnetospheric EMIC waves on low energy ions have also been studied. Cold protons experience a few eV heating at 90° pitch angles whereas He$^+$ displays significant heating, often greater than 100 eV, and at pitch angles intermediate between 0° and 90° indicating that the interaction occurs at lower altitudes where the wave frequency approaches the He$^+$ gyrofrequency. The preferential heating of He$^+$ by EMIC waves raises significant difficulties in determining the low energy ion composition during EMIC wave events.

Acknowledgements. Preparation of this manuscript was supported by NASA under the SR&T Program through grant NAGW-3052.

REFERENCES

Anderson, B. J., and D. C. Hamilton, Electromagnetic ion cyclotron waves stimulated by modest magnetospheric compressions, *J. Geophys. Res.*, *98*, 11369, 1993.

Anderson, B. J., and S. A. Fuselier, Response of thermal ions to electromagnetic ion cyclotron waves, *J. Geophys. Res.*, *99*, 19413, 1994.

Anderson, B. J., R. E. Erlandson, and L. J. Zanetti, A statistical study of Pc 1-2 magnetic pulsations in the equatorial magnetosphere: 1. equatorial occurrence distributions, *J. Geophys. Res.*, *97*, 3075, 1992.

Anderson, B. J., S. A. Fuselier, S. P. Gary, and R. E. Denton, Magnetic spectral signatures from 0.1 to 4.0 Hz in the Earth's magnetosheath and plasma depletion layer, *J. Geophys. Res.*, *99*, 5877, 1994.

Bossen, M., R. L. McPherron, and C.T. Russell, A statistical study of Pc 1 magnetic pulsations at synchronous orbit, *J. Geophys. Res.*, *81*, 6083, 1976.

Cornwall, J. M., Cyclotron instabilities and electromagnetic emission in the ultra low frequency and very low frequency ranges, *J. Geophys. Res.*, *70*, 61, 1965.

Denton, R. E., S. P. Gary, B. J. Anderson, S. A. Fuselier, and M. K. Hudson, Low-frequency magnetic fluctuation spectra in the magnetosheath and plasma depletion layer, *J. Geophys. Res.*, *99*, 5893, 1994.

Fairfield, D. H., Waves in the vicinity of the magnetopause, in *Magnetospheric Particles and Fields*, pp. 67, B. M. McCormac ed., D. Reidel Pub. Co., Dordrecht-Holland, 1976.

Fraser, B. J., and R. L. McPherron, Pc 1-2 magnetic pulsation spectra and heavy ion effects at synchronous orbit: ATS 6 results, *J. Geophys. Res.*, *87*, 4560, 1982.

Gary, S. P., and M. A. Lee, The ion cyclotron anisotropy instability and the inverse beta correlation between proton anisotropy and proton beta, *J. Geophys. Res.*, *99*, 11297, 1994.

Gary, S. P., S. A. Fuselier, and B. J. Anderson, Ion cyclotron instabilities in the magnetosheath, *J. Geophys. Res.*, *98*, 1481, 1993.

Gary, S. P., M. E. McKean, D. Winske, B. J. Anderson, R. E. Denton, S. A. Fuselier, Proton cyclotron anisotropy instability and the anisotropy/beta inverse correlation, *J. Geophys. Res.*, *99*, 5903, 1994.

Gendrin, R., M. Ashour-Abdalla, Y. Omura, and K. Quest, Linear analysis of ion cyclotron interaction in a multicomponent plasma, *J. Geophys. Res.*, *89*, 9199, 1984.

Horne, R. B. and R. M. Thorne, On the preferred source location for the convective amplification of ion cyclotron waves, *J. Geophys. Res.*, *98*, 9233, 1993.

Kangas, J., A. Aikio, and J. V. Olson, Multistation correlation of ULF pulsation spectra associated with sudden impulses, *Planet. Space Sci.*, *34*, 543, 1986.

Kennel, C. F., and H. E. Petschek, Limit on stably trapped particle fluxes, *J. Geophys. Res.*, *71*, 1, 1966.

Mauk, B. H., Helium resonance and dispersion effects on geostationary Alfven/ion cyclotron waves, *J. Geophys. Res.*, *87*, 9107, 1982.

Mauk, B. H., Frequency gap formation in electromagnetic cyclotron wave distributions, *Geophys. Res. Lett.*, *10*, 635, 1983.

Mauk, B. H., and R. L. McPherron, An experimental test of the electromagnetic ion cyclotron instability within the earth's magnetosphere, *Phys. Fluids.*, *23*, 2111, 1980.

Olsen, R. C., C. R. Chappell, D. L. Gallagher, J. L. Green, and D. A. Gurnet, The hidden ion population: revisited, *J. Geophys. Res.*, *90*, 12121, 1985.

Olson, J. V., and L. C. Lee, Pc 1 wave generation by sudden impulses, *Planet. Space Sci.*, *31*, 295, 1983.

Rauch, J. L., and A. Roux, Ray tracing of ULF waves in a multicomponent magnetospheric plasma: consequences for the generation mechanism of ion cyclotron waves, *J. Geophys. Res.*, *87*, 8191, 1982.

Roux, A., S. Perraut, J. L. Rauch, C. de Villedary, G. Kremser, A. Korth, and D. T. Young, Wave-particle interactions near Ω_{He^+} observed on board GEOS 1 and 2: 2. Generation of ion cyclotron waves and heating of He^+ ions, *J. Geophys. Res.*, *87*, 8174, 1982.

Saito, T., Geomagnetic pulsations, *Space Sci. Rev. 10*, 319, 1969.

B. J. Anderson, Johns Hopkins University Applied Physics Laboratory, Johns Hopkins Road, Laurel, Maryland, 20723-6099.

Aspects of Mesoscale Phenomena in the Middle Magnetosphere and Speculations on the Role of Microscale Processes

Barry H. Mauk

The Johns Hopkins University Applied Physics Laboratory, Laurel, Maryland

The role of mesoscale phenomena in the workings of the middle magnetosphere (e.g., $r < 8\ R_E$) is highly uncertain. Evidence will show that mesoscale phenomena play fundamental roles in the transport of energetic plasmas within the middle magnetosphere and in the coupling to different regions, including the ionosphere. Included will be discussions of magnetic field–aligned electrical discharge phenomena, the association of such phenomena with discrete auroral forms in the ionosphere, and the processes of energization and transport that give rise to injection boundary structures. Fundamental problems remain in our understanding of these phenomena that undoubtedly require the application of microscale processes. Although definitive answers do not exist, speculations on the role of microscale processes in the workings of these mesoscale phenomena are presented. This paper is not a comprehensive review of the topics addressed.

MESOSCALE PHENOMENA OF THE MIDDLE AND INNER MAGNETOSPHERES

Discussed here are mesoscale phenomena of the inner and middle magnetosphere and the possible role that microscopic processes play in moderating the mesoscale phenomena. I start by making the case that mesoscale phenomena—by which I mean processes involving scale sizes of, say, an Earth radius and substantially less—constitute fundamental aspects of how the inner and middle magnetospheric regions operate; that is, how plasmas are transported within these regions and how these regions couple to other regions [see also *Mauk and Meng*, 1991].

The bottom panel of Figure 1 shows a 2-h spectrogram of ion and electron plasma data sampled from the geosynchronous orbit. The display is centered on about 2000 local time. The top panel shows electrons for energies between ~1 eV and ~82 keV. The bottom panel shows ion data for the same energy range, but with the energy scale inverted. Particle intensity is proportional to the whiteness of the display.

Within the e⁻ display one sees electron distributions that appear to be quite unstructured, much like the way one might imagine central plasma sheet populations to look. One might also imagine that precipitation of this population would produce diffuse auroral structures, which is indeed the case. The top left portion of Figure 1 shows part of a study wherein electron distributions measured within the equatorial magnetosphere were compared with precipitating electron distributions measured simultaneously at low altitudes by a Defense Meteorological Satellite Program (DMSP) spacecraft. Those distributions matched each other to a remarkable degree. The dots are the geosynchronous measurements, the triangles are DMSP measurements sampled at the magnetic conjugate point of the geosynchronous satellite [according to the *Olson and Pfitzer*, 1974 model], and the bars are DMSP measurements sampled 0.5° equatorward of the calculated conjugate point. The DMSP spectrum indeed appears to have resulted from the diffusive precipitation of the trapped e⁻ population. By "diffuse precipitation" we mean pitch angle diffusion into the loss cone

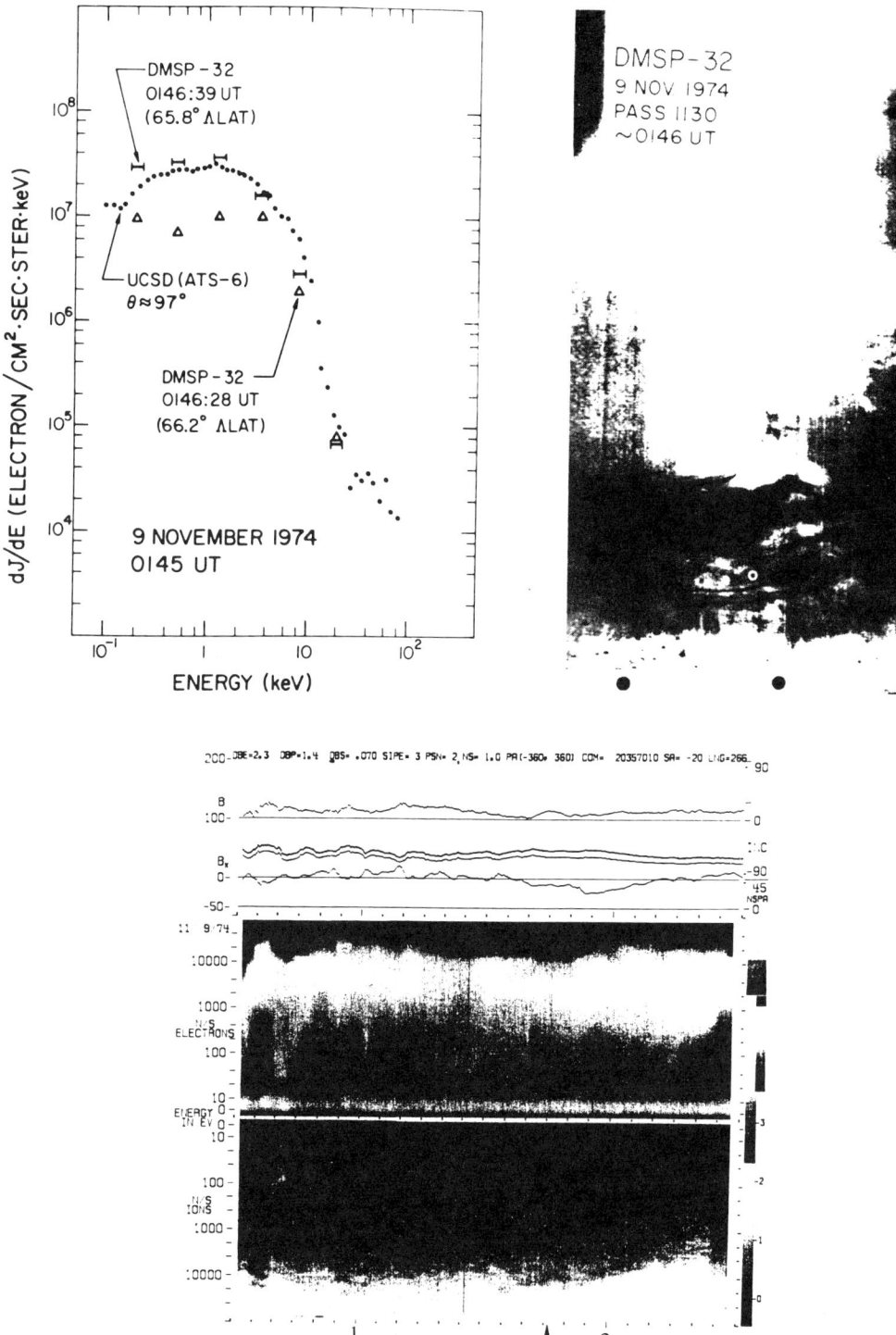

Fig. 1. Comparison between geosynchronous electron data (ATS-6) and DMSP electron precipitation data at the conjugate point. The geosynchronous spectrum was sampled at a magnetic latitude of about +10° and shows electrons moving away from the magnetic equator with pitch angles of about 35° [after *Meng et al.*, 1979].

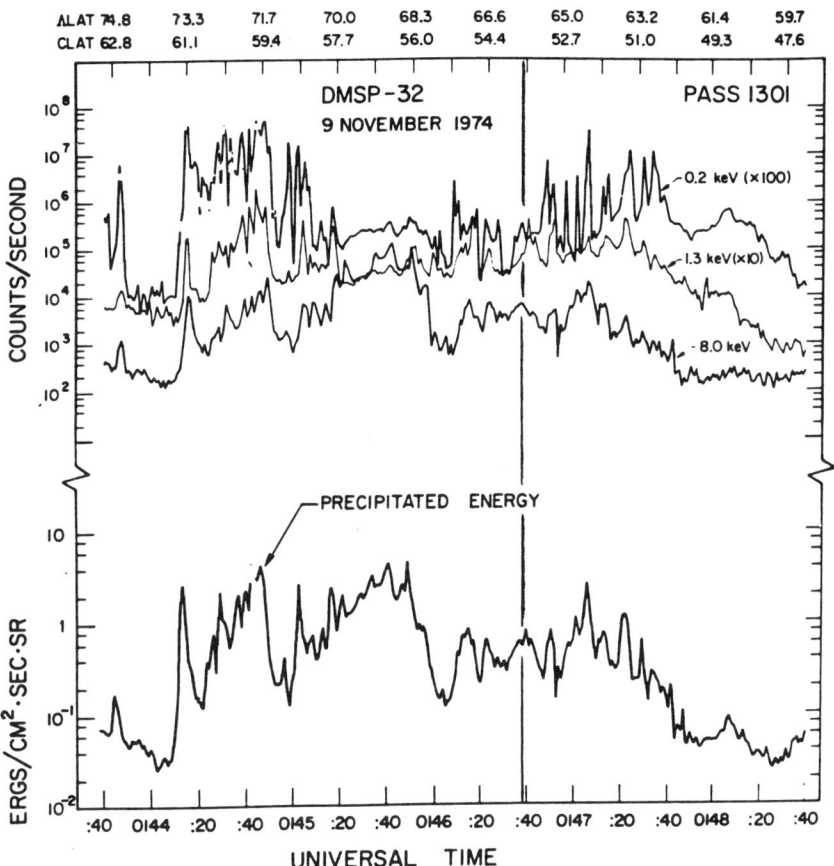

Fig. 2. More electron precipitation data sampled by the DMSP satellite for the period presented in Figure 1. The vertical line shows the time when DMSP crossed the magnetic conjugate position of the geosynchronous ATS-6 satellite [after *Meng et al.*, 1979].

by mechanisms that depend only weakly on the electron energy.

In this example, however, out of about five cases examined in detail [*Meng et al.*, 1979], something rather remarkable was noted: discrete auroral arc structures appeared equatorward of the geosynchronous magnetic footpoint. The top right portion of Figure 1 shows a DMSP image of the aurora in negative, and discrete auroral forms are apparent throughout the lower regions of the image. The circle shows the calculated footpoint of the geosynchronous satellite. What we see are these very distinct auroral arc structures equatorward of the geosynchronous footpoint. These arcs map to the equatorial magnetosphere to positions that are earthward of the geosynchronous orbit and earthward of these central plasma sheet populations. The *Olson and Pfitzer* [1974] model used to perform the magnetic mapping is a quiet time model, and the period in question was quite active. Thus, the magnetic mapping could be incorrect. However, as discussed more fully below, the matching electron distributions confirm that the magnetic mapping is approximately correct because the electron spectra measured equatorward of the auroral arcs differ dramatically from the spectra measured in the geosynchronous orbit.

Figure 2 shows additional aspects of the DMSP data. The time period is about 5 min and covers magnetic latitudes from about 75° on the poleward extreme (on the left) to about 58° on the equatorward extreme (on the right). The vertical line in the figure shows the position where DMSP crossed the magnetic footpoint of the geosynchronous satellite and where we found the striking agreement between the high- and low-altitude spectra. One finds a remarkable amount of discrete-like structured precipitation that maps to equatorial positions as close to the Earth as 5 R_E. The structure has scale sizes down to 10 km and less. This figure also shows that there is no mistaking the magnetic mapping. One might hypothesize, as discussed previously, that the geosynchronous footpoint actually maps to positions equatorward of the arcs. However, when one searches those regions in the particle data, one finds that the character of the spectra has changed completely, and that no precipitating spectra come even close to matching the geosynchronous spectra. Note that the close match between the near-equatorial and precipitating electron spectra would not necessarily be expected if the spacecraft resided at the time on field lines that threaded magnetic field–aligned auroral acceleration regions. However, an arc is not present in the DMSP data at the time when the spectra are shown to match.

This mesoscale structure is potentially a key aspect of transport within and coupling of these middle and inner magnetospheric regions. I would like to speculate on how this and other such middle region structures come about. I will start by addressing the processes by which the middle and inner regions of the magnetosphere become populated from populations that have been cycled through the magnetotail.

PLASMA TRANSPORT IN THE MIDDLE MAGNETOSPHERE

Magnetotail populations are known to convect toward the Earth owing to the so-called cross-tail electric field caused by the interactions between the solar wind and the magnetopause [e.g., *Kavanagh et al.*, 1968; *Stern*, 1977]. However, a problem associated with such convective transport was revealed, first by *Erickson and Wolf* [1980] and later by *Schindler and Birn* [1982]. The former considered the problem of convective flow from the magnetotail toward the Earth and showed that the pressure within the central plasma sheet rises so dramatically, based on lossless adiabatic motion, that the magnetic pressure of the tail lobes would be unable to contain the plasma sheet pressures. The different lines in Figure 3 correspond to the pressure calculations as applied to different magnetic field models. The conclusion reached by *Erickson and Wolf* [1980] and *Schindler and Birn* [1982] is that time-stationary convection could not be supported. This work has spawned a small industry. For example, *Kivelson and Spence* [1988] showed that time-stationary convection could be supported tailward of 30 R_E, but there is still a problem earthward of 30 R_E. Also, the importance of transient or bursty flow within the magnetotail has been recently recognized [*Baumjohann et al.*, 1990; *Angelopoulos et al.*, 1992, 1993].

In parallel with this theoretical work was empirical work on the spatial distributions of plasma populations

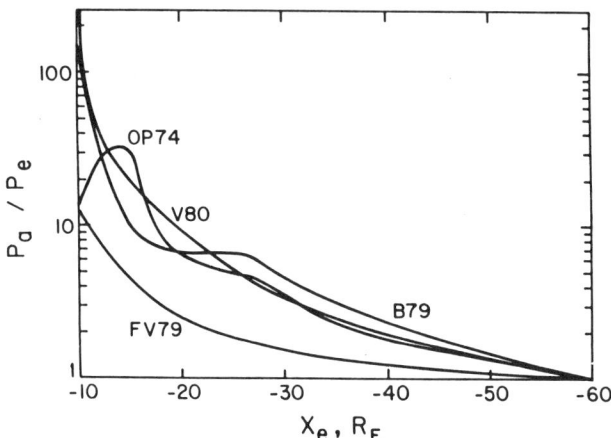

Fig. 3. Model results suggesting that steady-state convection may not be possible within the Earth's magnetotail [from *Erickson and Wolf*, 1980].

observed within the geosynchronous and other middle and inner regions of the magnetosphere. Some of us have concluded that the plasma distributions observed near the geosynchronous orbit cannot be explained by the concept of global convection, even during relatively quiet periods [e.g., see *Mauk and Meng*, 1983a, 1986]. The best evidence in favor of the global convection picture for sometimes populating the middle magnetosphere without localized transient processes is given by *Kerns et al.* [1994]. However, there is an unmodeled, higher-energy component to the data shown that has the appearance of structures known to be generated by dynamical processes, and thus I believe that this evidence is uncertain. To explain the observed geosynchronous distributions, an injection boundary model was developed [*McIlwain*, 1974; *Mauk and McIlwain*, 1974; *Konradi et al.*, 1975; *Mauk and Meng*, 1983b].

The injection boundary model is intrinsically time-dependent. At the initiation of a substorm expansion phase, plasmas are dynamically injected into the middle magnetosphere so that a very sharp boundary is formed between plasmas that are strongly disturbed and energized by the substorm initiation and the pre-existing populations that remain relatively undisturbed (see Figure 4, top). This boundary is quite sharp, thereby putting it into the class of small and/or mesoscale structures that are of concern here. After the formation of this boundary, global convection then takes over and disperses these plasmas in an energy- and species-dependent manner.

If one then flies a spacecraft through that dispersing pattern, one sees characteristic signatures that look very much like the plasma signatures observed in the geosynchronous orbit [*Mauk and Meng*, 1983a, 1983b]. The

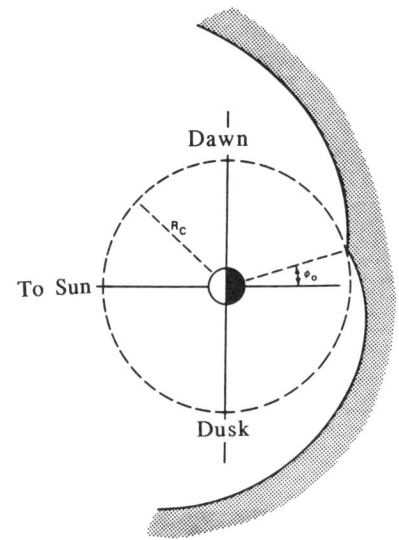

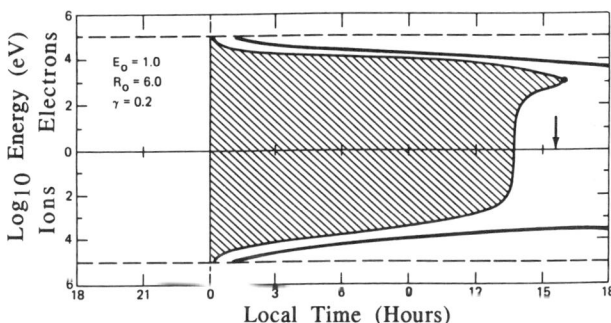

Fig. 4. (top) Schematic of the substorm injection boundary [after *Mauk and Meng*, 1983b]. (bottom) Geosynchronous charged particle spectrogram predicted by the injection boundary model [after *Mauk and Meng*, 1983a].

bottom panel of Figure 4 shows an example of such a signature predicted for the geosynchronous orbit. Here we show a simulated spectrogram for a 24-h period with electrons from 1 to 100 keV and ions with an inverted energy scale, also from 1 to 100 keV. The shaded region is where one expects to see enhanced particle fluxes. This pattern qualitatively matches patterns observed in the geosynchronous orbit for the conditions considered.

Most people in our field, however, do not appreciate the fact that success of the injection boundary model at explaining the many kinds of observed plasma patterns is not the most compelling evidence in favor of the injection boundary model. Rather, the most compelling evidence is that old and fresh plasma dispersion patterns seem to pass through each other without appearing to disturb each other [*McIlwain*, 1974; *Mauk and Meng*, 1986]. A good example is shown in Figure 5, top. Here, starting at about 0300 to 0400 universal time (UT), a relatively fresh plasma population appears, having been injected into the middle magnetosphere some hours before it was observed. Qualitatively the pattern resembles one of the many patterns predicted on the basis of the injection boundary model [Pattern 4 identified by *Mauk and Meng*, 1983b].

However, what is most interesting is what is happening in the background to the fresh feature. To the extreme left of the panel, a relatively old ion population from relatively old injection events appears to pass through the new populations as if the latter did not even exist; and a very sharp boundary in energy–time–space is obvious between the old and new plasmas.

A similar example can be seen during a much more active period in the bottom panel of Figure 5. At about 0730 UT a local, dispersionless substorm injection occurred with freshly energized ion and electron distributions; and subsequently, at the highest ion energies, we see ion echo events associated with that injection from ions traveling completely around the Earth and reappearing at the satellite. The feature of interest here appears within the very dark regions, say at 1100 to 1200 UT, and at ion energies of about 10 keV. This region was called the deep proton minimum by *McIlwain* [1972]. The faint traces of the pre-existing ion dispersion curves observed again appear to pass undisturbed through the newer plasmas. Obviously these curves are much dimmer than the dispersion curves seen before all of the new activity, suggesting that, in addition to the occurrences of plasma injections, the global convection electric field has intensified, causing the ring current plasmas to scale inward. Just as shown in the upper panel of Figure 5, an exceedingly sharp boundary is observed between the old and fresh dispersion curves.

What is going on here? All of those ions that appear at the interface between the older and newer populations on the spectrograms were lined up just along the injection boundary at the time of the injection (the inner edge boundary shown in Figure 4). The older plasmas were earthward of the injection boundary at the time of the injection and were left essentially undisturbed by the injection. The newer plasmas were tailward of the injection boundary at the time of the injection and were very much disturbed. The sharpness of the energy transition between the old and new populations provides a measure of the spatial sharpness of the boundary. A proper study of this transition has not been done, but if, for example, the transition on the spectrograms were to take about 5

206 MIDDLE AND INNER MAGNETOSPHERE

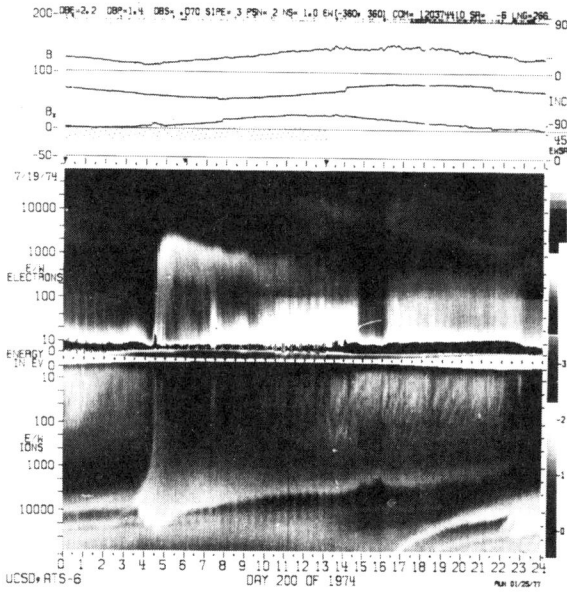

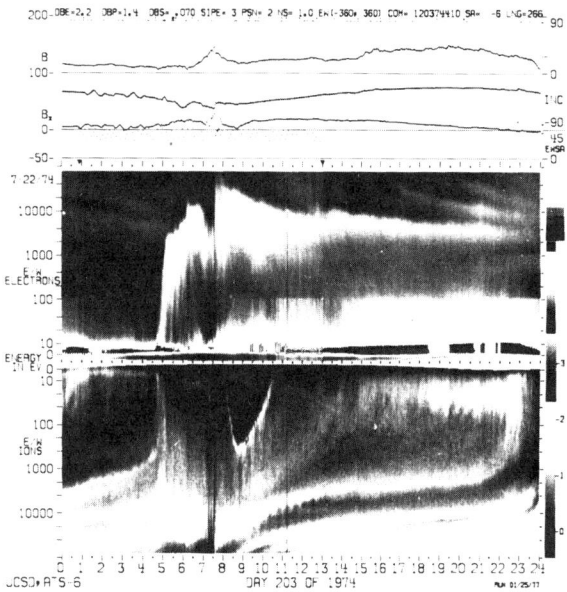

Fig. 5. Two 24-h charged particle spectrograms sampled by the geosynchronous satellite ATS-6. Electron and ion energies go from ~1 eV to 82 keV. Whiter regions correspond to more intense fluxes of particles.

min, the spatial scale size for the boundary would be less than 0.1 to 0.2 R_E [consistent with the scale sizes inferred by *Moore et al.*, 1981 for propagating injection fronts]. The spiral angle of the injection boundary has been used to make this estimate [*Mauk and McIlwain*, 1974].

The spatial sharpness utilized with the injection boundary is therefore not just a convenience for performing calculations. Rather, it reflects a spatial sharpness that appears to truly exist within the data. To date no one has convincingly explained the occurrence of these sharp spatial boundaries.

Moore et al. [1981] tried to explain the generation of this boundary on the basis of the so-called propagating injection fronts model. On the basis of timings of injection phenomena observed by two satellites that were radially displaced from one another, they concluded that injections were caused by a compressional wave that was launched from the base of the magnetotail during substorm expansions. That wave would propagate earthward, steepening as it goes, much like an ocean wave propagating toward a beach. Eventually, the wave would either break or dissipate, and the inner extreme of the wave's propagation would then constitute the injection boundary. Microscopic processes would, of course, be responsible for the breaking and/or dissipation of this mesoscale structure; however, the nature of those microscopic processes is unknown.

This model has, from my perspective, been pretty much ignored in recent years for several reasons, but in particular because discussions of the injection front model have become confused with discussions of the very exciting discovery by the AMPTE CCE (Active Magnetospheric Particle Tracer Explorers Charge Composition Explorer) spacecraft of current disruptions at about 9-R_E positions on the nightside [*Lui et al.*, 1988; *Lopez et al.*, 1989]. *Lopez et al.* [1989, 1990], in particular, have forcefully argued that these current disruptions initiate at an intermediate radial position and then propagate tailward. They have challenged the conclusion of *Moore et al.* [1981] that disturbances propagate Earthward. However, there would appear to be no inconsistency in having a current disruption region propagate tailward while the current disruption disturbance generates an MHD wave that propagates earthward. The current disruption, of course, relieves magnetic stress, causing the plasmas to surge earthward. This should be an ideal situation for launching a wave.

Microscopic processes must somehow be involved in establishing the scale size of the injection boundary. In the *Moore et al.* [1981] model the scale size is established by a balance between MHD wave steepening and microscopic dissipation. However, it remains unresolved whether the microscopic processes associated with current disruption, those associated with wave breaking and dissipations, or those associated with some additional mechanism are responsible for creating the very narrow mesoscale injection boundary structure.

COUPLING TO THE IONOSPHERE

What has been ignored so far in the discussion of the injection boundary formation is the role of ionospheric coupling. One of the most important discoveries about the injection boundary is that its creation appears to be associated with intense, magnetic-field–aligned electron beams [*McIlwain*, 1975; see review by *Mauk and Meng*, 1991].

Figure 6 shows the geosynchronous velocity distribution functions of electrons, plotted as a function of energy, for pitch angles perpendicular to the magnetic field $\bar{B}$ and for pitch angles parallel to $\bar{B}$. What one sees is a parallel electron beam that is qualitatively similar to downgoing electron beams measured in the vicinity of discrete auroral activity, with a positive slope in the intensity spectrum as well as in the velocity distribution spectrum. The beams are also associated with east–west magnetic perturbations that have been attributed to magnetic-field–aligned electric currents (upper right corner). These data have all the appearance of corresponding to a very energetic magnetic-field–aligned electrical discharge.

Beams of this sort, which show the character of auroral acceleration, are observed for just a few minutes in the vicinity of the leading edge of local substorm injections (e.g., at 0730 UT in the bottom panel of Figure 5). Electron beams are observed over much broader regions following the substorm injection, but those beams do not show the character of discrete-like auroral acceleration. The distributions with positive slopes appear only to be associated with the leading edges of the injections.

For the particular case shown in Figure 5 (bottom), the observed electron beam appeared just at the earthward edge of the injection boundary (Figure 4, top). This is deduced from the character of the dispersion curves, particularly as the ion injection at 0730 UT reaches above the 80-keV limit of the detector for only about 1 min [see further discussions in *Mauk and Meng*, 1991]. What we see, then, at the leading edge of local injections is the occurrence of two very distinct processes: (1) the generation of the isotropic injection, and (2) the generation of the electron beam and an apparent magnetic-field–aligned electrical discharge. I believe that the apparent electrical discharge observed here has a manifestation in the auroral ionosphere, as diagnosed in some auroral x-ray data [*Mauk and Meng*, 1991].

Now the question becomes how to generate an auroral-like electric discharge on magnetic field lines that

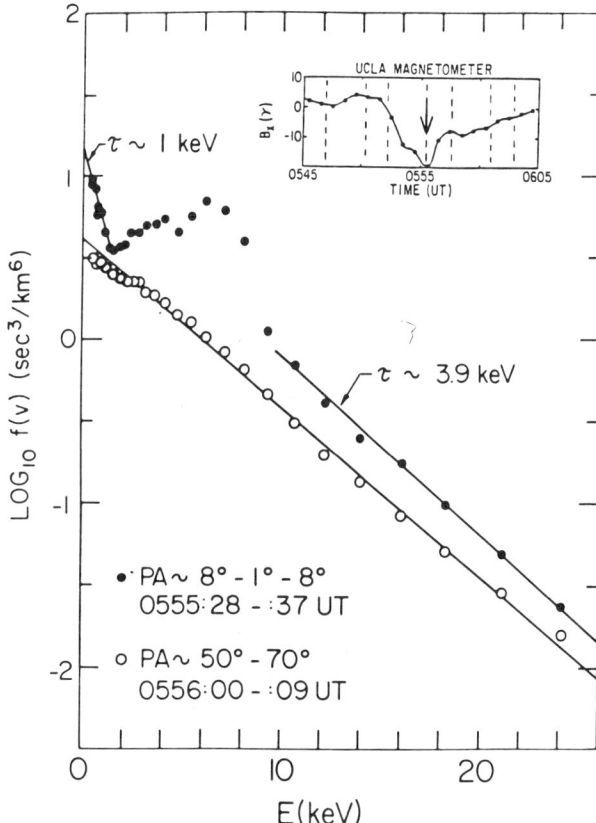

Fig. 6. Magnetic-field–aligned electron beam measured in the geostationary orbit in association with a substorm injection [after *McIlwain*, 1975 and *Mauk and Meng*, 1991].

are quasi-dipolar in character. Part of the answer, I believe, is in the response of plasma distributions to the dynamical dipolarizations associated with substorm injections.

The top panel of Figure 7 shows the configuration; a field line is allowed to go from a stretched configuration to a dipolar configuration over a period of 1 min, consistent with observations. When the response to this process is modeled, we find that the ion distributions end up dramatically field-aligned. In the second panel, we see the predipolarization ion intensity spectrum and the postdipolarization spectra. These spectra were calculated with a kinetic particle simulation model that preserves the first adiabatic invariant of the particles but not the second adiabatic invariant. The parallel distribution is dramatically enhanced over the perpendicular distribution [*Mauk*, 1986]. Since the electron populations respond very differently to the dipolarization, the dipolarization, in effect, tries to separate electric charges along the magnetic field lines. The right panel of Figure 8 shows a simulation of

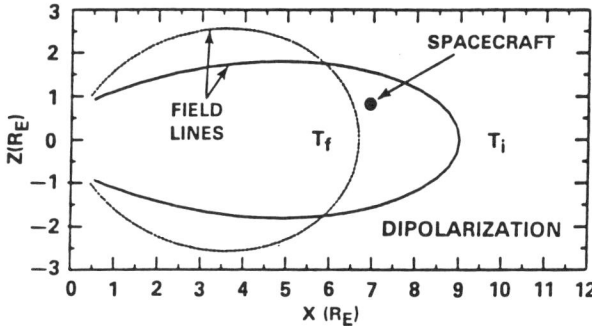

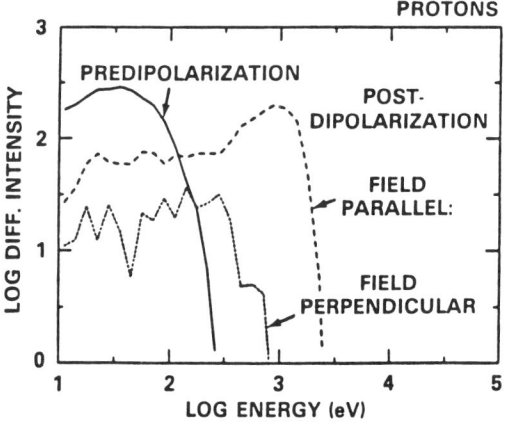

Fig. 7. (top) Schematic of a magnetic dipolarization associated with a substorm injection. (bottom) Results of a simulation of the response of an ion distribution to dynamic dipolarizations [after *Mauk*, 1986].

this effect, where the electric potentials are derived using the quasi-neutrality condition [*Mauk*, 1989].

Shown is electric potential plotted as a function of magnetic latitude along a single flux tube. The different profiles correspond to different times into the dipolarization process, from 0 to 220 s. One finds that an electric potential drop is formed near the magnetic equator, and that potential front propagates toward the ionosphere, spreading out as it goes. Eventually, a steady, macroscopic potential profile forms, in this case representing a potential drop of about 2 to 3 kV. Note that no ionospheric interactions have been included in this simulation.

This potential profile looks very much like the potential profiles achieved in a laboratory experiment by *Stenzel et al.* [1981], which was concerned with the generation of electrostatic double layers in auroral regions (left panel of Figure 8). In that experiment, an ion beam was fired toward the pole of a magnetic dipole, and the results were some electric potential profiles that looked like the potential profiles found during the simulations. However, to generate what we might consider a true electrical discharge with a detached double-layer–like potential drop, it was necessary to introduce a reflecting potential at the magnet's face in order to generate microturbulence in the near-magnet region associated with ion–electron counter-streaming. This microturbulence resulted in the needed generation of a new electron population trapped between the magnetic and electrostatic potential barriers. It is my hypothesis that by introducing the microturbulence that would be associated with the interactions between the ionospheric plasmas and the streaming plasmas associated with the dipolarizations, a true electrical discharge would result, possibly generating the kind of electron beams observed by *McIlwain* [1975] at the injection boundary. In this way, the introduction of the microscale processes associated with auroral beaming processes would be critical to establishing the character of the mesoscale features associated with the injection boundary.

The kind of electrical discharges diagnosed by *McIlwain* [1975] are highly transient in nature (1 min). We still cannot understand the mesoscale auroral features highlighted at the very beginning of this paper (Figures 1 and 2). I would guess that these features are more time-stationary than would be expected if they were associated directly with the transient electrical discharges observed in the geosynchronous orbit. One possibility is that the transient electrical discharge generates conditions whereby the arc-like structures can be sustained by other processes. For example, *Rothwell et al.* [1988] have developed a model whereby auroral arcs can be sustained on quasi-dipolar magnetic field lines, driven by the global convection. They presuppose the existence of a channel of enhanced conductivity within the ionosphere. I suggest that the transient electrical discharge observed by *McIlwain* [1975] at the injection boundary may well be ideal for generating the channels of enhanced conductivity that are needed as a starting point for the *Rothwell et al.* [1988] model. Injection boundary processes would then initiate the arc structure, but subsequently other processes would take over to sustain the arcs. Also, the arc structures could well be "images" of the positions of previous injection boundary structures. Interestingly, the scale size suggested for the width of the injection boundary, 0.1 to 0.2 R_E, maps to ionospheric scale sizes comparable to the arcs shown in Figure 2, i.e., 10 to 20 km. Regarding the coupling of mesoscale and microscale processes, the latter would appear to be critical to turning the magnetospheric dynamics into true electrical dis-

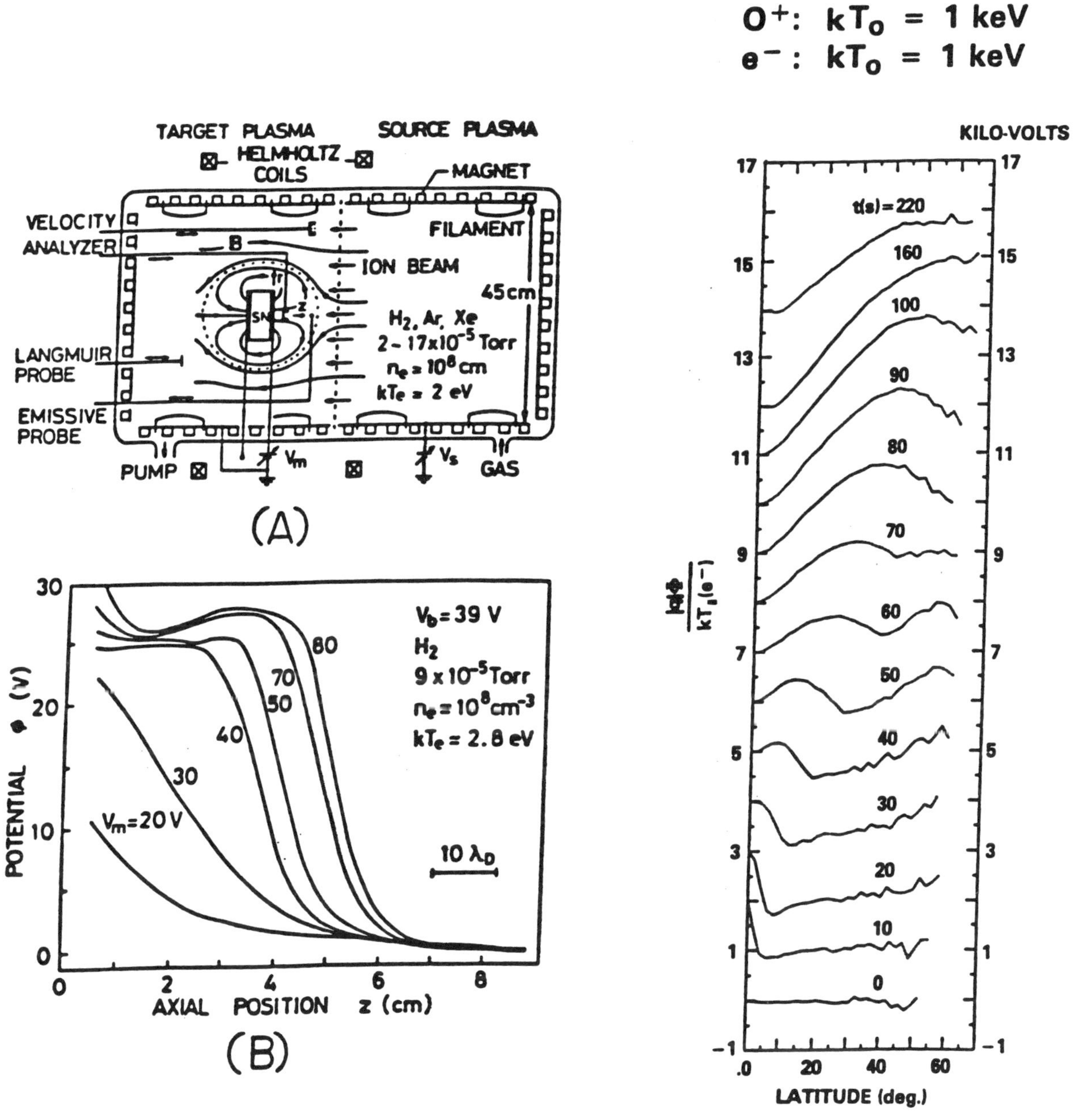

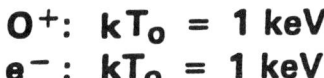

Fig. 8. (right) Simulation of the generation of magnetic-field–aligned electric fields in response to a substorm dipolarization [after *Mauk*, 1989]. (left) Laboratory study of the formation of electrostatic double layers using magnetic-field–aligned ion beams [from *Stenzel et al.*, 1981].

charges with the ionosphere, perhaps helping to establish the scale of the auroral manifestation of the magnetospheric dynamics.

Obviously, these ideas are all highly speculative.

CONCLUSIONS

Observations suggest that mesoscale processes (<1 R_E) appear to play critical roles in the transport of plasmas within the inner and middle magnetospheric regions and in the coupling of those regions to other regions (magnetotail, ionosphere). We have discussed the example of the narrow injection boundary structure and the auroral arcs that map to geosynchronous regions on quasi-dipolar field lines.

Microscale processes must play critical roles in moderating the actions and influences of the mesoscale processes. For the examples cited here, microscale processes should be critical toward establishing the spatial scale of the injection boundary and its ionospheric manifestation. The ionospheric coupling of the mesoscale structures via electrical discharge must also be moderated by microscale processes, as shown in the experiments of *Stenzel et al.* [1981]. The character of the coupling between the mesoscale phenomena highlighted here and the moderating microscale processes is not known and requires further study.

Acknowledgments. This work was supported by NSF Grant ATM-9108193 and NASA SR&T Grant NAGW-2583.

REFERENCES

Angelopoulos, V., W. Baumjohann, C. F. Kennel, F. V. Coroniti, M. G. Kivelson, R. Pellat, R. J. Walker, H. Lühr, and G. Paschmann, Bursty bulk flows in the inner central plasma sheet, *J. Geophys. Res.*, 97, 4027, 1992.

Angelopoulos, V., C. F. Kennel, F. V. Coroniti, R. Pellat, H. E. Spence, M. G. Kivelson, R. J. Walker, W. Baumjohann, W. C. Feldman, J. T. Gosling, and P. T. Russell, Characteristics of ion flow in the quiet state of the inner plasma sheet, *Geophys. Res. Lett.*, 20, 1711, 1993.

Baumjohann, W., G. Paschmann, and H. Lühr, Characteristics of high speed ion flows in the plasma sheet, *J. Geophys. Res.*, 95, 3801, 1990.

Erickson, G. M., and R. A. Wolf, Is steady convection possible in the Earth's magnetotail? *Geophys. Res. Lett.*, 7, 897, 1980.

Kavanagh, L. D., Jr., J. W. Freeman, Jr., and A. J. Chen, Plasma flow in the magnetosphere, *J. Geophys. Res.*, 73, 5511, 1968.

Kerns, K. J., D. A. Hardy, and M. S. Gussenhoven, Modeling of convection boundaries seen by CRRES in 120-eV to 28-keV particles, *J. Geophys. Res.*, 99, 2403, 1994.

Kivelson, M. G., and H. E. Spence, On the possibility of quasi-static convection in the quiet magnetotail, *Geophys. Res. Lett.*, 15, 1541, 1988.

Konradi, A., C. L. Semar, and T. A. Fritz, Substorm-injected protons and electrons and the injection boundary model, *J. Geophys. Res.*, 80, 543, 1975.

Lopez, R. E., A. T. Y. Lui, D. G. Sibeck, K. Takahashi, R. W. McEntire, L. J. Zanetti, and S. M. Krimigis, On the relationship between the energetic particle flux morphology and the changes in the magnetic field magnitude during substorms, *J. Geophys. Res.*, 94, 17105, 1989.

Lopez, R. E., D. G. Sibeck, R. W. McEntire, and S. M. Krimigis, The energetic ion substorm injection boundary, *J. Geophys. Res.*, 95, 109, 1990.

Lui, A. T. Y., R. E. Lopez, S. M. Krimigis, R. W. McEntire, L. J. Zanetti, and T. A. Potemra, A case study of a magnetotail current disruption and diversion, *Geophys. Res. Lett.*, 15, 721, 1988.

Mauk, B. H., Quantitative modeling of the 'convection surge' mechanism of ion acceleration, *J. Geophys. Res.*, 91, 3423, 1986.

Mauk, B. H., Generation of macroscopic magnetic-field–aligned electric fields by the convection-surge ion acceleration mechanism, *J. Geophys. Res.*, 94, 8911, 1989.

Mauk, B. H., and C. E. McIlwain, Correlation of Kp with the substorm-injected plasma boundary, *J. Geophys. Res.*, 79, 3193, 1974.

Mauk, B. H., and C.-I. Meng, Dynamical injections as the source of near geostationary quiet time particle spatial boundaries, *J. Geophys. Res.*, 88, 1011, 1983a.

Mauk, B. H., and C.-I. Meng, Characterization of geostationary particle signatures based on the 'injection boundary' model, *J. Geophys. Res.*, 88, 3055, 1983b.

Mauk, B. H., and C.-I. Meng, Macroscopic ion acceleration associated with the formation of the ring current in the Earth's magnetosphere, in *Ion Acceleration in the Magnetosphere and Ionosphere*, Geophysical Monograph *38*, edited by T. Chang, American Geophysical Union, Washington, D.C., 1986.

Mauk, B. H., and C.-I. Meng, The aurora and middle magnetospheric processes, in *Auroral Physics*, edited by C.-I. Meng, M. J. Rycroft, and L. A. Frank, p. 223, Cambridge University Press, Cambridge, England, 1991.

McIlwain, C. E., Plasma convection in the vicinity of the geosynchronous orbit, in *Earth's Magnetospheric Processes*, edited by B. M. McCormac, p. 268, D. Reidel, Hingham, Mass., 1972.

McIlwain, C. E., Substorm injection boundaries, in *Magnetospheric Physics*, edited by B. M. McCormac, p. 143, D. Reidel, Hingham, Mass., 1974.

McIlwain, C. E., Auroral electron beams near the magnetic equator, in *The Physics of Hot Plasma in the Magnetosphere*, edited by B. Hultqvist and L. Stenflo, p. 91, Plenum, New York, 1975.

Meng, C.-I., B. H. Mauk, and C. E. McIlwain, Electron precipitation of evening diffuse aurora and its conjugate electron fluxes near the magnetospheric equator, *J. Geophys. Res.*, 84, 2545, 1979.

Moore, T. E., R. L. Arnoldy, J. Feynman, and D. A. Hardy, Propagating substorm injection fronts, *J. Geophys. Res.*, 86, 6713, 1981.

Olson, W. D., and K. A. Pfitzer, A quantitative model of the magnetospheric magnetic field, *J. Geophys. Res., 79*, 3739, 1974.

Rothwell, P. L., L. P. Block, M. B. Silevitch, and C.-G. Fälthammar, A new model for substorm onsets: The pre-breakup and triggering regimes, *Geophys. Res. Lett., 15*, 1279, 1988.

Schindler, K., and J. Birn, Self-consistent theory of time-dependent convection in the Earth's magnetotail, *J. Geophys. Res., 87*, 2263, 1982.

Stenzel, R. L., M. Ooyama, and Y. Nakamura, Potential double layers in strongly magnetized plasmas, in *Physics of Auroral Arc Formation*, Geophysical Monograph, 25, edited by S.-I. Akasofu and J. R. Kan, p. 226, American Geophysical Union, Washington, D.C., 1981.

Stern, D. P., Large-scale electric fields in the Earth's magnetosphere, *Rev. Geophys. Space Phys., 15*, 156, 1977.

B. H. Mauk, The Johns Hopkins University Applied Physics Laboratory, Laurel, MD 20723-6099, (301) 953-6023; E-mail, Barry_Mauk@jhuapl.edu.

Relative Contribution of the Solar Wind and the Auroral Zone to Near-Earth Plasmas

Vahé Peroomian[1] and Maha Ashour-Abdalla

Institute of Geophysics and Planetary Physics, University of California at Los Angeles

We have carried out a three-dimensional large-scale kinetic simulation of the terrestrial electric and magnetic fields in order to gain a better understanding of the contribution of the solar wind and the ionosphere to the ring current during slightly disturbed times. Our results indicate that the ring current is populated by ions from all significant sources of ions for these periods, but that the near-Earth plasma sheet has distinct regions where ions from a single source dominate.

1. INTRODUCTION

Observations have shown that the Earth's magnetosphere is populated not only by solar wind ions, but also by ions of ionospheric origin [e.g. *Shelley et al.*, 1972; *Yau et al.*, 1985]. The ionospheric contribution to near-Earth plasma increases with geomagnetic activity, sometimes comprising up to 80% of the storm-time ring current energy density [*Hamilton et al.*, 1988].

Recently, in a three-dimensional magnetic and electric field model, *Delcourt et al.* [1989, 1990] used the cleft ion fountain as a source region to populate the plasma sheet and found that O^+ from the cleft can attain energies up to 60 keV and become trapped in the ring current. This study, however, used guiding center approximation, which is not entirely valid for the magnetotail regions considered. *Cladis and Francis* [1992] improved on existing particle tracing techniques by employing a more realistic magnetic and electric field model and a full-motion trajectory calculation to obtain number densities, pressures and plasma beta of O^+ ions for a limited region in the near-Earth magnetotail; these results led them to postulate that O^+ ions trigger substorms. *Peroomian* [1994] showed that during quiet and slightly disturbed times when the cross-tail electric field was relatively weak, ions from the dayside cleft region remained in the lobes and the nightside auroral zone was the most viable source of ionospheric ions in the near-Earth region. Accordingly, *Peroomian* [1994] used this source to obtain moments of the O^+ ion distribution in the near-Earth region. While the studies mentioned above have greatly increased our knowledge of the process of populating the magnetotail from the ionosphere, it is not completely clear how ionospheric particles combine with solar wind ions to result in the overall magnetospheric population. In this paper, we address the problem of populating the near-Earth magnetosphere by using both solar wind (plasma mantle and low latitude boundary layer (LLBL)) and auroral ionospheric sources and studying the relative abundances of ions from the sources in the near-Earth plasma sheet and ring current.

2. RESULTS

We followed the exact trajectories of ions through a static magnetic and electric field configuration representing Earth's magnetosphere. The position and velocity of each particle were recorded every time it crossed the "virtual" plane detector at z = 0; this information allowed us to calculate the local distribution functions and the distributions' moments there. The magnetic field model used in this study is the $K_p=1$ version of the *Tsyganenko*

---

[1] Present Address: The Aerospace Corp., M2-260, P.O. Box 92957, El Segundo, CA 90009-2957

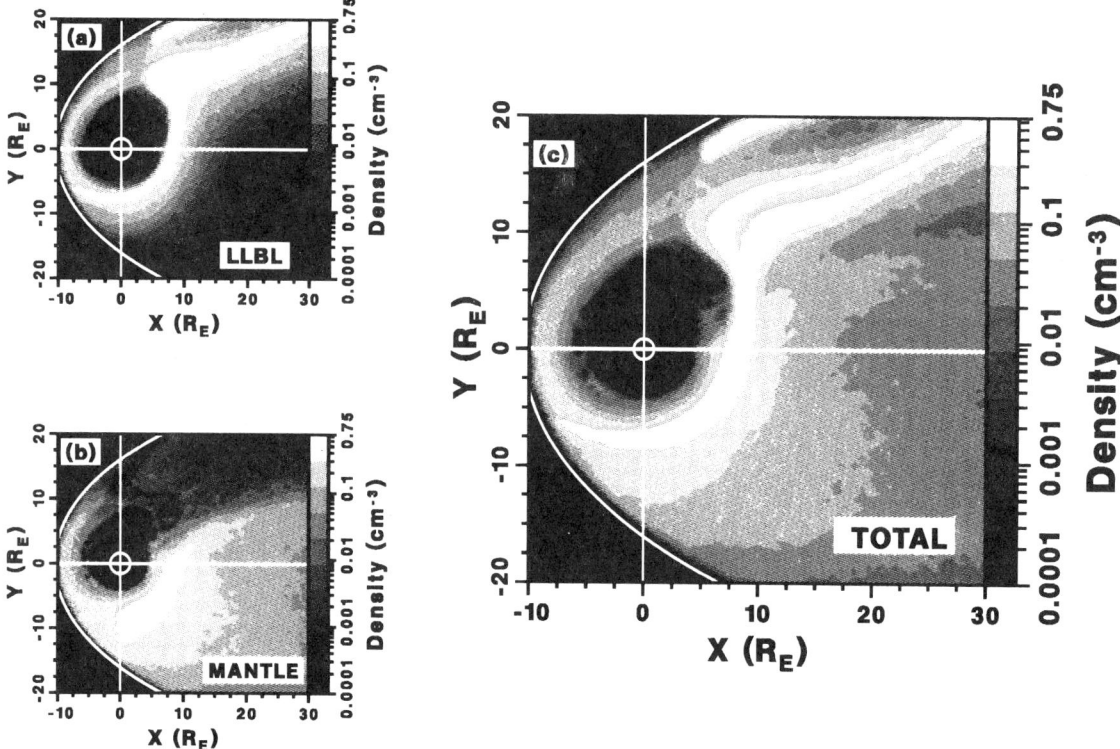

Fig. 1. a) Plot of LLBL H^+ ion density in the equatorial magnetosphere ($z = 0$ plane). The gray scale extends from 0.001 (black) to 0.75 (white) cm^{-3} on a logarithmic scale. b) Plot of mantle H^+ densities in the equatorial plane. c) Total solar wind contribution to equatorial H^+ density.

[1989] magnetic field model, and the electric field used is computed from the *Heppner and Maynard* [1987] ionospheric potential model "A" by assuming magnetic field lines to be equipotentials (see *Peroomian* [1994] for a thorough discussion of this model). The cross-polar cap potential drop chosen for this study is 40 kV. The solar wind contribution is modeled using 30 sources in the plasma mantle (distributed over invariant latitudes of 75°-82° and 1800-0600 MLT on the nightside) and 30 sources in the LLBL. Distributions of 1000 particles with a temperature of 300 eV and a streaming velocity of 200 km/s were launched from each solar wind source. The total solar wind contribution was normalized by assuming an influx of 1.0×10^{27} H^+ ions/s into the magnetosphere. The auroral ionospheric outflow was assumed to occur between invariant latitudes of 67° and 71° and between 1940 and 0440 local times. This area was then divided into 30 sectors and distributions of 1000 particles were launched with 200 eV temperature and 1.0 keV streaming energy. Ions were launched at an altitude of 20,000 km, above the auroral acceleration region. Each noninteracting ion trajectory was followed until it exited the system, either by precipitation or loss to the flanks. The total outflow from the auroral zone for $K_p=2$ consisted of 3.0×10^{25} H^+ ions/s, 1.0×10^{25} He^+ ions/s, and 3.0×10^{25} O^+ ions/s [*Yau et al.*, 1985].

a. Solar Wind. Figure 1 shows the density of solar wind H^+ ions in the equatorial plane. The horizontal (x) axis in each of the 3 panels is positive tailward and the vertical (y) axis is positive toward dawn. The gray scale on the right of each panel shows density in cm^{-3} on a logarithmic scale. The magnetopause is also shown. Figure 1a (upper left-hand panel) shows the density of ions originating in the LLBL. We see that ions from this source populate the dawn flanks and the ring current but don't cross into the dusk flanks of the plasma sheet. Conversely, ions from the plasma mantle (lower left-hand panel in Figure 1) travel far downtail and are convected duskward, filling the near-midnight and dusk flanks of the plasma sheet. In the absence of a time-varying electric field, mantle ions populate the partial (open) ring current and are excluded from the trapped ring current. Figure 1c

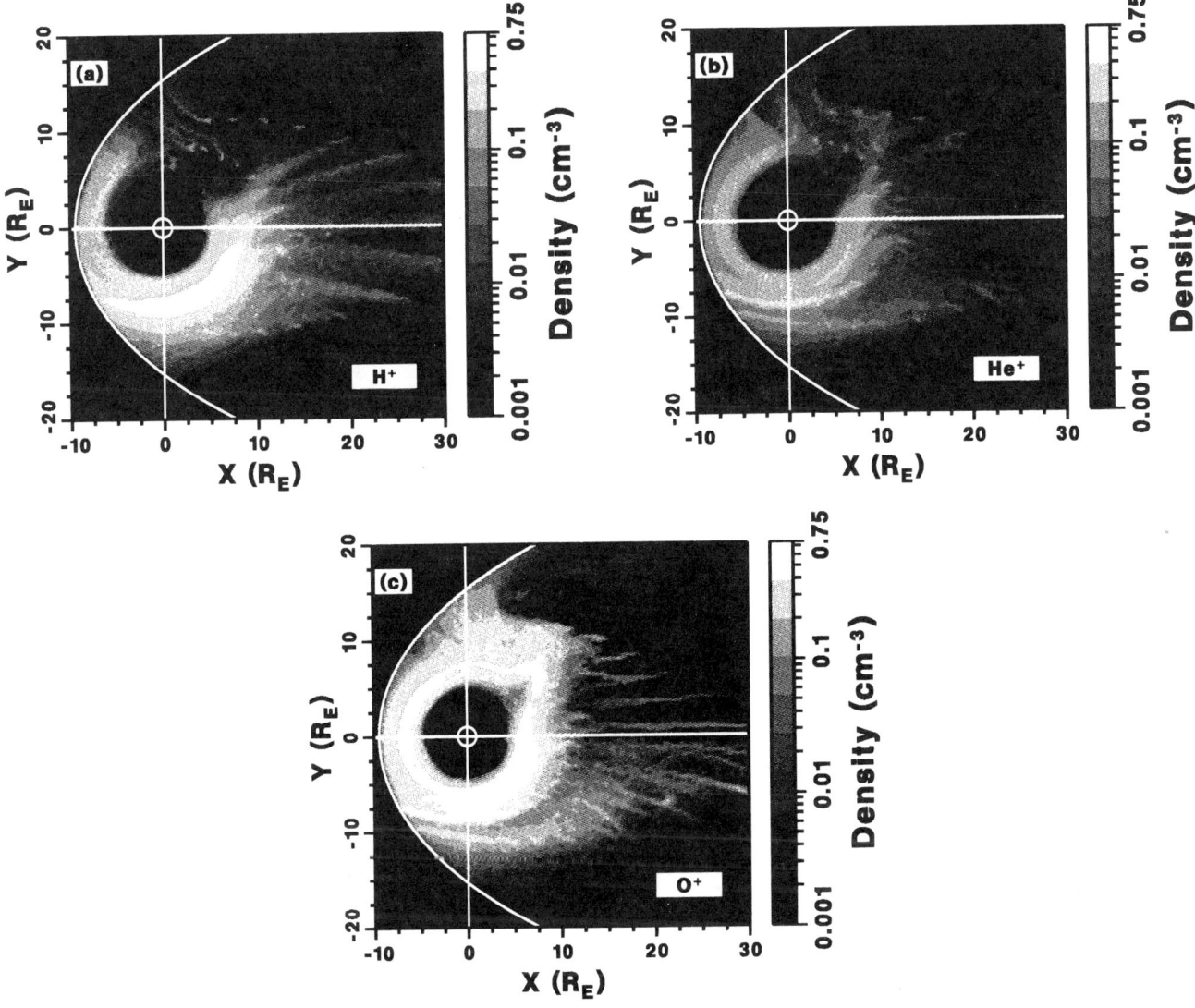

Fig. 2. a) Equatorial density plot of ionospheric (a) H^+, (b) He^+, and (c) O^+. Panels are in the same format as in previous figure.

(large right-hand panel) shows the combined (LLBL + mantle) solar wind contribution to the near-Earth density profile. This figure shows that the LLBL and mantle populate the near-Earth region in a complementary manner: the LLBL fills the dawn flanks and the ring current, while the mantle fills the midnight-dusk flanks. Both solar wind source regions contribute ions to the dusk-midnight ring current sector, resulting in density values of ~1 cm^{-3} in this region.

b. Auroral Ionosphere. Figure 2a shows the equatorial density profile of H^+ ions launched from the nightside auroral zone in a format similar to the panels of the previous figure. First, it is apparent that ions from this source do not travel far downtail, and that they are of meaningful numbers only earthward of ~12 R_E. Also, ionospheric H^+ ions launched in our calculations fail to form a complete ring current. This is because the magnetic field is more dipolar in the flanks as compared to the tail midplane, and whereas ions launched near midnight are stochastically trapped, ions launched toward the flanks precipitate in the opposite hemisphere and are lost. This leads to the density drop-out in the dawn flanks which is apparent in the figure. The highest H^+ density occurs in the dusk-midnight sector and is comparable to the density of solar wind ions there.

The formation of a partial ring current can also be seen in Figure 2b, showing the density of He^+ ions in our calculations. He$^+$ ions behave in much the same way as H$^+$ ions do, except in the dawn flanks where the density dropout area is smaller. The peak He$^+$ density is ~ 0.3 cm^{-3}, which is consistent with the smaller outflow rate of this species.

Figure 2c, the density of O$^+$ ions, shows that O$^+$ ions launched from the nightside auroral zone have succeeded in populating the ring current between r = 4 R$_E$ and r = 9 R$_E$. The maximum density of O$^+$ ions in this region is ~ 1 cm^{-3} and occurs at a radial distance of ~ 6 R$_E$. Observations place the location of maximum ring current density earthward of this location [*Lennartsson et al.*, 1979]. In a static electric field model, ions accessing regions earthward of this location have to be launched at lower latitudes than those considered here. However, our aim was not only to populate the ring current but also to study the energization of particles accessing this region. Particles launched at lower latitudes would simply execute trapped orbits and would not become as energized as do ions having access to the quasiadiabatic regions of the tail. Comparison of macroscopic parameters obtained in this study to observations is difficult because of the general lack of quiet-time ring current observations. However, we can obtain a ball-park figure for ring current parameters by comparing our results with observations from the late recovery phase of substorms. *Lennartsson et al.* [1979] studied one such period and found that O$^+$ ion densities ranged from ~ 2 - 4 cm^{-3} 3 hours after peak Dst to > 1 cm^{-3} 12 hours after peak Dst. During the storm, H$^+$ and O$^+$ ion densities were found to be comparable. O$^+$ densities were actually higher than H$^+$ densities at L < 4 R$_E$, and the H$^+$ ion densities were considerably higher than O$^+$ densities when measured 12 hours after peak Dst. We see that the peak value of 1 cm^{-3} obtained in our study corresponds well to the ISEE-1 measurements of *Lennartsson et al.* [1979], with the difference that the peak in density in our study occurs tailward of the measured storm-time and recovery phase values.

3. SUMMARY AND CONCLUSIONS

In this paper, we examined the density distribution of ions from various source regions in the near-Earth magnetosphere. Our findings can be summarized by the schematic shown in Figure 3 and are as follows:

1. Ions from the LLBL populate the dawn flanks and the ring current, but do not enter the near-midnight and duskside plasma sheet.

2. Mantle ions populate the midnight and duskside plasma sheet and the partial ring current.

3. Ionospheric ions are present in significant densities

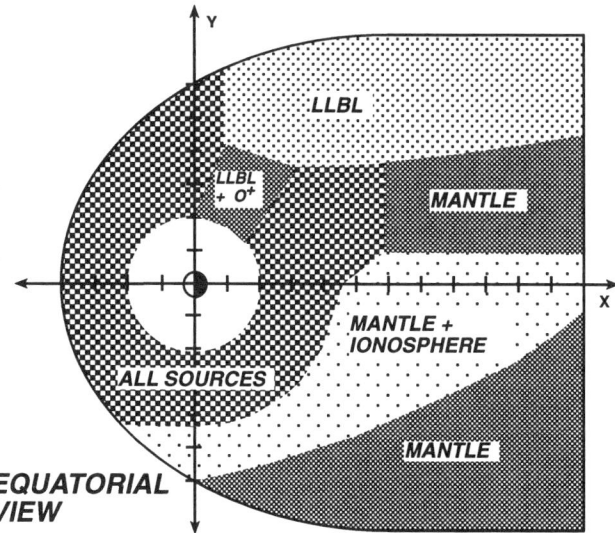

Fig. 3. Schematic showing equatorial regions where ions from each source are dominant.

only earthward of 12 R$_E$. Ionospheric H$^+$ and He$^+$ ions form a partial ring current while O$^+$ ions form a complete ring current.

4. The peak in density for all species occurs in the dusk-midnight sector of the ring current, between r = 4 R$_E$ and r = 9 R$_E$.

While many source regions and mechanisms have been postulated as supplying the ring current with energetic particles, these have been limited to either storm-time models with large transient electric fields or models with a short-tail configuration; these limitations restricted the study of ion orbits to within 20 R$_E$ of Earth. We have shown here that ions from many sources contribute to the overall plasma composition, and that areas of the magnetosphere exist in which all sources combine to give the observed plasma densities.

Acknowledgments. The authors are grateful to L. M. Zelenyi and R. J. Walker for helpful discussions. This work was supported by NASA ISTP grant NAG5-1100 and NSF grant ATM-9414760. Computing was performed at the San Diego Supercomputer Center and at the Office of Academic Computing at UCLA.

REFERENCES

Cladis, J. B. and W. E. Francis, Distribution in magnetotail of O$^+$ ions from cusp/cleft ionosphere: a possible substorm trigger, *J. Geophys. Res.*, *97*, 123, 1992.

Delcourt, D. C., C. R. Chappell, T. E. Moore, and J. H. Waite, Jr., A three-dimensional numerical model of ionospheric plasma in the magnetosphere, *J. Geophys. Res.*, *94*, 11,893, 1989.

Delcourt, D. C., J. A. Sauvaud, and T. E. Moore, Cleft contribution to ring current formation, *J. Geophys. Res.*, *95*, 20937, 1990.

Hamilton, D. C., G. Gloeckler, F. M. Ipavich, W. Stüdemann, B. Wilken, and G. Kremser, Ring current development during the great geomagnetic storm of February 1986, *J. Geophys. Res.*, *93*, 14343, 1988.

Heppner, J. P., and N. C. Maynard, Empirical high-latitude electric field models, *J. Geophys. Res.*, *92*, 4467, 1987.

Lennartsson, W., E. G. Shelley, R. D. Sharp, R. G. Johnson, and H. Balsiger, Some initial ISEE-1 results on the ring current composition and dynamics during the magnetic storm of December 11, 1977, *Geophys. Res. Lett.*, *6*, 483, 1979.

Peroomian, V., Large Scale Kinetic Modeling of Magnetospheric Plasma, PhD thesis, University of California at Los Angeles, March 1994.

Shelley, E. G., R. G. Johnson, and R. D. Sharp, Satellite observations of energetic heavy ions during a geomagnetic storm, *J. Geophys. Res.*, *77*, 6104, 1972.

Tsyganenko, N. A., A magnetospheric magnetic field model with a warped tail current sheet, *Planet. Space. Sci.*, *37*, 5, 1989.

Yau, A. W., E. G. Shelley, W. K. Peterson, and L. Lenchyshyn, Energetic auroral and polar ion outflow at DE 1 altitudes: magnitude, composition, magnetic activity dependence, and long-term variations, *J. Geophys. Res.*, *90*, 8417, 1985.

Vahé Peroomian, The Aerospace Corp., M2-260, P. O. Box 92957 M2-260, El Segundo, CA 90009-2957. (e-mail vahe_peroomian@qmail2.aero.org)

Maha Ashour-Abdalla, Institute of Geophysics and Planetary Physics, University of California, Los Angeles, CA 90024-1567. (e-mail mabdalla@igpp.ucla.edu)

Coupling Between Microscale and Mesoscale Processes in the Dayside Magnetosheath, Magnetopause, and Boundary Layer Regions

L.C. Lee and J. G. Hawkins

Geophysical Institute, University of Alaska Fairbanks

This paper reviews the following five scenarios in which coupling between microscale and mesoscale processes may occur in the dayside magnetosheath-magnetopause-boundary layer regions. (1) Downstream of the perpendicular bow shock, the ion temperature ratio $T_{i\perp}/T_{i\|}$ is large, and mirror waves and ion cyclotron waves can be generated. These waves in turn modify the particle velocity distribution, reduce the $T_{i\perp}/T_{i\|}$ ratio, and hence affect the evolution of the anisotropic magnetofluid. (2) As a result of the convection of the magnetosheath plasma to the dayside magnetopause, the magnetic fields are compressed and pile up at the dayside magnetopause, and a thin magnetopause current sheet with enhanced $T_{i\perp}/T_{i\|}$ can be formed. The enhanced ion temperature ratio at the thin current sheet may enhance the kinetic tearing instability and thus magnetic reconnection at the dayside magnetopause. As a result of magnetic reconnection, various MHD discontinuities and waves, including rotational discontinuities, Alfvén waves, slow shocks, and slow expansion waves may be formed in the magnetopause-boundary layer. (3) In the presence of density and magnetic field gradients near the Earth's magnetopause, the mesoscale MHD surface waves may generate kinetic Alfvén waves (KAWs), which lead to particle transport across the magnetopause and form boundary layer plasma. (4) Impinging dynamic pressure pulses or interplanetary Alfvén waves on the bow shock may result in the generation of fast shocks, slow shocks, and Alfvén waves in the magnetosheath. Transmission of fast-mode waves through the magnetopause may lead to the non-adiabatic heating of magnetospheric plasma, which generates ion cyclotron waves and thus Pc 1 waves in the magnetosphere. (5) Impulsive penetration may occur when irregularities in the solar wind with sufficient momentum impact the magnetopause. Plasma irregularities which lack the excess momentum to penetrate the magnetopause may still produce vorticity and field aligned currents within the magnetosphere.

1. INTRODUCTION

The interaction of the solar wind with the Earth's magnetosphere leads to the formation of the bow shock, magnetosheath, magnetopause, and boundary layer. The mesoscale features in these regions create several forms of free energy that drive microscale processes, which in turn are responsible for the transport of mass, momentum, and energy across the magnetopause boundary. Microscale processes are defined to have temporal scales that are on the order as the ion gyro period and spatial scales that are on the order of the ion gyro radius, while mesoscale features have temporal and spatial scales that are much greater than the ion gyro period and radius, respectively. The purpose of this paper is to review the coupling between the mesoscale features in these regions and various microscale processes. These interactions can be grouped into the following categories (references and detailed discussions for each type of interaction are given in subsequent sections).

1. *Bow Shock.* The bow shock is formed when the supersonic flow of the solar wind encounters the Earth's

magnetosphere. At locations along the bow shock where the interplanetary magnetic field (IMF) is nearly perpendicular to the normal direction of the bow shock surface (i.e., a so-called quasi-perpendicular shock), a temperature anisotropy is produced in the ions of the shocked plasma with $T_{i\perp} > T_{i\|}$. This anisotropy serves as the source of free energy that drives the mirror wave and electromagnetic ion cyclotron instabilities. At locations along the bow shock where the IMF is nearly parallel to the shock normal (quasi-parallel shock), the ions in the shocked plasma will exhibit a temperature anisotropy with $T_{i\|} > T_{i\perp}$, which serves as a source of free energy to generate Alfvén waves through the firehose instability.

2. *Slow Mode Structures and the Depletion Layer.* Since the magnetic flux is "frozen-in" to the solar wind plasma, it will tend to pile-up in front of the dayside magnetopause as the solar wind convects past the magnetosphere. The pile-up of magnetic flux in front of the magnetopause increases magnetic pressure, which pushes the plasma away from this region to form the plasma depletion layer. The ejection of plasma along the field lines leaves behind a thermal anisotropy with $T_{i\perp} > T_{i\|}$ in the plasma depletion layer, which can drive ion cyclotron instabilities. Magnetohydrodynamic (MHD) slow mode waves, in which the plasma and magnetic pressures are anti-correlated, are observed to form a "near" standing wave in front of the depletion layer and magnetopause. These slow mode waves can be generated by the interaction of interplanetary MHD waves with the bow shock.

3. *Magnetic Reconnection.* If the IMF and geomagnetic fields have components that are anti-parallel, then the pile-up of magnetic flux in front of the dayside magnetopause may provide a source of free energy to create off-diagonal terms in the plasma pressure tensor (particle inertia effects) and drive wave-particle interactions. Although collisional resistivity is insignificant near the magnetopause, particle inertias and plasma instabilities may provide a source of anomalous resistivity that allows the magnetic field to violate the "frozen-in" condition. In this case, the magnetic field may diffuse across the plasma and reconnect to form a more stable topology by converting magnetic energy into kinetic energy. Magnetic reconnection may remove the piled-up magnetic flux, and hence the depletion layer. Recent observations confirm that no depletion layer is observed when the IMF is southward ($B_z < 0$). As a result of magnetic reconnection, a layered structure consisting of various MHD discontinuities and waves is formed in the magnetopause boundary layer region.

4. *Viscous Interactions.* As the shocked solar wind plasma is diverted to flow around the magnetosphere, the resulting gradients in density, velocity, current, and magnetic field provide sources of free energy to drive a variety of instabilities (e.g., lower hybrid drift instabilities, whistler waves, Kelvin-Helmholtz, etc.). The resulting wave-particle interactions provide a source of anomalous viscosity that allow magnetosheath particles to random walk across the magnetopause boundary and enter the magnetosphere. In addition, fluctuations in the solar wind can produce large scale MHD waves at the surface of the magnetosphere that can be coupled to kinetic Alfvén waves (KAWs), which are Alfvén waves in which the finite ion gyroradius effects become important (i.e., $k_\perp \rho_i \approx 1$). In the presence of large gradients in the Alfvén velocity, KAWs produce a parallel electric field that allows magnetosheath particles to diffuse more efficiently into the magnetosphere.

5. *Dynamic Pressure Pulses and Impulsive Penetration.* The actual solar wind is not a steady flow, but composed instead of mesoscale variations in magnetic field and momentum density (variations in both plasma density and flow velocity). These dynamic pressure pulses form gusts in the solar wind that continually buffet the magnetosphere. Under very restricted conditions in which the IMF and geomagnetic fields are nearly parallel or antiparallel, it may be possible for a plasma irregularity in the solar wind with enhanced momentum density to impulsively penetrate the magnetopause and become trapped in the magnetospheric plasma. However, even if penetration does not occur, the

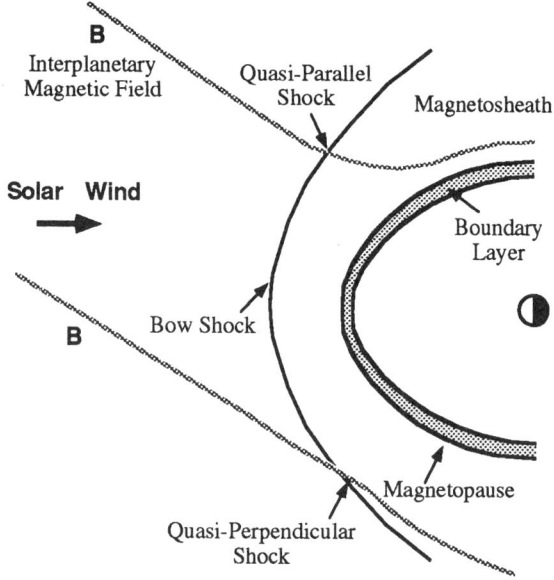

Fig. 1. Quasi-parallel and quasi-perpendicular shocks form at the bow shock, depending on the orientation of the interplanetary magnetic field (IMF) relative to the normal to the shock surface.

resulting impact of dynamic pressure pulses and interplanetary Alfvén waves can produce vorticity within the magnetosphere (which in turn produces field-aligned currents) and weak fast mode shocks (which produce a temperature anisotropy with $T_{i\perp} > T_{i\parallel}$ that can generate Pc 1 waves).

While the solar wind and IMF drive the geomagnetic activity, it is the plasma and magnetic field in the highly-turbulent magnetosheath that impinge directly onto the magnetosphere. Hence an understanding of the solar wind control of the magnetosphere must include the processes associated with the bow shock and magnetosheath.

2. QUASI-PERPENDICULAR SHOCKS AND MIRROR WAVES/LION ROARS IN THE MAGNETOSHEATH

The characteristics of the shocked plasma in the magnetosheath depends on the orientation of the IMF relative to the bow shock. Quasi-perpendicular shocks, in which the IMF is nearly perpendicular to the shock normal, results in a large ion temperature anisotropy in the shocked plasma with $T_{i\perp} > T_{i\parallel}$. This is due to preferential ion heating across the quasi-perpendicular shock [*Leroy et al.*, 1982; *Lee et al.*, 1986]. On the other hand, quasi-parallel shocks, where the IMF is nearly parallel to the shock normal, results in a large ion thermal anisotropy with $T_{i\parallel} > T_{i\perp}$. Due to the curvature of the bow shock, both types of shocks can occur simultaneously but at different locations, as shown in Figure 1.

Many theoretical and observational studies have expanded our knowledge of the structure of quasi-perpendicular shocks in the solar-terrestrial context [*Leroy et al.*, 1982; *Schopke et al.*, 1983; *Scudder et al.*, 1986]. One-dimensional hybrid simulations show that a large ion temperature anisotropy ($T_{i\perp}/T_{i\parallel} \approx 2-3$) is produced downstream of a quasi-perpendicular shock [*Lee et al.*, 1986; 1987b]. *Tsurutani et al.* [1982] suggested that magnetosheath mirror waves, which are characterized by an anti-correlation between the magnetic field strength and plasma density, may be excited by the mirror instability associated with an ion temperature anisotropy with $T_{i\perp} > T_{i\parallel}$. One dimensional hybrid simulations confirm that large amplitude mirror waves are produced downstream of quasi-perpendicular shocks with a large Alfvén Mach Number $M_A \geq 3$, as shown if Figure 2 [*Lee et al.*, 1988]. Near the bow shock in this figure, the positive correlation between the magnetic field strength and plasma density represents the signature of the fast shock, which decays downstream of the bow shock. Further downstream, the anticorrelation between these variables is the characteristic signature of mirror waves.

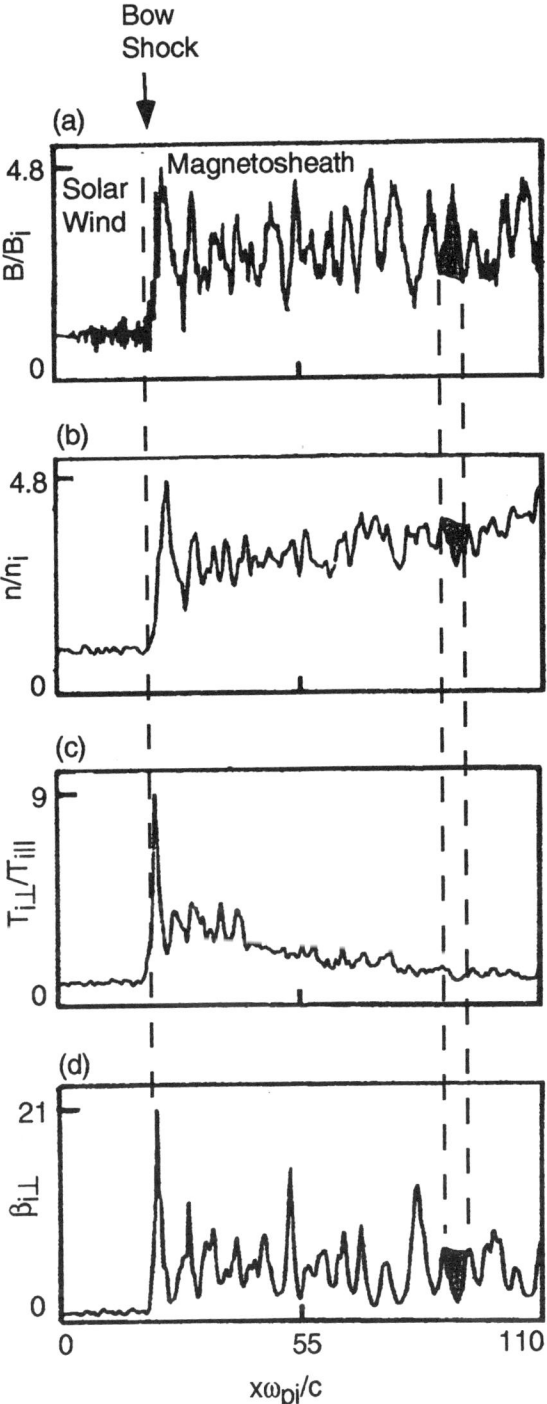

Fig. 2. Simulation results from a 1-dimensional hybrid code which shows the formation of mirror waves downstream from a quasi-perpendicular shock [*Lee et al.*, 1988]. Mirror waves are characterized by the anti-correlation between the magnetic field and plasma density.

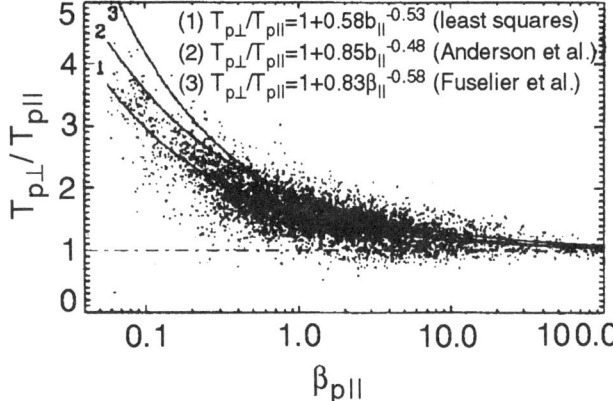

Fig. 3. Observations of temperature anisotropies of the protons and electrons in the magnetosheath region, plotted against plasma β [*Phan et al*, 1994]. The solid lines represent least squares fit of (1) all data, (2) downstream of quasi-perpendicular shocks, and (3) downstream of quasi-parallel shocks.

An analytical and simulation study of the generation of mirror waves found that the ion temperature anisotropy can also excite electromagnetic ion cyclotron waves which compete with non-oscillatory mirror modes [*Price et al.* 1986; *McKean et al.*, 1994]. While one-dimensional simulations cannot simulate the electromagnetic ion cyclotron mode, two-dimensional simulations predict that the electromagnetic ion cyclotron mode will dominate within and just behind the shock ramp [*Winske and Quest*, 1988; *Gary*, 1995]. The presence of minor ions in the solar wind may, however, reduce the growth rate of ion-cyclotron waves sufficiently for mirror modes to dominate [*Price et al.*, 1986].

Observations in the highly compressed terrestrial magnetosheath have shown the proton and electron anisotropies are strongly dependent on plasma β, as shown in Figure 3 [*Phan et al.*, 1994; *Denton et al.*, 1995]. *Anderson et al.* [1994] have shown that the proton temperature anisotropy varies inversely with the proton parallel β according to the following relationship

$$\frac{T_{\perp p}}{T_{\| p}} - 1 = \frac{0.85}{\beta_{\| p}^{0.48}} \qquad (1)$$

where $\beta_{\| p} \equiv 8\pi n_p T_{\| p} / B_0^2$ and $\mathbf{B}_0^2$ is the background magnetic field. Linear Vlassov theory predicts that the threshold for the electromagnetic proton cyclotron instability is consistent with this relationship [*Gary and Lee*, 1994] and hybrid simulations of this instability have further demonstrated that such a relation represents an upper bound on the proton temperature anisotropy [*Gary et al.*, 1994].

Observations downstream of a quasi-perpendicular shock in the Earth's magnetosheath have revealed characteristic signatures of mirror type modes [*Hubert et al.*, 1989], as first reported in the simulation by *Lee et al.* [1988]. Long duration lion roars (> 5 minutes) were often observed downstream of quasi-perpendicular shocks [*Rodriguez*, 1985]. Lion roars (electron whistler waves) are thought to associated with the electron cyclotron instability driven by an electron temperature anisotropy in the magnetosheath [*Thorne and Tsurutani*, 1981]. *Smith et al.* [1969] and *Tsurutani et al.* [1982] observed such electron temperature anisotropies with $T_{e\perp} > T_{e\|}$ and lion roars in regions where mirror waves provide a reduced magnetic field and a high plasma β factor. *Lee et al.* [1987a] discussed a theoretical explanation by which the free electron temperature ratio $T_{e\perp} / T_{e\|}$ may be enhanced in high β regions.

3. SLOW MODE WAVE STRUCTURES IN FRONT OF THE DAYSIDE MAGNETOPAUSE

The convection of the solar wind past the magnetosphere causes magnetic flux to pile up at the dayside magnetopause. Through this process, some of the kinetic energy associated with convection is converted and stored as magnetic energy. *Lees* [1964] and *Zwan and Wolf* [1976] used MHD calculations to predict that the pile-up of magnetic flux at the magnetopause would create a plasma depletion layer in front of the dayside magnetopause as the flux tubes are compressed and plasma is squeezed out the ends. The existence of the depletion layer was subsequently confirmed by observations [*Paschmann et al.*, 1978; *Crooker et al.*, 1979]. The slow-mode compression of draped magnetosheath fields and the squeezing of plasma along field lines can produce a temperature anisotropy in the ions with $T_{i\perp} / T_{i\|} > 1$ since the particles with a high parallel velocity will leave the region faster [*Crooker and Siscoe*, 1977]. This temperature anisotropy favors the generation of electromagnetic ion cyclotron waves since the plasma β is not very high in this region.

Recent analysis of ISEE 1 and ISEE 2 spacecraft data reveals that the plasma depletion layer is frequently accompanied by a complementary region with enhanced plasma density and decreased magnetic field strength [*Song et al.*, 1990; 1992; *Hubert*, 1994]. These plasma and magnetic field characteristics have been interpreted in terms of slow-mode wave structures, which can be summarized as follows:

1. The variations in plasma density and magnetic field strength are anti-correlated,

2. The observed waves do not convect with the magnetosheath plasma flow; they seem to stand against the magnetosheath flow,

3. The wavelength (λ_n) normal to the magnetopause is typically 2000-5000 km, while the wavelength (λ_t) tangent to the magnetopause is much larger.

4. A steepened plasma density profile with a shock-like structure also appears in the enhanced density region.

Properties (1) and (2) indicate that the observed waves are slow-mode waves. Observations also indicated that some of these slow-mode waves may come from the interaction of the bow shock and fluctuations in the solar wind.

Two-dimensional MHD simulations of the stagnation flow show that the plasma depletion region is often accompanied with a complementary region of enhanced plasma density and reduced magnetic field intensity if the IMF has a B_x component [*Lee et al.*, 1991]. *Southwood and Kivelson* [1992] suggested that slow shocks can be formed by disturbances near the magnetopause. MHD simulations show that the interaction of the interplanetary MHD waves with the bow shock may lead to the observed slow-mode waves in the inner region of the dayside magnetosheath [*Yan and Lee*, 1994]. The transmission of any MHD wave through the bow shock generally produces fast, Alfvén, and slow mode waves within the magnetosheath (see Figure 4). As the slow mode waves approach the magnetopause, their propagation speed is reduced. Both the flow speed and the slow-mode phase speed normal to the magnetopause decrease to zero at the magnetopause, which allows the slow-mode waves to stay in front of the magnetopause for a long time (over 15 minutes) before the wave energy is finally convected tailward. Since MHD waves are common in the solar wind, this mechanism may lead to the frequent appearance of slow-mode waves in front of the magnetopause. As shown in Figure 5, these waves will have the observed anti-correlated variations in plasma pressure P and magnetic field **B** along the sun-earth line. It is interesting to note that the interaction of Alfvén waves (no density variations) with the bow shock also leads to the generation of slow-mode waves with density variations in the downstream magnetosheath.

4. MAGNETIC RECONNECTION

When the IMF and geomagnetic fields have anti-parallel components, the magnetic energy may be efficiently converted into kinetic energy through magnetic reconnection. The basic geometry of reconnection is illustrated in Figure 6. The field line "a", originally located in domain 3, moves toward the separatrix surface and lies in that surface at location "b". When reconnection occurs, the field lines at the separator break into components "c" and "d", located in domains 1 and 2, respectively. The term "separator" is synonymous with "reconnection line", "merging line", or "X line" for the general case in which a component of the magnetic field may lie along the separator. The phrases "neutral line" and "null line" are also used for the special case in which there is no component of the magnetic field along the separator.

Mathematically, the magnetic reconnection process is governed by the magnetic induction equation

$$\frac{\partial \mathbf{B}}{\partial t} = \nabla \times (\mathbf{v} \times \mathbf{B}) + \frac{\eta}{\mu_0} \nabla^2 \mathbf{B} \qquad (2)$$

where **B** is the magnetic field strength, t is time, **v** is the plasma flow velocity, μ_0 is the magnetic permeability, and η is the resistivity. The basic magnetic field configuration for the reconnection process consists of two regions:

1) A convection region in which the first term on the right hand side of (2) dominates because the resistivity is nearly zero, which implies that the plasma is "frozen" to the magnetic field lines, and

2) A diffusion region with finite resistivity in which the "frozen-in" condition is violated and the magnetic field lines can diffuse through the plasma and change topology.

At the dayside magnetopause, collisions between particles are relatively infrequent and conventional resistivity is nearly zero. However, particle inertias or microscale instabilities at the magnetopause may produce sufficient "anomalous" resistivity at the magnetopause to allow magnetic reconnection to occur.

Several 2-D models have been proposed for steady-state reconnection ($\partial \mathbf{B}/\partial t = 0$). In the Sweet-Parker model, the oppositely-directed magnetic field lines are carried toward the diffusion region, which is represented as a current sheet as illustrated in Figure 7a [*Parker*, 1957; *Sweet*, 1958]. However, the reconnection rate predicted by this model is limited by the time required for the magnetic field to diffuse through the diffusion region. The Petschek model, shown in Figure 7b, removes this limitation by reducing the size of the diffusion region and relying on two pairs of slow shocks to accelerate the plasma from the inflow to the outflow regions [*Petschek*, 1964].

The vast literature on magnetic reconnection processes has been extensively reviewed over the past two decades [*Vasyliunas*, 1975; *Sonnerup*, 1979; *Forbes and Priest*, 1987; *Lee*, 1991, 1995]. MHD simulations have been used very successfully in the study of magnetic reconnection by explicitly specifying a value of anomalous resistivity. In

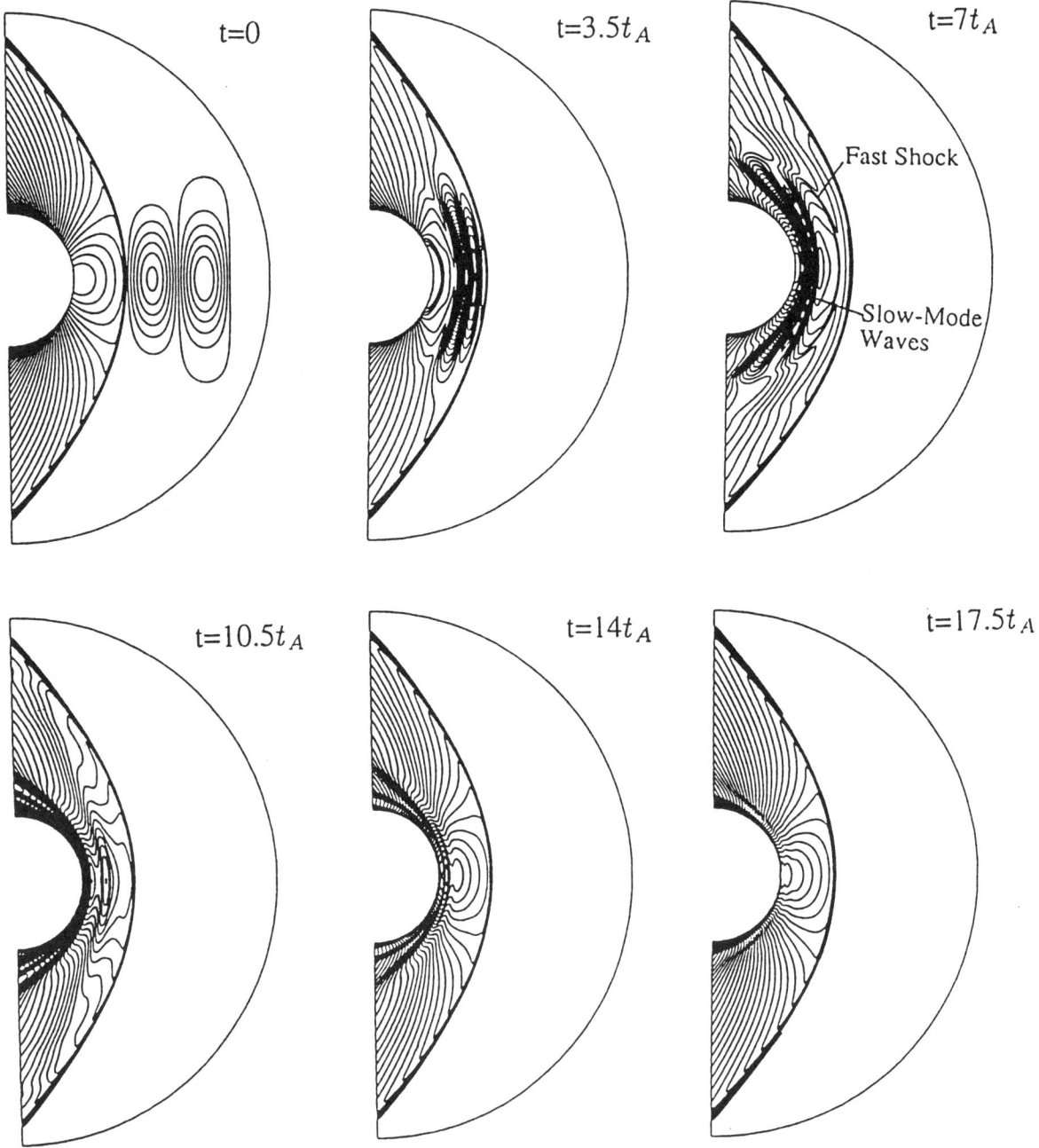

Fig. 4. The evolution of plasma density contours between the bow shock and magnetopause show that MHD waves in the solar wind will generate fast, Alfvén, and slow waves in front of the magnetopause [*Yan and Lee*, 1994].

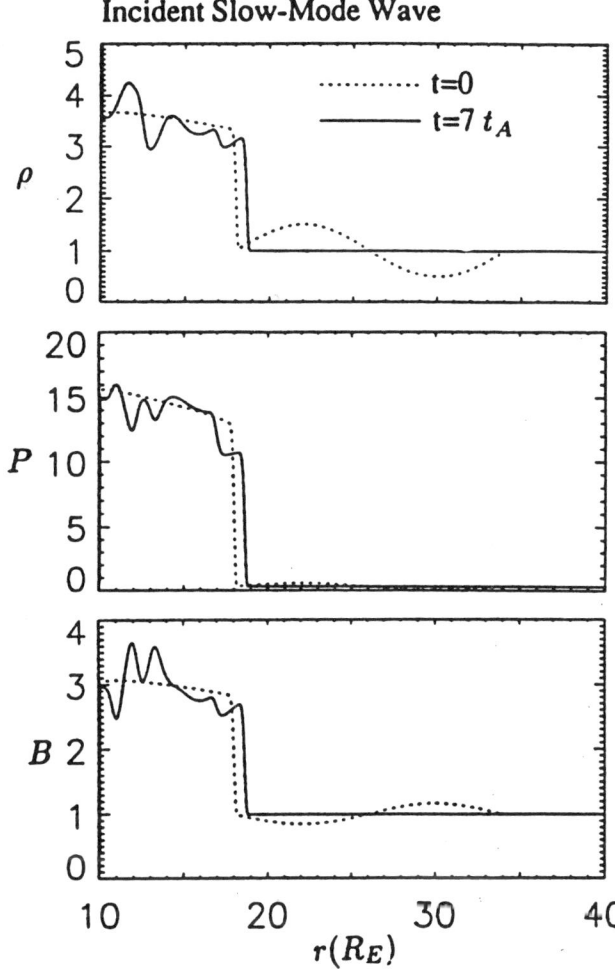

Fig. 5. Profiles of plasma pressure P, and magnetic field $\mathbf{B}$ along the sun-earth line at $t=0$ (dashed lines) and $t=7t_A$ (solid lines) for contours shown in Figure 4 [*Yan and Lee*, 1994].

this section, we emphasize the aspects of reconnection that are influenced by microscale processes which are better represented by hybrid and particle simulation codes.

In the Earth's magnetosphere, reconnection can take place at the dayside magnetopause and at the flank of the magnetopause. The reconnection layer is a layered structure containing several MHD discontinuities and expansion waves in the high speed outflow region during quasi-steady reconnection. These reconnection layers have been studied by solving the Riemann problem using the ideal MHD [*Heyn et al.*, 1988; *Biernat et al.*, 1989], resistive MHD, and hybrid models [*Lin and Lee*, 1994]. In these studies, Petschek's symmetric reconnection model is generalized by including an asymmetry in the magnetic field and plasma density, a finite guide-field (B_y), the ion kinetic effect, and a shear plasma flow in the magnetosheath. At the dayside magnetopause, the reconnection layer is bounded on the magnetosheath side by a large rotational discontinuity that can be identified as the magnetopause current sheet. Accelerated plasma flow is present in the boundary layer earthward of the magnetopause current sheet. Due to the mixing of plasmas from the magnetosheath and magnetosphere, an increase in the ion temperature ratio $T_{i\parallel}/T_{i\perp}$ is found in the boundary layer. At the flank of the magnetopause, the large magnetosheath plasma flow results in a thicker field transition region in the reconnection layer. These results are consistent with satellite observations, as shown in Figure 8.

The particle dynamics near a magnetic X line is an important issue in the study of collisionless reconnection. In the fluid description [e.g., *Krall and Trivelpiece*, 1986], the generalized Ohm's law can be written as

$$\mathbf{E} + \frac{\mathbf{v}}{c} \times \mathbf{B} = \eta \mathbf{J} + \frac{1}{nec} \mathbf{J} \times \mathbf{B} \\ + \frac{m_e}{ne}\left[\frac{\partial \mathbf{J}}{\partial t} + \nabla \cdot (\mathbf{vJ} + \mathbf{Jv})\right] - \frac{1}{ne}\nabla \cdot \mathbf{P}^{(e)} \quad (3)$$

where m_e and e are electron mass and charge respectively, c is the velocity of light, n, $\mathbf{v}$, and $\mathbf{J}$ are plasma density, flow velocity, and current density respectively, $\mathbf{P}$ is the electron pressure tensor, $\mathbf{E}$ and $\mathbf{B}$ are the electric field and magnetic field, and η is the resistivity. In resistive MHD theory, the generalized Ohm's law is simplified to include only the following terms: $\mathbf{E} + (\mathbf{v}/c) \times \mathbf{B} = \eta \mathbf{J}$. However, the resistivity due to binary collisions is negligible ($\eta \approx 0$) in collisionless plasmas. Therefore, in order to understand magnetic reconnection in collisionless plasmas, one has to search for dissipation mechanisms other than binary collisions. An effective anomalous resistivity may arise

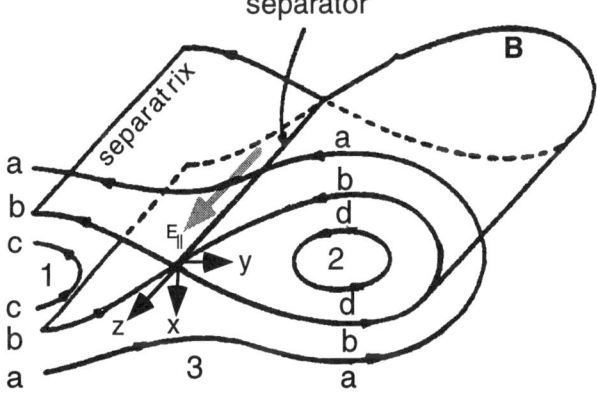

Fig. 6. Basic geometry for the magnetic field line reconnection process [*Sonnerup*, 1985].

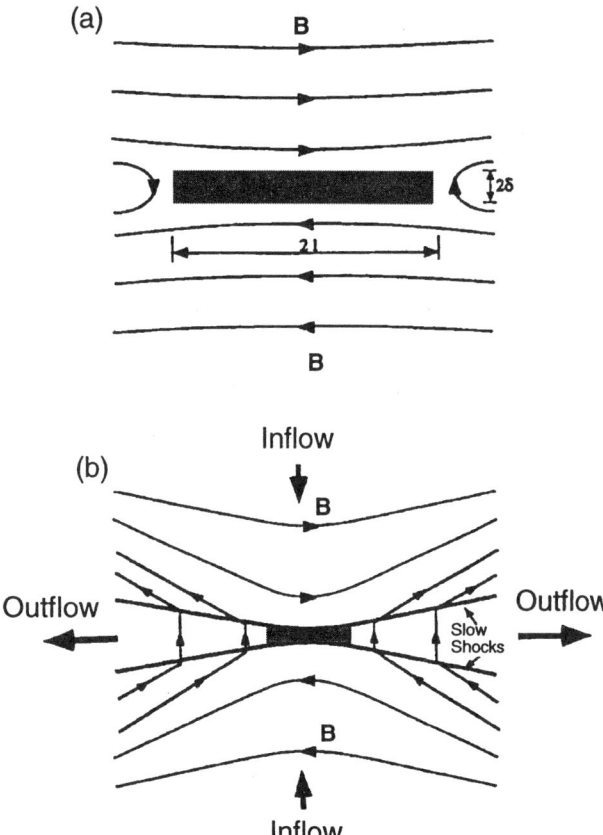

Fig. 7. (a) A schematic sketch of Sweet-Parker's reconnection model. (b) Petschek's symmetric reconnection model which consists of the inflow region, the outflow region, and the small central diffusion region. Pairs of slow shocks are present in each part of the outflow region.

from particle inertia effects, wave particle interactions, or nonlinear particle dynamics.

Using two dimensional particle simulations, *Cai et al.* [1994] recently demonstrated that the off-diagonal pressure terms provide an important inertia effect in the generalized Ohm's law. Along the X-line where $\mathbf{v} \times \mathbf{B} = 0$, the force balance equation can be written as

$$E_y = \frac{m_i}{e}\frac{\partial v_y}{\partial t} + \frac{1}{ne}\frac{\partial P_{xy}}{\partial x} + \frac{1}{ne}\frac{\partial P_{zy}}{\partial z} \qquad (4)$$

Figure 9 shows the time evolution of each term in the force balance equation at the X line. In Figure 9a, the dotted line corresponds to the y-momentum transport in the x direction $(1/ne)\partial P_{xy}/\partial x$, the dashed line corresponds to the y-momentum transport in the z direction $(1/ne)\partial P_{zy}/\partial z$, and the solid line is the inertial term $(m_i/e)\partial v_y/\partial t$. The solid line in Figure 9b shows the sum of these three terms, whereas the dashed line in Figure 9b shows the evolution of the reconnection electric field E_y. It can be clearly seen that the reconnection electric field is approximately balanced by the sum of the three terms in Figure 9a.

5. ANOMALOUS TRANSPORT ACROSS THE MAGNETOPAUSE

Axford and Hines [1961] proposed viscous interaction as a mechanism by which particles and momentum from the solar wind may be transferred across the magnetopause to the magnetosphere. The anomalous viscosity responsible for this diffusion process may arise from lower-hybrid turbulence [*Labelle and Treumann*, 1988; *Gary and Sgro*, 1990; *Treumann et al.*, 1992] or whistler mode waves driven by current gradients [*Drake et al.*, 1994]. Kelvin Helmholtz turbulence provides a mechanism for the transport of momentum across the boundary [*Miura*, 1984]. The streaming energy of the shocked solar wind flowing past the magnetospheric plasma, and the gradients in the density and fields, provide sources of free energy to drive the wave instabilities in this region. Observations of enhanced fluctuations in magnetic and electric fields in the magnetopause boundary layer region offer evidence for these processes [*Gurnett et al.*, 1979; *Tsurutani et al.*, 1981; *Anderson et al*, 1982]. However, the observed wave power at the ion cyclotron and lower hybrid frequencies seem to be insufficient to provide the necessary diffusion coefficient of 10^9 m²/s [*Treumman et al.*, 1992].

As the shocked solar wind with variable plasma density and magnetic field impinges on the dayside magnetopause, it is likely to generate large-scale Alfvén waves at the solar wind magnetosphere interface. However, large gradients in the density and magnetic field at the magnetopause boundary effectively couple large-scale Alfvén waves with kinetic Alfvén waves [*Hasegawa and Chen*, 1975; *Hasegawa*, 1976; *Eastman and Hones*, 1979]. The conversion of wave power into kinetic Alfvén waves may play an important role in plasma transport at the dayside magnetopause and in electron acceleration along field lines [*Lee et al.*, 1994]. The transport can occur because, unlike the magnetohydrodynamic (MHD) shear Alfvén wave, the kinetic Alfvén wave has an associated parallel electric field which breaks down the "frozen-in" condition and decouples the plasma from the magnetic field lines. This diffusion process can occur for a single KAW with a fixed frequency, which is distinct from the case of lower-hybrid and Kelvin-Helmholtz turbulence in which there is a broad spectrum of wave power at low frequencies. In Figure 10, we illustrate the displacement of the plasma and magnetic field. The

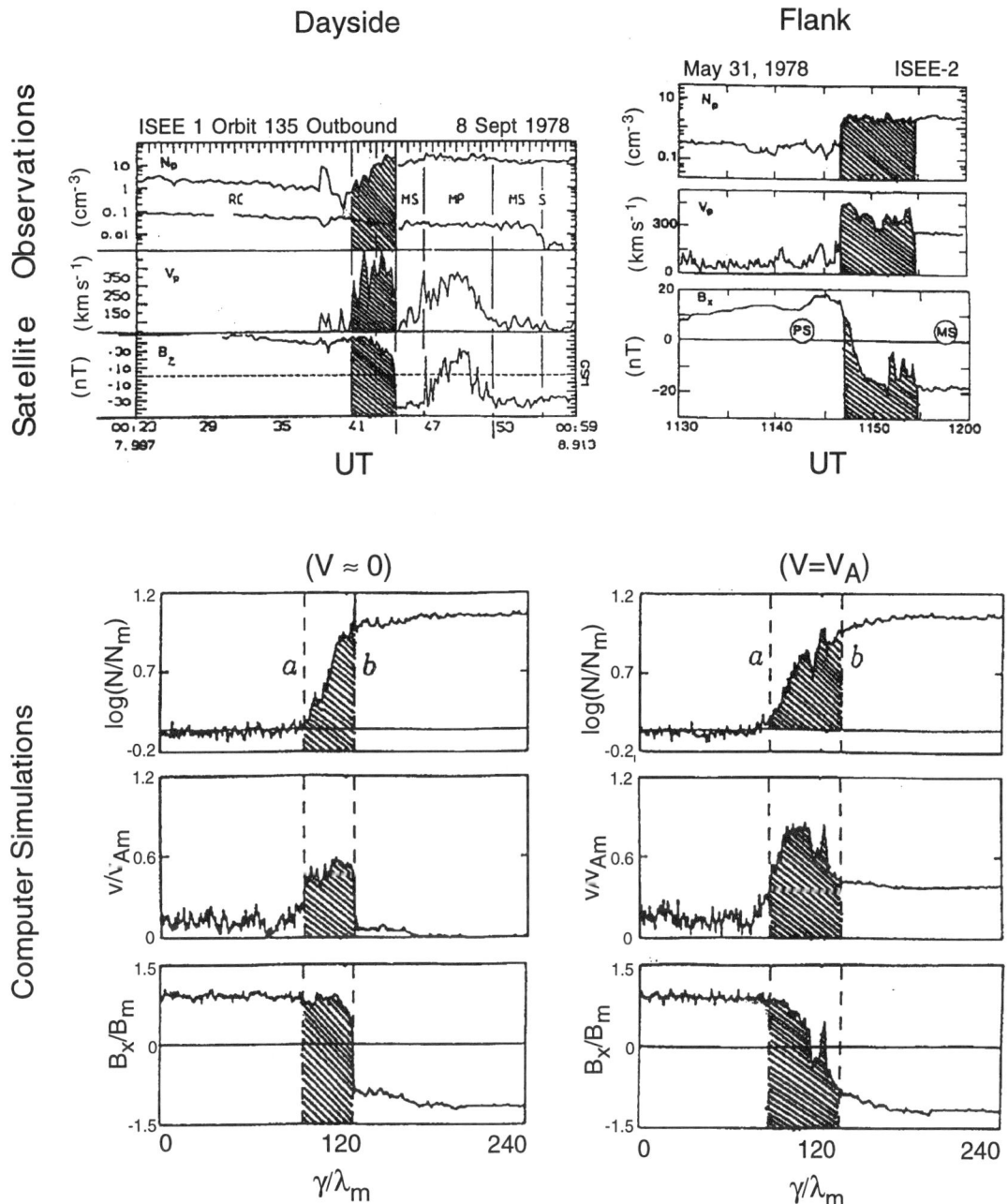

Fig. 8. Satellite observations [*Sonnerup et al.*, 1981; *Gosling et al.*, 1986] and hybrid simulations [*Lin and Lee*, 1994] of reconnection layers at (a) the dayside magnetopause, and (b) the flank of the magnetopause. The magnetospheric side and magnetosheath side of the boundary layer are indicated by "a" and "b", respectively, and v_{AM} is the Alfvén speed in the magnetosphere.

magnetic field is more deformed than the plasma so that, in effect, plasma is left behind on the magnetospheric side of the magnetopause. Because the scale length of the KAWs is the order of ρ_i, the waves can easily couple with particles to provide the diffusion coefficient of 10^9 m^2/s required to

explain the observed transport. As a result, filaments of plasma can enter the magnetosphere with a typical cross-field velocity given by the relative displacement between the field lines and plasma in bundles v_D which has a characteristic scale size Δx. An estimate for the diffusion

Force Balance and Momentum Transport

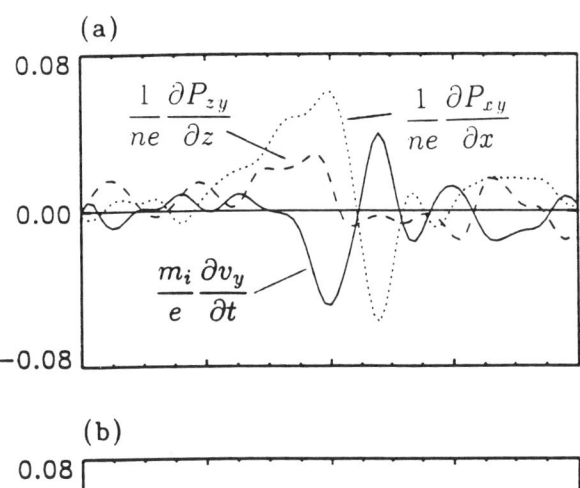

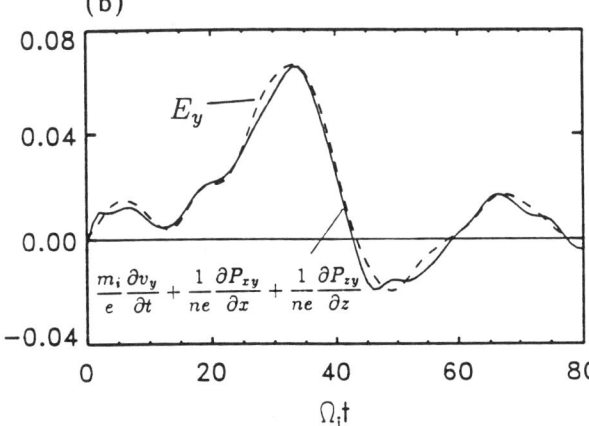

Fig. 9. (a) Time evolution of the terms in the force balance equation: the $(1/ne)\partial P_{xy}/\partial x$ term is represented by a dotted line, the $(1/ne)\partial P_{zy}/\partial z$ term is represented by a dashed line, and the $(m/e)\partial v_y/\partial t$ term is represented by a solid line. (b) Time history of the force balance at the X line: The sum of the three terms in (a) is plotted with a solid line and the reconnection electric field E_y is plotted with a dashed line [Cai et al, 1994].

coefficient D resulting from this process is

$$D \sim v_D \Delta x \sim 2\left(\frac{T_e}{T_i}k_x^2\rho_i^2\right)^2 \frac{v_A}{k_\parallel}\left(\frac{\delta B_x}{B_0}\right)^2 \quad (5)$$

where δB_x is the magnetic flux at the Alfvén frequency [Lee et al., 1994]. For the typical parameters discussed above, the diffusion is estimated to be of the correct order of magnitude (10^9 m^2/s) to explain the observed transport.

6. IMPULSIVE PENETRATION

The impulsive penetration model suggests that magnetosheath plasma may penetrate into the magnetopause when a filament or plasmoid with enhanced momentum impinges on the magnetopause [Lemaire et al., 1979; Lundin and Dubinin, 1984; Lemaire, 1985; Heikkila, 1982]. The geometry of the basic impulsive penetration process is shown in Figure 11 [Schindler, 1979]. If an IMF flux tube has sufficient excess momentum to penetrate deeply into the magnetosphere, then reconnection may be triggered in the current sheet behind the filament, thereby trapping it in the magnetosphere. Thus, impulsive penetration is dependent on the same microscale processes that govern magnetic reconnection, as discussed in section 4.

Two-dimensional MHD and hybrid simulations show that the penetration threshold is likely to be limited to cases when the magnetic fields are nearly parallel or anti-parallel [Ma et al., 1991; Savoini et al., 1994]. When the penetration threshold is exceeded, the simulation results show that twin vortices are generated both within the

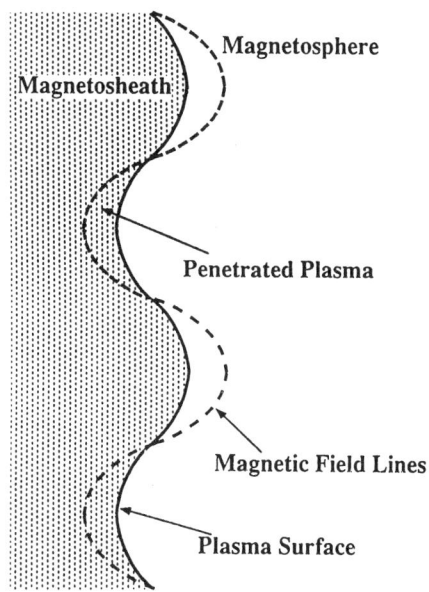

Fig. 10. Plasma penetration into the magnetosphere may occur when the magnetic field decouples from the plasma at the magnetopause boundary. As shown by the dashed line, the magnetic field displacement exceeds the plasma displacement and leaves plasma behind in the magnetosphere. Sudden damping of the wave or particle motion along the magnetic field can lead to plasma transport across field lines [Lee et al., 1994].

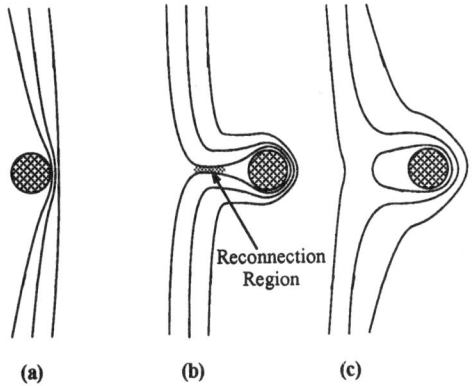

Fig. 11. A schematic of the 2-dimensional reconnection process: (a) The shaded circle represents the cross-section of an infinite filament (flux tube) of enhanced plasma density (momentum) that encounters a transverse magnetic field component; (b) If the excess momentum is sufficiently large, a current sheet (reconnection region) will form behind the filament, which (c) triggers reconnection between the antiparallel magnetic field lines (adapted from *Schindler* [1979]).

filament and in the magnetospheric plasma (see Figure 12). If the momentum of the impinging filament is not sufficient to penetrate into the magnetosphere, then the impact will still perturb the magnetopause and generate pairs of vortices within the magnetosphere. Although the irregularities rebound from the magnetopause surface, the generated vortices may twist the geomagnetic field lines and generate Alfvén waves inside the magnetosphere, which in turn may generate field aligned currents that propagate to the polar ionosphere. The simulations show that the threshold for penetration is given by

$$\frac{\frac{1}{2}\rho_F v_F^2}{\frac{B_{2y}^2}{2\mu_0}} = 50 \qquad (6)$$

where ρ_F is the plasma density of the filament, v_F is the initial velocity of the filament, and B_{y0} is the transverse component of the geomagnetic field that opposes penetration. Therefore, the threshold for penetration requires that the kinetic energy density of the filament must exceed the magnetic energy density of the transverse component of the magnetic field by a factor of 50.

The impact of a dynamic pressure pulse ($\Delta\rho v^2$) in the solar wind with the Earth's bow shock leads to the formation of a weak fast mode shock propagating into the magnetosheath [*Mandt and Lee*, 1991]. The shock can pass right through a tangential discontinuity at the magnetopause and into the magnetosphere without disturbing either structure. In a quasi-perpendicular geometry, the shock wave exhibits anisotropic heating of the ions with $T_{i\perp} > T_{i\parallel}$ as discussed in section 2. A contour plot of the expected anisotropy as a function of the Alfvén Mach number and plasma beta is shown in Figure 13. This anisotropy drives unstable ion cyclotron waves which may contribute to the generation of the Pc 1 waves [*Erlandson et al.*, 1990].

Table 1. Summary of Mesoscale-Microscale Coupling Processes in the Magnetosheath / Boundary Layer Regions

Category	Mesoscale Features	Microscale Features
Bow Shock	Quasi-Perpendicular Shock	$T_{i\perp} > T_{i\parallel}$ Mirror Waves / Lion Roars Ion Cyclotron Waves
Pileup of IMF	Slow Mode Structures Plasma Depletion Layer	$T_{i\perp} > T_{i\parallel}$ Ion Cyclotron Waves
Antiparallel Magnetic Fields	Magnetic Reconnection Reconnection Layers	Anomalous Resistivity and Diffusion
Viscous Interactions	Boundary Layers Kinetic Alfvén Waves	Plasma Transport and Diffusion
Dynamic Pressure Pulses	Impulsive Penetration Weak fast mode shocks	Anomalous Resistivity $T_{i\perp} > T_{i\parallel}$ --> Pc 1 Waves

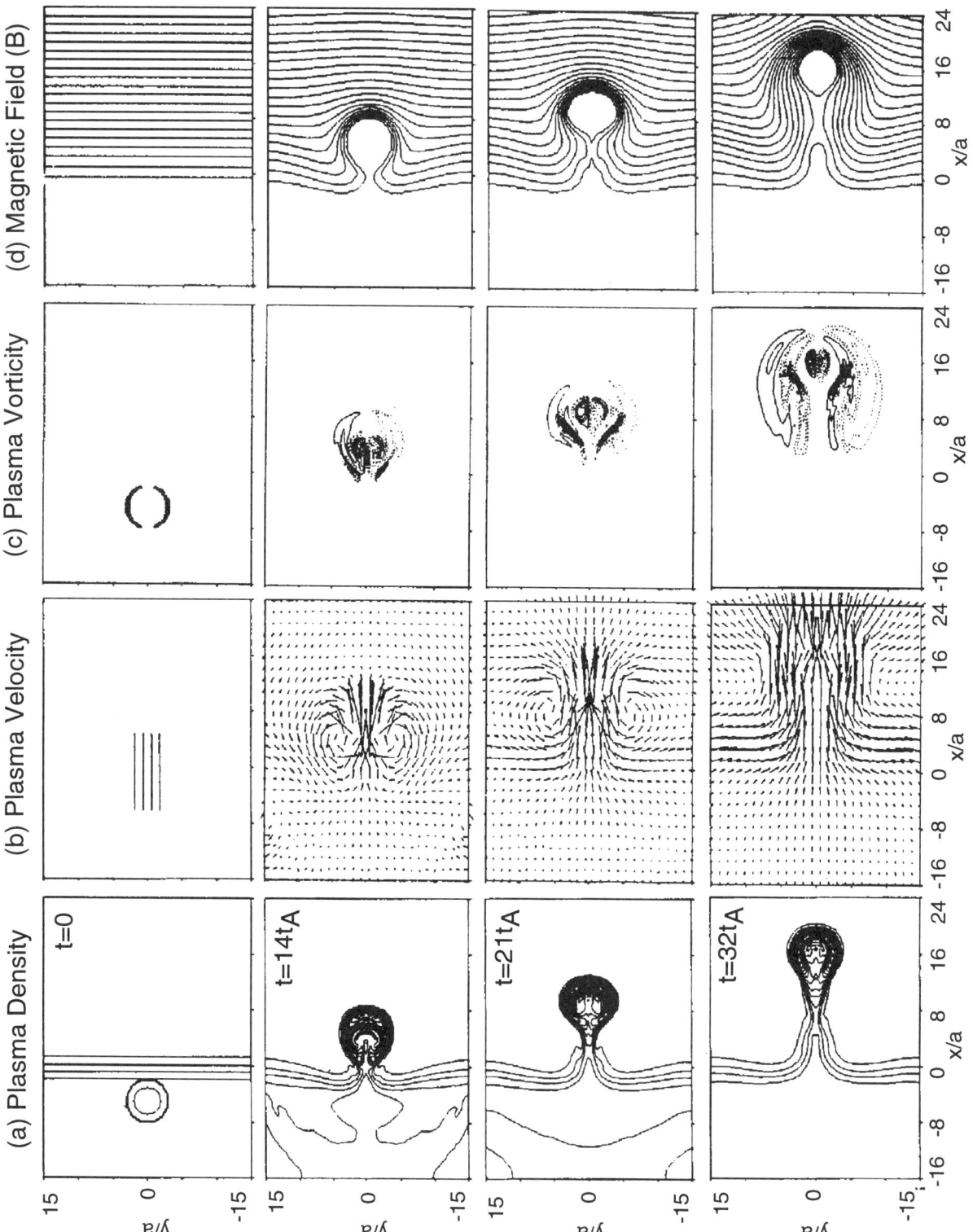

Fig. 12. Contours for impulsive penetration showing (a) plasma density, (b) plasma velocity, (c) vorticity, and (d) magnetic vector potential in the simulation plane [Ma et al., 1991].

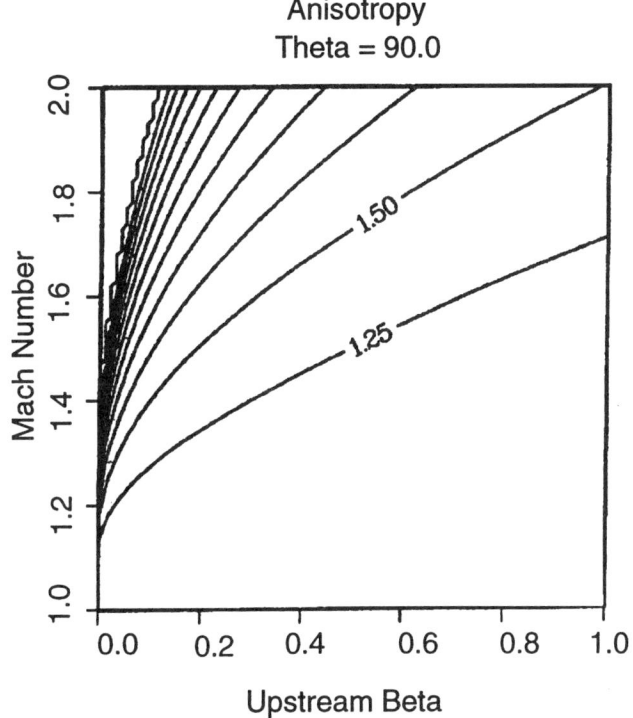

Fig. 13. Contour plot of the expected anisotropy when a dynamic pressure pulse of the given Alfvén Mach number and plasma beta impact with the Earth's bow shock. Contours are equally spaced at intervals of 0.25, and values of the anisotropy greater than 4.0 have been truncated to 4.0 [*Mandt and Lee*, 1991].

7. SUMMARY

Many features of the bow shock, magnetosheath, magnetopause, and boundary layer can be at least qualitatively understood from MHD theory and simulations. However, there are many additional features in which the mesoscale features of these regions drive microscale processes that are not well-represented by MHD theory. As summarized in Table 1, we have reviewed the processes that depend on this mesoscale/microscale coupling.

Acknowledgements. This work was supported by Department of Energy grant DE-FG06-91ER-13530 and National Aeronautics and Space Administration NASA/SPTP grant NAG5-1504 to the University of Alaska. Supercomputer resources were provided by grant ATM900002P from the Pittsburgh Supercomputing Center.

REFERENCES

Anderson, B. J., S. A. Fuselier, S. P. Gary, and R. E. Denton, Magnetic spectral signatures in the Earth's magnetosheath and plasma depletion layer, *J. Geophys. Res.*, *99*, 5877, 1994.

Anderson, R. R., C. C. Harvey, M.M. Hoppe, B. T. Tsurutani, T. E. Eastman, and J. Etcheto, Plasma waves near the magnetopause, *J. Geophys. Res.*, *87*, 2087, 1982.

Axford, W. I., and C. O. Hines, A unifying theory of high-latitude geophysical phenomena and geomagnetic storms, *Can. J. Phys.*, *39*, 1433, 1961.

Biernat, H. K., M. F. Heyn, R. P. Rijnbeek, V. S. Semenov, and C. J. Farrugia, The structure of reconnection layers: Application to the Earth's magnetopause, *J. Geophys. Res.*, *1*, 287, 1988.

Cai, H. J., D. Q. Ding, and L. C. Lee, Momentum transport near a magnetic X line in collisionless reconnection, *J. Geophys. Res.*, *99*, 35, 1994.

Crooker, N. U., and G. L. Siscoe, A mechanism for pressure anisotropy and the mirror instability in the dayside magnetosheath, *J. Geophys. Res.*, *82*, 185, 1977.

Crooker, N. U., T. E. Eastman, and G. S. Stiles, Observation of plasma depletion in the magnetosheath at the dayside magnetopause, *J. Geophys. Res.*, *84*, 869, 1979.

Denton, R. E., S. P. Gary, X. Li, B. J. Anderson, J. W. LaBelle, and M. Lessard, Low-frequency fluctuations in the magnetosheath near the magnetopause, *J. Geophys. Res.*, in press, 1995.

Drake, J. F., J. Gerber, and R. G. Kleva, Turbulence and transport in the magnetopause current layer, *J. Geophys. Res.*, *99*, 11117, 1994.

Eastman, T. E., and E. W. Hones Jr., Characteristics of the magnetospheric boundary layer and magnetopause layer as observed by IMP 6, *J. Geophys. Res.*, *84*, 2019, 1979.

Erlandson, R. E., L. J. Zanetti, T. A. Potemra, L. P. Block, and G. Homgren, Viking magnetic and electric field observations of Pc 1 waves at high latitudes, *J. Geophys. Res.*, *95*, 5941, 1990.

Forbes, T. G., and E. R. Priest, A comparison of analytical and numerical models for steadily driven magnetic reconnection, *Rev. Geophys.*, *25*, 1583, 1987.

Gary, S. P., An upper bound for the proton temperature anisotropy, in *Proceedings of the Workshop on "Coupling of Micro- and Mesoscale Processes in Space Plasma Transport"*, this issue, 1995.

Gary, S. P., and M. A. Lee, The ion cyclotron instability and the inverse correlation between proton anisotropy and proton beta, *J. Geophys. Res.*, *99*, 11297, 1994.

Gary, S. P., and A. G. Sgro, The lower hybrid drift instability at the magnetopause, *Geophys. Res. Lett.*, *17*, 909, 1990.

Gary, S. P., M. E. McKean, D. Winske, B. J. Anderson, R. E. Denton, and S. A. Fuselier, The proton cyclotron instability and the anisotropy / β inverse correlation, *J. Geophys. Res.*, *99*, 5903, 1994.

Gosling, J. T., M. F. Thomsen, S. J. Bame, and C. T. Russell, Accelerated plasma flows at the near-tail magnetopause, *J. Geophys. Res.*, *91*, 3029, 1986.

Gurnett, D. A., R. R. Anderson, B. T. Tsurutani, E. J. Smith, G. Paschmann, G. Haerendel, S. J. Bame, and C. T. Russell, Plasma wave turbulence at the magnetopause: Observations from ISEE 1 and 2, *J. Geophys. Res.*, *84*, 7043, 1979.

Hasegawa, A, Particle acceleration by MHD surface wave and formation of aurora, *J. Geophys. Res.*, *81*, 5083, 1976.

Hasegawa, A., and L. Chen, Kinetic process of the plasma heating by Alfvén wave, *Phys. Rev. Lett.*, *35*, 370, 1975.

Heikkila, W. J., Impulsive plasma transport through the magnetopause, *Geophys. Res. Lett.*, *9*, 159, 1982.

Heyn, M. F., H. K. Biernat, R. P. Rijnbeek, and V. S. Semenov, The structure of reconnection layer, *J. Plasma Phys.*, *40*, 235, 1988.

Hubert, D., Nature and origin of wave modes in the dayside magnetosheath, *Adv. Space Res.*, in press, 1994.

Hubert, D., C. Perche, C. C. Harvey, C. Lacombe, and C. T. Russell, Observation of mirror waves downstream of a quasi-perpendicular shock, *Geophys. Res. Lett.*, *16*, 159, 1989.

Krall, N. A., and A. W. Trivelpiece, *Principles of Plasma Physics*, San Francisco Press (reprint of original published by McGraw Hill), 1986.

Labelle, J., and R. A. Treumann, Plasma waves at the dayside magnetopause, *Space Sci. Rev.*, *45*, 175, 1988.

Lee, L. C., A review of magnetic reconnection, in *Proceeding of the 1994 Chapman Conference on Physics of the Magnetopause*, AGU Monograph, in press, 1995.

Lee, L. C., The magnetopause: a tutorial review, in the *Physics of Space Plasmas (1990), SPI Conference Proceedings and Preprint Series, Volume 10*, T. Chang, G. B. Crew, and J. P. Jasperse, eds., Scientific Publishers, Inc., Cambridge, MA, 1991.

Lee, L. C., C. S. Wu, and X. W. Hu, Increase of ion kinetic temperature across a collisionless shock: I. A new mechanism, *Geophys. Res. Lett.*, *13*, 209, 1986.

Lee, L. C., C. S. Wu, and C. P. Price, On the generation of magnetosheath lion roars, *J. Geophys. Res.*, *92*, 2343, 1987a.

Lee, L. C., M. E. Mandt, and C. S. Wu, Increase of ion kinetic temperature across a collisionless shock: II. A simulation study, *J. Geophys. Res.*, *92*, 13438, 1987b.

Lee, L. C., C. P. Price, and C. S. Wu, A study of mirror waves generated downstream of a quasi-perpendicular shock, *J. Geophys. Res*, *93*, 247, 1988.

Lee, L. C., M. Yan, and J. G. Hawkins, A study of slow-mode structures in the dayside magnetosheath, *Geophys. Res. Lett.*, *18*, 381, 1991.

Lee, L. C., J. R. Johnson, and Z. W. Ma, Kinetic Alfvén waves as a source of plasma transport at the dayside magnetopause, *J. Geophys. Res.*, *99*, 17405, 1994.

Lees, L., Interaction between the solar wind and the geomagnetic cavity, *AIAA J.*, *2*, 1576, 1964.

Lemaire, J., Plasmoid motion across a tangential discontinuity (with application to the magnetopause), *J. Plasma Phys.*, *33*, 425, 1985.

Lemaire, J., M. J. Rycroft, and M. Roth, Control of impulsive penetration of solar wind irregularities into the magnetosphere by the interplanetary magnetic field direction, *Planet. Space Sci.*, *27*, 47, 1979.

Leroy M. M., D. Winske, C. C. Goodrich, C. S. Wu, and K. Papadopoulos, The structure of perpendicular bow shocks, *J. Geophys. Res.*, *87*, 5081, 1982.

Lin, Y., and L. C. Lee, Structure of reconnection layers in the magnetosphere, *Space Sci. Rev.*, *65*, 59, 1994.

Lundin, R., and E. Dubinin, Solar wind energy transfer regions inside the dayside magnetopause, 1, Evidence for magnetosheath plasma penetration, *Planet. Space Sci.*, *32*, 745, 1984.

Ma, Z. W., J. G. Hawkins, and L. C. Lee, A simulation study of impulsive penetration of solar wind irregularities into the magnetosphere at the dayside magnetopause, *J. Geophys. Res.*, *96*, 15751, 1991.

Mandt, M. E., and L. C. Lee, Generation of Pc 1 waves by the ion temperature anisotropy associated with fast shocks caused by sudden impulses, *J. Geophys. Res.*, *96*, 17897, 1991.

McKean, M. E., D. Winske, and S. P. Gary, Two-dimensional simulations of ion anisotropy instabilities in the magnetosheath, *J. Geophys. Res.*, *99*, 11141, 1994.

Miura, A., Anomalous transport by magnetohydrodynamic Kelvin-Helmholtz instabilities in the solar wind-magnetosphere interaction, *J. Geophys. Res.*, *89*, 801, 1984.

Parker, E. N., Sweet's mechanism for merging magnetic field in conducting fluids, *J. Geophys. Res.*, *62*, 509, 1957.

Paschmann, G., N. Sckopke, G. Haerendel et al, ISEE plasma observations near subsolar magnetopause, *Space Sci. Rev.*, *22*, 717, 1978.

Petschek, H. E., Magnetic field annihilation, in AAS-NASA Symposium on the Physics of Solar Flares, *NASA Spec. Publ.*, *SP-50*, 425, 1964.

Phan, T.-D., G. Paschmann, W. Baumjohann, and N. Sckopke, The magnetosheath region adjacent to the dayside magnetopause: AMPTE/IRM observations, *J. Geophys. Res.*, *99*, 121, 1994.

Price, C. P., D. W. Swift, and L.-C. Lee, Numerical simulation of nonoscillatory mirror waves at the Earth's magnetosheath, *J. Geophys. Res.*, *91*, 101, 1986.

Rodriguez, P., Long duration lion roars associated with quasi-perpendicular bow shocks, *J. Geophys. Res.*, *90*, 241, 1985.

Savoini, P., M. Scholer, and M. Fujimoto, Two-dimensional hybrid simulations of impulsive plasma penetration through a tangential discontinuity, *J. Geophys. Res.*, *99*, 19377, 1994.

Schindler, K., On the role of irregularities in plasma entry into the magnetopause, *J. Geophys. Res.*, *84*, 7257, 1979.

Schopke, N., G. Paschmann, S. J. Bame, J. T. Gosling, and C. T. Russell, Evolution of ion distributions across the nearly perpendicular bow shock: specularly and non-specularly reflected gyrating ions, *J. Geophys. Res.*, *88*, 6122, 1983.

Scudder, J. D., A. Mangeney, C. Lacombe, C. C. Harvey, T. L. Aggson, R. R. Anderson, J. T. Gosling, G. Paschmann, and C. T. Russell, The resolved layer of a collisionless, high beta, supercritical, quasi-perpendicular shock wave: 1, Rankine-Hugoniot geometry, currents, and stationarity, *J. Geophys. Res.*, *91*, 11019, 1986.

Smith, E. J., E. Holzer, and C. T. Russell, Magnetic emissions in the magnetosheath at frequencies near 100 Hz, *J. Geophys. Res.*, *74*, 3027, 1969.

Sonnerup, B. U. O., Magnetic field reconnection, in *Solar System Plasma Physics, Vol. III*, p. 45, ed. by C. F. Kennel, L. T. Lanzerotti, and E. N. Parker, North Holland Publishing Company, 1979.

Sonnerup, B. U. O., p. 5 in *Unstable Current System and Plasma Instabilities in Astrophysics*, M. R. Kundu and G. D. Homan, eds, D. Reidel, Dordrecht, Holland, 1985.

Sonnerup, B. U. O., G. Paschmann, I. Papamastorakis, N. Sckopke, G. Haerendel, S. J. Bame, J. R. Asbridge, J. T. Gosling, and C. T. Russell, Evidence for magnetic reconnection at the Earth's magnetopause, *J. Geophys. Res.*, *86*, 10049, 1981.

Song, P., C. T. Russell, J. T. Gosling, M. Thomsen, R. C. Elphic, Observation of the density profile in the magnetosheath near the stagnation streamline, *Geophys. Res. Lett.* *17*, 2035, 1990.

Song, P., C. T. Russell, M. Thomsen, Slow mode transition in the front side magnetosheath, *J. Geophys. Res.*, *97*, 8295, 1992.

Southwood, D. J., and M. G. Kivelson, On the form of the flow in the magnetosheath, *J. Geophys. Res.*, *97*, 2873, 1992.

Sweet, P. A., The neutral point theory of solar flares, in *Electromagnetic Phenomena in Cosmic Physics*, edited by B. Lehnert, p.135, 1958.

Thorne, R. M., and B. T. Tsurutani, The generation mechanism for magnetosheath lion roars, *Nature*, *293*, 384, 1981.

Treumann, R. A., J. LaBelle, G. Haerendal, and R. Pottelette, Anomalous plasma diffusion and the magnetopause boundary layer, *IEEE Trans. Plasma Sci.*, *20*, 833, 1992.

Tsurutani, B. T., E. J. Smith, R. M. Thorne, R. R. Anderson, D. A. Gurnett, G. K. Parks, C. S. Lin, and C. T. Russell, Wave-particle interactions at the magnetopause: Contribution to the dayside aurora, *Geophys. Res. Lett.*, *8*, 183, 1981.

Tsurutani, B. T., E. J. Smith, R. R. Anderson, K. W. Ogilvie, J. D. Scudder, D. N. Baker, and S. J. Bame, Lion roars and nonoscillatory drift mirror waves in the magnetosheath, *J. Geophys. Res.*, *87*, 6060, 1982.

Vasyliunas, V. M., Theoretical models of magnetic field line merging, 1, *Rev. Geophys.*, *13*, 303, 1975.

Winske, D., and B. Quest, Magnetic field and density fluctuations at perpendicular supercritical collisionless shocks, *J. Geophys. Res.*, *93*, 9681, 1988.

Yan, M., and L. C. Lee, Generation of slow-mode waves in front of the dayside magnetopause, *Geophys. Res. Lett.*, *21*, 629, 1994.

Zwan, B. J., and R. A. Wolf, Depletion of solar wind plasma near a planetary boundary, *J. Geophys. Res.*, *81*, 1636, 1976.

L. C. Lee and J. G. Hawkins, Geophysical Institute, University of Alaska, Fairbanks, AK 99775-5900.

Micro/Mesoscale Phenomena in the Dayside Magnetopause: A Tutorial

Paul Song

High Altitude Observatory, National Center for Atmospheric Research, Boulder, Colorado

In this tutorial, we describe the micro/mesoscale phenomena which are often observed in and near the dayside magnetopause. The phenomena can be classified in general as structures and waves. We define a structure within or near the magnetopause as a plasma region which is significantly different from the macroscopic properties of the magnetosheath and magnetosphere. The low latitude portion of the dayside magnetopause is observed with layered structures with some temporal variations. The north-south component of the interplanetary magnetic field (IMF) appears to be the controlling factor of the properties of the structures. The flux transfer events (FTEs) are the most important transient mesoscale phenomenon. We provide some in depth discussion on our current understanding of it. Waves are defined as fluctuations within similar plasmas. The wave activity is particularly rich near and within the magnetopause covering a very broad scale range. The types or modes of the waves appear to depend on the upstream IMF and maybe solar wind conditions. While many macro and mesoscale phenomena are understood to some extent, we know much less about the properties and functions of microscale processes. We briefly describe the most basic data analysis methods.

1. INTRODUCTION

The magnetopause is defined as the outer boundary of the magnetosphere, or the transition region between the shocked solar wind and magnetosphere. From this definition, it is easy to understand its importance: a) Any mass, momentum and energy transfer between the solar wind and magnetosphere must cross the magnetopause and hence the amount of the transfer and the processes in the outer magnetosphere and dayside cusps largely depend on the processes at the magnetopause; b) Because the magnetopause is the interface between two distinct magnetized plasmas, which is extremely difficult to be established in a quasi-steady state in laboratories, understanding of the processes at such an interface is fundamentally important to general plasma physics and to understanding some astrophysical processes.

The magnetopause was first observed and identified from the magnetometer data as a current layer as shown in Figure 1, (see a recent historical review by *Cahill* [1995]). Later a layer within the transition on the earthward side of the current containing magnetosheath particles was identified from plasma measurements [*Akasofu et al.*, 1973; *Eastman et al.*, 1976; *Paschmann et al.*, 1976; *Haerendel et al.*, 1978]. The high latitude portion of this layer has been referred to as the high latitude boundary layer (HBL) or entry layer (EL), and the low latitude portion, the low latitude boundary layer (LLBL). Since most recent observations with advanced particle detectors, such as those onboard ISEE 1 and 2 and AMPTE/IRM, came from low latitudes, the LLBL has been a particularly popular topic in the past seventeen years. As the major missions in the International Solar-Terrestrial Physics (ISTP) program, such as the Cluster and Polar, will be on polar orbits, it is expected that much attention will be focused on the HBL. However, at the present time, the relationship between the LLBL and HBL is not all together clear, although one may guess one of them to be the source region of the other depending on where the major entry of the solar wind particles into the magnetosphere occurs. It may depend on the interplanetary magnetic field (IMF) orientation. All phenomena discussed in this tutorial have been observed in low latitudes.

Over the past more than twenty years, we have collected more than a thousand magnetopause crossings [*Fairfield*,

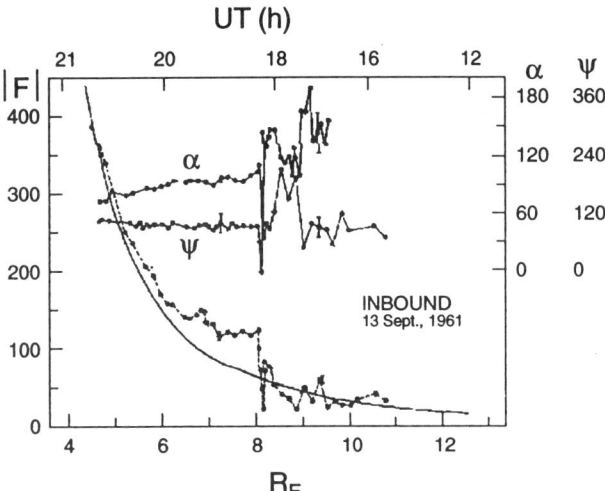

Fig. 1. One of the earliest magnetopause crossings recorded by Explorer 12 [*Cahill and Amazeen*, 1963]. The curve without the symbol is the prediction of the dipole field and the dashed curve with dots is the measured field strength. The magnetopause current layer crossing occurred near 8.2 R_E when the measured field changes its magnitude and direction.

1971; *Song et al.*, 1988; *Petrinec et al*, 1991; *Roelof and Sibeck*, 1993]. To identify a magnetopause crossing is by no means trivial. For example, sometimes in some regions the magnetosheath field is similar to the magnetospheric field in both strength and direction and thus the magnetopause current is extremely weak. In these cases, one may mistakenly pick up a field change upstream of the real magnetopause which is caused by the change in the IMF. The consequence of such a misidentification is that the LLBL would be very thick. However, an experienced data analyst could avoid such a mistake by looking, for example, at the change in the wave pattern at the magnetopause, as will be discussed in the next section. In these cases, confirmation from instruments other than the magnetometer becomes important. Figure 2 shows a set of the carefully identified crossings. One can see that the magnetopause was observed at varying locations. The variation in its radial distance and its openness in the nightside is caused by the solar wind and IMF conditions, and magnetospheric processes [*Petrinec et al.*, 1991; *Roelof and Sibeck*, 1993]. The location and the shape of the magnetopause determine the size of the magnetosphere. When the magnetopause moves to within the geostationary orbit, which is about 6.6 R_E, the high speed solar wind may cause severe damage to certain geosynchronous satellites. An important thing that one needs to keep in mind when looking at a magnetopause observation is that because the spacecraft moves relatively slowly, about 2 km/s near the magnetopause, compared with other characteristic speeds, Alfven speed and sound speed are about $10^2 \sim 10^3$ km/s near the magnetopause, the magnetopause is more often observed when it changes its motion, namely, it is not in a steady state to some extent when it is observed.

Although the phenomena observed near the magnetopause are as temporal variations in a spacecraft record, they can be described in general in terms of moving (relative to the spacecraft) structures and waves. These two different descriptions usually lead to using different theoretical languages and modeling methods, as well as observational techniques and interpretations. Structures could be defined as regions of plasmas with macroscopic properties significantly different from that in the magnetosphere and magnetosheath. Layers within the magnetopause and the slow mode structure in front of the magnetopause are examples of such structures. A structure can be spatial (with slow time evolution) or temporal (with 1-D, 2-D or 3-D spatial extension), depending on the frame of reference in which the structure is discussed. For example, a moving flux tube causes a temporal perturbation in the spacecraft frame and is a 2-D (3-D) spatial structure in the frame moving with it. Waves could be defined as fluctuations within similar plasmas and could be coherent or noncoherent, and quasi-periodic, solitary or random. According to these definitions, convecting mirror mode structures, which are created in the magnetosheath by the mirror mode instability and are characterized by the antiphase oscillation of the density and field strength, are classified as (nonpropagating) waves because the plasma condition is similar among these structures. The magnetopause surface wave is a special type of wave. If a spacecraft locates on one side of the magnetopause, it observes the perturbations the same as the waves that were defined above. If it crosses the magnetopause back and forth, it observes the motion of structures. To apply these definitions is not trivial in data interpretation because data

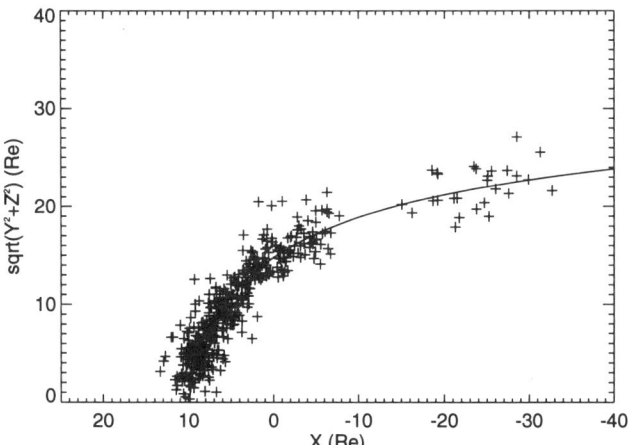

Fig. 2. Location of the magnetopause crossings observed from ISEE 1 and 2, AMPTE/IRM, and IMP 8 [*Shue et al.*, unpublished manuscript, 1995]. The locations of the crossings have been normalized by the solar wind dynamic pressure.

were recorded in only one or two points in space at a time. An example is magnetopause surface waves versus eddies. While both of them are observed as oscillations between distinct plasma regions, the physical implications are radically different: the former implies only minor transfer across the boundary but the latter results in transfer in all three moments. We define eddies as structures because they are entities significantly different from the plasmas on the two sides. A flux tube which may be associated with an FTE is the singular case of the eddies. Actually, the currently most controversial issue in the magnetopause studies which will be discussed in section 5 is rooted in this simple conceptual problem. The forthcoming four-spacecraft Cluster missions will significantly improve this situation.

In this tutorial, we will first take a look at the physical scales and phenomena near the magnetopause (section 2) and then describe the structures (section 3) and waves (section 4) at the magnetopause. In section 5, we discuss the magnetopause surface waves and flux transfer events (FTEs) as examples of the complexity of the magnetopause observations and interpretations. For recent reviews on each of the subjects mentioned above, one is referred to *Russell* [1995] and *Sonnerup et al.* [1995] (for structures), *Song* [1994] and *Anderson* [1995] (for waves), and *Elphic* [1995] (for surface waves and FTEs)

2. CHARACTERISTIC SCALES NEAR THE MAGNETOPAUSE

The tangential scale of the magnetopause, similar to the size of the magnetosphere, is 10^{1-2} R_E. This is referred to as the macro or global scale. The wavelength of large scale magnetopause surface waves is of the order of 10 R_E.

A variety of phenomena occur on a scale of $10^2 \sim 10^3$ km, or from a fraction to one Earth radius. This is about $10^1 \sim 10^2$ ion gyroradii and is usually referred to as the mesoscale. The thickness of the internal layers of the magnetopause, such as the current layer (or the sheath transition layer) and LLBL, is of such a scale. A frequently observed slow mode structure which stands upstream of the magnetopause current layer [*Song et al.*, 1992a] has a scale of about 1 R_E normal to the magnetopause. We will not discuss this phenomenon in this tutorial. Interested readers are referred to *Russell* [1995], *Lee et al.* [1991], *Southwood and Kivelson* [1992, 1995], *Yan and Lee* [1994], and *Omidi and Winske* [1995]. Observations have indicated that the wavelength of the compressional waves adjacent to the magnetopause in the magnetosheath and within the slow mode structure is also in this scale range [*Tsurutani et al.*, 1982; *Song et al.*, 1991]. These compressional waves are part of the semi-permanent features of the magnetosheath near the magnetopause and may play an important role in the magnetosheath-magnetosphere coupling. The scales of the FTE, a transient phenomenon near the magnetopause, in the direction normal to the magnetopause and the direction along its motion which is believed to be tangential to the magnetopause have been observed to be in this scale range. Its scale in the third dimension has not been determined observationally but could be very long in most models. There are some reported events which consist of perturbations that are difficult to interpret. They may be either interpreted as FTEs or eddies but different interpretations will lead to different global consequences.

The magnetosheath ion gyroradius is about 50-100 km and lower hybrid radius (geometric mean of the ion and electron gyroradii) 5 km. The phenomena of these scales or smaller are related to the kinetic processes and are referred to as the microscale phenomena. The sharp edges that separate sublayers within the magnetopause are of this scale range. There are a variety of electromagnetic fluctuations in this scale range. These waves should play key roles in the microprocesses in the magnetopause.

The magnetosheath electron gyroradius is about 500 m. The Debye lengths across the magnetopause are 20 ~ 500 m associated with the density change. Observationally there has been no convincing report on coherent structures in this scale range. Most signals appear to be waves with small coherent length [e.g., *Elphic*, 1988].

Table 1 summarizes the scales and phenomena.

3. STRUCTURES

The structure of the magnetopause is determined by the physical processes. For example, if diffusion is dominant in the transport processes, the magnetopause should appear as gradients in the density, temperature and field (see reviews by *Winske et al.* [1995] and *Treumann et al.* [1995]). Furthermore, one may even postulate the gradients to be different for different species and fields. If massive solar wind irregularities can cross the magnetopause directly as proposed by the so-called impulsive penetration models, one may expect to see isolated magnetosheath plasma blobs in the magnetosphere (see a review by *Roth* [1995]), some of which should be detached from the magnetopause. If reconnection is important at the magnetopause, one may find the phenomena to be sorted according to the IMF direction or local magnetic field shear (see a review by *Lee* [1995]). Here we should point out that because of the existence of the bow shock and magnetosheath, the local sheath field direction, in particular its x component, could be significantly different from the IMF direction. We should also note that because the magnetopause crossing may occur at different locations relative to the reconnection site at the time, the observed signals could be radically different even for the same reconnection model under a similar condition.

One should not be surprised if the interpretation of an observation is not unique. There are basically two approaches to narrow down the possible interpretations. One is statistics and the other detailed multiple spacecraft, multiple instrument case studies. The statistics which averages over many cases under the similar condition

TABLE 1. Characteristic Scales Near the Magnetopause

Length	Physical Scale	Phenomena
20~500 m	Debye Length	(affect E field measurement)
0.5 km	ele gyroradius	no coherent macrostructure. waves.
1 km	ele skin depth	no coherent macrostructure. waves.
5 km	lower hybrid radius	sharp edges? waves.
50~100 km	ion gyroradius	sharp edges, waves.
10^2~10^3 km (0.1~1 Re)	10~100 ion gyroradii	layers, FTEs, waves. eddies?
~10 Re	size of magnetosphere	tangential scale of MP surface waves

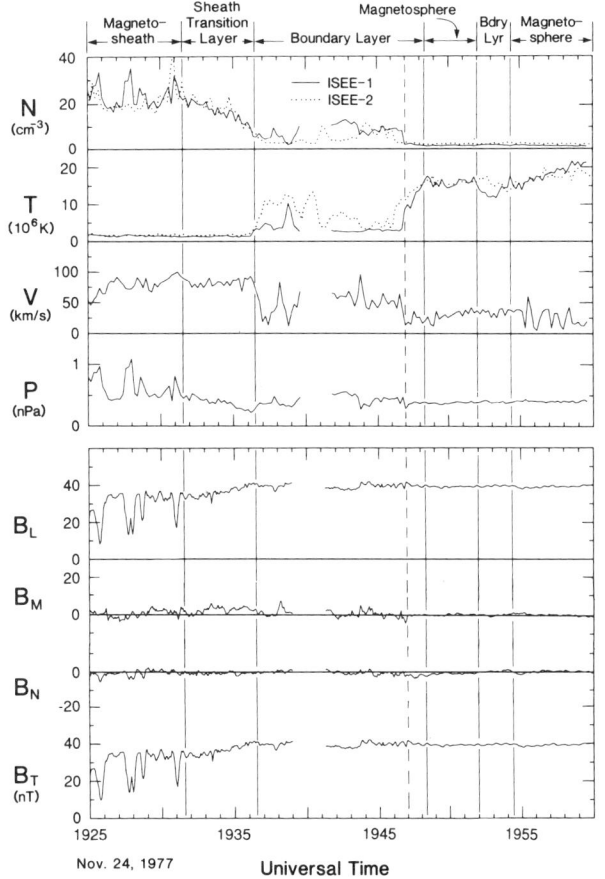

Fig. 3. A magnetopause crossing for strongly northward IMF [*Song et al.*, 1993a]. The crossing occurred at (9.5, -5.9, 5.0) R_E GSM. From the top are ion density, temperature, magnitude of xy 2-D velocity, thermal pressure, three components of the magnetic field and field strength.

provides the representative structural profile and value of a parameter with a setback that some important but sensitive features may be washed out. A multiple instrument case study could eliminate many speculations because they may be inconsistent among information gathered from different sources, but it could represent a rare accident. Both approaches have been and are being taken. The intercalibration between the results from the two is extremely important.

When looking at the data of dayside magnetopause crossings, one would not miss the major differences associated with IMF Bz component or the field shear across the magnetopause. Therefore, spontaneously some groups took the same approaches: study the magnetopause according to IMF Bz or field shear [*Russell and Elphic*, 1978; *Paschmann et al.*, 1978, 1986, 1990, 1993; *Song et al.*, 1989, 1990, 1993a; *Phan et al.*, 1994; *Phan and Paschmann*, 1995].

Figure 3 shows an example of the crossings that occurred for strongly northward IMF. In magnetopause studies, the magnetic field is usually presented in the local boundary normal coordinates, or so-called LMN coordinates [*Russell and Elphic*, 1978]. In this coordinate system, the N direction is outward along the local normal of the magnetopause, the L direction is along the tangential magnetospheric field, and the M direction (=NxL) completes the system. The normal direction is determined by one of the following techniques. The model normal direction is determined assuming that the magnetopause is an empirically determined ellipsoidal surface [e.g., *Fairfield et al.*, 1971; *Petrinec et al.*, 1991]. The tangential discontinuity method assumes that the normal component of the magnetic field on average is zero on both sides of the magnetopause. The minimum variance technique assumes that the magnetopause normal is in the direction along which the field variation is minimum within the magnetopause current layer [*Sonnerup and Cahill*, 1967]. One technique may be more accurate than others under a particular circumstance, but no one is superior all the time. A data analyst must very carefully examine for such differences. If these differences are significant, he/she must decide which technique(s) to use. For the crossing shown in Figure 3, the tangential discontinuity and minimum variance methods give the similar normal direction. The data shows that the field is nearly parallel on the two sides of the magnetopause, and a significant change is the increase in the field strength from the magnetosheath to the magnetosphere in the region we refer to as the sheath transition layer which contains most of the magnetopause current. It would be difficult to identify the magnetopause as a single point for such a crossing. The disappearance of the compressional field fluctuations can be used as a signature of entering the magnetopause region from the magnetosheath. The significant change in the plasma properties, as indicated by the change in the temperature, does not occur until the inner edge of the sheath transition layer in this crossing. Therefore, for this crossing, the magnetopause current is provided by the sheath plasma mainly through the density gradient. The boundary layer is the transition of the plasma from that of magnetosheath origin to that of magnetospheric origin. There are some differences between the measurements from the two spacecraft, which were separated by about 400 km along the magnetopause normal; indicating that some temporal processes may take place at the time. One of the most important features in this crossing is that the boundary layer is relatively uniform and the edge is sharp as indicated by the dashed vertical line. The timing difference between the two spacecraft indicates that the sharp edge is less than 120 km thick. The whole magnetopause transition is about a few thousand kilometers thick.

Song et al. [1993a] studied three crossings for strongly northward IMF with ten different instruments. These three cases show the similar structure and scales. The sheath transition layer starts with the disappearance of the large amplitude compressional waves, appears as a decrease in the density and an increase in the field, and occurs in the sheath plasma. The boundary layer consists of plasmas from both sides of the magnetopause, sometimes with slight heating and can be divided into sublayers. The number of the sublayers may not be definite. Each sublayer is relatively uniform and separated by sharp edges from others. *Paschmann et al.* [1993] and *Phan et al.* [1994] studied the structure statistically. They use the inner edge of the sheath transition layer, or the key time, to align different cases. This key time (more discussion will be given later in this section) marks the most significant change in the plasma properties and is best seen in the electron temperature anisotropy. They found that the sharp edge is about 50 km thick and that both the boundary layer and sheath transition layer are around 1000 km thick. The overall agreement between the two different approaches is remarkable. Even

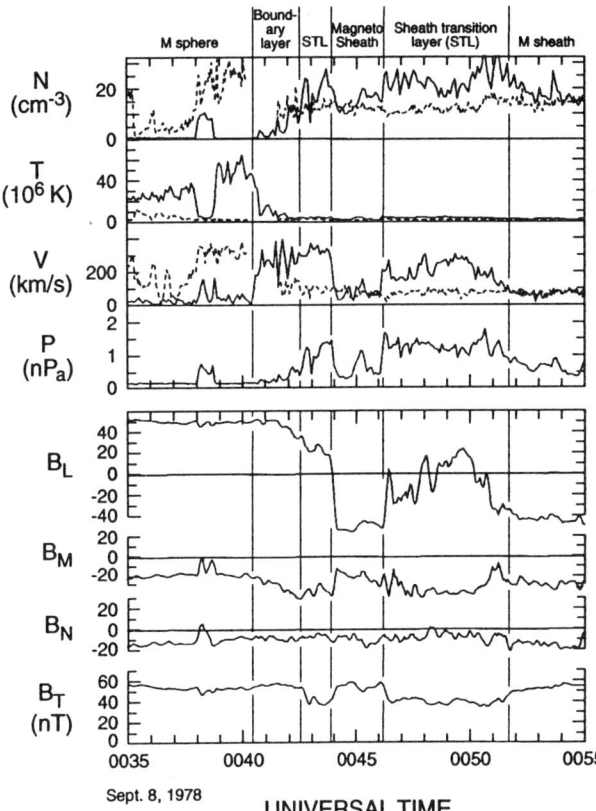

Fig. 4. A magnetopause crossing for strongly southward IMF in the same format as Fig. 3. The regions are defined according to ISEE 1 measurements (solid lines).

more interesting is that both groups are in favor of the interpretation that the LLBL is formed from reconnection at cusps for northward IMF [*Paschmann et al.*, 1990; *Song and Russell*, 1992] even though the techniques they used are drastically different. However, we caution the reader that one should not be surprised if he/she sees a northward IMF crossing that is significantly different from that in Figure 3, in particular if the crossing occurs in the flank. The variability of the structure has not yet been investigated systematically.

Figure 4 shows an example of magnetopause crossings for strongly southward IMF. The normal component of the field is not zero indicating that it is more like a rotational discontinuity than a tangential discontinuity. In discussions of observations, a rotational discontinuity usually refers to a rotation of the field across from the magnetopause while the normal component is significantly different from zero. This implies that the magnetic connection between the two sides of the magnetopause is strong but does not necessarily mean the magnetopause is a strict rotational discontinuity in a theoretical sense. A method of detailed tests is briefly discussed at the end of the section. The most remarkable feature of this crossing is the enhancements of the plasma

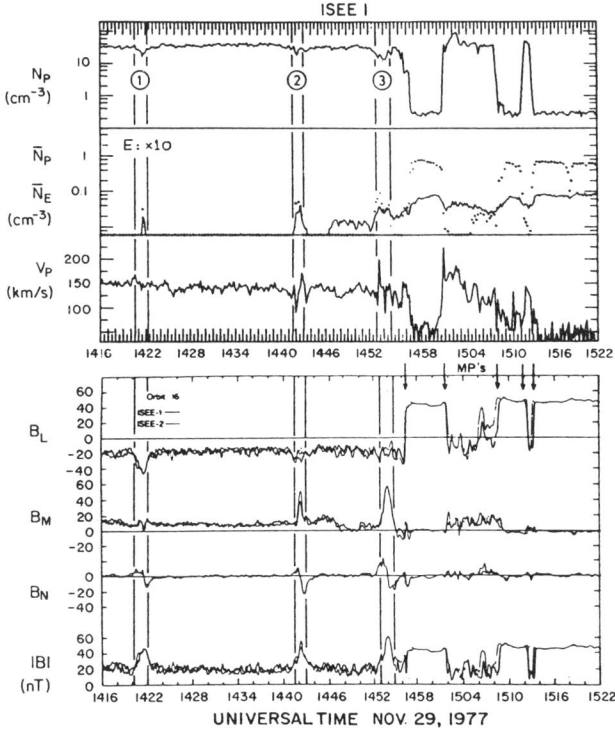

Fig. 5. A magnetopause pass for southward IMF with multiple crossing as indicated by vertical arrows [*Paschmann et al.*, 1982]. $\overline{N}_p$ and $\overline{N}_E$ are partial densities of energetic ions (solid line) and electron (x10, dotted line) respectively. Three FTEs are identified with the vertical lines.

velocity in the current layer and LLBL. The long duration (note that the same flow enhancement was observed by both ISEE 1 and 2) of such enhancements have been interpreted as the evidence of active reconnection. Further detailed analyses indicate that the plasma properties associated with the flow enhancements are consistent with those predicted by quasi-steady state reconnection models [*Sonnerup et al.*, 1981]. Readers interested in theoretical discussion on this topic are recommended to read *Lee* [this volume] and *Lin and Lee* [1994]. The magnetopause crossings with such a long duration of enhanced flow have often not been reported (less than 1/4 of the time) [*Cowley*, 1982; *Phan and Paschmann*, 1995]. More often, the enhanced flow is observed in a rather bursty manner, lasting a few minutes [*Gosling et al.*, 1990].

There are some differences in terminology used in magnetopause structure studies by different groups. These differences are unimportant when not dealing with the fine structures within the magnetopause. *Song et al.* [1990, 1993a] define the magnetopause structure according to physical phenomena. Two major physical phenomena at the magnetopause are a field change and a particle change from the magnetosheath to the magnetosphere. The field change is a current layer and the particle change a plasma layer. Song et al. termed the field change the "sheath transition layer." It was not termed the "current layer" because for northward IMF as seen in Figure 3 this layer contains little current and plasma (density) change seems more important. It was also not termed the "plasma depletion layer" because for southward IMF it does not show obvious density gradient as seen in Figure 4. The boundary layer is defined as the portion of the plasma layer that is excluded from the current layer. *Paschmann et al.* [1993], *Phan et al.* [1994], and *Phan and Paschmann* [1995] define the magnetopause structure according to a reference time, the "key time." The "key time" is defined differently for northward IMF (low shear) and southward IMF (high shear). For low shear cases, the key time is when proton and electron temperatures and their anisotropies have a major change. For high shear cases, it is when the field has its most rapid change. Since the definition of the key time is not based on the same quantity for differing IMF, one should not expect it to correspond to the same point in Song et al.'s definition. From Figures 3 and 4, one can see that the key time corresponds to the inner (outer) edge of the sheath transition layer for low (high) shears. For northward IMF the plasma depletion layer is the same as the sheath transition layer [*Song et al.*, 1990, 1993; *Paschmann et al.*, 1993; *Phan et al.*, 1994]. It does not occur for southward IMF [*Phan et al.*, 1994; *Phan and Paschmann*, 1995]. For low shear cases, the key time is about 50 km thick [*Paschmann et al.*, 1993] and is not the magnetopause used by *Berchem and Russell* [1982] who report the magnetopause to be a few hundred to a thousand kilometers thick.

Figure 5 shows another pass for southward IMF. The magnetopause was in rapid motion as indicated by the multiple crossings of its current layer. Because of such motion, to resolve its internal structure is difficult. There is unlikely a significant flow enhancement which should be in the range of the Alfven velocity, ~400 km/s, in this crossing. The magnetopause behaves more like a tangential discontinuity than a rotational discontinuity. Interestingly enough, there are several disturbances as indicated by the vertical lines in the magnetosheath. For these intervals, the normal component of the field first becomes positive and then goes negative before it goes back to zero. In the literature this is referred to as the bi-polar signature. The field strength increases during the events. If these fluctuations are caused by partial crossings of the current layer, one should expect B_L to go toward the positive and B_M toward zero. However, this is obviously not the case. Therefore they are not caused by partial magnetopause current layer crossings. *Russell and Elphic* [1978] first paid attention to such a phenomenon and referred to it as flux transfer events (FTEs). During the events, the cold plasma population, N_p drops from the magnetosheath level, while the magnetospheric hot population, $\overline{N}_p$ and $\overline{N}_E$ significantly present; indicates a linkage between the two sides of the magnetopause. Therefore, FTEs have been interpreted as the evidence for localized transient reconnection [*Russell and Elphic*, 1978; *Lee and Fu*, 1985; *Sonnerup*, 1987; *Southwood et al.*, 1988; *Scholer*, 1988]. The size of the FTE structure normal to the magnetopause has been determined independently by two spacecraft simultaneous observations [*Saunders et al.*, 1984] and by statistically the distance of the region where FTEs are observed across the magnetopause [*Rijnbeek et al.*, 1984]. The size of the FTE along its motion direction can be estimated by the duration of an event times the moving velocity. The FTE is on average about 1 R_E in these two directions.

The FTE has a clear correlation with the southward IMF [*Berchem and Russell*, 1984; *Rijnbeek et al.*, 1984]. There are usually more than one FTE to be observed in a magnetopause pass and the inter-FTE interval is on average 8 min, although this does not mean FTEs recur every 8 min [*Rijnbeek et al.*, 1984]. There may be some differences in the criteria to identify an FTE used in previous studies. Some results which were based on less strict definitions may have been contaminated by wave activities which are bounteous near the magnetopause. Some of these cases have recently been used to argue against the FTE model. *Kuo et al.* [1995] and *Le et al.* [1994] have done the FTE statistics using a more conservative definition and a more complete dataset which covers the whole 10 years of the ISEE mission. They found similar but much clearer results than that in previous studies.

Comparing Figure 5 with Figure 4, one can see that the magnetopause structure for southward IMF can be very different. The difference may be related to different reconnection processes. Because the FTE is one of the most important mesoscale phenomena at the magnetopause and is most challenging to data analyses and interpretation, we will discuss it further in section 5.

The most important methodological development in this area has been the determination of the so-called DeHoffmann-Teller (HT) frame and Walén relation test (see a review by *Sonnerup et al.* [1995]). This method provides a systematic way to test whether the magnetopause is a rotational discontinuity at the crossing. If it is, the motion of the magnetopause then can be derived. This technique has become a powerful tool to study reconnection-related phenomena.

4. WAVES

As the magnetosheath is the region in which the solar wind interacts with the magnetosphere in the form of waves, it fills with a variety of waves. Some of them are of solar wind origin but change their properties when crossing the bow shock [*Yan and Lee*, 1994]. Some of them are generated at or downstream of the bow shock and carried by the sheath flow. These waves may play a significant role in the magnetosheath-magnetosphere coupling. As the magnetopause is approached, because of the changes in the plasma properties, the wave properties also change. Figure

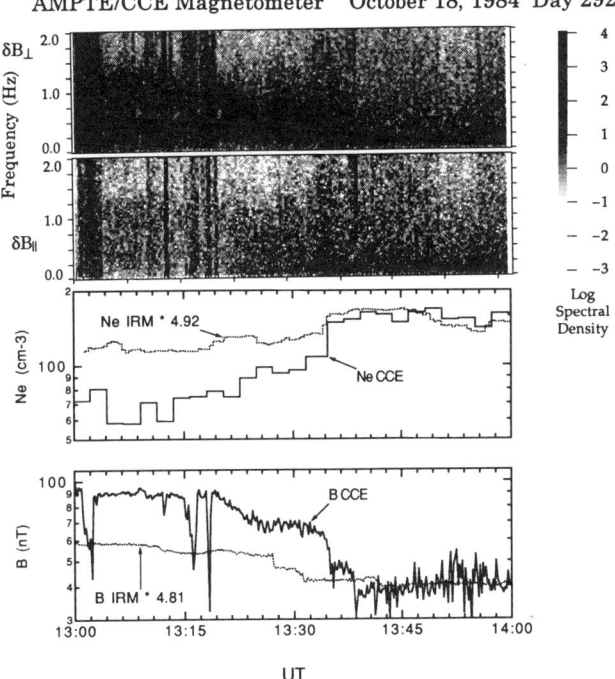

Fig. 6. An AMPTE/CCE magnetopause crossing [*Denton et al.*, 1994]. The crossing occurred at about 8.5 R_E, and 11.3 LT near the equator. From the top are the transverse and compressional components of the field fluctuation, electron density and field strength. The corresponding density and field measured in the solar wind by IRM are given as reference (dotted lines).

6 shows an example of the Fourier spectra near the magnetopause from AMPTE/CCE. CCE had an apogee of 8.8 Re. In the dayside, it observed the magnetosheath and magnetopause only when the magnetosphere was compressed with high solar wind pressure. To separate the temporal from spatial features is extremely difficult for CCE because its orbit was nearly along the magnetopause during the crossing. The magnetopause crossing occurred at about 1300 UT associated with a large field rotation. We are interested in the interval after 1300 UT. Similar to the crossing in Figure 3, the field increases while the density decreases toward the magnetopause in the magnetosheath. However, *Phan et al.* [1994] show that this feature has not been observed for large field shear, as shown in this case, from AMPTE/IRM. In this event this feature is at least partially caused by a solar wind density change as recorded by IRM. Nevertheless, the waves appear to be different in the region with high fields from that with lower fields. In the low field region, the waves are dominantly compressional and concentrated in the lowest frequencies. In the high field region, the waves are dominantly transverse and concentrated in the frequencies just below the proton gyrofrequency. Within the magnetopause current layer, the fluctuations are broadband associated with the field shear.

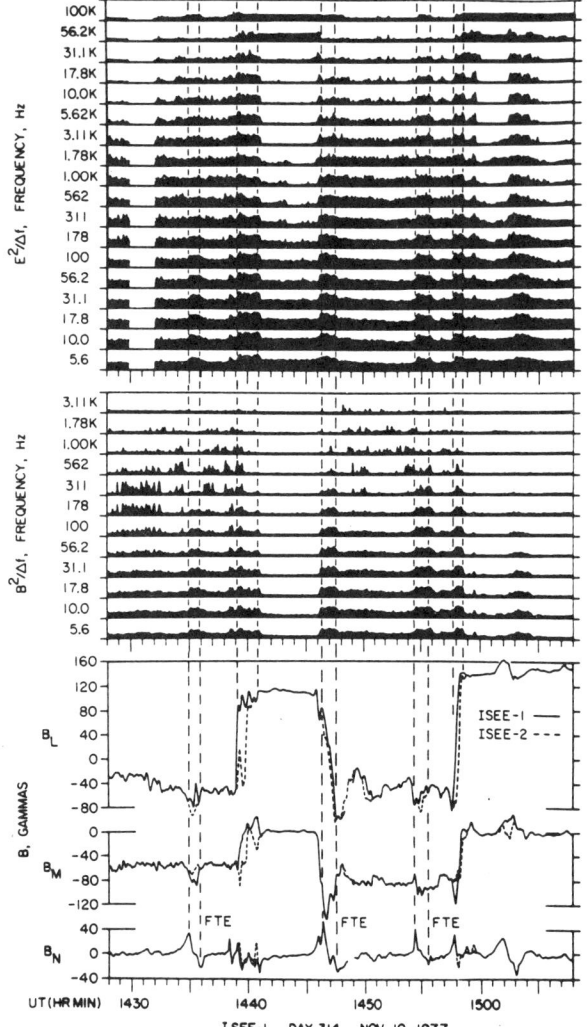

Fig. 7. Plasma wave measurements at the magnetopause [*Gurnett et al.*, 1979]. The bottom three panels show the magnetic field. The magnetopause current layer crossings are indicated by the major changes in B_L. The top panels show the wave power in the electric field received from each frequency channel. The middle panels are the wave power of the magnetic field. Electromagnetic waves are indicated by enhancements in both the electric and magnetic fields. Whereas, electrostatic waves have significant enhancements only in the electric field. A good example of the electrostatic waves is during the interval near 1503 UT.

Similar wave patterns have repeatedly been observed from other satellites as well. It is usually observed that the high field region, more precisely the low ß region, is of high ion temperature anisotropy [*Anderson et al.*, 1991; *Song et al.*, 1993b; *Anderson and Fuselier*, 1993]. A trough on the spectrum at the He^{++} gyrofrequency in the transverse component, which is the lighter regions, about 0.5 Hz, between the darker regions, is often observed in CCE data.

Furthermore, the upper branch is usually left handed and the lower branch linearly polarized. These features may provide very important clues on the processes which generate these waves. However, these features may not belong to the permanent features of the magnetopause because CCE observes the magnetopause only when the magnetopause is highly compressed and the cases from ISEE reported by *Song et al.* [1993b] do not show such a trough.

It is generally agreed that the temperature anisotropy is one of the main free energy sources for the transverse waves in the low-ß region and that these waves cause pitch angle diffusion which in turn forms the density gradient as the plasma convects toward the magnetosphere [*Song et al.*, 1993b; *Gary et al.*, 1994]. *Gary et al.* (see also *Gary*, this volume) have further developed a more quantitative model to understand an anti-correlation between the anisotropy and beta.

There are still some differences in the understanding of the functions and mode of the low frequency compressional waves [*Song et al.*, 1992b; *Denton et al.*, 1995; *Omidi and Winske*, 1995]. *Price et al.* [1986] and *Lee et al.* [1988] showed that the downstream condition of the bowshock is favorable to the growth of the mirror instability. *Denton et al.* suggest that these waves be mirror mode structures simply convected from the upstream magnetosheath. *Song et al.* and *Omidi and Winske* show from observation and computer simulation respectively that the mirror mode is a very important source of the waves and that the waves very close to the magnetopause propagate upstream. How the propagation occurs remains unclear, but it may be caused by the interaction of the mirror mode structures with the magnetopause. Whether the wave is upstream propagating or simply convecting downstream is an essential issue in the solar wind-magnetosphere coupling. The former implies that the magnetopause is at least partially responsible to some of the waves in the sheath, and the latter implies that the magnetopause is a passive recipient of the waves upstream. The measurements of the Poynting flux indicate that a fraction of the compressional wave energy can be transferred across the magnetopause current layer and hence plays a role in the energy coupling [*Song et al.*, 1993c]. The transmitted wave seems to experience a mode conversion at the current layer and become field-aligned, or Alfven modes. This conversion is significant to the consequent magnetospheric and ionospheric processes because the wave will couple along the field to the high latitude ionosphere and not radially to the low latitudes. However, that report was based on only one crossing, and it is not clear how often and to what spatial scale the wave transmission and mode conversion occur.

The waves of frequencies higher than the ion gyrofrequency have received much less attention, although they play the key role in any kinetic description of the processes. One possible reason is that the data analysis and the interpretation of these waves are more sophisticated. A simple-minded approach may not bring any convincing results. Figure 7 shows the plasma wave measurements from a pass of multiple crossings. Enhancements in both magnetic and electric fields appear in association with the current layer crossings and FTEs. However, a statistical study has not shown impressive correlations of the wave power with any parameters [*Tsurutani et al.*, 1989]. Recently, *Song* [1994] demonstrated that these high frequency waves can provide much more information about the physical conditions and processes. However, to extract the information correctly is by no means a simple task.

On the methodology front, the most interesting development is to identify the wave modes of low frequency fluctuations, (see a review by *Anderson* [1995]). The fluctuations of a wave are usually small, but the ratios of the fluctuations in different quantities may not be small. Based on theory, these ratios (the so-called transport ratios [*Gary and Winske*, 1992]), can be used as diagnostics of a mode. In practice, the observed transport ratios may not all be consistent with theoretical prediction and one has to make the decision to use some of the measured ratios and ignore the remaining. Different philosophies in this decision making lead to different schemes. *Song et al.* [1994a] proposed a hierarchical scheme in which more accurately measured ratios overwrite the less accurate ones. *Denton et al.* [1995] proposed a parallel scheme in which all measured ratios are treated equally and a mode is chosen for maximum agreement with theory.

5. FTE VERSUS SURFACE WAVE

To almost all experienced observationalists, FTEs and magnetopause surface waves are in general, two distinct phenomena, although in some particular cases, there is an uncertainty. However, they can be confusing in a modeler's cartoon. It is unfortunate that people have spent much time and energy in this debate [*Sibeck et al.*, 1989; *Lanzerotti*, 1989; *Elphic*, 1990; *Sibeck*, 1990, 1992; *Lockwood*, 1991; *Sckopke*, 1991; *Smith and Owen*, 1992; *Elphic et al.*, 1994; *Song et al.*, 1994b; *Hapgood and Lockwood*, 1995]. Nevertheless, from a purely observational viewpoint, this debate helps us to understand how important a quantitative observation is. In this section, we briefly review the several key arguments in this debate from a purely observational viewpoint. It is worth mentioning that the debate is not on whether or not upstream pressure pulses [*Song et al.*, 1988] cause magnetopause motion with which most people agree. The only controversy is whether the FTE is purely pressure related or reconnection related. The reader should be aware when reading literature on this subject that much of the complication has been caused by the confusion of many different phenomena which may be observed in the magnetosphere and in the ionosphere and assumed to be associated with FTEs but without justification.

When a perturbation propagates along the magnetopause

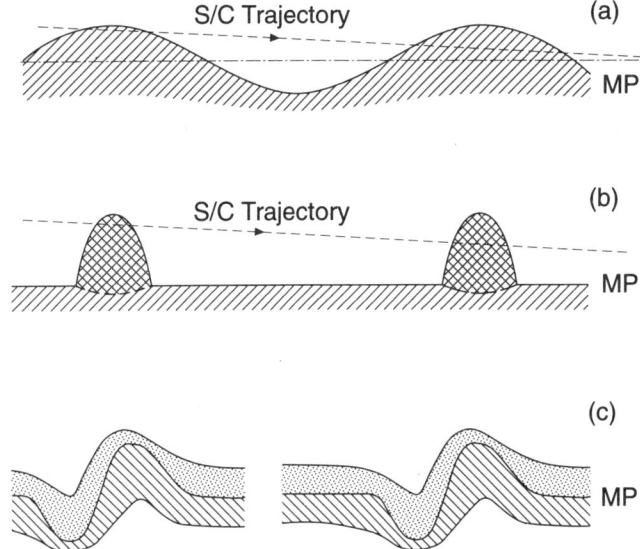

Fig. 8. Magnetopause perturbations associated with (a) a surface wave; (b) a convecting structure along the magnetopause; (c) a pressure-pulse-driven-surface-wave in Sibeck's model [*Sibeck*, 1990]. The magnetosheath is assumed to be above (unshaded) the magnetopause current layer (MP).

while a satellite crosses the magnetopause, the satellite trajectory in the frame of a reference rest to the perturbation is shown in Figure 8 (a) (for a surface wave) and (b) (for convecting structures). The differences between the two seem obvious. One can easily number several of them [e.g., *Song et al.*, 1994b]. Figure 8a is conventionally used to interpret surface waves generated by solar wind dynamic pressure fluctuations [*Elphic and Southwood*, 1987; *Song et al.*, 1988; *Elphic*, 1988] and Figure 8b to FTEs [*Russell and Elphic*, 1978]. *Sibeck* [1990] proposed the magnetopause perturbation which is assumed to be caused by a solar wind dynamic pressure pulse, as shown in Figure 8c. In this picture, the magnetopause perturbation is compacted into a small region or solitary form similar to that in Figure 8b, and layers consisting of plasmas partially different from either side of the magnetopause are added to explain the contents seen in an FTE. The physical differences, in the context of solar wind-magnetosphere coupling, between the FTE model and Sibeck's model are that in Sibeck's model, the coupling is mainly via momentum with little energy; but in the FTE model, the coupling is through electromagnetic fields and mass fluxes in addition to the momentum and energy, and that the coupling is controlled by IMF Bz in the FTE model and by IMF Bx in Sibeck's model.

The most direct and convincing way to observationally differentiate Figures 8b from 8c is to show the perturbations of the magnetopause seen on both sides simultaneously. *Elphic* [1990] was able to find such a rare occasion when FTEs occur and when ISEE 1 and 2 were separated by the

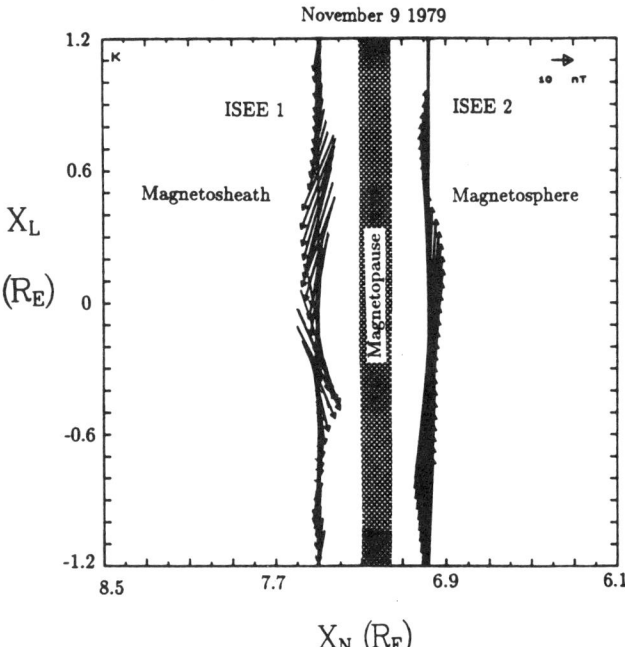

Fig. 9. Observed field perturbations near the magnetopause as an FTE passes by ISEE 1 and 2 [*Elphic*, 1990].

magnetopause with a distance to sample the boundary perturbation, see Figure 9. The perturbation shown in Figure 9 is consistent with Figure 8b and not 8c.

Similarities in the perturbations seen in Figures 8b and 8c would confuse the two if the spacecraft always crossed the perturbation with a path one wished, if one ignored the information from plasmas, total field, and the out-of-plane component of the field, if one had only one spacecraft, and if one discarded the statistical results that show clear correlation between the FTE and IMF Bz. Even so, to produce, by pressure variation alone, a magnetopause perturbation that is so similar to an FTE that is confused with the FTE, the pressure variation must satisfy some conditions. It is well-recognized that the strength of such a pressure pulse must be one or two times of the background pressure [*Sibeck*, 1990; *Song et al.*, 1994b]. As the FTE is repeatedly observed predominantly for southward IMF during a pass with an average inter-event interval of 8 min, although there is not a particular period between FTEs, in order to interpret the FTEs as pressure-pulse-driven surface waves according to Sibeck's model, the pressure pulses should be frequently produced with a large amplitude for southward IMF. On average the dynamic pressure fluctuations in the solar wind is about 10 to 20 percent of the background pressure on the time scale of less than 20 min. [*Song et al.*, 1988]. Therefore, the solar wind cannot be considered as a source for Sibeck's model. The perturbations associated with observed surface waves, Figure 8a, are consistent with 10 ~ 20% of the pressure variation.

The observational evidence used in Sibeck's model is the fluctuations upstream of the quasi-parallel bow shock [*Fairfield et al.*, 1990]. These fluctuations correlate with IMF Bx and not Bz. Many years after the proposal, the correlation between such upstream pressure pulses and FTEs has not been demonstrated observationally and there is no documented case that lends support to the model, whereas *Elphic et al.* [1994] show that there is no such correlation. Figure 10 shows the best example of the dynamic pressure variation based on which Sibeck's model was proposed, noting that neither FTE nor surface wave was documented at the magnetopause during the interval. The measurements come from IRM upstream of the bow shock when it appears to be quasi-parallel. There are fluctuations with periods of a few minutes. If one looks at the largest pulse, the amplitude is bigger than the average pressure. If one draws a qualitative conclusion that the pulses have an amplitude of one or two times of the background pressure and a period of 8 min, it may not sound too unreasonable. However, quantitatively, one can easily show that the average amplitude during the interval is less than 20% [*Song et al.*, 1994b] and only one pulse, not 22 pulses (which are required by the frequency of the FTEs if they were solely generated by these pulses), with an amplitude of one or two times of the background value, see the lower panel of Figure 10. This means that the above qualitative estimate is one order higher than it is either in amplitude or in frequency. With the actual observed value of pressure variations the pressure-pulse-driven surface wave is similar to that in Figure 8a and not in Figure 8c and should be easily distinguished from the FTE signatures even without consulting with other measurements and statistical features.

An important lesson we learned through this debate is that to quantify observations is becoming more and more crucial.

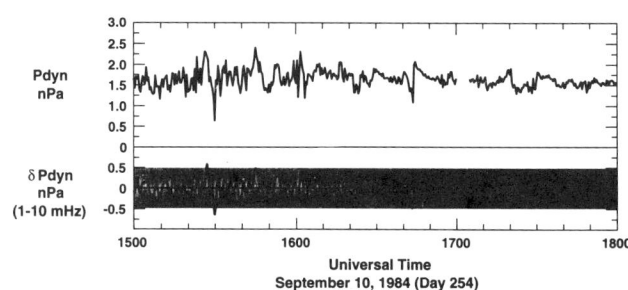

Fig. 10. Dynamic pressure measured by IRM upstream of a quasi-parallel bow shock (courtesy of G. Le). *Sibeck* [1990] assessed the perturbation as with periods of a few minutes and amplitude of one or two times the background pressure. The lower panel shows the fluctuations after a bandpass filter which takes away the average and high frequency noise. The shaded region is of peak-to-peak amplitude smaller than 67% and hence is not important in contribution to Sibeck's model. According to Sibeck's model, there should be at least 20 pulses above the shaded region.

People also recognize the importance of the methodology used in data analyses and the consistency between the methodology and interpretation. An important methodology development related to this subject is forwarded by the Dartmouth group [*Walthour et al.*, 1994] (see a review by *Walthour and Sonnerup* [1995]) to differentiate an FTE and a surface wave by reconstruction of the perturbations of the magnetopause using the observed field perturbations near the magnetopause.

6. CONCLUSIONS

Over more than three decades, we have accumulated much knowledge about the magnetopause from observations. Although we obtained the picture of the structure of the magnetopause under some simple circumstance, the variability of such a structure due to location and upstream conditions remain poorly understood. It is clear that the IMF Bz is the most important factor that controls the structure and processes of the magnetopause. Any one who is discussing a model of the magnetopause or citing a reference of magnetopause observations must first make it clear what IMF condition is in context. Significant progress has been made in recent years in observation of waves below the ion gyrofrequency. Active methodology development is leading this area to quantitative studies. There is strong potential for the development in analyses of the waves above the ion cyclotron frequency. It becomes clear that the pressure-pulse-driven surface wave model which was proposed to explain the FTE is not consistent with sound quantitative observations. As data analyses become more and more quantitative, we have started paying attention to the uncertainty and limit of an analysis. Uncertainties and limitation of a method or an instrument remain a subject of systematic investigations.

Acknowledgments. This effort at HAO was sponsored by the National Science Foundation and supported by NASA under research grant W-18,582

REFERENCES

Akasofu, S. I., E. W. Hones, Jr., S. J. Same, J. R. Asbridge, and A. T. Y. Lui, Magnetotail and boundary layer plasma at geocentric distance of 18 R_E: Vela 5 and 6 observations, *J. Geophys. Res., 78*, 7257, 1973.

Anderson, B. J., ULF signals observed near the magnetopause, in *Physics of the Magnetopause, AGU Monograph*, edited by P. Song, B. U. Ö. Sonnerup, and M. F. Thomsen, AGU, Washington, D. C., in press, 1995.

Anderson, B. J., and S. A. Fuselier, Magnetic pulsations from 0.1 to 4.0 Hz and associated plasma properties in the earth's subsolar magnetosheath and plasma depletion layer, *J. Geophys. Res., 98*, 1461, 1993.

Anderson, B. J., S. A. Fuselier, and D. Murr, Electromagnetic ion cyclotron waves observed in the plasma depletion layer, *Geophys. Res. Lett., 18*, 1955, 1991.

Berchem, J., and C. T. Russell, The thickness of the magnetopause current layer: ISEE 1 and 2 observations, *J. Geophys. Res., 87*, 2108, 1982

Berchem, J., and C. T. Russell, Flux transfer events on the magnetopause: Spatial distribution and controlling factors, *J. Geophys. Res., 89*, 6689, 1984.

Cahill, L. J., Jr., Early studies of the boundary of the geomagnetic field: A historical review, in *Physics of the Magnetopause, AGU Monograph*, edited by P. Song, B. U. Ö. Sonnerup, and M. F. Thomsen, AGU, Washington, D. C., in press, 1995.

Cahill, L. J., and P. G. Amazeen, The boundary of the geomagnetic field, *J. Geophys. Res., 68*, 1835, 1963.

Cowley, S. W. H., The causes of convections in the earth's magnetosphere: A review of developments during the IMS, *Rev. Geophys. and Space Phys., 20*, 531, 1982.

Denton, R. E., B. J. Anderson, S. A. Fuselier, S. P. Gary, and M. K. Hudson, Ion anisotroy-driven waves in the Earth's magnetosheath and plasma depletion layer, Geophysical Monograph 84, *Solar System Plasmas in Space and Time*, 111, 1994.

Denton, R. E., S. P. Gary, X. Lin, B. J. Anderson, J. W. LaBelle, M. Lessard, Low-frequency fluctuations in the magnetosheath near the magnetopause, *J. Geophys. Res., 100*, 5665, 1995.

Eastman, T. E., E. W. Hones, Jr., S. J. Bame, and J. R. Asbridge, The magnetospheric boundary layer: Site of plasma, momentum and energy transfer from the magnetosheath into the magnetosheath into the magnetosphere, *Geophys. Res. Lett., 3*, 685, 1976.

Elphic, R. C., Multipoint observations of the magnetopause: Results from ISEE and AMPTE, *Adv. Space Res., 8*, (9), 223, 1988.

Elphic, R. C., Observations of flux transfer events: Are FTEs flux ropes, islands, or surface waves?, in *Physics of Magnetic Flux Ropes, Geophys. Monogr. Ser.*, vol. 58, edited by C. T. Russell, et al., pp. 455-471, AGU, Washington, D. C., 1990.

Elphic, R. C., Observations of flux transfer events: A review, in *Physics of the Magnetopause, AGU Monograph*, edited by P. Song, B. U. Ö. Sonnerup, and M. F. Thomsen, AGU, Washington, D. C., in press, 1995.

Elphic, R. C., and D. J. Southwood, Simultaneous measurements of the magnetopause and flux transfer events at widely separated sites by AMPE UKS and ISEE 1 and 2, *J. Geophys. Res., 92*, 13,666, 1987.

Elphic, R. C., W. Baumjohann, C. A. Cattell, H. Luhr, and M. F. Smith, A search for upstream pressure pulses associated with flux transfer events: An AMPTE/ISEE case study, *J. Geophys. Res., 99*, 13,521, 1994.

Fairfield, D. H., Average and unusual locations of the earth's magnetopause and bow shock, *J. Geophys. Res., 76*, 6700, 1971.

Fairfield, D. H., W. Baumjohann, G. Paschmann, H. Luhr, and D. G. Sibeck, Upstream pressure variations associated with the bow shock and their effects on the magnetosphere, *J. Geophys. Res., 95*, 3773, 1990.

Gary, S. P., and D. Winske, Correlation function ratios and the identification of space plasma instabilities, *J. Geophys. Res., 97*, 3013, 1992.

Gary, S. P., M. E. McKean, D. Winske, B. J. Anderson, R. E. Denton, and S. A. Fuselier, Proton cyclotron anisotropy instability and the anisotropy/beta inverse correlation, *J. Geophys. Res., 99*, 5903, 1994.

Gosling, J. T., M. J. Thomsen, S. J. Bame, R. C. Elphic, and C. T. Russell, Plasma flow reversals at the dayside magnetopause and the origin of asymmetric polar cap convection, *J. Geophys. Res., 95*, 8073, 1990.

Gurnett, D. A., R. R. Anderson, B. T. Tsurutani, E. J. Smith, G. Paschmann, G. Haerendel, S. J. Bame, and C. T. Russell, Plasma waves turbulence at the magnetopause: Observations from ISEE 1 and 2, *J. Geophys. Res., 84*, 7043, 1979.

Haerendel, G., G. Paschmann, N. Sckopke, H. Rosenbauer, and P. C. Hedgecock, The frontside boundary layer of the magnetosphere and the problem of reconnection, *J. Geophys. Res., 83*, 3195, 1978.

Hapgood, M., and M. Lockwood, Rapid changes in LLBL thickness, *Geophys. Res. Lett., 22*, 77, 1995.

Kuo, H., C. T. Russell, and G. Le, Statistical studies of flux transfer events, *J. Geophys. Res., 100*, 3513, 1995.

Lanzerotti, L. J., Comment on "Solar wind dynamic pressure variations and transient magnetospheric signatures," *Geophys. Res. Lett., 16*, 1197, 1989.

Le, G., P. J. Chi, C. T. Russell, and C. A. Cattell, A study of the motion of flux transfer events, *EOS, Trans. AGU., 75*, No. 4, 555, 1994.

Lee, L. C., A review of magnetic reconnection: MHD models, in *Physics of the Magnetopause, AGU Monograph,* edited by P. Song, B. U. Ö. Sonnerup, and M. F. Thomsen, AGU, Washington, D. C., in press, 1995.

Lee, L. C., and Z. F. Fu, A theory of a magnetic flux transfer at the Earth's magnetopause, *Geophys. Res. Lett., 12*, 105, 1985.

Lee, L. C., C. P. Price, C. S. Wu, and M. E. Mandt, A study of mirror waves generated downstream of a quasi-perpendicular shock, *J. Geophys. Res., 93*, 247, 1988.

Lee, L. C., M. Yan, and J. G. Hawkins, A study of slow-mode structures in the dayside magnetosheath, *Geophys. Res. Lett., 18*, 381, 1991.

Lin, Y., and L. C. Lee, Structure of reconnection layers in the magnetosphere, *Space Sci. Rev., 65*, 59, 1994.

Lockwood, M., Flux transfer events at the dayside magnetopause: Transient reconnection to sheath dynamic pressure pulses?, *J. Geophys. Res., 96*, 5497, 1991.

Omidi, N., and D. Winske, Structure of the magnetopause inferred from the kinetic Riemann problem, *J. Geophys. Res.,* in press, 1995.

Paschmann, G., G. Haerendel, N. Sckopke, H. Rosenbauer, and P. C. Hedgecock, Plasma and magnetic field characteristics of the distant polar cusp near local noon: The entry layer, *J. Geophys. Res., 81*, 2883, 1976.

Paschmann, G., N. Sckopke, G. Haerendel, I. Papamastorakis, S. J. Bame, J. R. Asbridge, J. T. Gosling, E. W. Hones, Jr., and E. R. Tech, ISEE plasma observations near the subsolar magnetopause, *Space Sci. Rev., 22*, 717, 1978.

Paschmann, G., G. Haerendel, I. Papamastorakis, N. Sckopke, S. J. Bame, J. T. Gosling, and C. T. Russell, Plasma and magnetic field characteristics of magnetic flux transfer events, *J. Geophys. Res., 87*, 2159, 1982.

Paschmann, G., I. Papamastorakis, W. Baumjohann, H. Sckopke, C. W. Carlson, B. U. Ö. Sonnerup, and H. Luhr, The magnetopause for large magnetic shear: AMPTE/IRM observations, *J. Geophys. Res., 91*, 11049, 1986.

Paschmann, G., B. U. Ö. Sonnerup, I. Papamastorakis, W. Baumjohann, N. Sckopke, and H. Luhr, The magnetopause and boundary layer for small magnetic shear: Convection electric fields and reconnection, *Geophys. Res. Lett., 17*, 1829, 1990.

Paschmann, G., W. Baumjohann, N. Sckopke, T. -D. Phan, and H. Luhr, Structure of the dayside magnetopause for low magnetic shear, *J. Geophys. Res., 98*, 13,409, 1993.

Petrinec, S. P., P. Song, and C. T. Russell, Solar cycle variation in the size and shape of the magnetopause, *J. Geophys. Res., 96*, 7893, 1991.

Phan, T. -D., and G. Paschmann, The low-latitude dayside magnetopause and boundary layers for high shear: 1. Structure and motion, *J. Geophys. Res.,* submitted, 1995.

Phan, T. -D., G. Paschmann, W. Baumjohann, N. Sckopke, and H. Luhr, The magnetosheath region adjacent to the dayside magnetopause: AMPTE/IRM observations, *J. Geophys. Res., 99*, 121, 1994.

Price, C. P., D. W. Swift, and L. C. Lee, Numerical simulation of nonoscillatory mirror wave at the Earth's magnetosheath, *J. Geophys. Res., 91*, 101, 1986.

Rijnbeek, R. P., S. W. H. Cowley, D. J. Southwood, and C. T. Russell, A survey of dayside flux transfer events observed by the ISEE 1 and 2 magnetometers, *J. Geophys. Res., 89*, 786, 1984.

Roelof, E.C., and D. G. Sibeck, Magnetopause shape as a bivariate function of interplanetary magnetic field Bz and solar wind dynamic pressure, *J. Geophys. Res., 98*, 21,421, 1993.

Roth, M., Impulsive transport of solar wind into the magnetopause, in *Physics of the Magnetopause, AGU Monograph,* edited by P. Song, B. U. Ö. Sonnerup, and M. F. Thomsen, AGU, Washington, D. C., in press, 1995.

Russell, C. T., The structure of the magnetopause, in *Physics of the Magnetopause, AGU Monograph,* edited by P. Song, B. U. Ö. Sonnerup, and M. F. Thomsen, AGU, Washington, D. C., in press, 1995.

Russell, C. T., and R. C. Elphic, Initial ISEE magnetometer results: Magnetopause observations, *Space Sci. Rev., 22*,

681, 1978.

Saunders, M. A., C. T. Russell, and N. Sckopke, Flux transfer events: Scale size and interior structure, *Geophys. Res. Lett., 11,* 131, 1984.

Scholer, M., Magnetic flux transfer at the magnetopause based on single x line bursty reconnection, *Geophys. Res. Lett., 15,* 291, 1988.

Scholer, M., Models of flux transfer events, in *Physics of the Magnetopause, AGU Monograph,* edited by P. Song, B. U. Ö. Sonnerup, and M. F. Thomsen, AGU, Washington, D. C., in press, 1995.

Sckopke, N., Plasma structure near the low-latitude boundary layer: A rebuttal, *J. Geophys. Res., 96,* 9815, 1991.

Sibeck, D. G., A model for the transient magnetospheric response to sudden solar wind dynamic pressure variations, *J. Geophys. Res., 95,* 3755, 1990.

Sibeck, D. G., Transient events in the outer magnetosphere: Boundary waves or FTEs?, *J. Geophys. Res., 97,* 4009, 1992.

Sibeck, D. G., et al., The magnetospheric response to 8-minute period strong-amplitude upstream pressure variations, *J. Geophys. Res., 94,* 2505, 1989.

Smith, M. F., and C. J. Owen, Temperature anisotropies in a magnetospheric FTE, *Geophys. Res. Lett., 19,* 1907, 1992.

Song, P., Observations of waves at the dayside magnetopause, in *Solar Wind Sources of Magnetospheric Ultra-Low-Frequency Waves, AGU Monograph,* edited by M. J. Engebretson, K. Takahashi, M. Scholer, 159, 1994.

Song, P., and C. T. Russell, A model of the formation of the low-latitude boundary layer, *J. Geophys. Res., 97,* 1411, 1992.

Song, P., R. C. Elphic, and C. T. Russell, ISEE 1 and 2 observations of the oscillating magnetopause, *Geophys. Res. Lett., 15,* 744, 1988.

Song, P., C. T. Russell, N. Lin, R. J. Strangeway, J. T. Gosling, M. Thomsen, T. A. Fritz, D. G. Mitchell, and R. R. Anderson, Wave and particle properties of the subsolar magnetopause, in *Physics of Space Plasmas (1989), SPI Conf. Proc. and Reprint Ser.,* edited by T. Chang, G. Crew, and J. Jasperse, p. 463, Scientific Publishers, Inc., Cambridge, MA, 1989.

Song, P., R. C. Elphic, C. T. Russell, J. T. Gosling, and C. A. Cattell, Structure and properties of the subsolar magnetopause for northward IMF: ISEE observations, *J. Geophys. Res., 95,* 6375, 1990.

Song, P., C. T. Russell, R. J. Strangeway, and R. R. Anderson, Interrelationship between the magnetosheath ULF waves and VLF waves, *Phys. of Space Plasmas (1991), The SPI Conference Proceedings and Reprint Series,* edited by T. Chang, G. Crew, and J. Jasperse, p. 459, Scientific Publishers Inc., Cambridge, MA, 1991.

Song, P., C. T. Russell, and M. F. Thomsen, Slow mode transition in the frontside magnetosheath, *J. Geophys. Res., 97,* 8295, 1992a.

Song, P., C. T. Russell, and M. F. Thomsen, Waves in the inner magnetosheath: A case study, *Geophys. Res. Lett., 19,* 2191, 1992b.

Song, P., C. T. Russell, R. J. Fitzenreiter, J. T. Gosling, M. F. Thomsen, D. G. Mitchell, S. A. Fuselier, G. K. Parks, R. R. Anderson, and D. Hubert, Structure and properties of the subsolar magnetopause for northward interplanetary magnetic field: Multiple-instrument particle observations, *J. Geophys. Res,. 98,* 11,319, 1993a.

Song, P., C. T. Russell, and C. Y. Huang, Wave properties near the subsolar magnetopause: Pc 1 waves in the sheath transition layer, *J. Geophys. Res., 98,* 5907, 1993b.

Song, P., C. T. Russell, R. J. Strangeway, J. R. Wygant, C. A. Cattell, D. J. Fitzenreiter, and R. R. Anderson, Wave properties near the subsolar magnetopause: Pc 3-4 energy coupling for northward IMF, *J. Geophys. Res., 98,* 187, 1993c.

Song, P., C. T. Russell, and S. P. Gary, Identification of low frequency fluctuations in the terrestrial magnetosheath, *J. Geophys. Res., 99,* 6011, 1994a.

Song, P., G. Le, and C. T. Russell, Observational differences between flux transfer events and surface waves at the magnetopause, *J. Geophys. Res., 99,* 2309, 1994b.

Sonnerup, B. U. Ö., On the stress balance in flux transfer events, *J. Geophys. Res., 92,* 8613, 1987.

Sonnerup, B. U. Ö., and L. J. Cahill, Jr., Magnetopause structure and attitude from Explorer 12 observations, *J. Geophys. Res., 71,* 171, 1967.

Sonnerup, B. U. Ö., G. Paschmann, I. Papamastorakis, N. Sckopke, G. Haerendel, S. J. Bame, J. R. Asbridge, J. T. Gosling, and C. T. Russell, Evidence for magnetic field reconnection at the earth's magnetopause, *J. Geophys. Res., 86,* 10,049, 1981.

Sonnerup, B. U. Ö., G. Paschmann, T. -D. Phan, Fluid aspects of reconnection at the magnetopause: In situ observations, in *Physics of the Magnetopause, AGU Monograph,* edited by P. Song, B. U. Ö. Sonnerup, and M. F. Thomsen, AGU, Washington, D. C., in press, 1995.

Southwood, D. J., and M. G. Kivelson, On the form of the flow in the magnetosheath, *J. Geophys. Res., 97,* 2873, 1992.

Southwood, D. G., and M. G. Kivelson, The formation of slow mode fronts in the magnetosheath, in *Physics of the Magnetopause, AGU Monograph,* edited by P. Song, B. U. Ö. Sonnerup, and M. F. Thomsen, AGU, Washington, D. C., in press, 1995.

Southwood, D. J., C. J. Farrugia, and M. A. Saunders, What are flux transfer events?, *Planet. Space Sci., 36,* 503, 1988.

Treumann, R. A., J. LaBelle, T. Bauer, Diffusion processes: The observational perspective, in *Physics of the Magnetopause, AGU Monograph,* edited by P. Song, B. U. Ö. Sonnerup, and M. F. Thomsen, AGU, Washington, D. C., in press, 1995.

Tsurutani, B. T., E. J. Smith, R. R. Anderson, K. W.

Ogilvie, J. D. Scudder, D. N. Baker, and S. J. Bame, Lion roars and nonoscillatory drift mirror waves in the magnetosheath, *J. Geophys. Res., 87*, 6060, 1982.

Tsurutani, B. T., A. L. Brinca, E. J. Smith, R. T. Okida, R. R. Anderson, and T. E. Eastman, A statistical study of ELF-VLF plasma waves at the magnetopause, *J. Geophys. Res., 94*, 1270, 1989.

Walthour, D. W., and B. U. Ö. Sonnerup, Remote sensing of 2D magnetopause structures, in *Physics of the Magnetopause, AGU Monograph,* edited by P. Song, B. U. Ö. Sonnerup, and M. F. Thomsen, AGU, Washington, D. C., in press, 1995.

Walthour, D. W., B. U. Ö. Sonnerup, R. C. Elphic, and C. T. Russell, Double vision: Remote sensing of a flux transfer event with ISEE 1 and 2, *J. Geophys. Res., 99*, 8555, 1994.

Winske, D., V. A. Thomas, N. Omidi, Diffusion at the magnetopause: A theoretical perspective, in *Physics of the Magnetopause, AGU Monograph,* edited by P. Song, B. U. Ö. Sonnerup, and M. F. Thomsen, AGU, Washington, D. C., in press, 1995.

Yan, M., and L. C. Lee, Generation of slow-mode waves in the dayside magnetosheath, *Geophys. Res. Lett., 21*, 629, 1994.

P. Song, High Altitude Observatory, National Center for Atmospheric Research, P.O. Box 3000, Boulder, CO 80307-3000

Anomalous Plasma Diffusion Due to Kinetic Alfvén Wave Fluctuations at the Dayside Magnetopause

Manju Prakash

Department of Physics, State University of New York at Stony Brook

I calculate the anomalous diffusion coefficients of ions and electrons resonating with the Kinetic Alfvén Wave fluctuations (frequency $f \sim 0.01$Hz) observed near the dayside magnetopause. Following the quasilinear formalism, and, using the wave parameters those measured by GEOS-2 spacecraft, I obtain the values of the diffusion coefficients that are comparable to those estimated by *Sonnerup* [1980] to maintain the boundary layer thickness. The diffusion coefficient is studied as function of wave intensity and plasma β value at the magnetopause. Based on the diffusion coefficients, I infer the steady state boundary layer thickness as $\sim 10\rho_s$. The thickness is sensitive to the wave intensity but weakly dependent on the plasma β value. The significance of these calculations in the plasma entry at the magnetopause due to the observed nonlinear waves structures by *Rezeau et al.* [1989] in a high β plasmas will be discussed.

1. INTRODUCTION

The purpose of the present work is to investigate, the resonant wave particle interaction between the kinetic Alfvén wave (KAW) fluctuations and the magnetosheath particles, as a plausible mechanism of plasma entry at the magnetopause. Plasmas near the magnetopause is essentially collisionless, hence, the anomalous diffusion due to plasma waves can lead to a significant mass entry at the boundary layer at quiet times [*Treumann et al.*, 1992]. This is supported by the fact that the boundary layer thickness is insensitive to the direction of IMF [*Eastman and Hones*, 1979]. The exact analytical calculations of the anomalous transport coefficients is a difficult task, and, in general requires the simultaneous solutions of the coupled kinetic equations for the wave spectrum and the particle distributions. Considerable insight into particle diffusion can, however, be gained by examining the instability and the turbulence underlying the transport.

In the present work, I compute the ion-electron diffusion coefficients due to the KAW fluctuations using the quasi-linear theory [*Hasegawa and Mima*, 1978]. The cross-field diffusion coefficients are calculated using the wave parameters, those measured by GEOS-2 spacecraft. The present approach follows closely the work by *Lee et al.* [1994]. The diffusion coefficient is studied as function of the wave magnetic field intensity and the plasma β value at the magnetopause. Based on the diffusion coefficient, I infer the boundary layer thickness as $\sim 10\rho_s$ and its dependence on the wave and plasma parameters. The significance of these calculations in particle entry due to the KAW turbulence in a high β plasmas [*Rezeau et al.*, 1986] will be discussed.

The KAW spectrum can be excited at the magnetopause by the resonant mode conversion of the MHD surface wave excited by the solar wind of variable density and pressure [*Song et al.*, 1993 a; b; *Hasegawa and Chen*, 1976]. KAW with an enhanced amplitude (grea-

TABLE 1. Wave and Plasma parameters

T_e	T_i	ρ_s	ω_{ci}	ω	β	v_A
KeV	keV	km	Hz	Hz	—	[km s^{-1}]
2	10	100	2	0.01	1-4	700

ter than the surface wave) can become turbulent due to its large nonlinear coupling coefficients [*Hasegawa and Mima*, 1978]. The kinetic effects (finite ion gyroradius) result in a sizeable wave electric field parallel to the ambient magnetic field. The parallel electric field of the wave leads to the breakdown of the ideal MHD condition. This results in decoupling of plasma from the ambient field. Magnetosheath ions and electrons then undergo cross field diffusion.

The particle entry at low latitude boundary layer due to reconnection and impulsive penetration depends on the solar wind conditions. For quiet times plasma entry, microturbulence has been studied in the past using quasi-empirical approach with spacecraft parameters [*LaBelle and Treumann*, 1988; *Treumann et al.*, 1992]. Following an empirical approach *Thorne and Tsurutani* [1991] obtained an upper limit to the cross field diffusion in the ELF-VLF frequency range. However, it has been recently concluded by *Treumann* (*Proceedings of the Chapman conference on the physics of the magnetopause*, 1994) that the low frequency waves may be of importance in particle diffusion across the magnetopause, but currently available data is insufficient to conclude this point. Macroscopic processes such as magnetic field migration and stochastic $E \times B$ scattering can lead to particle diffusion coefficient 10^3km^{2}s^{-1} [*LaBelle and Treumann*, 1988; *Lee et al.*, 1994; *Wang and Abdalla*, 1994]. As of today, there is no observational data available in support of plasma entry due to the KAW fluctuations, but it is believed that the KAW turbulence can lead to a large scale spatial diffusion, which, is operative along the whole boundary of the magnetopause [*Hasegawa and Mima*, 1978; *Lee et al.*, 1994]. It is therefore, important to investigate the role of KAW turbulence in detail.

2. QUASILINEAR PARTICLE DIFFUSION

In this section we will calculate the diffusion coefficients of ions and electrons interacting with KAW due to the Cerenkov resonance $\omega - k_z v_z = 0$. The quantity ω denotes the dominant frequency, k_z the wave vector and v_z is the velocity of the resonant particle. The ambient magnetic field is along the z direction. We assume broadband spectrum of KAW at low frequencies ($\omega << \omega_{ci}$) near the magnetopause boundary. *Rezeau et al.* [1986] report that the GEOS-2 spacecraft has measured large amplitude electromagnetic fluctuations (frequency 0 - 10 Hz) with $\delta E = 10$ mV/m and $\delta B = 15$ nT in this region. The plasma β was greater than or nearly one. The electric and magnetic field of the KAW are given as follows [*Hasegawa*, 1988].

$$E_z \sim ik_z\lambda_s\phi \qquad (1)$$

$$B_x \sim i\frac{k_y k_z}{\omega}(1+\lambda_s)\phi \qquad (2)$$

The dispersion relation is given as,

$$\omega^2 = k_z^2 v_A^2 \left(1 + \lambda_i(3/4 + \frac{T_e}{T_i})\right) \qquad (3)$$

where, $\lambda_i = k_\perp^2 \rho_i^2$; $\lambda_s = \frac{T_e}{T_i}\lambda_i$. The above relations are valid under the approximation that $\lambda_i << 1$.

In Table 1 are listed some of the parameters that are relevant to the present work. The wave parameters are those measured by GEOS-2 spacecraft. These parameters will be used in calculating the cross field diffusion. The quantity B_x denotes the wave amplitude. Other notations have their usual meaning.

We note that the oscillatory magnetic component of the KAW wave field bends the ambient field lines. Particle motion follows the field lines, but Landau resonance leads to the dissipation of the particle velocity and electrons are decoupled from the field line. They undergo cross field diffusion due to the $v_z \frac{B_x}{B_0}$ drift velocity. The

TABLE 2. Diffusion coefficient as a function of β

β	D
—	km^{2}s^{-1}
1	128.34
2	149.62
3	144.32
4	135.85

quasilinear ion diffusion coefficient D is obtained as follows [*Hasegawa and Mima*, 1978; *Lee et al.*, 1994].

$$D = (\frac{\pi}{8})^{1/2} \sum_k \frac{B_{xk}^2}{B_0^2}(\frac{T_e}{T_i}\lambda_i)^2 \frac{v_A}{k_z}(\frac{2}{\beta})^{1/2} exp(-1/\beta) \quad (4)$$

Since the k- dependence of the power spectrum is not known [*Rezeau et al.*, 1989] we approximate the sum as,

$$\sum_k \frac{B_{xk}^2}{B_0^2} = \frac{B_x^2}{B_0^2} \quad (5)$$

Hence, the ion diffusion coefficient reduces to,

$$D = (\frac{\pi}{8})^{1/2} \frac{B_x^2}{B_0^2}(\frac{T_e}{T_i}\lambda_i)^2 \frac{v_A}{k_z}(\frac{2}{\beta})^{1/2} exp(-1/\beta) \quad (6)$$

Using the parameters B_0= 100 nT; B_x= 15 nT; λ_i= 0.25; $k_z = 1/10,000 km^{-1}$; ω= 0.01 Hz [*Rezeau et al.*, 1986], we study the diffusion coefficient as a function of β. The results are as shown in Table 2. .

We note that the maximum particle diffusion takes place when β is equal to 2. For β greater than 2 the diffusion coefficient is not very sensitive to the value of β. We next study the diffusion coefficient as a function of the wave magnetic field intensity. The results are shown in Table 3 and Table 4, for β values 1 and 2 respectively. We note that at low wave amplitudes, the diffusion coefficient is more sensitive to the wave amplitudes than at higher amplitudes. The diffusion coefficient is weakly dependent on the plasma β value. At the observable field amplitudes, (15 nT) the diffusion coefficient is comparable to that estimated by *Sonnerup* [1980] to account for the magnetopause boundary layer thickness.

TABLE 3. Diffusion coefficient function of wave intensity, $\beta=1$

| B_x^2 | D |
nT^2	km^{2}s^{-1}
100	57.04
225	128.34
400	228.16
625	356.50
900	513.36

TABLE 4. Diffusion coefficient function of wave intensity $\beta = 2$

| B_x^2 | D |
nT^2	km^{2}s^{-1}
100	66.498
225	149.62
400	265.99
625	415.61
900	598.48

We note from Eq. (6) that the diffusion coefficient decreases with an increase in frequency. In the same frequency range, the electron diffusion coefficient is the same as that of ions. If there is any difference in the ion-electron diffusion coefficients, then ambipolarity adjusts their values to a common value in the large scale magnetospheric plasmas [*Chen*, 1974].

We remark that the above formalism assumes ion temperature T_i is five times the electron temperature T_e. However, the temperature anisotropy along the perpendicular and the parallel directions has been ignored.

3. BOUNDARY LAYER THICKNESS

The magnetosheath plasma is collisionless, hence the boundary layer thickness δ_d can be defined in terms of the quasilinear diffusion coefficient D as follows,

$$\delta_d = (D\tau)^{1/2} \quad (7)$$

Here, τ is the correlation time of the KAW turbulence and is assumed to be the inverse of growth rate of the dominant mode in the KAW turbulence. Since the growth rate can atmost be the wave period (strong turbulence limit), we define the boundary layer thickness as,

$$\delta_d = (D\tau_w)^{1/2} \quad (8)$$

Here, τ_w is the wave period of KAW. The quantity δ_d is analogous to the diffusion length defined by *Sonnerup* [1980] in his viscous model. The quantity δ_d is sensitive to plasma β value, spectral wave intensity and the ion-electron temperatures. We find δ_d is of the order of $10\rho_s$

TABLE 5. Steady state boundary Layer Thickness as a function of β

β	δ_d
−	km
1	1133
2	1223
3	1201
4	1166

[*Thorne and Tsurutani*, 1991]. In Table 5, we study δ_d as a function of plasma β value. Based on the Tables (3) and (4) we deduce that the boundary layer thickness is sensitive to the wave amplitude but depends weakly on the plasma β value.

The equilibrium boundary layer thickness δ_d is determined by the balance between the gain and the loss of the particles. There are a number of loss mechanisms across the magnetopause, such as the azimuthal convection (due to a radial electric field) which sweeps the flux tube away from the source region. The particles can also be lost along the magnetic field [*Treumann et al.*, 1992].

4. ROLE OF HIGH β IN PARTICLE DIFFUSION

Recently, it has been pointed out by *Fu et al.* [1995] and *Johnson and Cheng* (*proceedings of the 1995 cambridge workshop*) that plasma β plays an important role in the particle diffusion at the magnetopause. We find that the dependence of particle diffusion on β value is very weak [Eq.6]. This is due to the fact that the electromagnetic effects which are important at higher β value do not play an important role because the magnetosonic mode (with frequency much larger than ω_{ci}) gets decoupled from KAW due to the finite ion gyroradius effects. The particle transport due to transit damping (which is proportional to β) can be ignored [*Hasegawa and Mima*, 1987; *Lee et al.*, 1994].

In view of this, the quasilinear formalism (Eq.6) can be used to calculate the plasma entry due to the nonlinear KAW fluctuations observed at the magnetopause with β greater than 1. Based on the correlation studies on turbulence between the ISEE 1 and ISEE 2 spacecrafts, it was inferred that these fluctuations are likely to be nonlinear solitary Alfvén waves with transverse dimension $L \simeq$ the ion gyroradius [*Rezeau et al.*, 1993]. The measurements on the isotropy in k space indicate that the spectrum is not of the Kolmogorov type but can be fitted with power law $f^{-\alpha}$, $\alpha \sim 3.0$ [*Rezeau et al.*, 1986; 1989]. The wave structures can trap particles and carry them to the auroral zone. Therefore, appropriate corrections arising from the effects due to the particle trapping should be incorporated. Since, the exact nature of the wave structures is not yet established, the quasilinear theory with appropriate β value will yield reasonable estimates of the diffusion coefficient.

5. RESULTS AND CONCLUSIONS

We have examined the resonant wave-particle interaction as a plausible mechanism of particle entry at the boundary layer. We conclude that the resonant wave-particle interaction between the KAW fluctuations and the magnetosheath particles can lead to a significant mass transfer (Kelvin-Helmholtz instability leads to momentum transfer only) across the magnetopause. Our estimates of the diffusion coefficients are comparable to those provided by *Sonnerup* [1980] with value $10^3 \text{km}^2 \text{s}^{-1}$. Based on the diffusion coefficient, we have inferred the boundary layer thickness and studied its dependence on the wave amplitude and the plasma β value. We remark that the nonresonant processes such as decay of KAW into ion-acoustic wave and another KAW can also lead to plasma entry across the magnetopause, if the wave-amplitude is of the order of the ambient field [*Hasegawa and Chen*, 1976; *Lee et al.*, 1994]. The ion acoustic wave is damped because the electron temperature is much less than ion temperature at the magnetopause.

Acknowledgements. I thank Prof. Akira Hasegawa for his encouragement in completing this work.

REFERENCES

Chen, F. F., *Introduction to Plasma Physics*, Plenum Press, New York, pp. 152, 1974.

Eastman, T. E., and E. W. Hones, Jr., Characteristics of the magnetospheric boundary layer and magnetopause layer as observed by IMP 6, *J. Geophys. Res.*, *84*, 2019, 1979.

Fu, S. Y., Z. Y. Pu, S. C. Guo and Z. X. Liu, Kinetic Alfvén wave instability and wave-particle interactions at the magnetopause, *AGU monograph, Space Plasmas: Coupling between small and medium scale processes*, 86, 73, 1995.

Hasegawa, A., Particle dynamics in low frequency electromagnetic waves in an inhomogeneous plasma, *Phys. Fluids*, 22, 483, 1988.

Hasegawa, A., and L. Chen, Kinetic Processes in plasma heating by resonant mode conversion of Alfvén wave, *Phys. Fluids*, 19, 1924, 1976.

Hasegawa, A., and K. Mima, Anomalous transport produced by kinetic Alfvén wave turbulence, *J. Geophys. Res.*, 83, 1117, 1978.

LaBelle, J., and R. A. Treumann, Plasma Waves at the dayside magnetopause, *Space Sci. Rev.*, 47, 175, 1988.

Lee L. C., J. R. Johnson, and Z. W. Ma, Kinetic Alfvén Waves as a source of plasma transport at the dayside Magnetopause, *J. Geophys. Res.*, 99, 17405, 1994.

Rezeau, L., A. Morane, S. Perraut, A. Roux, and R. Schmidt, Characterization of Alfvénic fluctuations in the magnetopause boundary layer, *J. Geophys. Res.*, 94, 101, 1989.

Rezeau, L., S. Perraut and A. Roux, Electromagnetic fluctuations in the vicinity of the magnetopause, *Geophys. Res. Lett.*, 13, 1093, 1986.

Rezeau, L., A. Roux, and C. T. Russell, Characterization of small scale structures at the magnetopause boundary from ISEE measurements, *J. Geophys. Res.*, 98, 179, 1993.

Song, P., C. T. Russell, R. J. Strangeway, J. R. Wygant, C. A. Cattell, R. J. Fitzenreiter, and R. R. Anderson, Wave properties near the subsolar magnetopause: Pc 3-4 energy coupling for northward interplanetary magnetic field, *J. Geophys. Res.*, 98, 187, 1993a.

Song, P., C. T. Russell, and C. Y. Huang, Wave properties near the subsolar magnetopause: Pc 1 waves in the sheath transition layer, *J. Geophys. Res.*, 98, 5907, 1993b.

Sonnerup, B. U. O, Theory of the low-latitude boundary layer, *J. Geophys. Res.*, 85, 2017, 1980.

Thorne, R. M., and B. T. Tsurutani, Wave particle interactions in the magnetopause boundary layer, in *Physics of Space Plasmas, SPI Conf. Proc. Reprint Ser., vol. 10*, edited by T. Chang, G. B. Crew, and J. R. Jasperse, Scientific, Cambridge, Mass, 1991.

Treumann, R. A., J. LaBelle, G. Haerendel, and R. Pottelette, "Anomalous plasma diffusion and magnetopause boundary layer, *IEEE Transactions on Plasma Science*, 20, 833, 1992.

Wang, J. Z., and Maha Ashour-Abdalla, Simulation of magnetic field line Stochasticity, *J. Geophy. Res.* 99, 2321, 1994.

Manju Prakash, Physics Department, State University of New York at Stony Brook, Stony Brook, New York 11794-3800.

Structure of Reconnection Layers at the Magnetopause and in the Magnetotail

Y. Lin

Physics Department, Auburn University, Auburn, Alabama

In the earth's magnetosphere, magnetic reconnection can take place at the dayside magnetopause, at the flank of the magnetopause, and in the magnetotail plasma sheet. As a result, reconnection layers which contain MHD discontinuities are formed at the magnetopause and in distant the magnetotail. In this paper, our recent one-dimensional hybrid simulations of the structure of these reconnection layers are summarized. In general, rotational discontinuities, intermediate shocks, slow shocks, and slow expansion waves may be present in the reconnection layer. At the dayside magnetopause, a thin rotational discontinuity bounds the reconnection layer from the magnetosheath side, and a high-speed accelerated plasma flow is present on the magnetospheric side of the rotational discontinuity. At the flank of the magnetopause, where a large plasma flow is present in the magnetosheath, the magnetic field transition region is thick, and the accelerated flow exists in the entire field transition region. The simulation results of the boundary layers on the dayside and at the flank are consistent with satellite observations in these regions of the magnetopause. In the magnetotail with symmetric magnetic fields and plasma densities in the two lobes, two switch-off slow shocks, whose intermediate Mach number $M_I=1$, are formed in the case with a zero guide field ($B_{y0}=0$) in the lobes. Each slow shock contains a circularly-polarized trailing wavetrain in magnetic field and propagates in either of the two lobe-plasma sheet boundary layers. In the cases with $B_{y0} \neq 0$, two rotational discontinuities and two non-switch-off slow shocks with $M_I<1$ are present in the reconnection layer. Our hybrid simulations show that the presence of the finite B_{y0} reduces the intermediate Mach number of slow shocks below a critical number $M_c \sim 0.98$ and thus destroys the coherent wavetrain in the magnetotail reconnection layer if the lobe $B_{y0} \geq 0.8\ B_{x0}$. The absence of a downstream wavetrain for the slow shocks observed in the magnetotail may be associated with the presence of non-switch-off slow shocks with $M_I<0.98$, which may be caused by a finite B_{y0} or the two-dimensional effects of magnetic reconnection.

1. INTRODUCTION

Magnetic reconnection can take place at a current sheet which separates two plasma regions with antiparallel magnetic field components [*Dungey*, 1961]. Through magnetic reconnection, magnetic energy can be efficiently converted into kinetic energy, leading to the ejection of high-speed plasma. A layered structure which contains several magnetohydrodynamic (MHD) discontinuities and expansion waves can be formed in the high-speed outflow regions in a quasi-steady reconnection [*Petschek*, 1964; *Heyn et al.*, 1988. *Lin and Lee*, 1994a]. This layered structure is referred to as the reconnection layer [e.g., *Heyn et al.*, 1988]. In the earth's magnetosphere, magnetic reconnection can take place at the dayside magnetopause for southward interplanetary magnetic field (IMF), at the flank of the magnetopause, and in the magnetotail plasma sheet [e.g., *Sonnerup et al.*, 1981; *Gosling et al.*, 1986; *Feldman et al.*, 1984]. Consequently, reconnection layers can be

Cross-Scale Coupling in Space Plasmas
Geophysical Monograph 93
Copyright 1995 by the American Geophysical Union

formed in these regions of the magnetosphere. Note that magnetic reconnection may also take place at the cusp for a northward IMF.

Recently, we carried out a systematic study of the structure of quasi-steady reconnection layers at the magnetopause and in the magnetotail. The ideal MHD formulation, resistive MHD simulations, and hybrid simulations are used. In this paper, we briefly summarize our recent hybrid simulations of reconnection layers at the magnetopause [Lin and Lee, 1993, 1994a, b] and in the magnetotail [Lin and Lee, 1995]. The results are compared with those obtained from the ideal MHD formulation and resistive MHD simulations. The structure of reconnectin layer in the ideal MHD has also been studied by other authors for quasi-steady reconnection [e.g., Heyn et al., 1988; Biernat et al., 1989] and time-dependent reconnection [e.g., Semenov et al., 1992]. In our hybrid model, ions are considered as individual particles and electrons as a massless fluid. Seven cases are shown in this paper. Cases 1-3 correspond to the reconnection layers at the dayside magnetopause and at the flank of the magnetopause, and Cases 4-7 are for the distant tail reconnection layer.

2. HYBRID SIMULATIONS OF THE DAYSIDE RECONNECTION LAYER

In our simulations of the dayside reconnection layer, the initial magnetopause current sheet separates two uniform plasma regions: the magnetosheath and the magnetosphere. These two regions have antiparallel magnetic field components in the z direction and a common guide magnetic field in the y direction. The normal of the magnetopause current sheet is in the x direction. Initially, the total pressure (thermal plus magnetic pressure) is assumed constant across the current sheet. The initial plasma flow velocity is assumed to be zero in the magnetosphere and magnetosheath. In the presence of a normal magnetic field component, which is caused by magnetic reconnection, the initial current sheet evolves to form the reconnection layer. In our study, the one-dimensional (1-D) Riemann problem is simulated for the evolution of an initial current sheet after the onset of magnetic reconnection. In the simulation, physical quantities are functions of spatial coordinate x and time t only.

The initial z-component magnetic field, magnetic field strength, and temperature are given by

$$B_{z0}(x) = (B_{zm}+B_{zs})/2 + [(B_{zm}-B_{zs})/2] \tanh(x/\delta) \quad (1)$$

$$B_0(x) = (B_m+B_s)/2 + [(B_m-B_s)/2] \tanh(x/\delta) \quad (2)$$

$$T_0(x) = (T_m+T_s)/2 + [(T_m-T_s)/2] \tanh(x/\delta) \quad (3)$$

where δ is the half-width of the initial current sheet, and the subscripts "s" and "m" indicate the quantities in the magnetosheath and magnetosphere, respectively. The thermal pressure profile is determined by the total pressure balance condition. The hybrid code used in this study is the one described by Swift and Lee [1983]. Two buffer zones are located at the two ends of the simulation domain. These boundaries are positioned far from the resulting reconnection layer. The simulation results shown below are only for the central part of the whole simulation domain.

The 1-D simulation results can be used to determine the 2-D quasi-steady reconnection layer in the xz plane, by relating the time t in the 1-D simulation to the coordinate z by $z=v_z t$, where v_z is approximately the flow speed near the initial current sheet [Lin and Lee, 1994a].

In Case 1, the magnetospheric plasma beta $\beta_m=0.2$ and ion number density $N_s=10N_m$. The tangential magnetic fields on the two sides of the initial current sheet are not exactly antiparallel, with the guide field $B_{ys}= B_{ym}= 0.1B_{zm}$.

The left column of Figure 1 shows the simulation result of Case 1. Plotted from the top are the spatial profiles of antiparallel magnetic field component B_z, ion number density N, plasma flow speed v, and temperatures perpendicular ($T_\perp$ by solid line) and parallel ($T_\parallel$ by dashed line) to the magnetic field. The vertical dashed lines a and b approximately bound the region of accelerated plasma flow. The system length shown in the figure is $240\lambda_m$, where λ_m is the ion inertial length in the magnetosphere.

A thin rotational discontinuity is located on the magnetosheath side (the righthand side) of the reconnection layer, and an accelerated high-speed flow appears on the magnetospheric side of the rotational discontinuity, as shown in the left column Figure 1. The electric current in the reconnection layer is concentrated mainly at this rotational discontinuity. In the boundary layer, the plasma density decreases from the magnetosheath to the magnetosphere, and both parallel and perpendicular temperatures increase. It is found that $T_\parallel>T_\perp$ because of the mixing of the accelerated magnetosheath ions and hot magnetospheric ions in the reconnection layer. The mixing of plasmas take place mainly along the magnetic field. A D-shaped ion velocity distribution is found in the boundary layer, which is associated with the transmitted magnetosheath ions. The presence of the D-shaped distribution has also been observed by satellites in the magnetopause-boundary layer [Gosling et al., 1990; Smith and Rodgers, 1991]. In addition, there exists a weak rotational discontinuity bounding the reconnection layer from the magnetospheric side, as seen from the small kink of B_z at the vertical line a.. The result of Case 1 is similar

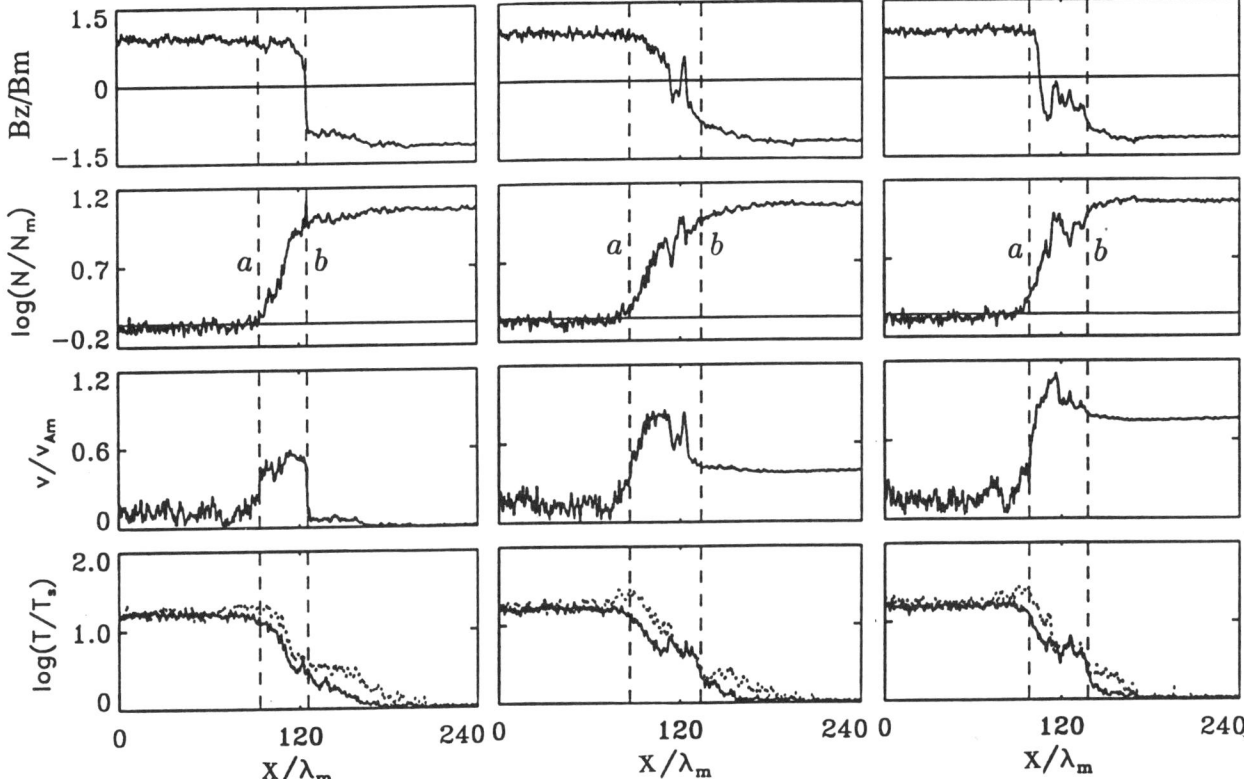

Fig. 1. Hybrid simulations of magnetopause reconnection layers. Case 1 with $v_s=0$ corresponds to the dayside reconnection layer, while cases 2 and 3 with $v_s=V_{As0}$ and $2V_{As0}$, respectively, are for the flank reconnection layer. Plotted from the top are the spatial profiles of tangential field B_z, ion number density N, flow speed v, and perpendicular (solid line) and parallel (dashed line) temperatures. The vertical dashed lines a and b in each case bound the region of accelerated plasma flows.

to the structure of the magnetopause-boundary layer observed at the dayside magnetopause [e.g., *Sonnerup et al.*, 1981].

In the ideal MHD formulation, a slow expansion wave, a slow shock, and a contact discontinuity exist between the two rotational discontinuities [e.g., *Lin and Lee*, 1994a]. In the hybrid simulation, however, the contact discontinuity cannot be identified because of the mixing of ions from the magnetosheath and the magnetosphere, and the slow expansion wave and slow shock are significantly modified.

In our resistive MHD simulation of the reconnection layer [e.g., *Lin and Lee*, 1993, 1994a], the steady rotational discontinuities do not exist. Instead, there exist two time-dependent intermediate shocks (TDISs) [e.g., *Wu*, 1990] bounding the reconnection layer in the cases with $B_y \neq 0$ on the two sides of the current layer. The width of the TDIS increases with time as $t^{1/2}$ and the strength decreases. As $t \to \infty$, the TDIS approaches a rotational discontinuity with an infinite width. In the case with exactly antiparallel magnetic fields in the magnetosheath and magnetosphere ($B_y=0$), the TDISs are replaced by an intermediate shock on the magnetosheath side and an Alfven wave pulse on the magnetospheric side. In the ideal MHD, however, the intermediate shock and time-dependent intermediate shock do not exist, and the rotational discontinuities bound the reconnection layer. Figure 2 summarizes the roles of rotational discontinuity, intermediate shock, and time-dependent intermediate shock obtained in the ideal MHD, resistive MHD, and hybrid simulations.

3. STRUCTURE OF THE RECONNECTION LAYER AT THE FLANK OF THE MAGNETOPAUSE

At the flank of the magnetopause, a large plasma flow is present in the magnetosheath and thus a large velocity shear exists across the magnetopause. We have studied the effects of the shear flow on the magnetopause-boundary layer, and found that the structure of the reconnection layer at the flank is very different from that on the dayside [*Lin and Lee*, 1994b].

Cases 2 and 3 correspond to the reconnection layer at the flank magnetopause, whose results are shown in the middle

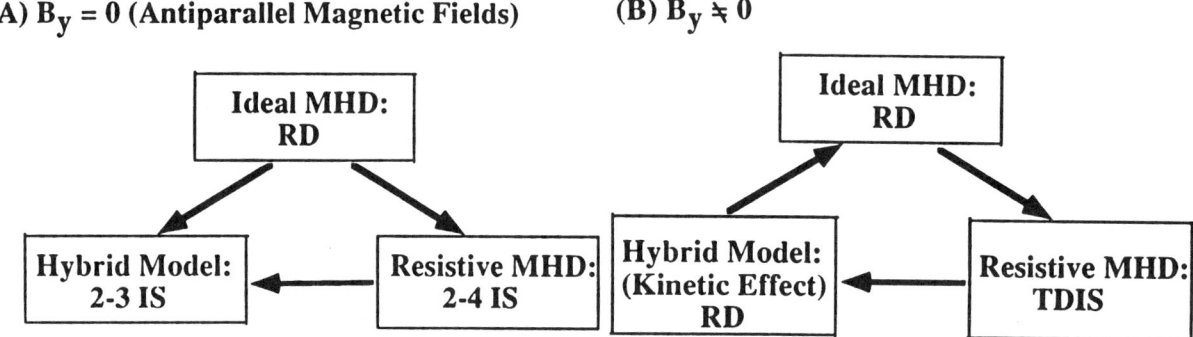

Fig. 2. Relations among the rotational discontinuity (RD), intermedaite shock (IS), and time-dependent intermediate shock (TDIS) in the ideal MHD, resistive MHD, and hybrid model.

and right columns of Figure 1, respectively. Note that here we still use the same coordinate system as in Case 1.

In Case 2, the magnetosheath flow speed $v_s = v_{As0}$, where v_{As0} is the Alfven speed in the magnetosheath. The magnetic field transition is thick, and the thin current layer no longer exists, as seen in Figure 1. The strength of B_z decreases from the value on each side to a small value at the center of the field transition region. The accelerated flow exists in the entire field transition layer. In addition, there exist large-amplitude oscillations in the magnetic field, density, and flow speed. As the magnetosheath flow becomes larger in Case 3 ($v_s = 2v_{As0}$), the field transition region is still thick, but a relatively sharp layer of magnetic field rotation is present on the magnetospheric side are the vertical line a, as shown in Figure 1. The simulation results of the flank reconnection layer appear to be consistent with the satellite observations at the flank magnetopause [e.g., *Gosling et al.*, 1986].

4. RECONNECTION LAYER IN THE DISTANT MAGNETOTAIL

In the magnetotail, the reconnection layer can be formed because of magnetic reconnection at the distant-tail X line between the two lobes. In the following, we show the simulations of the magnetotail reconnection layer, where the magnetic fields in the two lobes are assumed to be equal and so are the plasma densities. The antiparallel magnetic field components are in the x direction, and the normal of the initial current sheet is along the z direction. The presence of a finite guide field B_y in the lobes are considered in our simulation.

Cases 4-7 are for the magnetotail reconnection layer. In Case 4, the guide field $B_{y0}=0$ in the two lobes, and the lobe plasma beta $\beta_0=0.1$. In Cases 5-7, the guide field is assumed non-zero.

Figure 3 shows the hybrid simulation results of Case 4. The left column of the figure presents, from the top, spatial profiles of B_x, B_y, magnetic field magnitude B, and ion number density N. The right column shows the profiles of ion temperatures parallel ($T_\parallel$) and perpendicular ($T_\perp$) to the magnetic field, the x-component velocities of ion particles in the v_{ix}-z phase space, and the y-component velocities of ions in the v_{iy}-z phase space, where v_{ix} and v_{iy} represent, respectively, the x-velocity and y-velocity of ion particles. The hodogram of tangential magnetic field in Case 4 is shown in Figure 4a.

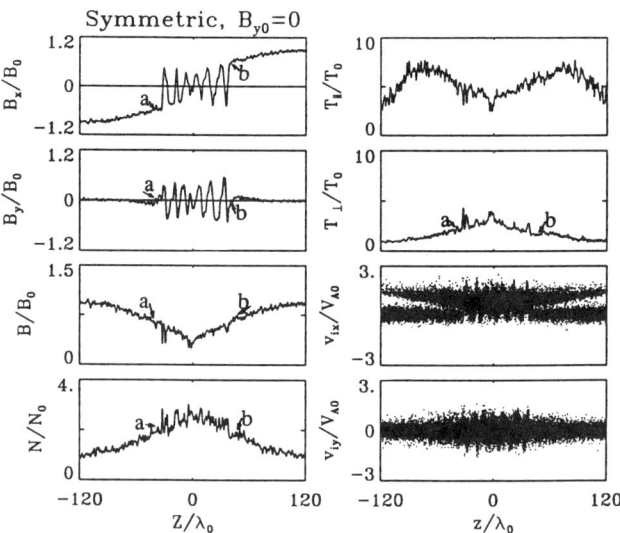

Fig. 3. Hybrid simulation result of Case 4. The left column presents, from the top, spatial profiles of B_x, B_y, magnetic field magnitude B, and density N. The right column shows the profiles $T_\parallel$, $T_\perp$, the x-component velocities of ions in the v_{ix}-z phase space, and the y-velocities of ions in the v_{iy}-z phase space. The shock fronts of two slow shocks are indicated by a and b, respectively.

Two switch-off slow shocks with $M_I=1$ are formed in the reconnection layer, where the M_I is the intermediate Mach number of the shock. The shock fronts of the two slow shocks are indicated in Figure 3 by a and b, respectively, with shock a propagating in the -z direction and shock b in the +z direction. The ion number density, perpendicular temperature, and flow speed increase across the slow shock, while the strength of magnetic field decreases. It is seen that there exists a large-amplitude, lefthand-polarized helical wavetrain of magnetic field in the downstream region, as seen from B_x and B_y profiles in Figure 3 and the magnetic field hodogram in Figure 4a. The fluctuation of the downstream magnetic field is around $\mathbf{B}_t=0$, where $\mathbf{B}_t$ is the tangential magnetic field. The presence of the coherent wavetrain in the slow shocks is consistent with the two-fluid theory of slow shocks [e.g., *Coroniti*, 1971].

The plasma temperature parallel to the magnetic field ($T_{||}$) is found to increase in most regions, except the center ($z\sim0$), of the reconnection layer. This is due to the interpenetrating of ions between the two lobes and the backstreaming of ions from downstream to upstream of the slow shock. It is seen from the phase-space plots of ion velocities in Figure 3 that there exists a beam of ions propagating upstream of each slow shock. The beam in each lobe contains approximately half of the backstreaming ions from the downstream of the slow shock and half of the transmitted ions from the other lobe.

In the presence of a finite B_{y0} in the lobes, the structure of the magnetotail reconnection layer is very different from that obtained in Case 4. As described for the results of the magnetopause reconnection layer, two rotational discontinuities bound the reconnection layer from the two sides when the guide field is non-zero. A non-switch-off slow shock propagates behind each rotational discontinuity. Figure 4 shows hodograms of magnetic field for Cases 4-7, with the lobe guide field $B_{y0}=0$, 0.1, 0.2, and $0.5B_{x0}$, respectively, where B_{x0} is the x-component magnetic field in the lobes. It is seen that two switch-off slow shocks with a large-amplitude rotational waves exist in Case 4 with $B_{y0}=0$, as described earlier. In Cases 5 and 6, two rotational discontinuities are present in addition to two slow shocks, and the slow shocks are non-switch-off shocks with $M_I<1$. In Case 7 with a larger B_{y0}, the slow shocks in the hybrid simulation are too weak to be identified.

One feature that is absent in the slow shocks in Cases 5-7 is the large-amplitude helical wavetrain, which appears in the switch-off shock in Case 1. It is found from the hybrid simulations that as B_{y0} increases from zero, the intermediate Mach number of the slow shocks in the reconnection layer decreases from $M_I=1$ in the switch-off shock to $M_I<1$ in the non-switch-off shocks. While B_{y0} increases to $0.08B_{x0}$ or

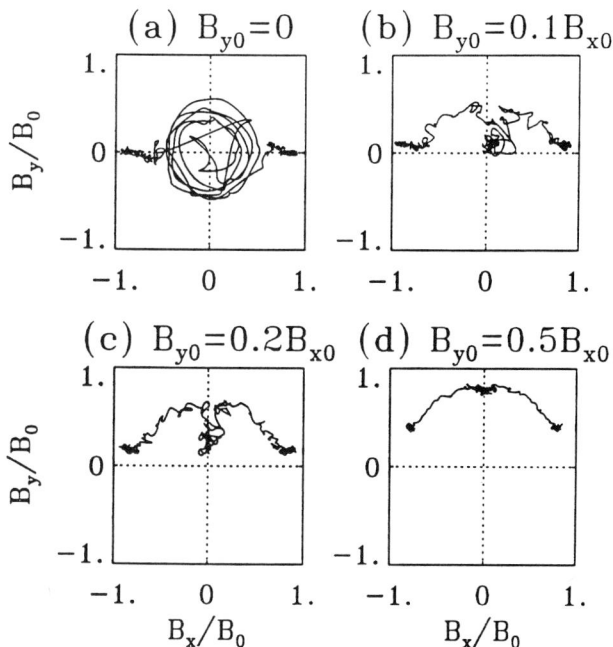

Fig. 4. Hodograms of tangential magnetic field for Cases 4-7, with the lobe guide field $B_{y0}=0$, 0.1, 0.2, and $0.5B_{x0}$, respectively. Large-amplitude rotational waves are present in the case with $B_{y0}=0$, while it does not exist in the other three cases with a finite B_{y0}.

larger, the intermediate Mach number M_I is smaller than a critical number $M_C\sim0.98$, and the coherent wavetrain disappears in the non-switch-off slow shocks. The present result is consistent with our earlier study which shows that the downstream wavetrain is absent in non-switch-off slow shocks with $M_I<M_C\sim0.98$ [*Lee et al.*, 1989; *Lin and Lee*, 1991].

We have also simulated the cases with unequal plasma densities in the two lobes [*Lin and Lee*, 1995]. It is found that a slight density asymmetry also reduces the intermediate Mach number of slow shocks to be below 0.98 and thus cause the absence of the coherent wavetrain in the reconnection layer. In addition, a 2-D reconnection which is not in a perfect steady-state may also lead to the presence of non-switch-off shocks with $M_I<0.98$ and therefore the disappearance of coherent wavetrains [*Lee et al.*, 1989].

The absence of a downstream wavetrain for the slow shocks observed in the magnetotail [e.g., *Feldman et al.*, 1984; *Smith et al.*, 1984] may be associated with the presence of non-switch-off slow shocks with $M_I<0.98$.

5. SUMMARY

In summary, our hybrid simulations show that at the dayside magnetopause, where the magnetosheath flow speed is small, a large-amplitude rotational discontinuity bounds the reconnection layer on the magnetosheath side, and a high-speed accelerated flow is present earthward of the rotational discontinuity. In the special case with antiparallel magnetic fields in the magnetosheath and magnetosphere ($B_y=0$), the large rotational discontinuity is replaced by an intermediate shock. At the flank of the magnetopause, where the magnetosheath flow is large, the magnetic field transition layer becomes thick, and the accelerated flow exists in the entire field transition layer. In the magnetotail with symmetric lobes, two switch-off slow shocks, whose intermediate Mach number $M_I=1$, are formed in the case with a zero guide field ($B_{y0}=0$) in the lobes. Each slow shock contains a circularly-polarized trailing wavetrain in magnetic field. In the cases with $B_{y0} \neq 0$, two rotational discontinuities and two non-switch-off slow shocks with $M_I<1$ are present in the reconnection layer. Our hybrid simulations show that the coherent wavetrain does not exist in the magnetotail reconnection layer if the intermediate Mach number of the slow shocks $M_I<0.98$, which may be caused by the presence of a finite B_{y0} ($\geq 0.08 B_{x0}$), a slight density asymmetry between the two lobes, or the 2-D effects of magnetic reconnection.

Acknowledgments. This work was supported by DOE grant DE-FG06-91ER 13530 and NASA SPTP grant NAG5-1504 to the University of Alaska. Computer resources were provided by the Pittsburgh Supercomputing Center.

REFERENCES

Biernat, H. K., M. F. Heyn, R. P. Rijnbeek, V. S. Semenov, and C. J. Farrugia, The structure of reconnection layers: Application to the earth's magnetopause, *J. Geophys. Res.*, 84, 287, 1989.

Coroniti, F. V., Laminar wave-train structure of collisionless magnetic slow shocks, *Nuclear Fusion*, 11, 261, 1971.

Dungey, J. W., Interplanetary magnetic field and the auroral zones, *Phys. Rev. Lett.*, 6, 47, 1961.

Feldman, W. C., S. J. Schwartz, S. J. Bame, D. N. Baker, J. Birn, J. T. Gosling, E. W. Hones, Jr., D. J. McComas, J. A. Slavin, E. J. Smith, and R. D. Zwickl, Evidence for slow-mode shock in the deep geomagnetic tail, *Geophys. Res. Lett.*, 11, 599, 1984.

Gosling, J. T., M. F. Thomsen, S. J. Bame, and C. T. Russell, Accelerated plasma flows at the near-tail magnetopause, *J. Geophys. Res.*, 91, 3029, 1986.

Gosling, J. T., M. F. Thomsen, S. J. Bame, R. C. Elphic, and C. T. Russell, Cold ion beams in the low latitude boundary layer during accelerated flow events, *Geophys. Res. Lett.*, 17, 2245, 1990.

Heyn, M. F., H. K. Biernat, R. P. Rijnbeek, and V. S. Semenov, The structure of reconnection layer, *J. Plasma Phys.*, 40, 235, 1988.

Lee, L. C., Y. Lin, Y. Shi, and B. T. Tsurutani, Slow shock characteristics as a function of distance from the X-line in the magnetotail, *Geophys. Res. Lett.*, 16, 903, 1989.

Levy, R. H., H. E. Petschek and G. L. Siscoe, Aerodynamic aspects of the magnetospheric flow, *AIAA J.*, 2, 2065, 1964.

Lin, Y. and L. C. Lee, Chaos and ion heating in a slow shock, *Geophys. Res. Lett.*, 18, 1615, 1991.

Lin, Y. and L. C. Lee, Structure of the dayside reconnection layer in resistive MHD and hybrid models, *J. Geophys. Res.*, 98, 3919-3934, 1993.

Lin, Y. and L. C. Lee, Structure of reconnection layers in the magnetosphere, *Space Sci. Rev.*, 65, 59-179, 1994a.

Lin, Y. and L. C. Lee, Reconnection layer at the flank magnetopause in the presence of shear flow, *Geophys. Res. Lett.*, 21, 855, 1994b.

Lin, Y. and L. C. Lee, A simulation study of the reconnection layer in the magnetotail, *J. Geophys. Res.*, submitted, 1995.

Petschek, H. E., Magnetic field annihilation, in *AAS-NASA Symposium on the Physics of Solar Flares*, NASA Spec. Publ., SP-50, 425-439, 1964.

Semenov, V. S., I. V. Kubyshkin, V. V. Lebedeva, M. V. Sidneva, H. K. Biernat, M. F. Heyn, B. P. Besser, and R. P. Rijnbeek, Time-dependent localized reconnection of skewed magnetic fields, *J. Geophys. Res.*, 97, 4251, 1992.

Sonnerup, B. U. O., G. Paschmann, I. Papamastorakis, N. Sckopke, G. Haerendel, S. J. Bame, J. R. Asbridge, J. T. Gosling, and C. T. Russell, Evidence for magnetic reconnection at the Earth's magnetopause, *J. Geophys. Res.*, 86, 10049, 1981.

Smith, E. J., J. A. Slavin, B. T. Tsurutani, W. C. Feldman and S. J. Bame, Slow mode shocks in the Earth's magnetotail: ISEE-3, *Geophys. Res. Lett.*, 11, 1054, 1984.

Smith, M. F. and D. J. Rodgers, Ion distributions at the dayside magnetopause, *J. Geophys. Res.*, 96, 11617, 1991.

Swift, D. W. and L. C. Lee, Rotational discontinuities and the structure of the magnetopause, *J. Geophys. Res.*, 88, 111, 1983.

Wu, C. C., Formation, structure, and stability of MHD intermediate shocks, *J. Geophys. Res.*, 95, 8149, 1990.

Y. Lin, Physics Department, 206 Allison Lab, Auburn University, Auburn, AL 36849-5311.

Micro/Mesoscale Coupling in Magnetotail Current Sheet: Observations

A. T. Y. Lui

Johns Hopkins University Applied Physics Laboratory, Laurel, Maryland

Observations from the Charge Composition Explorer are presented to illustrate the general features of current disruptions and the condition prior to their onsets. As current disruption onset is approached, the plasma pressure becomes enhanced and the pressure gradient at the neutral sheet becomes large along the tail axis. The deduced local current densities prior to onset range from ~27 to ~80 nA/m^2 and from ~85 to 105 mA/m when integrated over the sheet thickness. During disruption, there are large changes in the local magnetic field strength, enhancements of magnetic noise over a broad frequency range, magnetic field-aligned counter-streaming electron beams, ion energization perpendicular to the magnetic field, and reduction in current similar to the amount built up during the growth phase. Remote sensing capability from energetic particle measurements confirms that current disruptions can occur earthward of the downstream distance of ~9 R_E and are not due to an energetic particle front propagating earthward over the spacecraft. Reversal from northward to southward magnetic field or vice versa at the neutral sheet is not necessarily a site of particle energization related to an X-type neutral line. The observed features of current disruption compare favorably with the expected signatures of the cross-field current instability. It is emphasized that current disruptions modify the local current density and magnetic field, altering the force equilibrium to give rise to plasma and energy transport on a global scale.

1. INTRODUCTION

In our daily lives, we are accustomed to electrical currents delivering power to remote sites. It is therefore no foreign concept to us that magnetospheric currents are agents by which energy and momentum are transported between different plasma regions in the magnetosphere in spite of the vast distances between them. Electrical current is therefore a primary factor responsible for the often amazingly tight coupling exhibited by such a large-scale system as the magnetosphere.

It is natural that the first efforts in magnetospheric studies concentrate on large-scale phenomena and processes. This preference led to a promotion of using magnetohydrodynamics (MHD) equations to treat magnetospheric dynamics. A number of useful concepts were developed and significant progress was achieved from this approach, such as the notion of magnetospheric convection, the general shape of the magnetosphere, and their dependence on incoming solar wind condition. The simplifying assumptions adopted by MHD equations have both advantages and disadvantages. On the one hand, they allow us to gain some understanding of magnetospheric phenomena even with rather scanty survey of magnetospheric regions and sometimes with rather imprecise knowledge of the physical parameters involved. On the other hand, they exclude more advanced understanding on the exact nature and the intricacies involved in a number of magnetospheric processes which have global consequences.

At the present stage of magnetospheric research, we are often confronted with problems for which solutions lie beyond the MHD description. Good illustrative examples on the inadequacy of MHD formalism in understanding magnetospheric phenomena may be found in substorm problems. For a long time, magnetic reconnection is invoked as the process by which energy is released during substorms. The occurrence of magnetic reconnection requires the existence of a diffusion region which breaks down the frozen-in condition governing the ideal MHD behavior of the medium. However, there is very limited effort to address the underlying physical process for the diffusion region which may act as the "control valve" for magnetic reconnection. Some MHD modelers use a coarse numerical scheme to accomplish the magnetic diffusion required, while others adopt an ad hoc resistivity in a limited region in the magnetosphere. Both approaches essentially decouple the dynamics of the diffusion region from the large-scale evolution of the magnetosphere.

Detail evaluation of magnetospheric development before the onset of substorm expansions indicates that the basic simplifying assumptions adopted by MHD treatment are grossly violated prior to the initiation of dynamic activities found during substorm expansions [*Lui*, 1992]. For example, an intense current sheet with thickness comparable to the thermal ion gyroradius is observed near the inner edge of the cross-tail current. This development violates the MHD approximation of the system's length-scale being much longer than the thermal ion gyroradius. Furthermore, the relative drift between ions and electrons for this intense current sheet exceeds several times the Alfvén speed. This situation is anticipated to lead to the emergence of current-driven instabilities. Therefore, the precondition for substorm expansion onset is highly suggestive of an important role in substorms being played by a kinetic process which cannot be treated with MHD equations. It is also recognized that comprehensive knowledge of plasma dynamics in the magnetosphere requires understanding of not only macroscopic and microscopic processes individually but also the interplay between them.

In this paper, we examine in detail some observational features of the current sheet in the near-Earth magnetotail around the dynamic episodes of current disruptions. We point out the implications of these observations to the applicability of MHD treatment as well as suggest a plausible kinetic process which may be responsible for the dynamic activity of current disruption. We conclude by elaborating the scenario by which the onset of this microscale kinetic process may affect the global equilibrium of the magnetosphere and how plasma momentum and energy may be transported efficiently between different magnetospheric regions as a consequence.

2. CURRENTS IN THE MAGNETOSPHERE

Electrical currents are generated by differential motions between ions and electrons. In the collisionless state of the near-Earth space plasmas, forces are often set up to impart differential motions among different particle species, resulting in currents being a ubiquitous character of magnetospheric plasmas. The main current systems in the magnetosphere are illustrated in Plate 1. Currents are found in the magnetopause, the boundary layers, the central plasma sheet, and the ring current region. Some currents flow mainly along the magnetic field while others flow mainly perpendicular to the magnetic field. They are denoted by subscripts ∥ and ⊥, respectively. These current systems have the general shape of sheets which can have disparate thicknesses. For instance, the thickness of the current sheet at the dayside magnetopause lies approximately in the range of 400-2000 km while that of the cross-tail current sheet is typically 6000-18000 km. A key element in substorms is the existence of the substorm current wedge. This current wedge is often construed as a partial diversion or disruption of the cross-tail current which joins with an ionospheric westward electrojet through field-aligned currents, downwards to the ionosphere in the morning sector and upwards to the magnetosphere in the evening sector.

3. OBSERVATIONAL FEATURES OF CURRENT DISRUPTIONS

A major discovery from the Charge Composition Explorer (CCE) mission is the occurrence of large magnetic field fluctuation in the near-Earth current sheet at substorm expansion onsets. The disturbance is accompanied by a reduction of the fringe field associated with the cross-tail current buildup prior to substorm expansion onsets. The north-south component of the magnetic field is often found to reverse its sign momentarily and repeatedly. This feature has been interpreted as the turbulent characteristic of current disruption detectable by a spacecraft embedded within the disruption site [*Lui et al.*, 1988].

Let us examine two individual events to illustrate some general features found for current disruptions. Magnetic field fluctuation during current disruptions is exemplified by the event on August 28, 1986 in Figure 1 [*Takahashi et al.*, 1987]. The dipole VDH coordinate system is used to display the magnetic field measurements. The spacecraft was located at a geocentric distance of ~8.1 R_E and slightly before the magnetic local midnight. Prior to disruption onset, the field magnitude was extremely small, ~8 nT, in comparison with the Earth's dipole field of ~58 nT at that distance. The low field strength indicates the buildup of an intense cross-tail current such that the fringe field from that current canceled about six-sevenths of the dipole field. At ~1152:40 UT, large magnetic field fluctuation began with $\delta B \sim B$. The time scale of the field fluctuation is much shorter than the ion gyroperiod, thus violating one of the basic approximations adopted by the MHD theory. The activity onset coincides within 1 min of the onsets of magnetic activity and Pi 2 micropulsation on the ground in the CCE sector.

Plate 2 shows another current disruption event which was detected closer to the current sheet than the previous event. The close proximity of the spacecraft to the current sheet center during this event is indicated by the latitude angle of the magnetic field being near 90°. For the bottom panel which shows the anisotropy of 31-43 keV ions measured by the Medium Energy Particle Analyzer (MEPA), the "−" and "+" signs mark the sectors at which particles entered the detector from the dawn and dusk

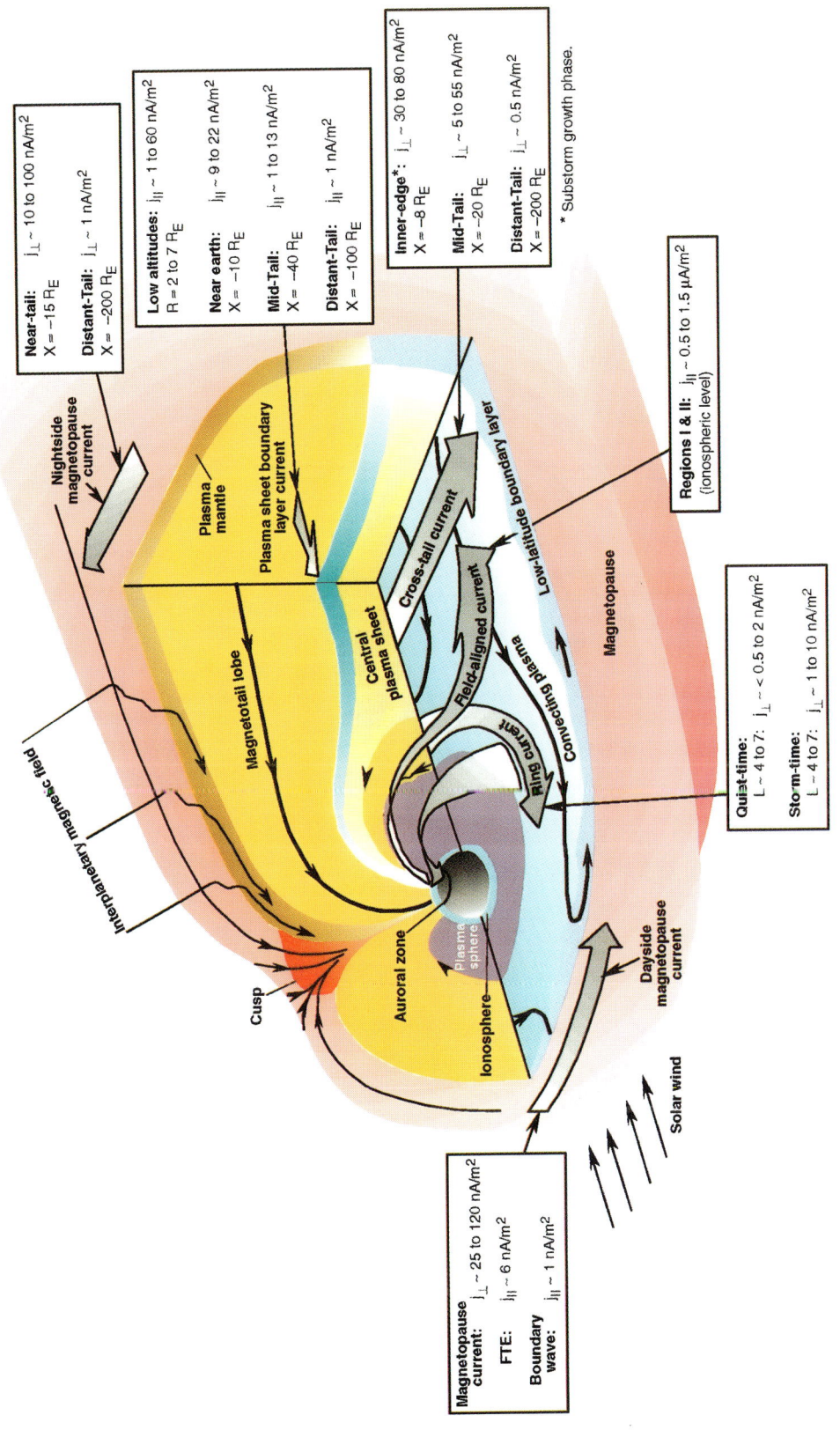

Plate 1. A schematic diagram to illustrate the current systems within the magnetosphere. The inferred current densities are also provided.

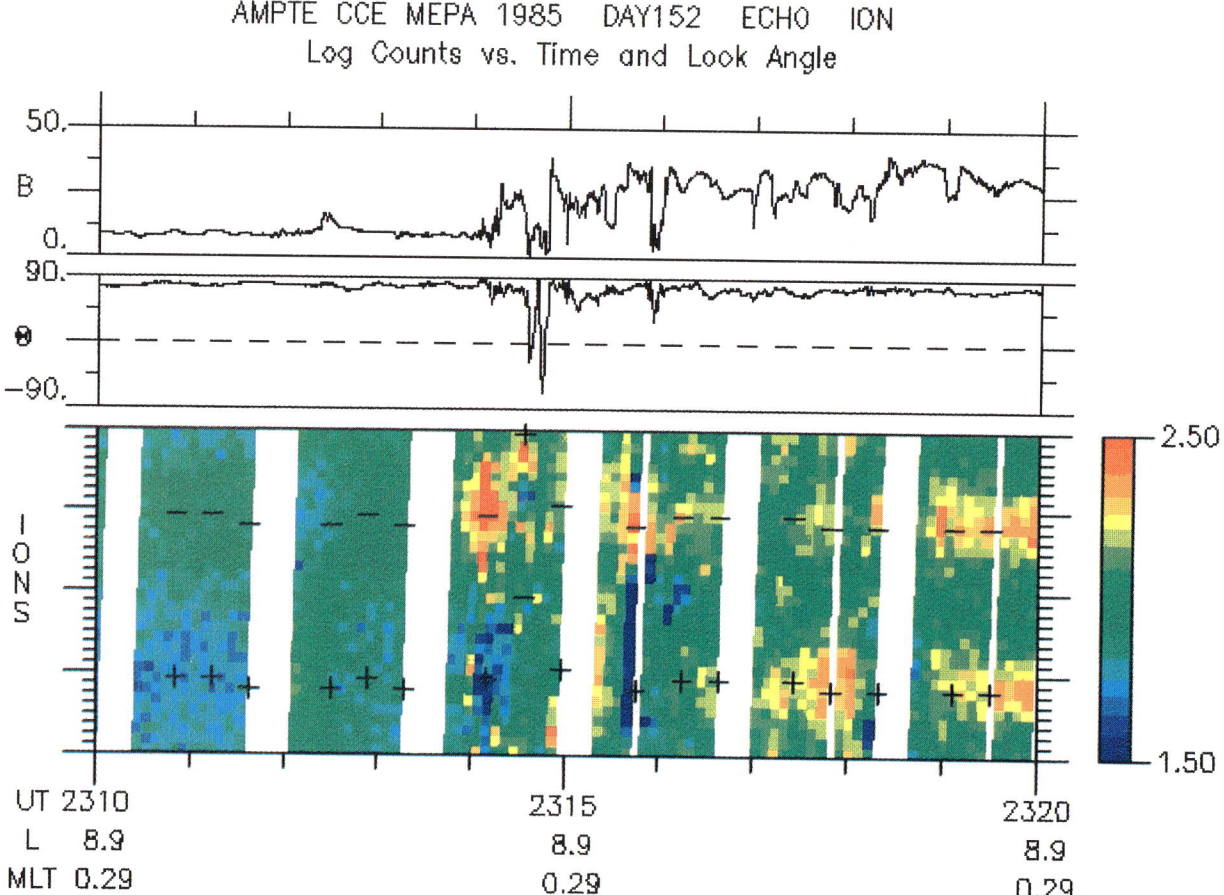

Plate 2. Magnetic field and energetic ion measurements on CCE during the current disruption event on June 1, 1985. The magnitude and latitude angle of the magnetic field are shown in the top two panels whereas the angular anisotropy of 31 - 43 keV energetic ions are given in the bottom panel. The first sign of particle energization came from particles with gyrocenters earthward of the spacecraft and it occurred before the increase of magnetic field strength.

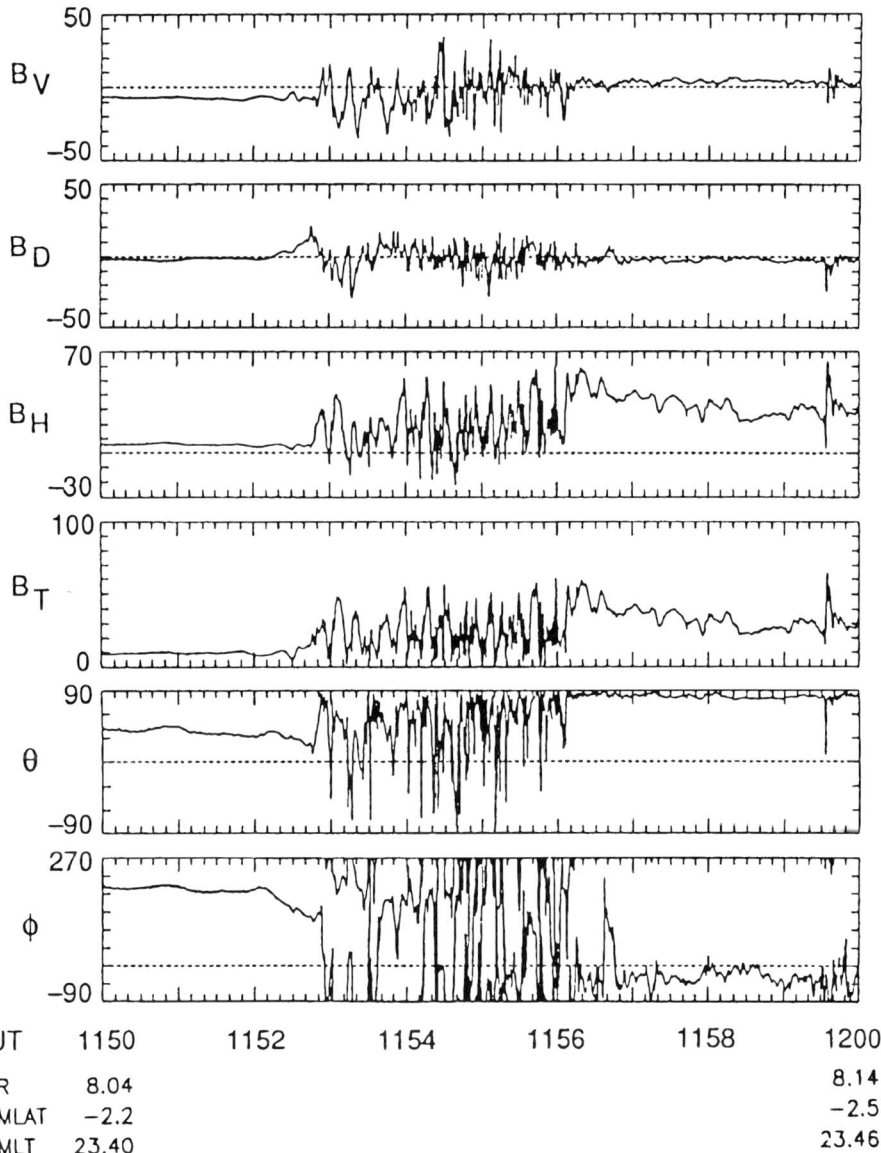

Fig. 1. CCE magnetic field measurements at high time resolution on August 28, 1986. The field strength was depressed to about one-seventh of the dipolar field value before the start of the large field fluctuation at ~1153 UT.

directions, respectively, at 90° with respect to the local magnetic field. Thus, for a northward field, the "−" and "+" sectors show the intensities of particles with gyrocenters at one gyroradius distance earthward and tailward, respectively, of the spacecraft location. Figure 2 illustrates this remote sensing capability from energetic particle measurements. The spinning motion of the spacecraft allows the detector to sense remotely the region earthward and tailward of the spacecraft within one gyroradius, albeit at different pitch angles. When the detector pointed toward west (east), it sampled particles with gyrocenters earthward (tailward) of the spacecraft. This remote sensing capability was first

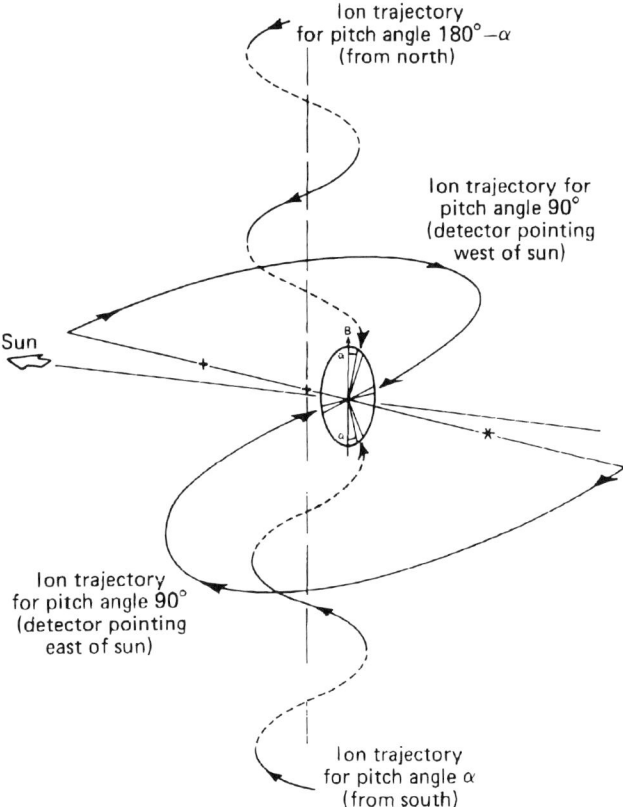

Fig. 2. A schematic diagram to illustrate the remote sensing capability of energetic ions from MEPA. This capability allows one to deduce that the energetic particle activity was not due to an earthward propagating front of energetic particles passing over the spacecraft.

used by *Kaufmann et al.* [1972] in the study of the motion of an energetic particle boundary.

Returning to Plate 2, the first indication of enhancement in particle intensity for this event came from the "−" sector, implying that the first sign of particle energization was earthward of the spacecraft. This occurred before the increase of magnetic field strength at ~23:14:08 UT (which coincided closely with substorm westward electrojet intensification at Syowa at ~23:15 UT; see *Lui et al.* [1992] for details) and well before the H-component reversed in direction at the spacecraft. The peak AE index (not shown) for this substorm activity is ~500 nT. Although only measurements of the low energy ions are shown here, the higher energy ions behaved similarly. If the activity were propagating from tailward of the spacecraft, then ion energization would have been seen first in the "+" sector, then in sectors between "+" and "−" sectors, and finally in the "−" sector. That this temporal sequence was not seen demonstrates clearly that the observation is consistent with activation first occurring earthward of the spacecraft at ~8.8 R_E and not consistent

with earthward propagation of energization across the spacecraft from the tailward direction. This finding is extremely important because it rules out substorm models that predict the near-Earth activity being originated from disturbances well beyond the downstream distance of 10 R_E. After ~2315 UT, the site of particle intensification shifted tailward and earthward of the spacecraft repeatedly. This continual jump of intensification site relative to the spacecraft strongly suggests that the impulsive ion energization occurred all around the spacecraft and is not due to a single acceleration site moving back and forth the spacecraft as the observation was first interpreted by *Takahashi et al.* [1987].

Does the north-south magnetic field reversal seen during current disruption signify the encounter of an X-type magnetic neutral line where particle energization takes place? The north-south component of the magnetic field reversed sign four times between 2314:30 and 2315:00 UT, which may be interpreted as an X-line crossing over the spacecraft four times. However, a look at the simultaneous particle intensity indicates that there is no dramatic enhancement during this interval of field reversals in comparison with the initial energization prior to this interval. As will be shown later (Fig. 4), the electrons also did not show any dramatic energization in association with this field reversal. Therefore, if the north-south field reversal were related to an X-line encounter, then there is no evidence that the X-line is an important site for particle energization. An interpretation more consistent with the observation than the X-line interpretation is that the field reversal is a part of the manifestation of field turbulence bearing no special significance except for a larger field perturbation.

CCE was in the neutral sheet region for an extended period, which permits a close examination on the temporal development of pressure components surrounding the disruption onset. The evolution of particle pressures, pressure anisotropy, and plasma beta are shown in Figure 3. The ion pressures are calculated from measurements of MEPA and the Charge-Energy-Mass (CHEM) spectrometer with a combined energy range of ~1 keV to ~4 MeV while the electron pressures are based on measurements from the Hot Plasma Composition Experiment (HPCE) covering an energy range of ~50 eV to ~25 keV. For typical thermal energies of ~10 keV for ion and ~2 keV for electrons at this distance, the energy ranges provided a coverage of ~90.5% and ~97.5% for ions and electrons, respectively, of the entire population with a stationary Maxwellian distribution. The percentage of population covered is even higher if the population were Kappa distributions. The perpendicular ($P_\perp$) and parallel ($P_\parallel$) components of pressure for protons, oxygen ions, and electrons are given in panels (a) and (b), respectively. The total plasma pressure components shown in panel (c) are the sum from these three species. The plasma beta is displayed in panel (d). These parameters are calculated in

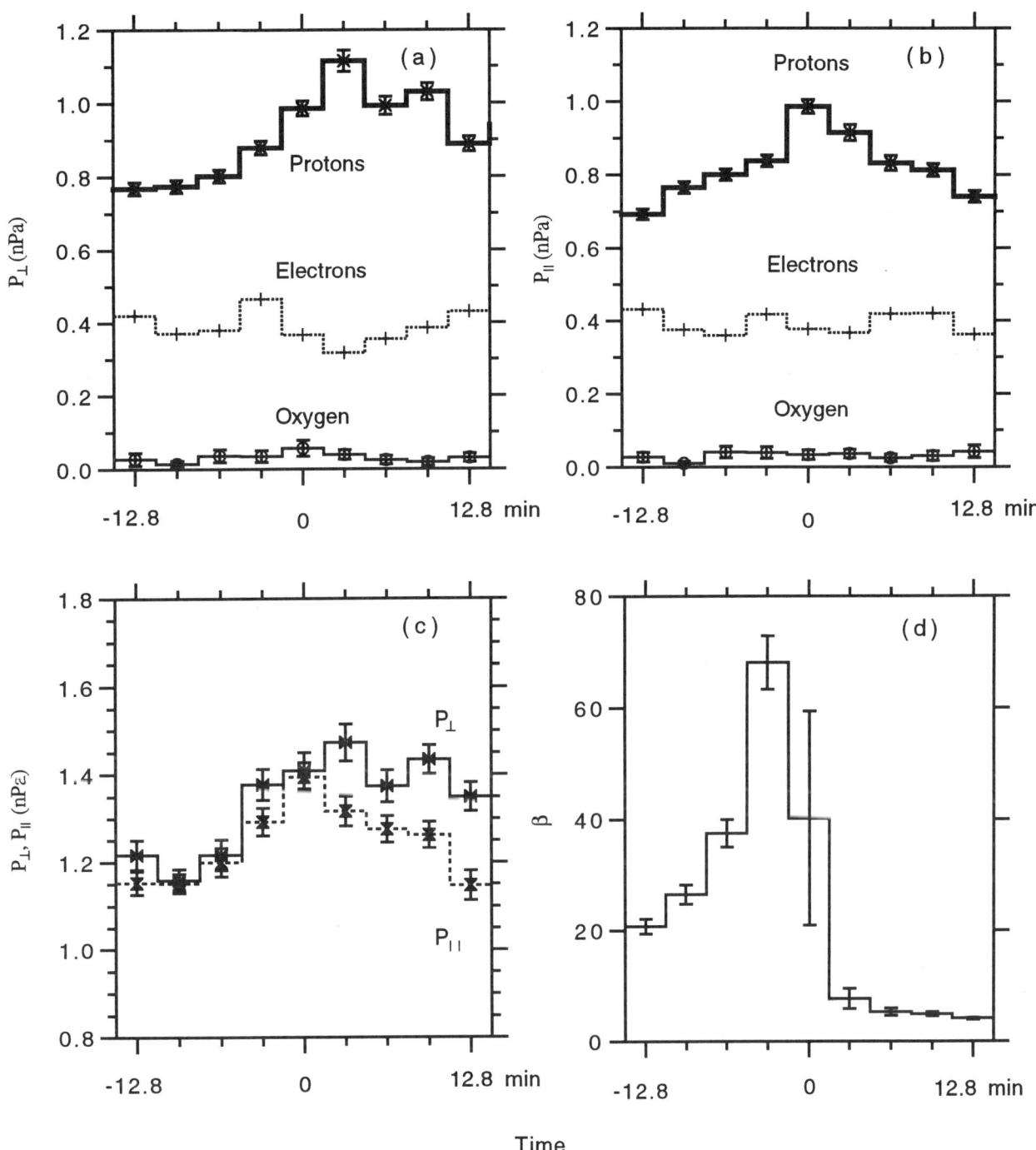

Fig. 3. Time development of pressure components during the current disruption event on June 1, 1985. The pressure components from protons were highest, followed by electrons and then by oxygen ions. The total plasma pressure was nearly isotropic prior to the current disruption onset.

intervals of ~3.2 min, corresponding to the temporal resolution based on the CHEM instrument cycle. Times shown are relative to the disruption onset time. It is important to bear in mind that although the level of magnetic field fluctuation was high during disruption, the field latitude angle was mostly near 90° so that the pitch angle of particles sampled in each sector remained relatively unchanged.

It can be seen from Figure 3 (panels (a) and (b)) that the perpendicular and parallel pressures from protons were highest among the particle species. Next were that of electrons. Oxygen ions contributed least to the pressure components among these three species. There was a general trend of increasing pressure in both perpendicular and parallel components toward the disruption onset. The total plasma pressure was nearly isotropic prior to the onset of activity. Near the end of the disturbance, protons developed a pancake pitch angle distribution ($P_\perp > P_\parallel$) while the electrons were observed to show field-aligned bidirectional anisotropy ($P_\parallel > P_\perp$) as reported by *Klumpar et al.* [1988] and will be discussed later (Figure 4). Furthermore, the total pressure components at the activity onset (panel (c)) were definitely enhanced relative to the level well before the activity onset. The plasma beta (panel (d)) was extremely large throughout the interval, ranging from ~6 to ~70. Its maximum value was reached just prior to the activity onset when the plasma energy density was high and the ambient magnetic field was low. Its minimum value occurred at the end of the activity when the magnetic field strength was highest.

Figure 4 shows an 2-min interval of electron data at the highest time resolution. The 2.9-4.9 keV channel sampled electrons near the thermal energy. The large increase in B_H, signifying current disruption, started at ~23:14:08 UT. Prior to the current disruption, there is an indication that electron fluxes showed a spin modulation with peak fluxes near 90° pitch angle. Electrons thus had a pancake distribution prior to the current disruption, which persisted well past the current disruption onset time till ~23:14:30 UT. This was followed by ~15s of nearly isotropic distribution when the magnetic field turbulence level was high and B_H reversed in sign for two brief intervals as noted before. As pointed out earlier, the north-south field reversals were not accompanied by any significant electron energization. After this period of isotropy, spin modulation of electron pressure reappeared with enhanced values occurring near 0° and 180° pitch angles. This is due mostly to the development of field-aligned counter-streaming electron beams at the higher energies as shown by 15-25 keV electrons. The electron intensities at the thermal and lower energies remained low. Similar field-aligned electron beams have been observed previously at the geosynchronous altitude [*McIlwain*, 1975; *Kremser et al.*, 1988].

The magnetic field fluctuation during current disruptions is generally fully developed such that the frequency band from which fluctuation originates is not readily discernible. However, for weaker events, structures in the power spectra of magnetic field fluctuation can be seen as shown in Figure 5 for the event on August 30, 1986. The power spectra are given in all three components for three separate time intervals of pre-disruption, early disruption, and late disruption. Details on the procedure of extracting the power spectra and the times associated with these intervals were given in *Lui et al.* [1992]. The slight leveling off of power beyond 1 Hz frequency for the pre-disruption spectra is due to digitization sensitivity. The dashed line adjacent to the early disruption spectrum in the B_x panel is drawn to represent the slope of the late disruption spectrum. This reference line indicates the enhanced power to be in the frequency band of ~0.1-1 Hz seen during early disruption. The enhanced power was seen at frequencies near and above the proton gyrofrequency since the averaged proton gyrofrequency was ~0.59 Hz with a low value of ~0.09 Hz. The power spectral index for field fluctuation varied from about –2.3 to –2.5.

A puzzling dilemma noted by *Kaufmann* [1987] is that particle fluxes measured by geosynchronous satellites are typically low when the near-Earth current is intense prior to substorm onset. This raises the question of what constitutes the intensified current during the substorm growth phase. One should recall that a decrease in particle flux may be due to thinning of the plasma sheet in the near-Earth tail prior to the substorm onset: particle flux decrease is a result of the observing spacecraft exiting from the plasma sheet region rather than due to an actual decrease in plasma pressure within the plasma sheet. This indicates that the dilemma is best addressed by observations made near the neutral sheet without the complication brought about by plasma sheet thinning. Such a situation is realized for the event shown in Figure 3 and provides an answer to the above dilemma. The plasma pressure was observed to increase as substorm onset approached. Therefore, the current intensification prior to substorm onset is associated with plasma pressure enhancement. This trend of increasing pressure towards current disruption onset is a common feature of current disruption events.

In a stationary plasma with anisotropic pressures $P_\perp$ and $P_\parallel$, the perpendicular current density is related to two terms involving these pressure components. The first term is associated with the pressure gradient perpendicular to the field (which can be approximated as $P_\perp/(BL_g)$, where B is the local magnetic field and L_g is the scale-length for the pressure gradient) while the second term is associated with the product of the pressure anisotropy and the inverse of the radius of field line curvature (which can be approximated by $(P_\parallel - P_\perp)/(BL_c)$, where L_c is the radius of field line curvature). If $P_\perp > P_\parallel$, then this second term opposes the first term. For this current disruption event, the plasma pressure anisotropy just prior to the activity shows the typical behavior of a predominance of $P_\perp$ over $P_\parallel$. Therefore, a physical process which simply isotropizes the pitch angle distribution of these particles will in fact enhance the current density by eliminating the negative contribution from the curvature current term. This result contradicts the suggested scenario that the current intensification prior to substorm onset could be due to a positive feedback mechanism in which $P_\parallel$ is enhanced relative to $P_\perp$.

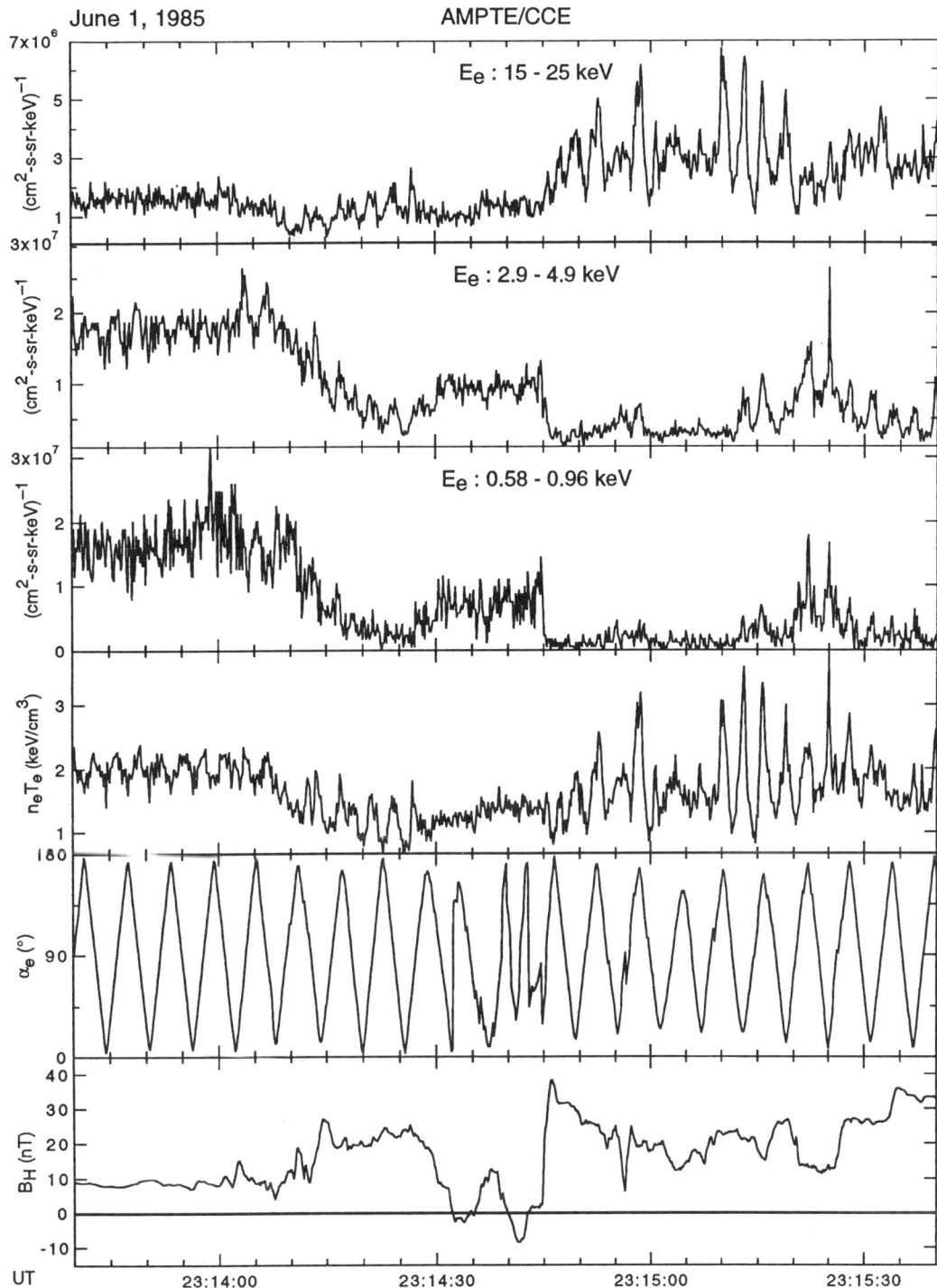

Fig. 4. Electron measurements at their highest time resolution surrounding the current disruption onset. Note that trapped pancake pitch angle distribution of thermal electrons (represented by the 2.9 - 4.9 keV channel) persisted well after the onset. Nearly isotropic electron distribution occurred when the magnetic field fluctuation was high and gave rise to very brief negative B_Z excursions. Afterwards, counterstreaming electron beams along the magnetic field were detected.

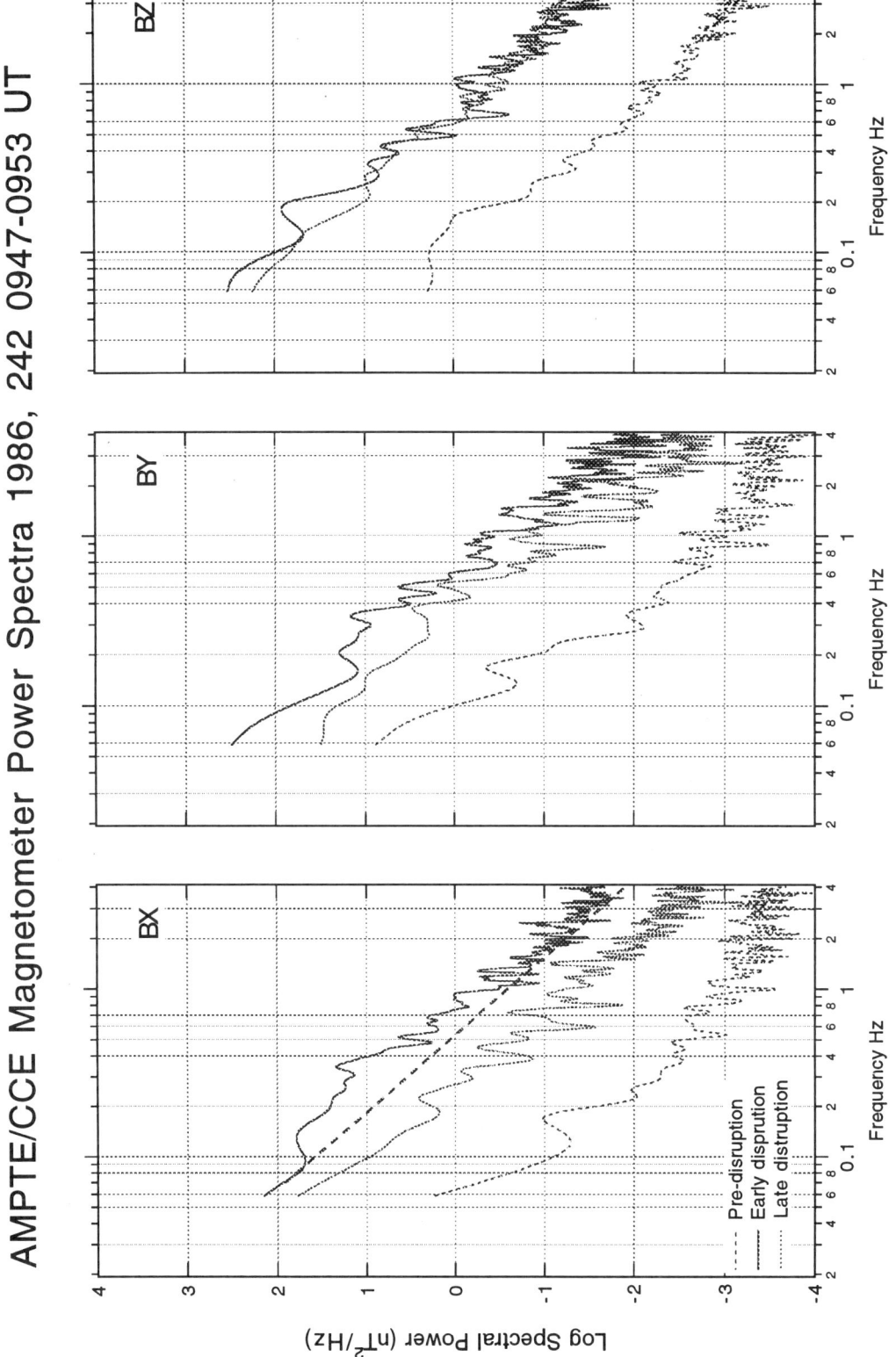

Fig. 5. Power Spectra of magnetic field fluctuation during three intervals of the current disruption event on August 30, 1986. Wave powers were generally two to three orders of magnitude above the pre-onset period.

We can further estimate the development of current density and other relevant parameters just prior to current disruption. Using the ion-sounding technique on MEPA data (>10 keV ion channel which has a temporal resolution of ~24s; note also the ion thermal energy was ~12 keV, close to the energy threshold of this channel), we estimate the pressure gradient scale-length L_g perpendicular to the magnetic field to be ~1400 km just prior to the large magnetic field fluctuation. This pressure gradient scale-length is along the tail axis rather than in the north-south direction since the field latitude angle was close to 90°. This short gradient length-scale implies that the strong radial gradient must be rather localized along the tail axis. The perpendicular current density associated with this pressure gradient is $j_y \approx P_\perp/(BL_g) \approx 71$ nA/m^2. The number density was determined to be ~0.6 cm^{-3}, implying the net relative drift between the ions and the electrons for this case to be ~750 km/s. The proton thermal speed was ~1300 km/s. Thus, the relative drift speed between protons and electrons was a sizable fraction of the proton thermal speed.

The magnetic field in the tail lobe needed to preserve the pressure balance in the north-south direction just prior to current disruption is ~57 nT and the north-south integrated current density is ~91 mA/m. One should note that the current densities may be underestimated because (1) the temporal resolution of measurements is not high enough to capture the highest current density developed seconds before activity onset (note that *Ohtani et al.* [1991] reported evidence for explosive enhancement of current just prior to disruption) and (2) the current density is obtained in most cases not at the center of the neutral sheet and is thus expected not to be the peak value in the current density profile. Nonetheless, the deduced current density values are quite consistent with other earlier estimates [*Lui*, 1978; *Kaufmann*, 1987; *Tsyganenko*, 1989; *Mitchell et al.*, 1990]. Together with the current density of 71 nA/m^2 as deduced above, we estimate the north-south current sheet thickness L_z to be ~1400 km, about twice the local gyroradius of the thermal protons. The short length scale revealed by this observation is another violation of a basic assumption adopted in MHD theory that the length scale of the system is much longer than the thermal ion gyroradius.

Does current disruption phenomenon represent simply a broadening of the current sheet or an actual reduction of north-south integrated current intensity? This can be answered from plasma pressure profile given in Figure 3 also. Just prior to disruption, the tail lobe field B_0 required to balance the total plasma pressure in the north-south direction is ~57 nT, while at the end of the disruption the required tail lobe field is ~43 nT. With these values and the Ampere's law, we obtain a reduction of the north-south integrated current density J_y (=$2B_0/\mu_0$) from ~91 mA/m to ~68 mA/m., i.e. ~25% reduction, in agreement with results from other earlier independent analyses [*Lui*, 1978; *Jacquey et al.*, 1991]. Furthermore, the north-south plasma pressure well before and well after disruption are about the same (Figure 3c), indicating that current (related to north-south plasma pressure) was indeed enhanced during the substorm growth phase and was reduced to nearly the pre-growth phase value after disruption. From this, one can conclude that current disruption (or dipolarization) events in the near-Earth region is not simply due to thickening of the current sheet without current reduction. Furthermore, the amount of current reduction in the near-Earth region after current disruption is approximately the same as the amount of current buildup during the substorm growth phase.

The observation also indicates that current change is enormous in the near-Earth region. No similarly large current density change is reported in the mid-tail or distant-tail regions during the growth phase. The lack of observational evidence that the cross-tail current is reduced in the mid-tail or distant-tail prior to substorm onset therefore suggests that current intensification in the near-Earth region is due to an enhancement of the solar-wind-magnetosphere dynamo in the near-Earth region as deduced from stress balance argument by *Siscoe and Cummings* [1969] and not due to a re-routing of the cross-tail current from further downstream in the tail.

4. A PLAUSIBLE THEORY FOR CURRENT DISRUPTIONS

One proposed mechanism for current disruptions is a current-driven instability known as the cross-field current instability (CCI) [*Lui et al.*, 1991, 1993]. The cross-tail current is visualized to intensify during the substorm growth phase to a point that exceeds the unstable threshold for this instability. A pre-requisite for this process is the drift associated with the current in the neutral sheet to be a substantial fraction of the ion thermal speed. Note that this concept of a current threshold for dynamic activity onset is foreign to the MHD theory since the MHD theory imposes no upper limit for current density and therefore predicts no dramatic consequence for intense current. Using kinetic theory, *Lui et al.* [1991] formulated the dispersion equation relevant to this environment and showed that at least two modes can be excited. One mode is the ion Weibel instability (IWI) which causes current filamentation and electromagnetic turbulence. For the near-Earth plasma parameters, waves are excited over a broadband near and above the ion gyrofrequency. The other mode is analogous to the modified two stream instability (MTSI) which gives rise to ion acceleration perpendicular to the magnetic field and electron acceleration parallel to the field [*McBride et al.*, 1972]. The excited waves for the near-Earth plasma parameters are also broadband at frequencies about one order of magnitude lower than the lower hybrid frequency. Numerical calculations demonstrate that both modes can be excited with a high growth rate at an e-folding time-scale of ~5 to 50s for the near-Earth parameters, quite comparable to the current disruption and substorm onset time-scales.

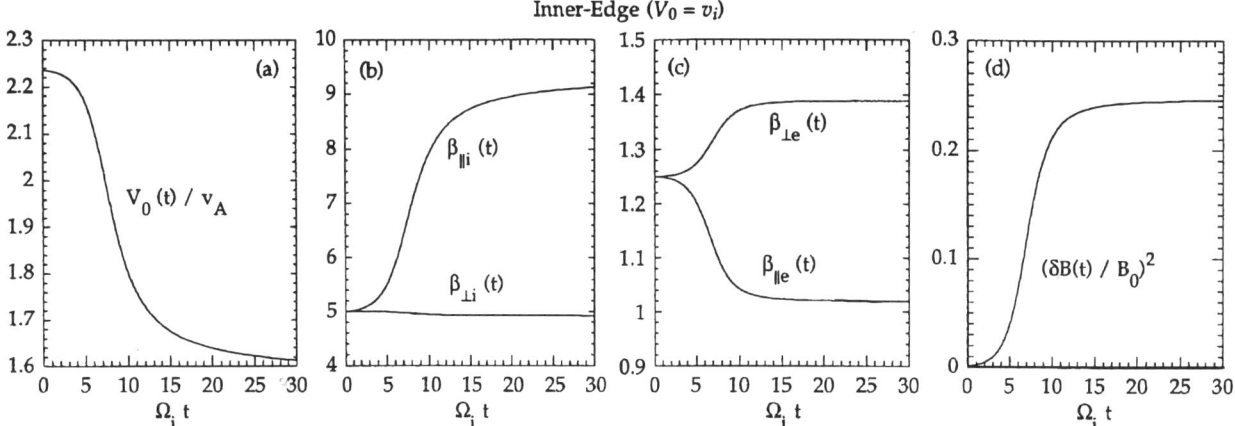

Fig. 6. Results from a quasilinear calculation of the ion Weibel instability showing the temporal development of (a) the normalized ion drift speed, (b) the normalized ion temperatures, (c) the normalized electron temperatures, and (d) the normalized magnetic component of the excited waves.

Figure 6 shows some results from a numerical calculation on the quasi-linear development of the IWI [*Lui et al.*, 1993]. Shown in the figure are the time evolution of (a) the ion drift speed normalized to the Alfvén speed, (b) the perpendicular and parallel ion temperatures relative to the magnetic field energy density, (c) the perpendicular and parallel electron temperatures relative to the magnetic field energy density, and (d) the amplitude of the magnetic component of the excited waves normalized to the initial ambient field strength. The parameters used by the run are $n_e = n_i = 0.6$ cm^{-3}, $T_i/T_e = 4$, $T_i = 12$ keV, and $B_z = 25$ nT. Here, the ion and electron quantities are denoted by subscripts i and e, respectively. The initial drift speed V_0 is assumed to be the ion thermal speed. The initial normalized magnetic fluctuation adopted is 0.001 for the entire unstable range of wavenumber. The normalized time $\Omega_i t = 30$ corresponds to ~12s. Figure 6a indicates that the drift speed departs from the linear prediction in less than half the ion gyroperiod. The drift speed decreases dramatically at time $\Omega_i t = 5$ and reaches close to a limit at $\Omega_i t = 15$ with ~28% reduction of its original value. The ion temperature is seen in Figure 6b to increase by ~86% for its perpendicular component but no significant change occurs in its parallel component. The electron temperatures change very slightly (Figure 6c). The growth of the unstable wave amplitude essentially ends at $\Omega_i t = 15$ (Figure 6d). Summarizing these findings, the free energy associated with the current creates waves which then reduce the current by heating the ions mainly along the magnetic field direction. The quasilinear saturation arises from the decrease in the drift speed and the hotter ion parallel temperature.

The above numerical results should be treated as preliminary because (1) only the parallel propagating mode is considered and (2) local approximation is assumed. Nonetheless, there are a number of observational features for current disruption presented in the previous section which are quite consistent with these preliminary predictions. First, the ion drift speed detected at the onset of current disruptions is quite consistent with the onset threshold predicted by CCI. The onset time scale of current disruption corresponds well to the onset time scale for the instability. There appears to be enhanced wave power near the ion gyrofrequency and above, consistent with the expected frequencies and the broadband nature of the predicted excited waves. Although the theoretical results on IWI in Figure 6 indicates no significant electron heating along the magnetic field, which is different from the observation shown in Figure 4 that indicates occurrence of field-aligned electron acceleration, this difference can be attributed to the fact that the MTSI mode, which is known to give rise to electron acceleration along the magnetic field [*McBride et al.*, 1972], is not included in the theoretical calculation in Figure 6.

5. COUPLING FROM A LOCAL SCALE TO A GLOBAL SCALE

Current disruptions are rather localized activities [*Lui et al.*, 1988, 1992; *Lopez and Lui*, 1990; *Ohtani et al.*, 1991]. One may wonder how these localized activities manifest themselves in the global scale. One immediately evident coupling between these different scales based on the previous discussion of CCI is that the general condition for current disruption is set up locally by the global magnetospheric processes. More specifically, the buildup of an intense cross-tail current during the substorm growth phase is controlled by the enhanced solar wind dynamo as a consequence of southward interplanetary magnetic field [*Siscoe and Cummings*, 1969].

Another cross-scale coupling comes from the ability of the CCI to alter the local current, which then affects the large-scale force equilibrium and gives rise to very efficient global transport of mass, momentum, and energy over an extended region within the magnetosphere. The equilibrium in the neutral sheet is maintained by the balance between the pressure force $\nabla \cdot \mathbf{P}$, where $\mathbf{P}$ is the plasma pressure tensor, and the $\mathbf{j} \times \mathbf{B}$ force. The pressure force acts to push the plasma tailward but is counter-balanced by the earthward $\mathbf{j} \times \mathbf{B}$ force. One may estimate the change in the $\mathbf{j} \times \mathbf{B}$ force in the neutral sheet region due to the nonlinear evolution of the CCI. The change in this force is not simply the reduction in current density since part of the magnetic field is contributed by the "local" current. This is clearly demonstrated by the observation that local current disruption at the neutral sheet is always immediately accompanied by a large increase in the local magnetic field [Lui et al., 1992]. The ambient magnetic field B_0 prior to current disruption can be separated into two parts as $B_0 = B_q + B_l$, where B_l represents the contribution of the local current and B_q represents the quiet time magnetic field at the current disruption region from the more distant sources. Just prior to current disruption, the quiet time magnetic field at the inner edge can be depressed by as much as a factor of 7 [Lui et al., 1992]. If we take the quiet time field to be reduced to one-third of its value prior to activity, which may be considered as representative (i. e., $B_0 = B_q/3$), then $B_l = -2B_q/3$.

The Biot-Savart law indicates the magnetic field from a current system to be proportional to the current density. A current reduction, say, of 25% of its initial value leads to $B_l = -B_q/2$, thus $B_0 = B_q/2$. The $\mathbf{j} \times \mathbf{B}$ force is therefore increased by ~12% above its pre-current-reduction value, providing a net earthward force to transport plasma earthward rapidly from the current disruption region. More generally, it can be readily shown that if the ambient field is suppressed from its quiet time value by a factor g prior to activity and the current is reduced by a factor h afterwards, i. e.,

$$B_0 = B_q/g$$

$$j_{\text{final}} = j_{\text{initial}}/h$$

then the final $\mathbf{j} \times \mathbf{B}$ force will be increased as long as

$$\frac{g}{h}\left[1 - \frac{1}{h}\left(1 - \frac{1}{g}\right)\right] > 1$$

If the left hand side of the inequality is less than 1, then the $\mathbf{j} \times \mathbf{B}$ force will be reduced, resulting in tailward transport of plasma from the current disruption region. It is possible that plasma is injected earthward in a part of the current disruption region and ejected tailward in another part. The forces within the disruption may not be coherent.

The plasma transport resulting from current disruption will also produce field-aligned currents locally as well as at other locations where the plasma is slowed down. Several previous theoretical studies show that magnetic field-aligned current can be generated whenever the cross-field current has a component in the direction of the magnetic field gradient [Sato and Iijima, 1979; Hasegawa and Sato, 1979; Vasyliunas, 1984]. Evaluation of this mechanism for generating field-aligned current is recently conducted with the result that the direction and magnitude of field-aligned currents from this mechanism are in good agreement with observations [Lui, 1995]. Another way by which magnetic field-aligned currents may be generated is by the longitudinal deflection of earthward injected plasma producing plasma flow vorticities. The generation of these field-aligned currents further leads to a transport of momentum and energy to other distant parts of the magnetosphere and the ionosphere.

6. SUMMARY AND CONCLUSIONS

Several features on current disruption are found from CCE observations which allow one to draw several important inferences:

1. The current disruption phenomenon exhibits a time-scale faster than ion gyroperiod and a length-scale near the thermal ion gyroradius, violating the basic assumptions adopted by the MHD theory.
2. The current intensification during the substorm growth phase is associated with an enhancement in the particle pressure at the neutral sheet. Therefore, the observed decreases in particle fluxes at the geosynchronous orbit during this substorm phase arise from thinning of the near-Earth plasma sheet at that time.
3. The lack of dominance of parallel plasma pressure over the perpendicular pressure prior to current disruption onsets provides a strong evidence that an extremely thin current sheet requiring parallel pressure exceeding the perpendicular pressure does not develop at these distances prior to current disruption. Therefore, any current disruption theory requiring parallel pressure exceeding perpendicular pressure prior to disruption onset is not applicable to these events.
4. Remote sensing capability of energetic particle measurements indicates that current disruptions can occur earthward of the downstream distance of ~9 R_E and are not due to disturbances propagating earthward from further downstream distances.
5. The occasional reversals from northward to southward magnetic field or vice versa seen during current disruptions are not necessarily the acceleration site of energetic particles, contrary to the expectation from the near-Earth neutral line substorm model.
6. Current disruption events in the near-Earth region are associated with a current reduction similar to the amount of current buildup during the growth phase. These events are not simply due to a broadening of the current sheet.
7. The observed parameters in the neutral sheet region indicate that the intense cross-tail current developed prior to current disruption is unstable to the cross-field current instability. The observed wave frequencies and the broadband nature in the magnetic field fluctuation spectra are

consistent with the characteristics of the excited waves predicted by that theory. The current disruption time-scale is in agreement with the predicted growth rate. The characteristics of particle energization are compatible with the anticipated interaction between the excited wave modes and the particles. However, the theory needs to be extended to non-local and additional nonlinear analyses in order to provide a more definitive comparison with observations.

8. Localized current disruptions can lead to large-scale magnetospheric changes by altering the local force equilibrium, allowing plasmas to be injected earthward or ejected tailward from the current disruption region. Field-aligned currents are also generated as a result, providing further means of transporting momentum and energy from localized regions associated with current disruption to distant locations in the magnetosphere and the ionosphere.

Acknowledgments. This work was supported in part by the Atmospheric Sciences Section of the National Science Foundation, grant ATM-9114316 and in part by the Space Physics Division of National Aeronautics and Space Administration under grant NAGW-3449 to the Johns Hopkins University.

REFERENCES

Hasegawa, A., and T. Sato, Generation of field aligned current during substorm, *Dynamics of the Magnetosphere*, ed. by S.-I. Akasofu, D. Reidel Publ. Co., Hingham, MA, USA, 529, 1979.

Jacquey, C., J. A. Sauvaud, J. Dandouras, Location and propagation of the magnetotail current disruption during substorm expansion: analysis and simulation of an ISEE multi-onset event, *Geophys. Res. Lett., 18*, 389, 1991.

Kaufmann, R. L., Substorm currents: growth phase and onset, *J. Geophys. Res., 92*, 7471, 1987.

Kaufmann, R. L., J.-T. Horng, and A. Konradi, Trapping boundary and field line motion during geomagnetic storms, *J. Geophys. Res., 77*, 2780, 1972.

Klumpar, D. M., J. M. Quinn, and E. G. Shelley, Counter-streaming electrons at the geomagnetic equator near 9 R_E, *Geophys. Res. Lett., 15*, 1295, 1988.

Kremser, G., A. Korth, S. L. Ullaland, S. Perraut, A. Roux, A. Pedersen, R. Schmidt, and P. Tanskanen, Field-aligned beams of energetic electrons (16 keV $\leq$ E $\leq$ 80 keV) observed at geosynchronous orbit at substorm onsets, *J. Geophys. Res., 93*, 14453, 1988.

Lopez, R. E., and A. T. Y. Lui, A multi-satellite case study of the expansion of a substorm current wedge in the near-Earth magnetotail, *J. Geophys. Res., 95*, 8009, 1990.

Lui, A. T. Y., Estimates of current changes in the geomagnetotail associated with a substorm, *Geophys. Res. Lett., 5*, 853, 1978.

Lui, A. T. Y., Role of cross-field current instability in substorm onsets and intensifications, *Proc. of the International Conference on Substorms, ESA SP-335*, 213, 1992.

Lui, A. T. Y., Formation of the substorm current wedge, submitted to *J. Geophys. Res.*, 1995.

Lui, A. T. Y., R. E. Lopez, S. M. Krimigis, R. W. McEntire, L. J. Zanetti, and T. A. Potemra, A case study of magnetotail current sheet disruption and diversion, *Geophys. Res. Lett., 15*, 721, 1988.

Lui, A. T. Y., C.-L. Chang, A. Mankofsky, H.-K. Wong, and D. Winske, A cross-field current instability for substorm expansions, *J. Geophys. Res., 96*, 11389, 1991.

Lui, A. T. Y., R. E. Lopez, B. J. Anderson, K. Takahashi, L. J. Zanetti, R. W. McEntire, T. A. Potemra, D. M. Klumpar, E. M. Greene, and R. Strangeway, Current disruptions in the near-Earth neutral sheet region, *J. Geophys. Res., 97*, 1461, 1992.

Lui, A. T. Y., P. H. Yoon, and C.-L. Chang, Quasi-linear analysis of ion Weibel instability in the Earth's neutral sheet, *J. Geophys. Res., 98*, 153, 1993.

McBride, J. B., E. Ott, J. P. Boris, and J. H. Orens, Theory and simulation of turbulent heating by the modified two-stream instability, *Phys. Fluids, 15*, 2367, 1972.

McIlwain, C. E., Auroral electron beams near the magnetic equator, in *Physics of the Hot Plasma in the Magnetosphere,* ed. by B. Hultqvist and L. Stenflo, p. 91, Plenum, New York, 1975.

Mitchell, D. G., D. J. Williams, C. Y. Huang, L. A. Frank, and C. T. Russell, Current carriers in the near-Earth cross-tail current sheet during substorm growth phase, *Geophys. Res. Lett., 17*, 583, 1990.

Ohtani, S., K. Takahashi, L. J. Zanetti, T. A. Potemra, R. W. McEntire, and T. Iijima, Tail current disruption in the geosynchronous region, *Magnetospheric Substorms*, ed. by J. R. Kan, T. A. Potemra, S. Kokubun, and T. Iijima, AGU, Washington, DC, p.131, 1991.

Sato, T., and T. Iijima, Primary sources of large-scale Birkeland currents, *Space Sci. Rev., 24*, 347, 1979.

Siscoe, G. L., and W. D. Cummings, On the cause of geomagnetic bays, *Planet. Space Sci., 17*, 1795, 1969.

Takahashi, K., L. J. Zanetti, R. E. Lopez, R. W. McEntire, T. A. Potemra, and K. Yumoto, Disruption of the magnetotail current sheet observed by AMPTE/CCE, *Geophys. Res. Lett., 14*, 1019, 1987.

Tsyganenko, N. A., On the re-distribution of the magnetic field and plasma in the near nightside magnetosphere during a substorm growth phase, *Planet. Space Sci., 37*, 183, 1989.

Vasyliunas, V. M., Fundamentals of current description, *Magnetospheric Currents, Geophys. Monogr. Ser., 28*, ed. by T. A. Potemra, 63, AGU Washington D. C., 1984.

A. T. Y. Lui, The Johns Hopkins University Applied Physics Laboratory, Laurel, MD 20723-6099.

The Role of Microprocesses in Macroscale Magnetotail Dynamics

Joachim Birn

Los Alamos National Laboratory, Los Alamos, New Mexico

Michael Hesse

Electrodynamics Branch, NASA Goddard Space Flight Center, Greenbelt, Maryland

S. Peter Gary

Los Alamos National Laboratory, Los Alamos, New Mexico

Microphysical processes influence the behavior of macroscopic variables and hence the dynamical processes on larger scales primarily through the isotropization of the pressure tensor and through anomalous transport processes which affect the conservation and transport of momentum, energy, and magnetic flux. Using MHD simulations of the dynamic evolution of the magnetotail with deviations from ideal, isotropic, and isentropic MHD, it is demonstrated that each of these deviations may have drastic effects on the stability and dynamic behavior of the magnetotail. As is well known, the occurrence of anomalous resistivity, violating ideal MHD, destabilizes the tail and enables tearing instability and magnetic reconnection. A variation of the value of γ, the ratio of specific heats, in the plausible range from about 1 to 4 has no drastic effect on the evolution of the tail. In the extreme case of isobaric thermodynamic conditions, $dp/dt = 0$, however, the tail dynamics is drastically altered: the tail becomes unstable even in the absence of resistivity, leading to the rapid formation of a thin current sheet, and the dynamic time scale is much shorter than that of ordinary tearing modes. Plasma anisotropy and its reduction through microscopic processes can affect the tail dynamics in two ways. As demonstrated by Birn et al. (1995), the formation of a thin current sheet in the substorm growth phase, as a consequence of magnetic flux added to the tail and the corresponding electric field, is more pronounced and hence will sooner lead to instability if the plasma model is closer to isotropy. Double adiabatic constraints, leading to mirror type anisotropy, tend to stabilize the resistive tearing instability, so that isotropization through microprocesses again produces a destabilizing effect.

1. INTRODUCTION

The stability and dynamics of the Earth's magnetosphere, like that of many other space plasma configurations, is governed not only by the large-scale conservation laws but also by microscopic physics that may crucially affect the behavior of the large-scale structure. A prototype of such an effect is the process of magnetic reconnection, which is usually invoked to explain large-scale energy releases from free energy stored in the magnetic fields, for instance, as a mechanism associated with magnetospheric substorms. Critical elements in this process are localized, small-scale particle scattering mechanisms which have not yet been identified. Large-scale dynamic models of the Earth's magnetotail typically are based on the MHD approximation, which incorporates various microphysics assumptions. The purpose of this paper is to identify some of the places where microphysics assumptions enter MHD and to discuss effects of modifications of these approximations within the framework of large-scale dynamic models that simulate substorm effects in the magnetotail. Obviously, the restriction to large scales does not permit a direct comparison of the MHD model and its modifications with a particle model. The results presented in this paper therefore mainly serve to illustrate possibilities of how microphysics might affect the macroscale dynamics and to demonstrate improved mechanisms to include microscale effects in large-scale fluid models.

2. BASIC EQUATIONS AND APPROXIMATIONS

Microscopic phenomena in space plasmas are usually investigated on the basis of the Vlasov description or by a full particle approach, which become essentially equivalent when the number of particles within a Debye sphere is large and particle-particle interactions are negligible. (The differences between these two approaches are not the subject of this paper.) The Vlasov description consists of Maxwell's equations

$$\frac{\partial \mathbf{B}}{\partial t} = -\nabla \times \mathbf{E} \quad (1)$$

$$\nabla \cdot \mathbf{B} = 0 \quad (2)$$

$$\nabla \times \mathbf{B} = \mu_o (\mathbf{j} + \epsilon_o \frac{\partial \mathbf{E}}{\partial t}) \quad (3)$$

$$\nabla \cdot \mathbf{E} = \frac{1}{\epsilon_o} \sigma \quad (4)$$

combined with the kinetic equations

$$\frac{\partial f_j}{\partial t} + \mathbf{u} \cdot \frac{\partial f_j}{\partial \mathbf{x}} + \frac{e_j}{m_j}(\mathbf{E} + \mathbf{u} \times \mathbf{B}) \cdot \frac{\partial f_j}{\partial \mathbf{u}} = 0 \quad (5)$$

where $f_j(\mathbf{x}, \mathbf{u}, t)$ denotes the phase space distribution of species j with mass m_j and charge e_j, σ the total charge density, and mks units are used. The macroscopic or fluid description of the plasma is usually derived by taking velocity moment integrals of Equation (5), using suitable assumptions. For more details we refer to appropriate text books [e.g., *Gartenhaus*, 1964; *Krall and Trivelpiece*, 1986]. Here we point out some of the major assumptions that could affect the macroscopic behavior, using generalized forms of the moment equations. For simplicity we restrict the discussion to two (singly charged) particle species, ions and electrons. In that case each velocity moment integration of (5) yields two equations, one for each particle species, which are typically added together, using m_j or e_j as multipliers.

The lowest order (in terms of velocity powers) equations are the continuity equations, representing conservation of mass and charge

$$\frac{\partial \rho}{\partial t} + \nabla \cdot \rho \mathbf{v} = 0 \quad (6)$$

$$\frac{\partial \sigma}{\partial t} + \nabla \cdot \mathbf{j} = 0 \quad (7)$$

where ρ is the mass density and $\mathbf{v} = (m_i \mathbf{v}_i + m_e \mathbf{v}_e)/(m_i + m_e)$ denotes the total plasma flow velocity. For large-scale phenomena with phase speeds considerably less than the speed of light it is justified to assume quasineutrality

$$n_i \approx n_e \quad \text{or} \quad \sigma \approx 0 \quad (8)$$

In that case the first term in (7) can be dropped, which is consistent with also dropping the displacement current term in (3), while (4) is not needed because of the smallness of σ.

In next higher order we obtain the momentum equation

$$\rho \frac{d\mathbf{v}}{dt} \equiv \rho \left(\frac{\partial \mathbf{v}}{\partial t} + \mathbf{v} \cdot \nabla \mathbf{v} \right) = -\nabla \cdot \underline{P} + \mathbf{j} \times \mathbf{B} \quad (9)$$

and the generalized Ohm's law

$$\mathbf{E} + \mathbf{v} \times \mathbf{B} = \eta \mathbf{j} + \frac{1}{n_e e} \mathbf{j} \times \mathbf{B} + \frac{1}{n_e e} \nabla \cdot \left(\frac{m_e}{m_i} \underline{P_i} - \underline{P_e} \right) + \frac{m_e}{n_e e^2} \left[\frac{\partial \mathbf{j}}{\partial t} + \nabla \cdot (\mathbf{j}\mathbf{v} + \mathbf{v}\mathbf{j}) \right] \quad (10)$$

where $\underline{P}$ is the total pressure tensor and $\underline{P_i}$ and $\underline{P_e}$ are the ion and electron pressure tensors, respectively, all evaluated in the plasma rest frame. In (9) and (10) quasineutrality has already been used, and in some places the electron mass has been neglected against the ion mass.

We note that the ohmic term $\eta \mathbf{j}$ in (10) follows directly from the kinetic equation only if a collision term is added on the right-hand side of (5) and momentum exchange is assumed to be proportional to the velocity difference of colliding particles. The reason for adding this term in (10) here is that it is the easiest way of approximating nonideal MHD effects that may result from one or more of the other terms on the right-hand side of (10), in order to incorporate it in a fluid model. It satisfies the heuristic argument that microscopic wave-particle scattering generates anomalous resistivity and the expectation that such processes are most likely concentrated in regions of strong gradients, that is, in thin current sheets. A further argument that is often used is that nonideal processes in space plasmas tend to be highly localized, so that it does not matter too much for the large-scale structure and behavior which process actually operates in such regions, as long as it provides approximately the correct nonideal electric field contribution. This argument appears to be most suitable for cases of driven evolution. We note further that the ohmic term also satisfies the generally accepted property that the main induced electric field associated with the magnetic field reconfiguration at and after substorm onset is oriented from dawn to dusk, that is, in the direction of the main current.

The explicit inclusion of any of the other terms on the right-hand side of the generalized Ohm's law (10) typically changes the character of the waves in the system, or rather adds smaller-scale, higher-phase-speed waves not included in MHD, which typically deals with wave periods of 1 min or more, scale lengths of 1 R_E or more, and phase speeds of several hundred km/s. For instance, the addition of the Hall term, which is typically the dominant one, extends the wave regime from the Alfvén into the whistler branch where phase speeds increase with increasing wave number. These effects become important when the characteristic scales become comparable to an ion skin depth c/ω_{pi}, where ω_{pi} is the ion plasma frequency [e.g., *Vasyliunas*, 1975]. For typical magnetotail parameters the ion skin depth is a few hundred km, which is smaller than the MHD scales investigated in our code. A correct treatment of these waves is certainly necessary for questions of stability and internal structure of thin current sheets, which appear to be crucial in the initiation of substorms in the tail [e.g., *McPherron et al.*, 1987; *Lui et al.*, 1990; *Schindler and Birn*, 1993; *Pritchett and Coroniti*, 1994]. We note, however, that the Hall term by itself does not enable reconnection. At present these terms have not been included in large-scale fluid models. It is therefore not clear yet whether and how they influence the large-scale dynamic tail evolution after the initiation of instability.

Other major microphysics assumptions concern primarily the properties of the pressure tensor, which is usually assumed to be isotropic

$$P_\perp = P_\| \tag{11}$$

or gyrotropic

$$\underline{P} = P_\perp \underline{I} + (P_\| - P_\perp)\frac{\mathbf{BB}}{B^2} \tag{12}$$

Finite Larmor radius effects become important when the scale size approaches the ion Larmor radius, which is typically a few hundred km. In that case, deviations from the gyrotropic form (12) may occur which are often put in the form of gyroviscous effects [e.g., *Macmahon*, 1965]. Such approaches have been used in the context of equilibrium models [e.g., *Stasiewicz*, 1987, 1989; *Hau and Sonnerup*, 1991], primarily for the magnetopause, but not yet for studies of tail dynamics. For the case of tail dynamics, *Hesse and Winske* [1994] showed by means of a modified hybrid model that nongyrotropic electron pressure tensors in (10) can lead to the growth of a collisionless tearing instability. This mechanism, however, requires very thin (of the order of ion Larmor radius) current sheets, and its details are therefore outside the framework of MHD.

A closure of the dynamic fluid equations requires equations for the pressure tensor, which are derived from the next higher moment integrations of (5) using further approximations. These approximations are typically ad hoc and rarely based on microphysics of the plasma configuration under consideration. However, in view of the uncertainties of how microphysics affects the macrophysics, it may be justified to explore these effects by using a wide range of closure relations. Among such relations for isotropic pressure p are an isothermal assumption

$$p/\rho = const \tag{13}$$

which requires a fast energy exchange with some thermal reservoir. In the other extreme of negligible heat conduction one can derive the adiabatic law

$$\frac{dp}{dt} + \gamma p \nabla \cdot \mathbf{v} = (\gamma - 1)\eta j^2 \tag{14}$$

where $\gamma = 5/3$ and the ohmic term has been included to be consistent with its inclusion in (10).

A value of $\gamma = 5/3$ in (14) corresponds to the absence of heat flux. Using different values of γ in (14), other approximations can be considered; for instance, the incompressible limit can be obtained for $\gamma \to \infty$ and negligible ohmic heating. Isothermal conditions, obtained for large heat conductivity, are modeled by $\gamma = 1$. Alternatively, an opening of field lines at the ionosphere may also allow for incompressible modes [*Erickson and Heinemann*, 1992] or some other process may lead to fast energy exchange, while maintaining approximate pressure balance in contact with some pressure reservoir. In such cases an isobaric law may also be possible [*Birn et al.*, 1994a]:

$$\frac{dp}{dt} = 0 \tag{15}$$

which formally corresponds to $\gamma = 0$ in (14) when the ohmic term is neglected. Observationally it is not clear which value of γ is the most appropriate. The main reason is that observations show only statistical relationships between p and ρ instead of following the changes within a moving plasma element. These statistics indicate values somewhat [*Baumjohann and Paschmann*, 1989] or even significantly smaller than 5/3 [*Huang et al.*, 1989].

Anisotropic models have been considered in the gyrotropic approximation [e.g., *Hesse and Birn*, 1992b], where one obtains, again including effects of finite resistivity

$$\frac{dP_\|}{dt} + P_\| \nabla \cdot \mathbf{v} + 2P_\|(\nabla \mathbf{v})_\| - 2\eta j_\|^2 = 2R \tag{16a}$$

$$\frac{dP_\perp}{dt} + 2P_\perp \nabla \cdot \mathbf{v} - P_\perp(\nabla \mathbf{v})_\| - \eta j_\perp^2 = -R \tag{16b}$$

with

$$(\nabla \mathbf{v})_\| = \frac{\mathbf{B}}{B} \cdot (\nabla \mathbf{v}) \cdot \frac{\mathbf{B}}{B}$$

Neglecting the heat flux leads to $R = 0$ in (16), that is, the double adiabatic generalization of (14), which again corresponds to the absence of heat flux. In this case (16a) and (16b) are uncoupled. Consequently, during the course of a dynamic evolution, $P_\|$ and $P_\perp$ change differently and anisotropies develop. If the ansiotropy exceeds some instability threshold, microinstabilities are expected to reduce it and hence couple the two pressure components. This is modeled by the term R on the right-hand sides of (16), which will be specified below. It is based on a new, limited closure relation for the temperature anisotropy of ions in plasmas of moderate β (ratio of plasma and magnetic pressure), proposed recently by *Gary et al.* [1994a]. *Denton et al.* [1994] used a similar pair of equations to study the consequences of an anisotropy upper bound in the magnetosheath. The anisotropy relation was discovered by *Anderson et al.* [1994] in the magnetosheath and has been interpreted by *Gary et al.* [1994a] as arising due to pitch-angle scattering of hot magnetospheric protons by the electromagnetic proton cyclotron instability. In the magnetosphere *Gary et al.* [1994b] express this closure as an upper bound on the hot proton temperature anisotropy given by

$$\frac{T_{\perp h}}{T_{\parallel h}} - 1 = \frac{S'}{\beta_{\parallel h}^{0.44}} \left(\frac{n_h}{n_e}\right)^{0.3} \quad (0.05 \leq \frac{n_h}{n_e} \leq 1.0) \quad (17)$$

where S' is taken as a constant with values somewhat less than 0.5, $\beta_{\parallel h} \equiv 2\mu_o p_{\parallel h}/B_o^2$, and the subscript h denotes the hot proton component of the magnetosphere. Because a fluid approach does not discern the difference between the hot and cool magnetospheric ions, and because the energy density of the former is typically substantially greater than approximate pressure balance in contact with some pressure reservoir. In such cases an isobaric law may also be possible [*Birn et al.*, 1994]:

$$\frac{dp}{dt} = 0 \quad (15)$$

which formally corresponds to $\gamma = 0$ in (14) when the ohmic term is neglected. Observationally it is not clear which value of γ is the most appropriate. The main reason is that observations show only statistical relationships between p and ρ instead of following the changes within a moving plasma element. These statistics indicate values somewhat [*Baumjohann and Paschmann*, 1989] or even significantly smaller than 5/3 [*Huang et al.*, 1989].

Anisotropic models have been considered in the gyrotropic approximation [e.g., *Hesse and Birn*, 1992b], where one obtains, again including effects of finite resistivity

$$\frac{dP_\parallel}{dt} + P_\parallel \nabla \cdot \mathbf{v} + 2P_\parallel (\nabla \mathbf{v})_\parallel - 2\eta j_\parallel^2 = 2R \quad (16a)$$

$$\frac{dP_\perp}{dt} + 2P_\perp \nabla \cdot \mathbf{v} - P_\perp (\nabla \mathbf{v})_\parallel - \eta j_\perp^2 = -R \quad (16b)$$

with

$$(\nabla \mathbf{v})_\parallel = \frac{\mathbf{B}}{B} \cdot (\nabla \mathbf{v}) \cdot \frac{\mathbf{B}}{B}$$

Neglecting the heat flux leads to $R = 0$ in (16), that is, the double adiabatic generalization of (14), which again corresponds to the absence of heat flux. In this case (16a) and (16b) are uncoupled. Consequently, during the course of a dynamic evolution, $P_\parallel$ and $P_\perp$ change differently and anisotropies develop. If the anisotropy exceeds some instability threshold, microinstabilities are expected to reduce it

and hence couple the two pressure components. This is modeled by the term R on the right-hand sides of (16), which will be specified below. It is based on a new, limited closure relation for the temperature anisotropy of ions in plasmas of moderate β (ratio of plasma and magnetic pressure), proposed recently by *Gary et al.* [1994a]. *Denton et al.* [1994] used a similar pair of equations to study the consequences of an anisotropy upper bound in the magnetosheath. The anisotropy relation was discovered by *Anderson et al.* [1994] in the magnetosheath and has been interpreted by *Gary et al.* [1994a] as arising due to pitch-angle scattering of hot magnetospheric protons by the electromagnetic proton cyclotron instability. In the magnetosphere *Gary et al.* [1994b] express this closure as an upper bound on the hot proton temperature anisotropy given by

$$\frac{T_{\perp h}}{T_{\parallel h}} - 1 = \frac{S'}{\beta_{\parallel h}^{0.44}} \left(\frac{n_h}{n_e}\right)^{0.3} \quad (0.05 \leq \frac{n_h}{n_e} \leq 1.0) \quad (17)$$

where S' is taken as a constant with values somewhat less than 0.5, $\beta_{\parallel h} \equiv 2\mu_o p_{\parallel h}/B_o^2$, and the subscript h denotes the hot proton component of the magnetosphere. Because a fluid approach does not discern the difference between the hot and cool magnetospheric ions, and because the energy density of the former is typically substantially greater than that of the latter, we here use this equation in the form [*Birn et al.*, 1995]

$$\frac{P_\perp}{P_\parallel} - 1 = \frac{S'}{\beta_\parallel^{0.5}} \quad (18)$$

where the exponent 0.44 in (17) has been approximated by 0.5. If we assume that anisotropies that exceed the treshold given by (18) are reduced to the marginal limit on a fast time scale τ_{micro}, we can represent the quantity R in (16) by

$$R = \text{Max}\left(\frac{1}{3}\frac{P_\perp - P_\parallel - SB\sqrt{P_\parallel}}{\tau_{micro}}, 0\right) \quad (19)$$

where $S = S'/\sqrt{2\mu_o}$.

Equations (16) with (19) describe the expected local effects of an anisotropy-driven microinstability. In addition there may be remote effects, through which local scattering reduces the anisotropy at other places along a field line as well. In this case one expects an earlier isotropization, which may be approached by reducing the value of S in (19). In the following sections we will primarily focus on the effects of different approximations of the energy or pressure equations in unstable models of the magnetotail where a tearing type instability is triggered by the occurrence of anomalous resistivity, using both isotropic and anisotropic models. The initial state and the numerical simulation model will be discussed in Section 3, the effect of various values of γ in isotropic models in Section 4, and anisotropic models with various forms of anisotropy reduction in Section 5.

3. INITIAL STATE AND NUMERICAL MODEL

As an initial state we use a realistic isotropic three-dimensional equilibrium model of the magnetotail developed by *Birn* [1987], which was used previously for dynamic simulation studies [e.g., *Birn and Hesse*, 1991; *Hesse and Birn*, 1992a,b]. It is illustrated in Figure 1 in the form of a magnetic flux surface emanating from an assumed circular cross section at its near-Earth end. The configuration includes flaring of the tail in y and z and an increase in plasma sheet thickness from midnight toward the flanks by a factor of about 2 and a somewhat weaker increase downtail. The length units in Figure 1 are scaled by the current sheet half-thickness L_z at $y = 0$ and the near-Earth end $x = 0$ of the tail section considered. Here and in the following we use dimensionless units, normalized at the near-Earth boundary by the initial current sheet half-thickness L_z, a lobe magnetic field strength B_L, a characteristic Alfvén speed v_A, defined by the characteristic lobe magnetic field strength and the plasma sheet density, and suitable combinations of these quantities. For illustration we use dimensional units $L_z \approx 12000$km, $B_L = 40$nT, and $v_A = 1000$km/s. This yields, for instance, a time scale $t_A = L_z/v_A = 12$s.

The boundary conditions consist of solid ideally conducting walls at each of the boundaries $x=0$, $y=y_{max}=10$, and $z=z_{max}=10$ and an open boundary at $x = -60$. All velocity components are set to zero except at the distant boundary in x, where a free outflow condition is assumed. Neumann boundary conditions ($\partial/\partial n = 0$) are imposed on density, pressure, and the tangential magnetic field components, while the normal magnetic field is held fixed. Symmetry conditions are used at $y = 0$ and $z = 0$. The code consists of an explicit finite difference scheme which solves the time-dependent resistive MHD equations; it has been described earlier in more detail [e.g., *Birn and Hesse*, 1991]. Only constant uniform resistivity is considered here, primarily to avoid prescribing where reconnection and neutral line formation occurs.

4. UNSTABLE EVOLUTION OF ISOTROPIC MODELS

For all but the isobaric model, which will be discussed later, the tail configuration is stable within our simulation period of a few hundred Alfvén times (comparable to about 1-2 hours) when the ideal MHD constraint $\mathbf{E} + \mathbf{v} \times \mathbf{B} = 0$ is used and no external electric field is imposed. The unstable evolution of the magnetotail hence is initiated by imposing finite resistivity. The uniform value chosen here is $\eta = 0.005$, corresponding to a Lundtquist number (magnetic Reynolds number) $R_m = 1/\eta = 200$. The occurrence of finite resistivity typically leads, after some initial diffusion, to the evolution of a generalized tearing instability [e.g., *Birn and Hesse*, 1991] with neutral line formation and plasmoid ejection into the far tail and dipolarization and wedge type

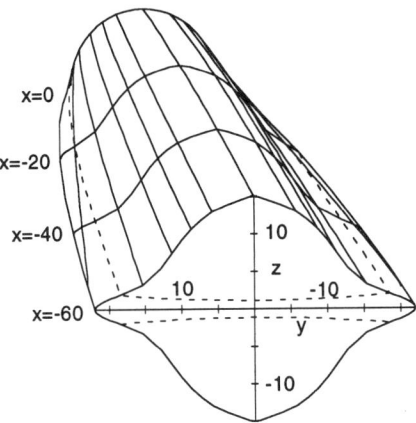

Fig. 1. Initial configuration of the simulations, showing a magnetic flux surface emanating from a circle of 10 units radius at $x = 0$ (which corresponds to a distance of 10-20R_E from the Earth). The dashed lines indicate the plasma sheet (current sheet) width. Length units are scaled with the current sheet half-width ($\approx 2R_E$) at $x = 0$, $y = 0$.

current diversion in the near tail.

Details of the magnetic field and flow characteristics have been presented earlier [e.g., *Birn and Hesse*, 1991; *Hesse and Birn*, 1992a,b] and are not shown here. We focus on a comparison between the different plasma models demonstrated by the temporal evolution of different local or global parameters which are characteristic of the unstable evolution. Figure 2 shows a comparison between isotropic models with various values of γ, ranging from 2/3 to 4. The value $\gamma = 5/3$ corresponds to the absence of heat flux, while $\gamma = 1$ represents infinite heat conductivity, $\gamma = 4$ approaches the incompressible limit, and $\gamma = 2/3$ was chosen to simulate nonadiabatic losses of higher energy particles and/or energies which may even go beyond the isothermal limit [e.g., *Huang et al.*, 1986]. Figure 2 shows (from top to bottom) (a) the reconnected magnetic flux, obtained by integrating $|B_z|$ in the equatorial plane over the region where it has become negative, as an indication of the start and rate of reconnection, (b) the maximum tailward speed v_x, (c) the total field-aligned current of region 1 signature, integrated at the near-Earth boundary $x = 0$, which is part of the substorm current wedge, and (d) the magnetic field component B_z at $x=-5.6$, $y=0$, $z=0$ earthward from the reconnection site, as a representation of the dipolarization of the field in the near tail. The figures are taken from the paper by *Hesse and Birn* [1992a].

All quantities shown in Figure 2 behave quite similarly for different values of γ. There is, however, a tendency for near-Earth effects, dipolarization and substorm current wedge signatures, to be stronger for smaller values of γ. This effect becomes extreme when the limit $\gamma \to 0$ or the isobaric case is considered. As shown by *Birn et al.* [1994a] through linear

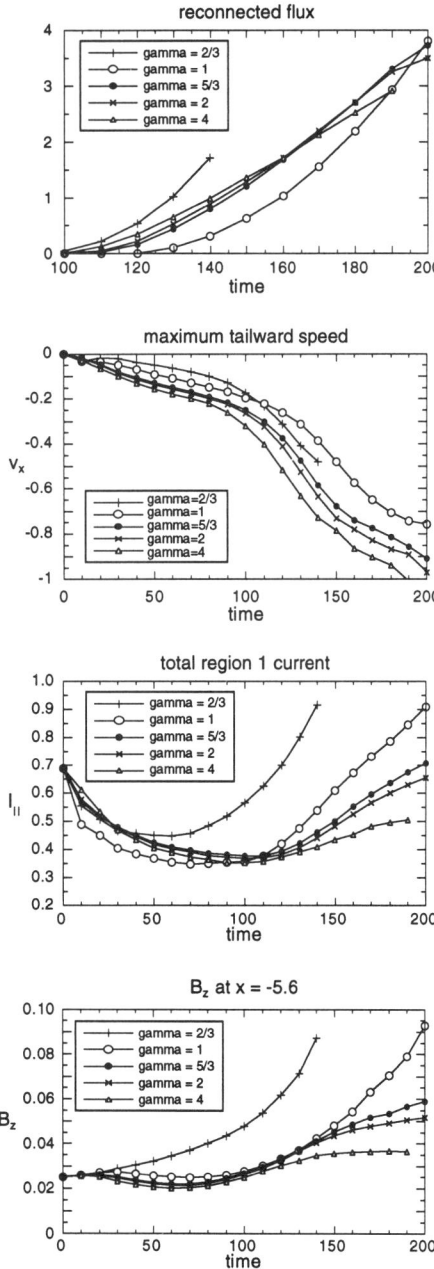

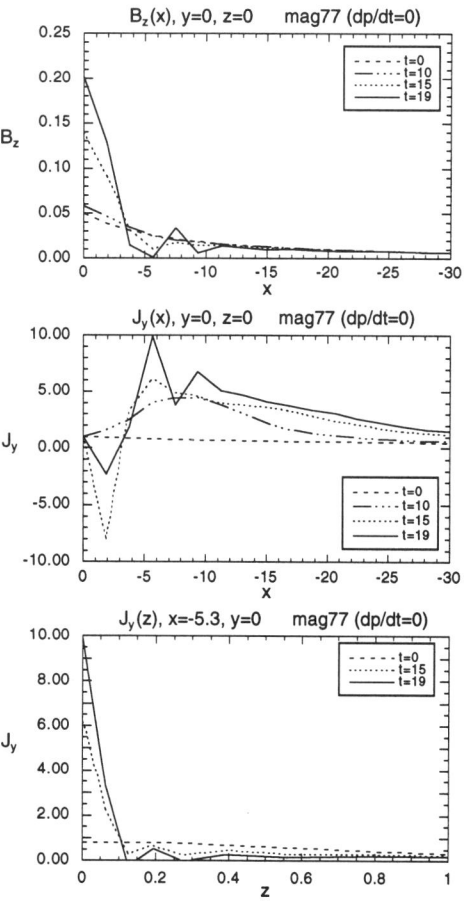

stability analysis and through two-dimensional and three-dimensional numerical simulations [*Birn et al.*, 1994b], this case becomes unstable even in the absence of resistivity. In that case the nonlinear evolution leads to the rapid formation of a thin current sheet in the near tail but also some dipolarization effects further earthward. This is demonstrated by Figure 3 taken from *Birn et al.* [1994b]. The figure shows (from top to bottom) the magnetic field component B_z, the current density j_y, both along the x axis, and j_y as a function of z at $x=-5.6$, $y=0$ for various times indicated in the figure. Note that the changes in B_z and j_y occur within about 5 to 10 Alfvén times, corresponding to about 1 to 2 min. In the absence of resistivity the evolution does not lead to reconnection. However, when a small resistivity $\eta = 0.005$ is added, reconnection and plasmoid formation is initiated on the same fast time scale.

Fig. 2. Temporal evolution of characteristic parameters for isotropic resistive simulations with various values of γ, showing from top to bottom: (1) the total reconnected flux, calculated from integrating B_z at $z=0$ over the area where $B_z < 0$, (2) the maximum tailward speed, (3) the total field-aligned current of region 1 signature integrated at $x=0$, and (4) the magnitude of B_z at $x=-5.6$, $y=0$, $z=0$ as an indicator of dipolarization in the near tail. The different values of γ, the ratio of specific heats, are indicated in the figures.

Fig. 3. Spatial and temporal variation of characteristic parameters for an isobaric nonresistive simulation, showing from top to bottom: (1) B_z as a function of x at $y=0$, $z=0$, (2) j_y as a function of x at $y=0$, $z=0$, and (3) j_y as a function of z at $x=-5.6$, $y=0$. Different times are indicated in the figures.

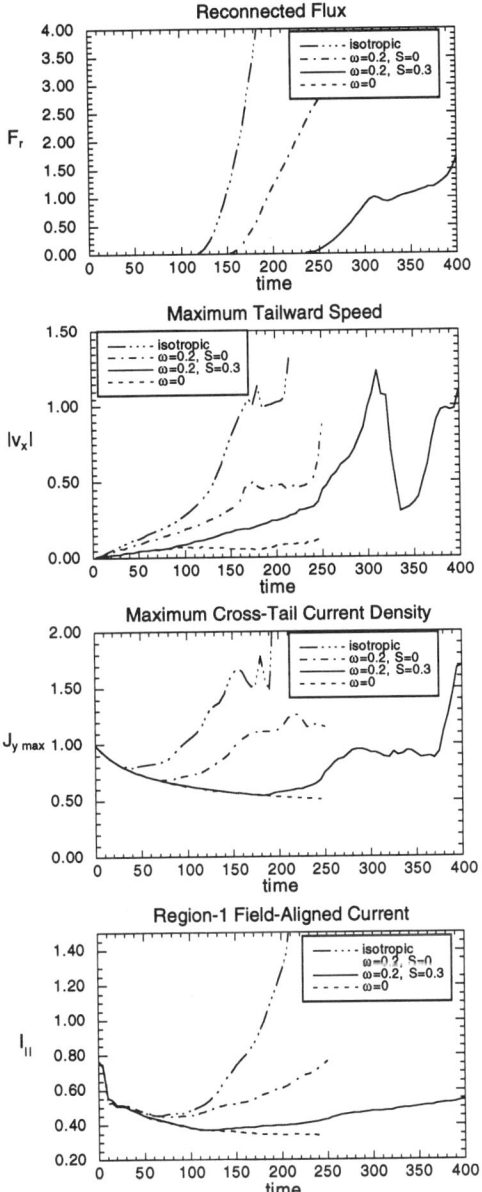

Fig. 4. Temporal evolution of characteristic parameters for nondriven resistive simulations with various anisotropic plasma models, showing from top to bottom: (1) the total reconnected flux, calculated from integrating B_z at $z = 0$ over the area where $B_z < 0$, (2) the maximum tailward speed, (3) to maximum cross-tail current density, and (4) the total field-aligned current of region 1 signature integrated at $x = 0$. The four plasma models indicated in the figures correspond to (A) a fully isotropic model, (B) a model with isotropization of arbitrarily small anisotropy ($S = 0$), (C) a model that is based on the marginal proton cyclotron stability threshold ($S = 0.3$), and (D) a model without any isotropization ($\omega = 0$), representing the double-adiabatic approximation.

5. EFFECTS OF ANISOTROPY AND ISOTROPIZATION

The effects of anisotropy and its reduction through microscale processes are studied within four dynamic models [Birn et al., 1995]: (A) an isotropic model with an adiabatic index $\gamma = 5/3$, as discussed in section 4, (B) an anisotropic model with isotropizing terms as in (19) but $S = 0$, that is, reducing any anisotropy, (C) an anisotropic model described by (19), which reduces anisotropies to the marginal proton cyclotron limit if the instability criterion is satisfied, and (D) an anisotropic model without any isotropization, $R = 0$ in (16), representing the double adiabatic approximation modified by ohmic heating. Again we use constant resistivity $\eta = 0.005$ and the initital state discussed in Section 3 for our three-dimensional simulations. For a better comparison, we have neglected the heat flux in all cases, although in general microscopic processes may also affect heat conduction. The models span the range from instantaneous isotropization (model A) to the absence of isotropization (model D) with C representing the local effects of proton cyclotron anisotropy instability and B simulating more efficient isotropization through remote effects. A time scale $\tau_{micro} = 20s$ of the anisotropy reduction in (19) was chosen to be comparable to, but slightly larger than, the ion cyclotron time scale, leading to a dimensionless value $\omega = 1/3\tau_{micro} = 0.2$.

Figure 4 shows the evolution of characteristic parameters in the four models, (from top to bottom) the total reconnected flux, the maximum tailward speed, the maximum current density j_y, and the integrated region 1 type field-aligned current at the near-earth end $x = 0$. All quantities in Figure 4 demonstrate clearly that the isotropic case is the most unstable while the double adiabatic case is the most stable; in fact it does not even lead to reconnection within the simulation period. (Note, however, that a similar two-dimensional run with a resistive double adiabatic model led to instability, although at a significant delay.) Consistent with our earlier results [Hesse and Birn, 1992b] we find that the closer the model is to isotropy, the more unstable it is. Isotropization, even if it does not lead to full isotropy, has a significant destabilizing effect.

In addition, we find that the anisotropy model can also have a significant effect on the structure of the mode and, most significantly, on the location of neutral line formation. This is demonstrated by Figure 5, which shows the variation of B_z along the x axis for models B (which is close to the fully isotropic case) and C (which reduces anisotropies only to the marginal instability limit). Neutral line formation in model C not only occurs much later than in model B but also much further tailward.

Typical anisotropies are demonstrated for model C in Figure 6, which shows the variation of parallel and perpendicular temperatures with z at two locations in x at $y = 0$. For an indication of the thickness and extent of the plasma/current sheet, the variation of the cross-tail cur-

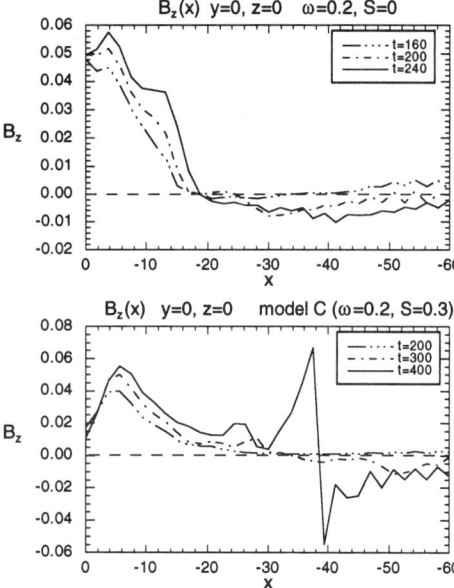

Fig. 5. Variation of B_z as a function of x at $y = 0$, $z = 0$, and different times as indicated for two of the four anisotropic resistive simulations: plasma model B (top), which is close to isotropy, and model C (bottom), based on the marginal proton-cyclotron instability criterion.

activity, implying the occurrence of reconnection, the process which generates resistivity or an equivalent deviation from ideal MHD has not been clearly identified, despite more than 20 years of effort.

In contrast to the effects of anomalous resistivity or other possible deviations from ideal MHD, the role of microphysics elsewhere within a fluid plasma model of the magnetotail has received much less attention. In this paper we have therefore concentrated on several models of energy transport and pressure closure relations and their effects on the dynamic evolution of the magnetotail.

Isotropic models were varied by the value of the ratio of specific heats γ, which is equivalent to the polytropic index in the absence of nonadiabatic (here, ohmic) heating. Large values of γ represent nearly incompressible conditions, $\gamma = 5/3$ the adiabatic case, valid for an isotropic plasma in the absence of heat flux, $\gamma = 1$ the isothermal case, and $\gamma \to 0$ or $dp/dt = 0$ the extreme, isobaric case, which requires fast energy and/or mass exchange with some external reservoir while maintaining approximate pressure balance. All unstable resistive cases are characterized by the growth of a generalized tearing mode with neutral line formation and plasmoid ejection into the far tail, while the near-Earth region exhibits dipolarization of the magnetic field and the current reduction and diversion associated with the substorm current wedge [McPherron et al., 1973].

rent density j_y is also shown. The location $x = -7.5$ represents a region of plasma sheet thickening and dipolarization, whereas the location $x = -30$ is close to the x-type neutral line where the plasma/current sheet has thinned considerably. At both locations (and others as well) the central part of the plasma sheet has been heated (the initial temperature in the chosen units was uniformly 0.5) but remained approximately isotropic. Large mirror-type anisotropies ($p_\perp > p_\parallel$) have developed in the boundary regions of the plasma/current sheet, extending into the lobes. A comparison with model D (not shown here) and with the marginal stability criterion (18) shows that these anisotropies are similar to those in the fully anisotropic model, but reduced in magnitude, primarily in the boundary regions, consistent with the marginal stability limit.

6. SUMMARY AND DISCUSSION

Using isotropic and anisotropic resistive magnetofluid simulations, we have investigated the influence of microphysical processes on the large-scale dynamic evolution of the magnetotail. Obviously, the occurrence of anomalous resistivity is one of the most drastic possible consequences of microscopic processes, as it changes the stability of the tail and enables magnetic reconnection. Although observations have provided ample evidence that flux transfer from closed to open field line regions and vice versa and plasmoid formation and release are an important part of magnetospheric

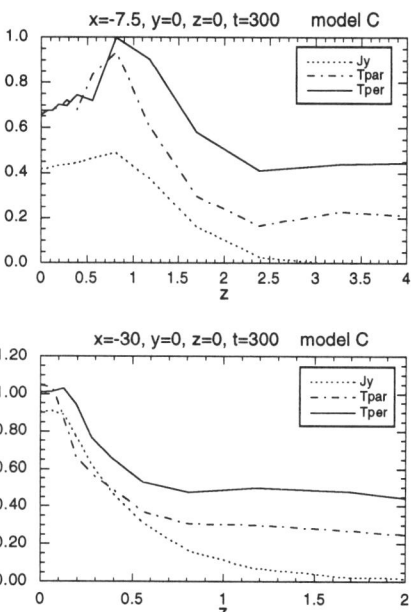

Fig. 6. Variation of the parallel and perpendicular temperatures with z at $x = -7.5$ (top) and $x = -30$ (bottom), $y = 0$, $t = 300$ for the anisotropic plasma model C, based on the marginal proton-cyclotron instability criterion. The current density J_y (dotted line) is shown also to indicate the extent of the plasma/current sheet.

A variation of the value of γ in the plausible range from about 1 to 4 has no drastic effect on the evolution of the tail. Reconnection proceeds similarly, and dipolarization and the generation of field-aligned currents associated with the substorm current wedge is qualitatively similar for different values of γ. There is, however, a tendency for the latter, near-tail effects to be more pronounced for smaller values of γ. In the extreme case of isobaric thermodynamic conditions, $dp/dt = 0$, the tail dynamics is drastically altered: the tail becomes unstable even in the absence of resistivity, leading to the rapid formation of a thin current sheet, and the dynamic time scale is much shorter [*Birn et al.*, 1994a,b]. Although this model is included here mostly to illustrate plasma behavior under extreme conditions, there may be some application to the late substorm growth phase when gradual externally driven current sheet thinning has led to a configuration in which the ions become unmagnetized, releasing previous energy constraints, while electrons are still magnetized and prevent reconnection of magnetic field lines.

Anisotropic simulations were based on three different plasma models, the double adiabatic model in which perpendicular and parallel pressure components are uncoupled, a model based on the marginal ion cyclotron instability limit, in which anisotropies that exceed this limit are reduced on a short time scale [*Gary et al.*, 1994a,b; *Birn et al.*, 1995], and a model in which any anisotropy is reduced on such short time scale. Consistent with earlier results by *Hesse and Birn* [1992b], we find that the more isotropic models tend to be more unstable: reconnection starts sooner, the reconnection rate is larger, and region-1 type field-aligned currents associated with the substorm current wedge increase faster when the model is closer to isotropy. As in driven nonresistive simulations [*Birn et al.*, 1995] there is a strong influence of the anisotropy reduction model on the spatial variation of B_z in the neutral sheet $z = 0$. As a consequence, the reconnection site and the location where a magnetic neutral line forms can vary significantly from model to model. This indicates that one has to be cautious in applying the results on the reconnection site from one particular model to the actual magnetotail. The sensitivity of the location of the reconnection site to changes in the plasma model may also mean that this site can vary considerably in the actual magnetotail, as suggested by *Cowley* [1991].

The effects of the anisotropy reduction on the stability of the tail in the resistive dynamic simulations are consistent with effects found in nonresistive simulations driven by an externally applied electric field [*Birn et al.*, 1995]. In these simulations a thin current sheet forms inside the near-tail plasma sheet combined with a reduction of B_z in the neutral sheet as a consequence of the external driving. Both the current density enhancement and the reduction of B_z are stronger the closer the model is to isotropy. This means that a more isotropic model will more likely, or sooner, lead to instability than a strongly anisotropic model (such as a double adiabatic model) when the instability threshold depends on the magnitude of the current density [e.g., *Lui et al.*, 1990] or on the magnitude of B_z in the neutral sheet [e.g., *Schindler*, 1974].

The anisotropic dynamic magnetotail models develop mirror-type anisotropies ($p_\perp > p_\parallel$) in the (low beta or low current density) boundary regions of the plasma sheet and the adjacent lobe regions, while the (high beta or high current density) center region of the plasma/current sheet remains close to isotropy. The type of anisotropy and its relation with the plasma beta are consistent with observations of *Lui et al.* [1992] after substorm onset.

Our results on the effects of anisotropy appear to differ from earlier studies of the collisionless tearing instability with anisotropic pressure which have been carried out using both analytic techniques [e.g., *Chen and Palmadesso*, 1984] and computer simulations [e.g., *Ambrosiano et al.*, 1986; *Pritchett and Coroniti*, 1990]. These investigations have shown that a mirror-type anisotropy present in the initial configuration or caused by quiet-time convection can significantly enhance the growth of this mode. However, these results cannot be compared directly to the conclusions drawn here and in *Hesse and Birn* [1992b] because the models are quite different. We first note that the above mentioned investigations concerned the collisionless tearing mode, while we investigated the anisotropy effects within the framework of resistive tearing. Chen and Palmadesso and Ambrosiano et al. have investigated the effects of anisotropy of the initial state under the same collisionless dynamic model, whereas our work has addressed different dynamical models using the same isotropic equilibrium state. Whereas anisotropy of a particular state represents a possible additional source of free energy and hence may lead to faster growth of an instability, the double adiabatic invariants constrain the flow and conversion of energy, so that a release of these constraints by pitch angle scattering may in fact destabilize.

The investigation by *Pritchett and Coroniti* [1990] is perhaps more similar to ours, as they start from a rather similar isotropic initial state. In their simulation, however, driven and unstable effects appear combined and they do not compare different dynamic models under the same initial and boundary conditions. Nevertheless, there is the possibility that anisotropy affects a collisionless tearing mode, as studied by the aforementioned authors, and a resistive tearing mode, studied in our paper, in different ways. We note that there is also a significant difference in the temperature dependence of the instability threshold. Lowering the ion temperature reduces the threshold of a current driven microinstability, which presumably is the cause of anomalous resistivity, and hence enables more easily the initiation of the resistive tearing mode, while it tends to stabilize the collisionless ion tearing mode.

We have shown that the large-scale dynamics of the magnetotail may strongly depend on microphysical processes incorporated in various approximations. This influence shows most clearly in the time scale of the evolution and factors that control the onset of tail instability, whereas the qualitative spatial features of the instability remain unchanged.

The reason for this fact is that MHD governs the vast majority of the magnetospheric plasma system, while deviations from MHD become important only in localized regions [e.g., *Vasyliunas*, 1975]. Dissipative processes in these localized regions mainly determine nonideal electric fields in unstable nondriven cases, which control in particular the time scale of the large-scale dynamics. A simple fluid model obviously cannot correctly represent features which depend critically on these localized processes. Examples include the onset time, details of the reconnection rate, and the exact growth rate of a large-scale tail instability. On the other hand, the similarities in the large-scale evolution of various modifications of the fluid model demonstrate that, through adoption of an appropriate dissipation model, MHD is able (and presently the only tool) to correctly represent the large-scale evolution on a qualitative and, depending on the quality of the dissipation model, possibly semi-quantitative level.

Acknowledgments. This work was supported by the U.S. Department of Energy's Office of Basic Energy Sciences through its Geosciences Research Program and by NASA's Space Physics Theory and SR&T Programs.

REFERENCES

Ambrosiano, J., L. C. Lee, and Z. F. Fu, Simulation of the collisionless tearing instability in an anisotropic neutral sheet, *J. Geophys. Res.*, *91*, 113, 1986.

Anderson, B. J., S. A. Fuselier, S. P. Gary, and R. E. Denton, Magnetic spectral signatures in the Earth's magnetosheath and plasma depletion layer, *J. Geophys. Res.*, *99*, 5877, 1994.

Baumjohann, W., and G. Paschamnn, Determination of the polytropic index in the plasma sheet, *Geophys. Res. Lett.*, *16*, 295, 1989.

Birn, J., Magnetotail equilibrium theory: The general three-dimensional solution, *J. Geophys. Res.*, **92**, 11,101, 1987.

Birn, J., and M. Hesse, The substorm current wedge and field-aligned currents in MHD simulations of magnetotail reconnection, *J. Geophys. Res.*, *96*, 1611, 1991.

Birn, J., K. Schindler, L. Janicke, and M. Hesse, Magnetotail dynamics under isobaric constraints, *J. Geophys. Res.*, *99*, 14,863, 1994a.

Birn, J., K. Schindler, and M. Hesse, Magnetotail dynamics: MHD simulations of driven and spontaneous dynamic changes, in *Substorms 2*, Proc. 2nd Int. Conf. on Substorms, Fairbanks, Alaska, 7-11 March 1994, edited by J. R. Kan, J. D. Craven, and S.-I. Akasofu, p. 135, Geophys. Inst., Univ. Alaska Fairbanks, 1994b.

Birn, J., S. P. Gary, and M. Hesse, Microscale anisotropy reduction and macroscale dynamics of the magnetotail, *J. Geophys. Res.*, in press, 1995.

Chen, J., and P. Palmadesso, Tearing instability in an anisotropic neutral sheet, *Phys. Fluids*, *27*, 1198, 1984

Cowley, S. W. H., The role and location of magnetic reconnection in the geomagnetic tail during substorms, in *Magnetospheric Substorms, Geophys. Monogr. Ser.*, vol. 64, edited by J. R. Kan, T. A. Potemra, S. Kokobun, and T. Iijima, p. 401, AGU, Washington, D.C., 1991.

Denton, R. E., B. J. Anderson, S.P. Gary, and S. A. Fuselier, Bounded anisotropy fluid model for ion temperatures, *J. Geophys. Res.*, *99*, 11,225, 1994.

Erickson, G. M., and M. Heinemann, A mechanism for magnetospheric substorms, in Substorms 1, *Eur. Space Agency Spec. Publ., ESA SP-335*, 587, 1992.

Gartenhaus, S., Elements of plasma physics, *Holt, Rinehart & Winston*, New York, 1964.

Gary, S. P., B. J. Anderson, R. E. Denton, S. A. Fuselier, and M. E. McKean, A limited closure relation for anisotropic plasmas from the Earth's magnetosheath, *Phys. Plasmas*, *1*, 1676, 1994a.

Gary, S. P., M. B. Moldwin, M. F. Thomsen, D. Winske, and D. J. McComas, Hot proton anisotropies and cool proton temperatures in the outer magnetosphere, *J. Geophys. Res.*, *99*, 23,603, 1994b.

Hau, L.-N., and B. U. Ö. Sonnerup, Self-consistent gyroviscous fluid model of rotational discontinuities, *J. Geophys. Res.*, *96*, 15,767, 1991.

Hesse, M., and J. Birn, Three-dimensional MHD modeling of magnetotail dynamics for different polytropic indices *J. Geophys. Res.*, *97*, 3965, 1992a.

Hesse, M., and J. Birn, MHD modeling of magnetotail instability for anisotropic pressure, *J. Geophys. Res.*, *97*, 10,643, 1992b.

Hesse, M., and D. Winske, Hybrid simulations of collisionless reconnection in current sheets, *J. Geophys. Res.*, *99*, 11177, 1994.

Huang, C. Y., and L. A. Frank, A statistical study of the central plasma sheet: Implications for substorm models, *Geophys. Res. Lett.*, *13*, 652, 1986.

Huang, C. Y., C. K. Goertz, L. A. Frank, and G. Rostoker, Observational determination of the adiabatic index in the quiet time plasma sheet, *Geophys. Res. Lett.*, *16*, 563, 1989.

Krall, N. A. and A. W. Trivelpiece, Principles of Plasma Physics, *McGraw Hill*, New York, 1986.

Lui, A. T. Y., A. Mankofsky, C.-L. Chang, K. Papadopoulos, and C. S. Wu, A current disruption mechanism in the neutral sheet: a possible trigger for substorm expansions, *Geophys. Res. Lett.*, *17*, 745, 1990.

Lui, A. T. Y., R. E. Lopez, B. J. Anderson, K. Takahashi, L. J. Zanetti, R. W. McEntire, T. A. Potemra, D. M. Klumpar, E. M. Greene, and R. Strangeway, Current disruptions in the near-Earth neutral sheet region, *J. Geophys. Res.*, *97*, 1461, 1992.

Macmahon, A., Finite gyro-radius corrections to the hydromagnetic equations for a Vlasov plasma, *Phys. Fluids*, *8*, 1840, 1965.

McPherron, R. L., C. T. Russell, and M. A. Aubry, Satellite studies of magnetospheric substorms on August 15, 1968, 9, Phenomenological model for substorms, *J. Geophys. Res.*, *78*, 3131, 1973.

McPherron, R. L., A. Nishida, and C. T. Russell, Is near-Earth current sheet thinning the cause of auroral substorm onset?, in *Quantitative Modeling of Magnetosphere-Ionosphere Coupling Processes*, edited by Y. Kamide and R. A. Wolf, p. 252, Kyoto Sangyo University, Kyoto, Japan, 1987.

Pritchett, P. L., and F. V. Coroniti, Plasma sheet convection and the stability of the magnetotail, *Geophys. Res. Lett., 17*, 2233, 1990.

Pritchett, P. L., and F. V. Coroniti, Convection and the formation of thin current sheets in the near-Earth plasma sheet, *Geophys. Res. Lett., 21*, 1587, 1994.

Schindler, K., A theory of the substorm mechanism, *J. Geophys. Res., 79*, 2803, 1974.

Schindler, K., and J. Birn, On the cause of thin current sheets in the near-Earth magnetotail and their possible significance for magnetospheric substorms, *J. Geophys. Res., 98*, 15,477, 1993.

Stasiewicz, K. A gyroviscous model of the magnetotail current layer and the substorm mechanism, *Phys. Fluids, 30*, 1401, 1987.

Stasiewicz, K. A fluid finite ion larmor radius model of the magnetopause layer, *J. Geophys. Res., 94*, 8827, 1989.

Vasyliunas, V. M., Theoretical models of magnetic field line merging, 1, *Rev. Geophys., 13*, 303, 1975.

J. Birn and S. P. Gary, Space and Atmospheric Sciences Group, M.S. D466, Los Alamos National Laboratory, Los Alamos, NM 87545. (e-mail: jbirn@lanl.gov, pgary@lanl.gov)

M. Hesse, Electrodynamics Branch, Code 696, NASA Goddard Space Flight Center, Greenbelt, MD 20771. (e-mail: hesse@kapaa.gsfc.nasa.gov)

Irreducible Cross-Scale Coupling in the Magnetotail Current Sheet: A Tutorial

J. B. Harold[1] and J. Chen

Beam Physics Branch, Naval Research Laboratory, Washington, D.C.

The magnetotail current sheet plays a central role in the global dynamics of the magnetosphere, representing storage of magnetic energy that can be released during large-scale reconfiguration of the magnetic field. An important question is what controls the current sheet structure and particle distribution functions and how they may be coupled to the surrounding regions. A key factor in determining the current sheet properties is that the charged particle motion is collisionless, with negligible classical resistivity or diffusion. It has been found that local particle distribution function properties (microscale), the current sheet structure (mesoscale), and the physical processes on larger (macro) scales are inseparably coupled and that the properties on a given (e.g., macro) scale are not reducible to those on other (e.g., micro) scales. The present paper provides a tutorial review of recent research results in the underlying charged particle motion and the properties of the magnetotail current sheet.

1. INTRODUCTION

A long-standing question in plasma physics is whether and how "anomalous" transport coefficients can be defined to incorporate the average microscale properties into a macroscopic description. (See, for example, *Papadopoulos* [1977] and *Manheimer and Boris* [1977].) This question is particularly relevant to space physics because of the important energetic phenomena such as magnetospheric substorms and solar eruptions that occur in these highly collisionless media. In such systems, large-scale properties, on the scale of the dimensions of the systems of interest, are often modeled by means of fluid approaches in which average moments with suitable closure are used. One well-established macroscopic framework is that of magnetohydrodynamics (MHD). This powerful approach, however, neglects kinetic effects which are important for energy exchange between fields and particles. In a highly collisionless system such as the magnetosphere, the classical transport coefficients (e.g., resistivity and diffusion) result in time scales that are longer and lengths scales that are shorter by many orders of magnitude than those inferred from observations. To remedy this shortcoming, kinetic effects on microscales are often invoked to justify enhanced, "anomalous" resistivity and diffusion to replace those determined by classical collisionality. Implicit in this approach is the premise that physical processes occurring on macroscales are reducible to a description parameterizable by phenomena on microscales or vice versa. In the context of the present paper, we refer to properties of distribution functions as local, or microscale, attributes while average fluid quantities are macroscopic properties. We will also use the term "nonlocal" to denote influences over large distances, including particle motion between the current sheet and its surrounding regions. These processes typically involve distances over an Earth radius (R_E) and larger.

The physical system of interest in this paper is the magnetotail and its associated current sheet. Its importance lies in its role as a magnetic energy reservoir; when this energy is released, the entire magnetosphere can undergo significant reconfiguration and disturbances. In this paper we discuss the ways in which current sheet properties on different scales are determined, with the focus on quasi-stationary properties. In the context of anomalous transport, the issue can be stated as whether or not one can define a "collisionless" con-

---

[1] NRC/NRL Postdoctoral Research Associate.

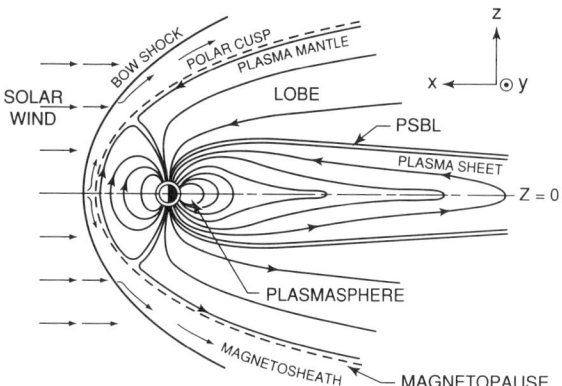

Fig. 1. A schematic illustration dipicting various regions of the Earth's magnetosphere. The coordinate system used in this paper is shown. (From *Chen* [1992].)

ductivity [*Speiser*, 1970; *Martin*, 1986; *Horton and Tajima*, 1990] so that one can determine the current distribution by $\mathbf{J}(\mathbf{x}) = \sigma \mathbf{E}(\mathbf{x})$, where σ is reducible to local plasma properties such as density and temperature with no reference to the electric field or macroscopic constraints.

Figure 1 is a schematic of the magnetosphere which identifies various regions. The so-called magnetospheric coordinate system is indicated. The magnetosphere is organized by large-scale electric and magnetic fields which govern the motion of charged particles. The central plasma sheet (CPS) contains a thin current sheet at the midplane, $z = 0$, where the B_x component reverses sign. Outside the current sheet, the particle motion is adiabatic. Inside the current sheet itself, however, the motion is nonadiabatic and can be stochastic [*Chen and Palmadesso*, 1986]. Figure 2 gives a conceptual diagram of a region of the magnetotail, with the current sheet indicated by the box. The shaded regions represent particle populations (f_1 and f_2) streaming toward the current sheet. The short arrows indicate that an incoming particle population consists of components originating from different spatial and temporal sources. Each component can be modified by different physical processes remote from the current sheet. These components can be mapped over long distances to the current sheet where they influence its structure. Thus, the colllisionless particle motion couples the current sheet to distant regions of the magnetosphere through the incoming distribution function. Figure 2 illustrates that the current sheet is in continual communication with the surrounding region; charged particles enter and leave the current sheet [*Speiser*, 1965] so that what happens outside the current sheet, perhaps tens of R_E away, can affect the current sheet through the nonlocal particle motion.

In this tutorial, we will discuss (1) how the collisionless particle dynamics control both the local distribution functions (microscale) and structure of the entire (mesoscale) current sheet (cross-scale coupling) and (2) how physical processes remote from the current sheet can determine the current profile (macroscale, nonlocality and spatial coupling). We find that the magnetotail current sheet cannot be reduced to a description parameterized by local processes because of the intrinsically nonlocal particle motion.

The remainder of this tutorial is organized in the following way. In section 2 we briefly discuss the particle dynamics in the current sheet region. Although this subject has been reviewed extensively [*Chen*, 1992; *Lakhina*, 1994], some salient dynamical properties will be summarized here for use in later sections. In section 3 we discuss cross-scale coupling as it relates to the effects of phase space partitioning on the distribution function. In section 4 we discuss the dependence of the quasi-equilibrium current sheet structure on the plasma distribution. In section 5 we discuss the role of nonlocal ("remote") processes in determining the current sheet structure. Section 6 presents numerical results modelling such effects.

2. PARTICLE DYNAMICS

In its simplest form, the basic particle motion in the CPS is described by the solutions of

$$m\frac{d\mathbf{v}}{dt} = \frac{e}{c}\mathbf{v} \times \mathbf{B}, \qquad (1)$$

with a magnetic field of the form

$$\mathbf{B}(x, z) = B_0 f(z)\hat{\mathbf{x}} + B_n \hat{\mathbf{z}}, \qquad (2)$$

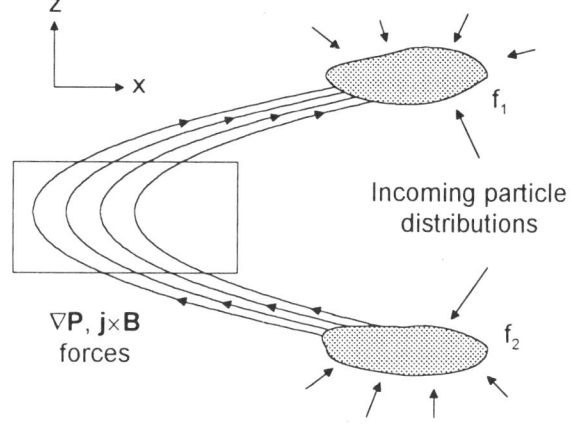

Fig. 2. Illustration of an idealized section of the current sheet. The incoming plasma distribution consists of a collection of disparate populations originating from different regions of the magnetosphere. Both the local $\mathbf{J} \times \mathbf{B}$ forces and the nature of the incoming distribution influence the structure of the current sheet.

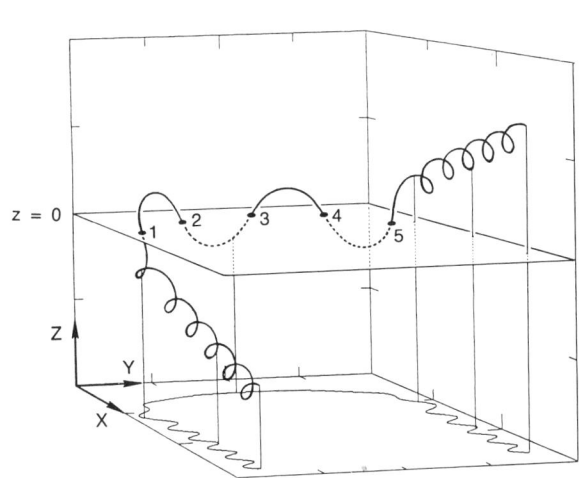 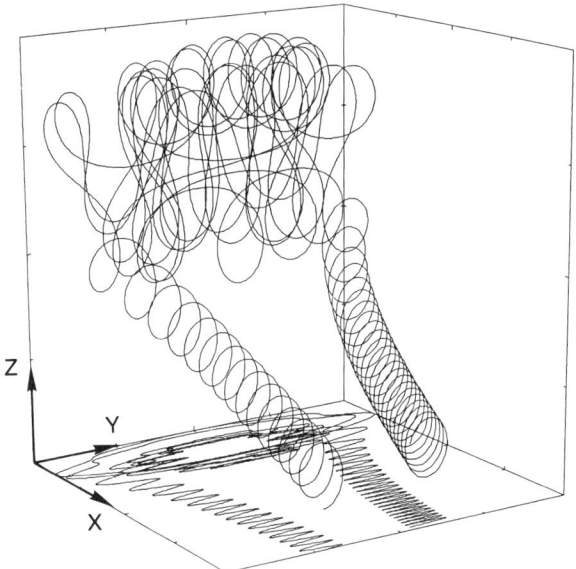

Fig. 3. (a) A characteristic Speiser-type orbit and its x-y projection. (b) A stochastic orbit. The x axis is compressed by ~10. (Adapted from *Chen* [1992].)

where m and e are the particle mass and charge and B_0 and B_n are constants. The coordinate system is shown in Figure 1. A frequently used field configuration is the modified Harris field given by

$$f(z) = \tanh(z/\delta), \quad (3)$$

where δ is the characteristic half-thickness of the current sheet so that $B_x \simeq B_0$ in the asymptotic region ($z/\delta \gg 1$). In general, f can also depend on x, and B_n can be nonuniform. In addition, there can be a nonzero magnetic field B_y across the tail. For our purposes, it is sufficient to consider the basic dynamical properties of the particle motion described by (1) and (2). We will only touch on some of these more realistic refinements as appropriate.

We refer to the field configuration (2) as a quasi-neutral sheet. In the magnetotail there exists, in addition, a cross-tail electric field, E_y. If E_y and B_n are uniform, then E_y can be transformed away by shifting into a coordinate frame moving with $\mathbf{V}_D \equiv (cE_y/B_n)\hat{\mathbf{x}}$, the so-called de Hoffman-Teller frame. In reality, of course, E_y and B_n vary in space. The criterion for application of one-dimensional models given by (2) (i.e., translationally invariant in x and y) is that $\Delta \ll L$, where Δ is the excursion distance of particles in x in the current sheet and L is the gradient scale length in x and y [*Burkhart and Chen*, 1993]. This condition is typically satisfied in the current sheet in the distant tail, say, $|x| \gtrsim 20$ R_E, where $\rho_n \simeq 1$–$2\,R_E$ while $L \gtrsim 20\,R_E$ [*Behannon*, 1968]. The other aspect of spatial dependence is the long-distance ($\gg R_E$) motion outside the current sheet. There, the motion

is magnetized, and adiabatic invariance can be used to map particles and distribution functions.

The basic motion of particles in the vicinity of the current sheet is a fast oscillation in the B_x component with the characteristic frequency $\Omega_0 \equiv eB_0/mc$ and a slow rotation about B_n with a frequency $\Omega_n \equiv eB_n/mc$ [*Speiser*, 1965; *Sonnerup*, 1971]. It was first shown by *Chen and Palmadesso* [1986] that the particle motion is nonintegrable, i.e., there are only two constants of the motion in involution, and that the phase space describing the particle motion is partitioned into distinct regions corresponding to three basic classes of trajectories: stochastic, transient, and regular. They suggested that the phase space partitioning, i.e. the existence of different classes of orbits, can influence the distribution functions and lead to the generation of observable non-Maxwellian features. This process was referred to as "differential memory." *Büchner and Zelenyi* [1986] proposed to model the stochastic motion by a mapping theory based on jumps ΔI in the action $I \equiv (2\pi)^{-1} \oint \dot{z}\,dz$, and *Büchner and Zelenyi* [1989] and *Brittnacher et al.* [1991] calculated ΔI using the so-called slow separatrix crossing technique [*Cary et al.*, 1986]. *Büchner and Zelenyi* [1987] then argued that chaotic electron motion may destabilize the collisionless tearing mode. More recently, *Delcourt et al.* [1994] suggested that the magnetic moment variation due to chaotic motion can be modeled using an impulsive centrifugal force. For a detailed review of the basic nonlinear dynamics of charged particles in the quasi-neutral sheet geometry, see *Chen* [1992].

Figure 3 shows two examples of particle trajectories in

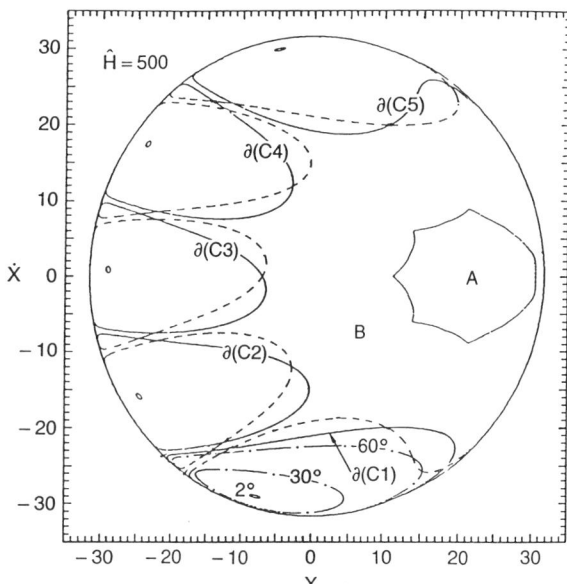

Fig. 4. Phase space boundaries in a Poincaré surface of section for $b_n = 0.1$ and $\hat{H} = 500$. Solid lines denote the entry region and its mappings $\partial(C1)$–$\partial(C5)$. Dashed lines denote the exit region and its time reversed mappings $\partial(C1')$–$\partial(C5')$. The contours of particles with asymptotic pitch angles of $2°$, $30°$, and $60°$ are shown.

the modified Harris field. These two orbits have the same kinetic energy, with different initial phase and pitch angles. Figure 3a is a transient orbit, which is sometimes referred to as a Speiser orbit, and Figure 3b shows a stochastic orbit. Examples of stochastic orbits were first discussed by *West et al.* [1978]. Such orbits exhibit extreme sensitivity to initial conditions [*Chen et al.*, 1990b] so that one orbit can have very different orbital characteristics from those with slightly different initial conditions.

The sensitivity to initial conditions makes it virtually impossible to fully understand the particle motion using a small number of selected orbits. Physically observable quantities, on the other hand, are determined by ensembles of particles. Thus, it is the properties of families, or classes, of orbits rather than those of individual orbits that one must understand accurately. A well-established technique for obtaining accurate properties of families of solutions of a dynamical system is the Poincaré surface of section method [e.g., *Lichtenberg and Liebermann*, 1983]. This technique provides a powerful tool for analyzing low-dimensional systems in which the solutions can be chaotic, defying analytic representation or simple characterization. In our case, the particles in the system described by equations (1)–(3) move in a six dimensional phase space. Two constants of the motion may be normalized away, leaving four initial conditions to describe a trajectory. If the energy is specified and a surface chosen, we are left with two initial conditions to completely specify the motion of the particle. Thus each point in a two-dimensional surface of section uniquely specifies a solution, i.e., an entire particle orbit. Applying this method, we can take a cut (not necessarily planar) through the phase space. For our purposes, it is convenient to choose the midplane, $z = 0$, and plot the coordinates $(x, \dot{x})$ of the midplane crossings. Such surfaces of section have been discussed extensively in the literature. Figure 4 shows a number of underlying structures in a representative surface of section for $\hat{H} = 500$. (See *Chen and Palmadesso* [1986] for the full plot.) The boundaries are shown for distinct phase space regions with A designating the integrable region and B the stochastic region. The curves $\partial(C1)$–$\partial(C5)$ bound the corresponding transient regions $C1$–$C5$ (not labeled explicitly). The integrable orbits remain trapped in the current sheet indefinitely. The transient orbits enter the current sheet through region $C1$, mapping from there to region $C2$, then to $C3$, and so on until they leave the current sheet. The exit region, where escaping trajectories cross the midplane for the last time, is simply the mirror image of $C1$ (the mirror images of each of the 5 transient regions are marked with dashed lines). Stochastic particles enter through $C1$, map through regions $C2$–$C5$, but rather than leaving the current sheet immediately, these particles undergo repeated traversals of the midplane in region B before finally exiting through $C1' \cap \overline{C5}$, where Cj' denotes the mirror image of Cj, $\overline{Cj}$ the complement, and $\cap$ is the intersection of the two regions.

As a particle traverses the midplane (Figure 3a shows one traversal), its contribution to the cross-tail current due to the shift in the guiding center is

$$j_y \simeq \frac{e\Delta y}{\Delta t}, \qquad (4)$$

where Δy is the total displacement in y and Δt is the amount of time the particle spends in the current sheet. (See section 4 for a more precise discussion of Δy.) For transient orbits, we have $\Delta y \simeq 2\rho_n$ and $\Delta t \simeq \pi\Omega_n^{-1}$. As can be seen from Figures 3 and 4, particles executing stochastic orbits have approximately the same Δy, while Δt is much greater, typically 30–100 times longer, than that for transient orbits for most values of $\hat{H}$ [*Burkhart and Chen*, 1991]. As a result, the stochastic contribution to the total current is much less significant than the transient contribution. In an ensemble of particles, the relative population of transient versus stochastic particles is a key factor in determining the total cross-tail current.

Much work had been done prior to the realization that the particle motion is nonintegrable. The most important and qualitatively new understanding from the nonlinear dynamics point of view is the phase space structure, i.e., the partitioning of the phase space. The phase space partitioning is generic to and depends on the quasi-neutral sheet magnetic topology, rather than on the details of the field. Physically, each of the phase space regions identified earlier results from a specific feature of the current sheet: (1) the current sheet has a finite thickness, so that there is a distinction between the inside and outside of the current sheet, giving rise to the transient orbits (regions C1–C5), and (2) the existence of two frequencies of the motion, Ω_0 and Ω_n, implies that some orbits are quasi-periodic, closing onto themselves (regular orbits, region A), while others are stochastic (region B), not closing onto themselves. These phase space properties are often thought of as part of chaos. However, chaos *per se*, defined by exponential divergence of nearby orbits, refers to only one aspect of the particle motion and should be clearly distinguished from the existence of distinct classes of trajectories. In the remainder of the paper, we will discuss the role that the collisionless particle motion plays in coupling properties on micro-, meso-, and macroscales and over long spatial distances. We will find that the phase space partitioning plays a central role in connecting the dynamical properties of the particles and physical observables.

3. PARTICLE DISTRIBUTION FUNCTIONS

In studying kinetic physics, the fundamental quantity is the particle distribution function $f(\mathbf{x}, \mathbf{v})$ where $\mathbf{x}$ and $\mathbf{v}$ are the spatial and velocity coordinates of particles. Various wave-particle interactions can take place locally where f deviates from a Maxwellian. In this context, the spatial extents of interest are generally of the order of the wavelengths in relation to local spatial gradients, and wave frequencies are compared with time scales of variation of the system. However, in a collisionless plasma, particles detected at a given time and spatial location can have different temporal and spatial histories so that the particle distribution functions are not simply determined by local processes, as would be the case in highly collisional systems. As a result, particle distribution functions should in general be non-Maxwellian. This also implies that non-Maxwellian features carry information regarding large-scale properties of the system. In this section, we illustrate this point using one example of non-Maxwellian signature that was predicted based on the phase space structure and that can yield information on the current sheet structure from *in situ* particle measurements.

Return for a moment to Figure 2. Suppose a population of particles is streaming toward the current sheet. The individual particles can enter the current sheet and then exit either toward the side from which they came or toward the other side of the current sheet. In the former (latter) case, particles undergo back (forward) scattering by the current sheet. It was found that the scattering by the current sheet is chaotic but has coherent resonance [*Chen et al.*, 1990a; *Burkhart and Chen*, 1991]. The resonance condition is given by

$$N = \hat{H}^{1/4} - 0.6, \quad (5)$$

where N is an integer, and $\hat{H}$ is defined by

$$\hat{H} \equiv \frac{H}{mb_n^2 \Omega_n^2 \delta^2}. \quad (6)$$

Here $H = mv^2/2$, Ω_n is the cyclotron frequency about the normal field component, $b_n \equiv B_n/B_0$, and δ is the current sheet half-width. For values of $\hat{H}$ corresponding to odd (even) integer values of N, the majority of particles entering the current sheet from $z = +\infty$ will escape to $z = +\infty$ ($-\infty$) giving rise to coherence. For $\hat{H}$ between two successive resonance values, the incoming ensemble of particles from either side are forward and backscattered with approximately equal probability.

Consider, then, a situation in which the populations f_1 and f_2 in Figure 2 are unbalanced: for instance, $f_1 > f_2$. Then the outgoing distribution measured at $z = +\infty$ will contain peaks at odd values of N, where most of detected particles will be f_1 particles which have been backscattered. Similarly, it will contain valleys at even values of N, where most of the f_1 particles escape through the opposite side of the sheet, and the measured population consists primarily of forward scattered f_2 particles ($f_2 < f_1$). Such structures have been observed in both test particle simulations and observed distribution functions [*Chen et al.*, 1990a]. Similar structures ("beamlets") have been obtained in test-particle simulations by *Ashour-Abdalla et al.* [1991]. Such non-Maxwellian features arise because different classes of orbits occupy distinct regions of phase space; a distribution of particles consists of subpopulations with very different dynamical properties (i.e., differential memory). In this application, it is the coherent scattering of the transient orbits and chaotic scattering of stochastic orbits that produce the predicted distribution function features.

The $\hat{H}^{1/4}$ scaling was originally derived to calculate the number of transient regions [*Chen and Palmadesso*, 1986], and the coherence in forward and backscattering was discussed in the 1D modified Harris model by *Burkhart and Chen* [1991] and in a 2D magnetic field by *Ashour-Abdalla et al.* [1991, 1993]. It has been further studied by *Kaufmann et al.* [1993]. It is interesting to examine the origin of this resonance effect. Physically, this scaling law can be traced to the property that $B_x \propto z$ near the midplane. The thickness of the region where the linear field profile applies is of the order of $d \equiv (2\rho_0 \delta)^{1/2}$ [*Sonnerup*, 1971]. This scaling law has been extended to field dependences of the form $B_x \propto z^s$ where s takes on small odd integer values, yielding a resonance scaling law of

$$N \sim \hat{H}^\chi,$$

where $\chi \equiv (1/2)s/(s+1)$. This scaling has been verified numerically [*Chen*, 1993]. Thus, the resonance structure in locally measured distribution functions is nonlocally determined by the average field geometry of the entire current sheet, a mesoscale property.

From a practical point of view, if the locally measured particle distributions depend on the field profile of the current sheet, then one should be able to infer the current sheet thickness from *in situ* particle data. Possible methods have been suggested along this line [*Chen et al.*, 1990a; *Lui*, 1993]. Note also that distribution function features detected at a point may have been generated tens of R_E away.

Another significant implication of this example of resonance in distribution functions is that it results from some asymmetry in the northern and southern incoming distribution functions, which are subjected to different remote processes as illustrated in Figure 2. This is a clear example of the fact that local distribution functions are not determined by local processes, in contrast to phenemena describable in a fluid picture or a collisional system. In the next section, we will discuss how the current sheet structure is in turn determined by the incoming distribution functions.

4. EQUILIBRIUM CURRENT SHEET STRUCTURE

We now consider current sheet properties under quasi-equilibrium conditions that may apply to the quiet-time magnetosphere or to slowly evolving conditions such as the substorm growth phase (slow in comparison with the time scale of particle motion through the current sheet, $\Delta t \simeq \pi \Omega_n^{-1}$, which is typically a few minutes or shorter). In particular, we will illustrate specific relationships between the incoming particle distributions (e.g., f_1 and f_2 in Figure 2), the current sheet structure, and the particle distributions inside the current sheet.

The dependence of the current sheet structure on the cross-tail E_y is a long-standing problem [*Eastwood*, 1972, 1974]. Recent studies have included the full nonlinear particle dynamics, rather than being limited to the Speiser orbits, and can be grouped into two broad types: self-consistent and non-self consistent simulations. By "self-consistent," we mean that the magnetic field satisfies $\nabla \times \mathbf{B} = (4\pi/c)\mathbf{J}$, where $\mathbf{J} = e \int \mathbf{v} f d^3 \mathbf{v}$. The non-self-consistent simulations include global test particle calculations, in which particles are injected into fixed field distributions and the particle trajectories, pressures, etc. are examined. These models seek to relate different regions of the magnetosphere by mapping particles using large-scale fields. Some studies have adopted empirical magnetic fields using electric fields computed from ionospheric models [*Delcourt et al.*, 1989a] or imposed uniform cross-tail electric field in the magnetotail [*Ashour-Abdalla et al.*, 1991, 1994; *Horton et al.*, 1993]. In these approaches, the electric and magnetic fields are not necessarily self-consistent for the global system. More recently, *Joyce et al.* [1995] have performed test particle simulations using field data acquired from global MHD simulations. This technique provides electric and magnetic fields self-consistently within a model, and the Poynting flux, $\mathbf{S} = (c/4\pi) \mathbf{E} \times \mathbf{B}$, gives energy flux and bulk flow that are globally consistent, although the test-particle results do not satisfy $\nabla \times \mathbf{B} = (4\pi/c)\mathbf{J}$.

Self-consistent simulation studies of current sheets have been carried out, including iterative test particle simulations with no electron dynamics [*Burkhart et al.*, 1992; *Pritchett and Coroniti*, 1993; *Holland and Chen*, 1993], hybrid simulations with kinetic ions and fluid electrons [*Pritchett and Coroniti*, 1993; *Burkhart et al.*, 1993; *Cargill et al.*, 1994], and full particle simulations (kinetic ions and electrons) [*Pritchett and Coroniti*, 1994]. These approaches result in magnetic field configurations which are consistent with the particle flows. They have been used to examine the structure and stability of the current sheet and its dependence on a uniform cross-tail electric (E_y) and normal magnetic (B_z) fields. In such a simple field geometry, E_y can be set to zero by transforming to the so-called de Hoffman-Teller frame which moves at $v_D \equiv cE_y/B_z$ relative to the original frame. If the incoming distribution f_{in} (e.g., f_1 or f_2 in Figure 2) is a Maxwellian whose only nonthermal motion is the $\mathbf{E} \times \mathbf{B}$ drift in the original frame, then f_{in} in the de Hoffman-Teller frame is a drifting Maxwellian with v_x replaced by $v_x - v_D$ and $E_y = 0$. In this case, the system may be characterized by b_n and v_D, where typically $b_n = B_n/B_0 \ll 1$ for thin current sheets. Extensive work has been carried out using such simulations. A salient conclusion from these studies is that a drifting Maxwellian with large v_D/v_{th} (i.e., a Maxwellian f_{in} with a large E_y) leads to a current sheet with a highly peaked density [*Eastwood*, 1972; *Burkhart et al.*, 1992; *Pritchett and Coroniti*, 1993], while a drifting Maxwellian with $v_D/v_{th} \ll 1$, typical for the quiet-time magnetotail, produces a current sheet with an essentially flat density distribution across the current sheet [*Holland and Chen*, 1993]. This flat density profile is consistent with those of observed current sheets during quiet times [*McComas et al.*, 1986], and corresponds to broader current sheets than highly peaked profiles.

Note that while one can choose to characterize the system using **E**, describing the system as an isotropic Maxwellian shifted into a frame with $v_D = cE_y/B_n$, this choice is not unique. The effect of moving into the de Hoffman-Teller frame is to make f_{in} more field aligned. Thus, the system could be characterized by the pitch angle distribution (PAD) rather than by the value of E_y or v_D/v_{th}. In the case of a simple, single-component Maxwellian f_{in}, these two approaches are equivalent. For more complex distributions, however, they are not. The cross-tail current depends on the form of f_{in} so that it cannot be reduced to a simply $J_y \sim E_y$ scaling in which the proportionality constant depends only on local quantities.

The precise dependence of J_y on E_y is a matter of great interest, since it is at the heart of the question of whether or not one can obtain the current sheet structure in the form of $\mathbf{J} = \sigma \mathbf{E}$, where σ is independent of **E**. The current contribution resulting from the shift in the guiding center of an individual particle is simply $j_y \sim \Delta y/\tau$, where Δy is the total cross-tail shift undergone by the particle during its traversal of the current sheet, and τ is the crossing time. It can be shown that [*Cowley*, 1978]

$$\Delta y = \left(\frac{2\hat{H}}{1+b_n^2}\right)^{1/2} (|\cos\beta_1| + |\cos\beta_2|), \qquad (7)$$

where β_1 and β_2 are the asymptotic incoming and outgoing pitch angles, respectively, and $\hat{H}$ has been previously defined in equation (6). Although there is no simple relation between β_1 and β_2, both transient and stochastic orbits have the typical value $\Delta y \sim 2\rho_n$ (e.g., Figure 4). Note that it is important to distinguish between chaotic versus regular pitch angle scattering: transient orbits are pitch angle scattered but have no sensitivity to initial conditions.

It has been suggested [*Martin*, 1986] that the exponential divergence rate of nearby chaotic orbits (the Lyapunov exponent) can serve as an effective decorrelation time scale, allowing one to define a collisionless conductivity σ. More recently, *Horton and Tajima* [1990, 1991] calculated the two-time velocity autocorrelation function and defined from it a zero-frequency chaotic conductivity. However, it has been argued [*Holland and Chen*, 1992] that intrinsic ambiguities exist in this definition of conductivity. This is because the particle motion is chaotic only inside the current sheet, and the calculated autocorrelation functions depend on the starting and ending points of the time integration. Earlier, *Speiser* [1970] suggested that a collisionless inertial conductivity of the form $\sigma = (ne^2/m)\tau$ can be defined, where $\tau \simeq \pi\Omega_n^{-1}$ is the time the particle spends in the current sheet for transient orbits. Returning to equation (4) and substituting $\rho_n = v/\Omega_n$ and $\Delta t = \pi\Omega_n^{-1}$, we obtain $j_y \propto v$ for a single particle. For $v_D \gtrsim v_{th}$, this implies $j_y \propto v_D \propto E_y$, consistent with the collisionless conductivity of Speiser. However, *Burkhart et al.* [1992] found that the current sheet density is dependent on E_y, scaling as $n \propto E_y^{4/3}$. This implies that σ is a function of E_y with $\sigma \propto E_y^{4/3}$. Furthermore, in the $v_D \ll v_{th}$ regime, $j_y \propto v_{th}$, so that the current is determined by the precise form of the velocity distribution function [*Holland and Chen*, 1993]. In this regime, v and hence j_y become essentially independent of E_y, and a well-defined current sheet can exist in the limit of $E_y = 0$ (i.e., $v_D = 0$). Thus, **J** depends on f_{in}, and one cannot write a linear relationship $J_y(z) = \sigma(z)E_y(z)$ where σ is dependent on local quantities and independent of the electric field. As discussed in connection with (4), J_y critically depends on the population of transient relative to the stochastic particles. In earlier studies in the X-line geometry, [*Martin*, 1988; *Burkhart et al.*, 1990] also found that the simple relation $J_y \propto E_y$ (or $\sigma \propto \tau$) is generally not applicable.

5. NONLOCAL EFFECTS ON THE INCOMING PLASMA DISTRIBUTION

So far, we have discussed the effects of the current sheet on the plasma distribution function, and the effects of the large scale fields (i.e., E_y) on the structure of the current sheet. Each of these is an example of large (meso, macro) to small (micro) scale coupling. Conversely, in this section we discuss coupling from the micro to the meso scale, in the form of the effects of the plasma distribution function on the structure of the current sheet.

In the previous section it was noted that in the simplified one-dimensional models, increasing E_y was equivalent to increasing the field alignment of f_{in}. However, in the case of more complex distributions no single frame of reference exists that would reduce the system to one described solely by an E_y and an isotropic Maxwellian. In the general case both E_y and the distribution function must be specified to define the system. It has already been demonstrated that the current sheet can be quite sensitive to the details of the incoming distribution: even apparently minor modifications to the distribution, such as replacing a Maxwellian with a κ function, have been found to alter the stability of the current sheet structure in numerical simulations [*Holland and Chen* 1993]. This suggests that a closer examination of both the processes that control the plasma distribution and its effects on the current sheet is warranted.

In all of the simulations mentioned previously, f_{in} is chosen by *fiat* to be a drifting Maxwellian whose only (non-thermal) motion in the Earth frame is that given by $\mathbf{E} \times \mathbf{B}$ drifts. Certainly we do not expect such a simple description of the distribution to be complete. Indeed, observational evidence suggests that plasma distributions in the far tail

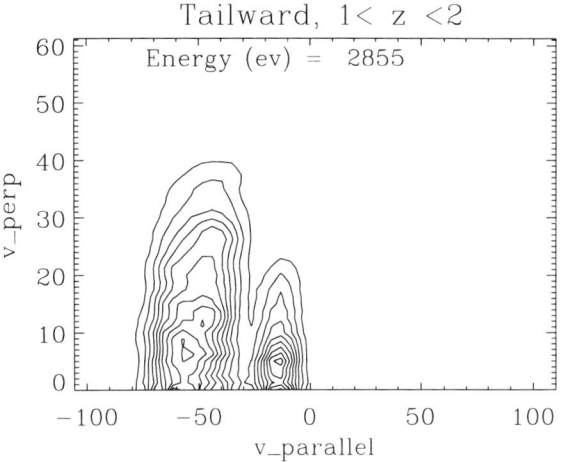

Fig. 5. Two component distribution function generated by a particles interacting with the X line. The distribution includes particles which left the system on the tailward side with $1 R_E < z < 2 R_E$ (from *Joyce et al.* [1995]).

($x \lesssim -35 R_E$) can be highly structured, with the distribution broken into multiple components [*Frank et al.*, 1994]. Observations nearer the Earth have found distributions which are more accurately described by kappa distributions than Maxwellians [*Christon et al.*, 1989], consisting of a nearly Maxwellian component with a power-law high-energy tail.

Various magnetospheric processes can produce multi-component distributions. For example, recent simulation work by *Joyce et al.* [1995] suggests one possible mechanism by which such distributions might form. Global MHD simulations were used to generate magnetic and electric fields, and these fields were then used in a (non-self consistent) two-dimensional test particle code to examine the energization effects of the X-line on particle distributions entering from the lobes. For an incoming low-energy (200 eV) Maxwellian distribution, they found that the outgoing distribution consisted of two distinct components: a low energy population similar to the lobe plasma, and a high energy population which had been energized through crossings of the neutral sheet (Figure 5). The explanation is simply that the cold component consists of particles that do not cross the midplane prior to detection while particles that cross the midplane and undergo current sheet acceleration [*Cowley*, 1978; *Lyons and Speiser*, 1982] comprise the hot component. These distributions bear a qualitative similarity to the observations of *Frank et al.* [1994], and reinforce the assertion that the plasma distribution in the vicinity of the current sheet cannot be well-described by a simple, single component drifting Maxwellian.

The complexity of these distributions is a reflection of the processes which affect the plasma as it moves through the magnetosphere. Understanding the influences of these processes is critical to understanding the behavior of the current sheet, since this represents a mechanism by which the current sheet may be controlled by processes in remote parts of the magnetosphere. In the next section we discuss a simplified model with which we can probe some of these effects.

6. EFFECTS OF REMOTE PROCESSES ON THE CURRENT SHEET

Recently, we have used one-dimensional iterative test-particle simulations incorporating multi-component distributions in an attempt to model the possible effects of non-local processes on the structure of the current sheet [*Harold and Chen*, 1995]. The simulation is performed in the de Hoffman-Teller frame (as described previously), with the system described by the normal component of the magnetic field, $b_n \equiv B_z/B_x$, evaluated at the boundary, and the temperature T and drift velocity v_D/v_{th} of the incoming ion distribution. It is beyond the scope of this work to perform a global modeling of specific remote processes in order to determine the appropriate distribution function for the incoming plasma. Instead, we examined the generic effects of physical processes that influence plasma distributions. One is localized acceleration (i.e., electric field) that modifies the pitch angle distribution (PAD) and energy distribution of the plasma passing through the region in relation to other components that do not. Another is the modification of PAD caused by certain magnetic field geometry (i.e., a magnetic mirror). Thus, particle interactions with electric and magnetic fields can be thought of as consisting of elemental processes that modify PAD and energy distribution. In this paper, we investigate a simple parameterization of elemental remote processes by considering two-component incoming distributions using self-consistent test particle simulations.

The general approach of the iterative test-particle simulation is to perform repeated runs of test-particles through static fields and use the resulting particle currents to calculate a self-consistent field. Initially, several thousand particles are injected at the boundaries of the simulation and followed through a static field configuration which is initialized to a modified Harris field (equations (2) and (3)). The particle currents are accumulated, and at the end of the run this information is used to calculate a new B_x. This calculated field is averaged with the original field (in order to reduce noise) and used as the input field for the next iteration. The process continues until the change in the calculated current from one iteration to the next becomes appropriately small, satisfying $\nabla \times \mathbf{B} = (4\pi/c)\mathbf{J}$.

Consider now an incoming ion population consisting of two independently specified components. The primary distribution represents the bulk of the incoming plasma and consists of a drifting Maxwellian. The secondary population represents a plasma which has been modified by a "generic" remote process, in this case through restriction of the particle PAD. This may occur if the second component is accelerated by an $E_\parallel$ along its way or if it is mapped from a region with stronger magnetic field. The two components of f_{in} remain unchanged throughout the iterative process. The current contributions of the two groups are combined at each iteration in order to calculate the field for the next iteration.

Figure 6 shows the results of a series of runs performed for varying pitch angles of the secondary distribution, which constitutes 10% of the total particle density. Each point represents a single converged, self-consistent current sheet solution for a different value of the maximum pitch angle of the secondary distribution (pitch angle θ is plotted in radians normalized to $\pi/2$). The ratio of the density (n_2) of the secondary distribution to the total (n_T) evaluated in the uniform magnetic field region at the simulation boundary ($z \gg \delta$) is $n_2/n_T = 0.1$. The temperatures of the two distributions are

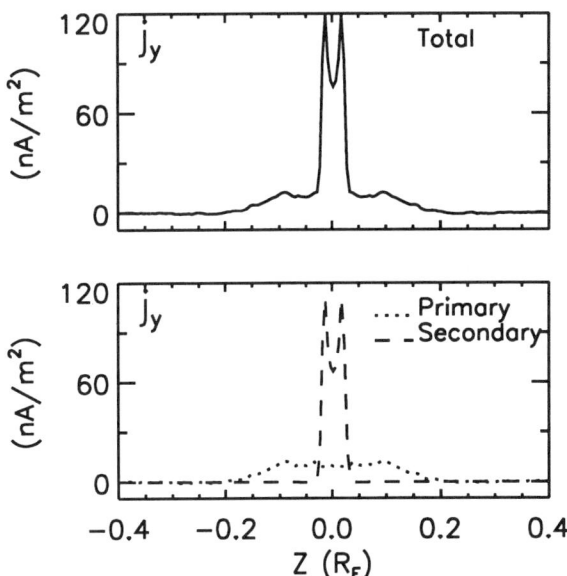

Fig. 7. Current densities for a run in which the secondary distribution has a pitch angle of approximately zero, i.e., field aligned, representing 10% of the total density. The top panel displays the total current as a function of z, while the bottom panel displays the current broken up into the contributions from the primary distribution (dotted line) and the secondary distribution (dashed line).

equal, and each has an initial drift velocity of $v_D/v_{th} = 0.2$, with b_n held at 0.1 for all runs. The figure shows that as the maximum pitch angle is varied from $\pi/2$ to 0, the density changes from an essentially flat profile to one peaked by a factor of almost two over the background. The sheet thins by a factor of just under eight, and the asymptotic field increases by 50%. This demonstrates the strong influence that such modifications to the distribution can have on the overall current sheet structure, even if only a small fraction (10%) of the total plasma is affected. Note that it has been found that B_y becomes significant in the thin current sheet regime [*Pritchett and Coroniti*, 1993; *Burkhart et al.*, 1993; *Cargill et al.*, 1994]. Therefore, the details of the current sheet structure in the $\theta \to 0$ limit discussed here are not necessarily physical because B_y is neglected. However, the overall trend demonstrates that the current sheet structure has a strong dependence on the form of the incoming distribution function. The neglect of B_y does not affect the basic conclusion that the current sheet can be moved from one regime to another (e.g., thick to thin) by modifying f_{in}.

A better understanding of the relative contributions of the two ion populations can be obtained from Figure 7, in which the current profile for a current sheet with an asymptotic ($z \gg \delta$) pitch-angle of 0 (field-aligned), is separated into components. The dotted line represents the contribution from the primary population (90% of the asymptotic density),

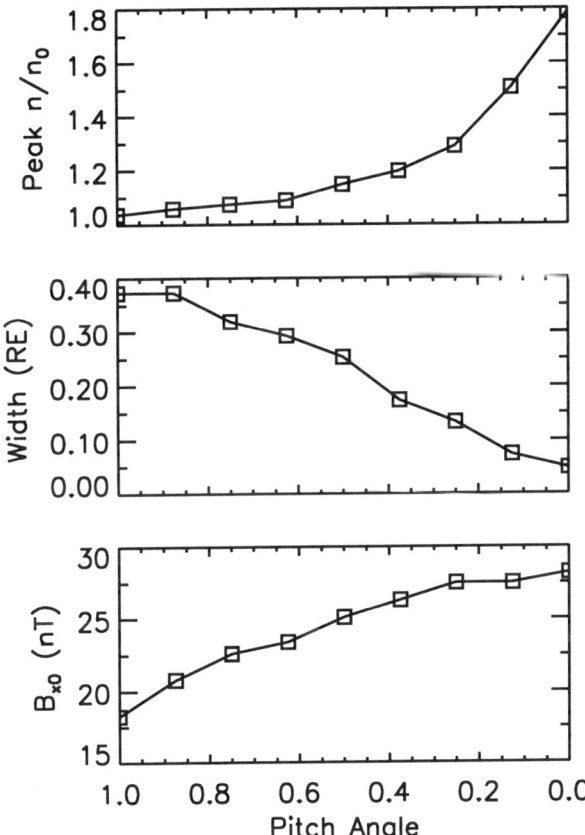

Fig. 6. Peak density, width and asymptotic B_x for runs in which the pitch angle of the secondary distribution was varied. The x axis measures the pitch angle in radians normalized to $\pi/2$.

while the dashed line marks the contribution of the secondary (10%) population. The contributions are quite distinct: the primary component provides a wide, shallow current sheet, while the field aligned secondary contributes a very narrow, highly peaked current structure. This is consistent with the previously discussed, single component simulations, in the sense that in those cases highly field aligned distributions (associated in that context with large E_y) generated thin current sheets. The results can be understood in the same general way: low pitch angle particles deposit their current in a narrower region around the midplane, leading to a thinner, more highly peaked current sheet. For example, the 2° contours in the transient regions in Figure 4 show that such particles traverse the midplane near the outer circle where $v_z \simeq 0$.

In addition, the field aligned particles also tend to contribute more current on an individual basis. The surface of section in Figure 4 shows the contours for particles entering the midplane which have pitch angles at $+\infty$ of of 2°, 30°, and 60°. [Note that Figure 4 describes one constant-$\hat{H}$ surface in phase space: the details of the entry region and its internal structure depend on the particle energy and field profile, though the general structure does not.] It can easily be seen that the field aligned particles generally have greater y excursions as they cross the midplane, thereby contributing more current. Furthermore, we note that for $\rho = v_\parallel/\Omega_n$, and $\Delta t \sim \pi\Omega_n$, equation (4) reduces simply to $j_y \propto v_\parallel$. Hence, the increase in the average $v_\parallel$ in a field aligned distribution also contributes to an increase in the net current per particle. A final contribution, quite significant in this case, stems from diamagnetic effects. It has been shown [*Holland and Chen*, 1993] that in current sheets with small v_D/v_{th} diamagnetic effects become significant, and oppose the current produced by the motion of the particles guiding centers (equation (4)). In the case of the two-component distribution, making one population field aligned serves to remove the competing diamagnetic component of this population, significantly enhancing the net field. Thus, even if the particle current were not enhanced as described above, the observed current would still increase due solely to the elimination of part of the diamagnetic component.

The previous results are based on a simple model, but they serve to demonstrate an important property: relatively minor modifications to the (local, microscale) plasma distribution function can lead to significant changes in the (mesoscale) current sheet structure. Remote processes can modify the plasma distribution, which then travels to the current sheet and affects its local structure. This is a purely nonlocal kinetic effect separate from any MHD processes in which local properties (e.g., the current sheet thickness) are determined by local forces ($\mathbf{J} \times \mathbf{B}$ and ∇p: see Figure 2). This is particularly relevant to magnetospheric dynamics in the context of ionospheric outflows. Significant concentrations of O^+ ions have been measured in the near-equatorial magnetosphere, with O^+/H^+ ratios typically being larger during quiet times [*Lennartsson*, 1982]. Numerical calculations support the idea that O^+ ions can be a significant component of the near-Earth CPS plasma [*Delcourt et al.*, 1989b]. (Also see *Moore* [1991] for a review of magnetospheric plasma of ionospheric origin.) We expect such plasmas to be more field aligned if they are mapped to the current sheet due to the lower magnetic field strengths in the tail as compared to the ionosphere. They would then form field-aligned components of the total distribution containing plasmas from other sources. These components can consist of different chemical species with different mass and charge state. While our simulations included only a single ion species, we can comment on some aspects of the more complete, multi-species problem. In particular, it can readily be shown [*Sonnerup*, 1971] that the turning points $\pm z_t$ for an ion in a simple linearly varying field $B_x(z) = B_0(z/\delta)$ are given by

$$z_t \propto (\rho_0\delta)^{1/2},$$

where $\rho_0 = v_\perp/\Omega_0$. By requiring that $z_t = \delta$ for a self-consistent field, we find immediately that the self-consistent current sheet width should scale as $\delta \propto \rho_0$, where ρ_0 is calculated for the average energy of the distribution. This shows that, for a given temperature, an ion species of charge Ze tends to form a current distribution of thickness $\delta \propto (m^{1/2}/Z)$. Note that the typical particle energy in a self-consistent field corresponds to $\hat{H} = \hat{H}_T = (1/2)b_n^{-4}$, independent of mass and charge state. Thus, a non-interacting O^+ (O^{++}) component would produce a current sheet four (two) times the width of an H^+ sheet, with essentially the same particle dynamics. In reality, of course, the ions of each species are coupled through the fields, and the resulting current sheet would reflect this. The net result of these combined effects remains to be determined. However the sensitivity of the current sheet to changes in the incoming distribution function demonstrated in this paper suggests a specific mechanism by which ionospheric plasmas can play a significant role in determining the current sheet structure.

7. SUMMARY

The recent studies of kinetic properties of the magnetotail have progressed from the nonlinear dynamics of single particles, to generation of non-Maxwellian distributions, to current sheet structure, and to particle signatures in large-scale electric and magnetic fields. We have discussed the structure of the magnetotail current sheet and its coupling to the larger magnetosphere in two contexts: (1) the local dependence on the large scale electric and magnetic fields, and

(2) the nonlocal coupling to remote processes via the plasma distribution. The first aspect has been discussed by a number of authors, and the particle (microscale) to field (mesoscale) coupling examined. While efforts have been made to describe this process using a chaotic conductivity term, these approaches cannot be applicable in all regimes. In particular, for non-Maxwellian distribution functions, the dependence of the cross-tail current J_y on the distribution function cannot be reduced to a dependence on E_y (or on $\mathbf{E} + \mathbf{v} \times \mathbf{B}$) via a local conductivity.

The sensitivity of the current sheet to the form of the incoming distribution function further suggests that processes remote from the current sheet can critically influence its structure. The plasma populations arriving at the current sheet have been affected by processes in different parts of the magnetosphere, and carry information about these processes within their distribution functions. Their interaction with the current sheet serves to couple it to remote parts of the magnetosphere over a range of spatial and temporal scales. Numerical results demonstrating the sensitivity of the current sheet structure to small changes in the incoming distribution function, combined with evidence of complicated distribution functions in the vicinity of the neutral sheet, reinforce the view that the magnetotail cannot be understood in isolation from these remote processes.

In the last three decades, much has been learned in magnetospheric physics, observationally and theoretically, by isolating phenomena on single spatial and temporal scales. The fact that physical processes on one scale are irreducibly coupled to those on other scales is an indication that the magnetosphere is a "complex" system, in the sense that the global behavior is qualitatively different from the properties of the constituent components. Complexity of physical systems has received much recent attention [e.g., *Garrido and Mendes*, 1992] because of the realization that many important physical systems cannot be fully analyzed by focusing on individual components, the traditional reductionist approach. A unique aspect of collisionless plasma systems, such as the magnetosphere and some astrophysical objects, is that large-scale magnetic (and electric) fields provide the organization over large distances, and nonlocal particle motion can couple processes on all scales within each system. We have illustrated these ideas using a number of specific current sheet properties. Many more questions remain to be asked and answered for a better understanding of global magnetospheric properties.

Acknowledgments. This work was supported by the Office of Naval Research and National Aeronautics and Space Administration (W-16,991).

REFERENCES

Ashour-Abdalla, M., J. Berchem, J. Büchner, and L. M. Zelenyi, Large and small scale structures in the plasma sheet: A signature of chaotic motion and resonance effects, *Geophys. Res. Lett.*, *18*, 1603, 1991.

Ashour-Abdalla, M., J. P. Berchem, J. Büchner, and L. M. Zelenyi, Shaping of the magnetotail from the mantle: global and local structuring, *J. Geophys. Res.*, *98*, 5651, 1993.

Ashour-Abdalla, M., Lev M. Zelenyi, Vahe Peroomian and Robert L. Richard, Consequences of magnetotail ion dynamics, *J. Geophys. Res.*, *99*, 14,891, 1994.

Behannon, K. W., Mapping of the Earth's bow shock and magnetic tail by Explorer 33, *J. Geophys. Res.*, *73*, 907, 1968.

Brittnacher, M. J. and E. C. Whipple, Chaotic jumps in the generalized first adiabatic invariant in current sheets, *Geophys. Res. Lett.*, *18*, 1599, 1991.

Büchner, J., and L. M. Zelenyi, Deterministic chaos in the dynamics of charged particles near a magnetic field reversal, *Phys. Lett. A*, *118*, 395, 1986.

Büchner, J. and L. M. Zelenyi, Chaotization of the electron motion as the cause of an internal magnetotail instability and substorm onset, *J. Geophys. Res.*, *92*, 13,456, 1987.

Büchner, J., and L. M. Zelenyi, Regular and chaotic charged particle motion in magnetotaillike field reversals, 1, Basic theory of trapped motion, *J. Geophys. Res.*, *94*, 11,821, 1989.

Burkhart, G. R., and J. Chen, Differential memory in the Earth's magnetotail, *J. Geophys. Res.*, *96*, 14,033, 1991.

Burkhart, G. R., and J. Chen, Particle motion in x-dependent Harris-like Magnetotail models, *J. Geophys. Res.*, *98*, 89, 1993.

Burkhart, G. R., J. F. Drake, and J. Chen, Magnetic reconnection in collisionless plasmas: Prescribed fields, *J. Geophys. Res.*, *95*, 18,833, 1990.

Burkhart, G. R., J. F. Drake, P. B. Dusenbery and T. W. Speiser, A Particle Model for Magnetotail Neutral Sheet Equilibria, *J. Geophys. Res.*, *97*, 13,799, 1992.

Burkhart, G. R., P. B. Dusenbery, T. W. Speiser, and R. E. Lopez, Hybrid simulations of thin current sheets, *J. Geophys. Res.*, *98*, 21373, 1993.

Cargill, P. J., J. Chen and J. B. Harold, One-dimensional hybrid simulations of current sheets in the quiet magnetotail, *Geophys. Res. Lett.*, *21*, 2251, 1994.

Cary, J. R., E. F. Escande, and J. L. Tennyson, Adiabatic-invariant change due to separatrix crossing, *Phys. Rev. A*, *34*, 4256, 1986.

Chen, J., Nonlinear Dynamics of Charged Particles in the Magnetotail, *J. Geophys. Res.*, *97*, 15,011, 1992.

Chen, J. and P. J. Palmadesso, Chaos and nonlinear dynamics of single-particle orbits in a magnetotaillike magnetic field, *J. Geophys. Res.*, *91*, 1499, 1986.

Chen, J., G. R. Burkhart, and C. Y. Huang, Observational signatures of nonlinear magnetotail particle dynamics, *Geophys. Res. Lett.*, *17*, 2237, 1990a.

Chen, J., J. L. Rexford, and Y. C. Lee, Fractal boundaries in magnetotail particle dynamics, *Geophys. Res. Lett.*, *17*, 1049, 1990b.

Christon, S. P., D. J. Williams, D. G. Mitchell, L. A. Frank and C. Y.

Huang, Spectral Characteristics of Plasma Sheet Ion and Electron Populations During Undisturbed Geomagnetic Conditions, *J. Geophys. Res.*, 94, 13,409, 1989.

Cowley, S. W. H., A note on the motion of charged particles in one-dimensional magnetic current sheet, *Planet. Space Sci.*, 26, 539, 1978.

Delcourt, D. C., T. E. Moore, J. H. Waite, Jr., and C. R. Chappell, Polar wind ion bands after neutral sheet acceleration, *J. Geophys Res.*, 94, 3773, 1989a.

Delcourt, D. C., C. R. Chappell, T. E. Moore, and J. H. Waite, Jr., A three-dimensional numerical model of ionospheric plasma in the magnetosphere, *J. Geophys. Res.*, 94, 11,893, 1989b.

Delcourt, D. C., R. F. Martin, F. Alem, A simple model of magnetic moment scattering in a field reversal, *Geophys. Res. Lett.*, 21, 1543, 1994.

Eastwood, J. W., Consistency of fields and particle motion in 'Speiser' model of the current sheet, *Planet. Space Sci.*, 20, 1555, 1972.

Eastwood, J. W., The warm current sheet model, and its implications on the temporal behaviour of the geomagnetic tail, *Planet. Space Sci.*, 22, 1641, 1974.

Frank., L. A., W. R. Paterson and M. G. Kivelson, Observations of nonadiabatic acceleration of ions in Earth's magnetotail, *J. Geophys. Res.*, 99, 14,877, 1994.

Garrido, M. S., and R. V. Mendes, *Complexity in Physics and Technology*, World Scientific, Singapore, 1992.

Harold and Chen, Kinetic Thinning in Self-Consistent Current Sheets, to be submitted, *J. Geophys. Res.*, 1995.

Holland, D. L. and J. Chen, Effects of collisions on the nonliner particle dynamics in the magnetotail, *Geophys. Res. Lett.*, 18, 1579, 1991.

Holland, D. L. and J. Chen, On chaotic conductivity in the magnetotail, *Geophys. Res. Lett.*, 19, 1231, 1992.

Holland, D. L., and J. Chen, Self-Consistent Current Sheet Structures in the Quiet-Time Magnetotail,*Geophys. Res. Lett.*, 20, 1775, 1993.

Horton, W. and T. Tajima, Decay of correlations and the collisionless conductivity in the geomagnetic tail, *Geophys. Res. Lett.*, 17, 123, 1990.

Horton, W. and T. Tajima, Collisionless conductivity and stochastic heating of the plasma sheet in the geomagnetic tail, *J. Geophys. Res.*, 96, 15,811, 1991.

Horton, W., L. Cheung, J.-Y. Kim, and T. Tajima, Self-consistent plasma pressure tensors from the Tsyganenko magnetic field models, *J. Geophys. Res.*, 98, 17327, 1993.

Joyce, G., J. Chen, S. Slinker and D. Holland, Particle energization near an X line in the magnetotail based on global MHD fields, *J. Geophys. Res.*, accepted, 1995.

Kaufmann, R. L. and C. Lu, Cross-tail current: resonant orbits, *J. Geophys. Res.*, 98, 15,447, 1993.

Lakhina, G. S., Solar wind-magnetosphere-ionosphere coupling and chaotic dynamics, *Surveys in Geophysics*, 15, 703, 1994.

Lennartsson, W., and R. D. Sharp, A comparison of the 0.1-17 keV/e ion composition in the near equatorial magnetosphere beween quiet and disturbed conditions, *J. Geophys. Res.*, 87, 6109, 1982.

Lichtenberg, A. J. and M. A. Lieberman, *Regular and Stochastic Motion*, Springer-Verlag, New York, 1983.

Lui, A. T. Y., Inferring global characteristics of current sheets from local measurements, *J. Geophys. Res.*, 98, 13,423, 1993.

Manheimer, W. and J. P. Boris, Marginal stability analysis – A simpler approach to anomalous transport in plasmas, *Comments Plasma Phys. Cont. Fusion*, 3, 15, 1977.

Martin, R. F., Chaotic particle dynamics near a two dimensional magnetic neutral point with applications to the geomagnetic tail, *J. Geophys. Res.*, 91, 11985, 1986.

Martin, R. F., Self-consistent neutral point current and fields from single particle dynamics, in *Modeling Magnetospheric Plasma*, Geophys. Monogr. Ser., Vol. 44, edited by T. E. Moore and J. H. Waite, Jr., p. 55, AGU, Washington, D. C., 1988.

McComas, D. J., C. T. Russell, R. C. Elphic, and S. J. Bame, The near-earth cross-tail current sheet: Detailed ISEE 1 and 2 case studies, *J. Geophys. Res.*, 91, 4287, 1986.

Moore, T. E., Origins of magnetospheric plasma, *U. S. Natl. Rep. Int. Union Geod. Geophys. 1987–1990, Rev. Geophys.*, 29, 1039–1048, 1991.

Papadopoulos, K., A review of anomalous resistivity for the ionosphere, *Rev. Geophys. Spac. Phys.*, 15, 113, 1977.

Pritchett, P. L. and F. V. Coroniti, A radiating one dimensional current sheet, *J. Geophys. Res.*, 98, 15355, 1993.

Pritchett, P. L. and F. V. Coroniti, Convection and the formation of thin current sheets in the near-Earth plasma sheet, *Geophys. Res. Lett.*, 21, 1587, 1994.

Rich, F. J., V. M. Vasyliunas and R. A. Wolf, On the Balance of Stresses in teh Plasma Sheet, *J. Geophys. Res.*, 77, 4670, 1972.

Sonnerup, B. U. Ö., Adiabatic particle orbits in a magnetic null sheet, *J. Geophys. Res.*, 76, 8211, 1971.

Speiser, T. W., Conductivity without collisions or noise, *Planet. Space Sci.*, 18, 613, 1970.

Speiser, T. W., Particle trajectories in model current sheets, 2, Applications to auroras using a geomagnetic tail model, *J. Geophys. Res.*, 70, 4219, 1965.

West, H. I., R. M. Buck, and M. G. Kivelson, On the configuration of the magnetotail near midnight during quiet and weakly distrubed periods: Magnetic field modeling, *J. Geophys. Res.*, 83, 3819, 1978.

J. Chen and J. B. Harold, Code 6790, Naval Research Laboratory, Washington, DC 20375.

Ion Energization and Cross-Scale Coupling During Magnetotail Reconnection

G. R. Burkhart

NOAA Space Environment Laboratory, Boulder, Colorado

During a plasma energization process, such as magnetic reconnection, it is likely that cross-scale coupling becomes important. Here, cross-scale coupling is studied, beginning with the ion scales and then proceeding to the global scales and electron scales. For the ion scales, I appeal to previous work that emphasizes the motion of ions near neutral lines. For coupling to global scales, previous work on the structure of thin, one-dimensional current sheets is put into the context of magnetotail reconnection. Bringing these different results together allows the conception of a model of magnetotail reconnection that includes thin current sheets deep in the tail (-10 $R_E > x_{\text{GSM}}$) and current disruption events such as were observed by AMPTE/CCE. Coupling to electron scales is also discussed, a simple model that can be used to study this coupling is presented, and some conjectures are given as to what the results might be.

1. INTRODUCTION

One of the most interesting aspects of magnetotail dynamics is the abrupt conversion of magnetic energy into plasma heat and bulk kinetic energy during the reconfiguration of the magnetotail magnetic field. Most investigators study the role these magnetotail reconfiguration events play in auroral substorms, but many are interested in these events because of the importance of the processes that can convert magnetic energy into plasma heat and acceleration. Plasma energization events are also interesting because, during energization, cross-scale coupling is likely to be very important. First, global scales are coupled to the energization scale for ions (i.e., for example the ion gyroradius) because the energized ions can transport energy and momentum across great distances. Second, ion scales are coupled to electron scales through electrostatic electric fields. During energization, the two species will generally respond in different ways, thereby setting up charge separation.

In this paper, I discuss kinetic theory for one possible energization process, magnetic reconnection. Particular emphasis is placed upon the motion of individual particles near neutral lines and within current sheets – not because I believe wave-particle interactions are unimportant, but because I feel that the configuration formed by particle motion in the large scale fields is the proper framework for studying instabilities and wave-particle interactions.

While the linear phase of reconnection may be important for understanding the expansion phase onset, the conversion of field energy to plasma energy occurs only during the non-linear phase (by definition). Unfortunately, the strongly non-linear regime of kinetic reconnection is difficult to study with particle simulations because they must accommodate widely disparate length and time scales; therefore, the non-linear regime for kinetic reconnection has, in the past, been only studied extensively with the test-particle approach and with self-consistent, steady-state calculations. In the remainder of this paper, these studies of non-linear kinetic reconnection will be discussed.

2. THE DISSIPATION REGION

Here we consider the region near a neutral line that is commonly called the dissipation region, where it is thought that the slippage between the plasma and the magnetic field lines occurs. The aim here is to describe a picture of proton energization, based upon the motion of individual particles near a neutral line, and to determine the observational consequences of such a picture. Although wave-particle interactions within the dissipation region are not included here, they are not irrelevant. Wave-particle interactions are likely to occur within the context of the picture presented here, which is only a starting point for a future theory.

Figure 1, shows the results of a self-consistent, steady-state calculation of neutral-line reconnection by *Burkhart et al.* [1991]. Here, the x-line is in the lower left-hand corner of each plot, and symmetry is assumed about both the $z = 0$ and the $x = 0$ axes. The units of the axes are the ion-inertial length evaluated at the top left corner, λ_b: If l_b is 718 km (0.1 particle/cm^3), the vertical axis (z) is about 7,000 km and the horizontal axis (x) is about 14,000 km in length. An electric field E_y drives flow into and out of the region near the x-line, and it accelerates particles near the x-line; this results in the large current in that region. In panels (a) and (b), the grey-tone scaling is such that the most intense values are represented by black and the least intense by white. Stretching from the x-line to about $x = 6$ we find a long, thin current sheet (the half-width is around 300 km and the half-length is around 5,000 km) coinciding with a region of enhanced flow and density.

If it is assumed that the variation of the magnetic field in y can be ignored, particle motion conserves the quantity $P_y = v_y + \frac{q}{mc} A_y(x, z, t)$ exactly, where $A_y(x, z, t)$ is the y component of the vector potential that labels flux surfaces in a time-dependent, two-dimensional geometry. Because P_y differs from A_y by the particle velocity, particles are confined to a region of one gyroradius around flux surfaces. Field lines can move with respect to the plasma only while they are within one gyroradius of a neutral line. Thus, we define the dissipation region as the area within one gyroradius of the neutral line.

To limit the current within the dissipation region, particles must escape from the dissipation region on some finite time scale, τ. The assumption that the particles are freely accelerated by the electric field while they are within the dissipation region allows one to calculate an effective resistivity, which results from the finite trapping time $\eta_{\text{eff}} = 1/(\sum_{e,i} qn\tau/m)$, where n is the number density in the dissipation region and τ is the trapping time. Here the contributions from electrons and ions are summed and it is assumed that charge separation is not relevant.

Near the neutral line, the particle motion is composed of two weakly coupled components exhibiting different time scales. These two components are (1) the fast motion in the z direction and (2) the slow motion in the x-y plane. During one oscillation of the fast motion, the particle has progressed only a small amount along the slow motion, and we can approximate the slow motion by a constant velocity and position. The fast motion is identical to motion in a neutral sheet (i.e. $B_z = 0$) [*Sonnerup*, 1971]. If we assume that the half-thickness of the current sheet is equal to the average amplitude of the oscillations of the ions in z, the half-thickness can be evaluated using the constancy of the adiabatic invariant $J_z = \oint dz\, v_z$. From the average J_z the half-thickness of the current sheet is estimated to be about $a \sim (v_{Th}/V_y)^{4/3} \lambda_0$, where V_y is the characteristic or average y velocity of ions within the current sheet and λ_0 is the ion-inertial length c/ω_{pi} just outside the current sheet [*Francfort and Pellat*, 1976; *Burkhart et al.*, 1992a]. Since the fast motion in this limit does not depend upon whether the current sheet is near a neutral line (where the B_z scale length is shorter than the ion gyroradius) or in a region with weakly varying B_z, this estimate of the current sheet thickness applies in both cases. The estimate of the current sheet thickness given above was calculated for a current sheet in a region of constant B_z, and the result was found to agree with numerical solutions for $V_y \gtrsim v_{Th}$. For $V_y < v_{Th}$ the general behavior (that the current sheet is thicker for smaller V_y), remains [*Burkhart et al.*, 1992a].

The trapping time is determined by the slow motion near a neutral line. A simple model of this slow motion includes (1), the acceleration of particles in the y-direction by the electric field E_y, and (2), the conversion of the resulting, large v_y into a large v_x by the $v_y B_z$ force. B_z is zero only at the x-line. A particle infinitesimally far from the x-line will eventually develop a v_x velocity, and this v_x velocity allows the particle to exit the neutral line region on a time $\tau_c \sim \Omega_z'^{-2/3}(cE_y/B_z')^{-1/3}$, where $B_z' = \partial B_z/\partial z$ is assumed to be constant and $\Omega_z' = qB_z'/mc$ [*Burkhart et al.* 1990].

This simple model of the slow motion can be used to derive an analytic, self-consistent theory of the dissipation region [*Burkhart et al.*, 1991]. The principle feature of this solution is the tendency for a long, thin current sheet to form in the dissipation region. The thickness of this current sheet is less than the ion gyroradius, and the length becomes infinite near the maximum reconnection rate. The reason for the singular result is the theory assumes that the dissipation region length is much smaller than global scale sizes. *Burkhart et al.* [1991] suggested that in reality the dissipation region length would be limited to macroscopic scales.

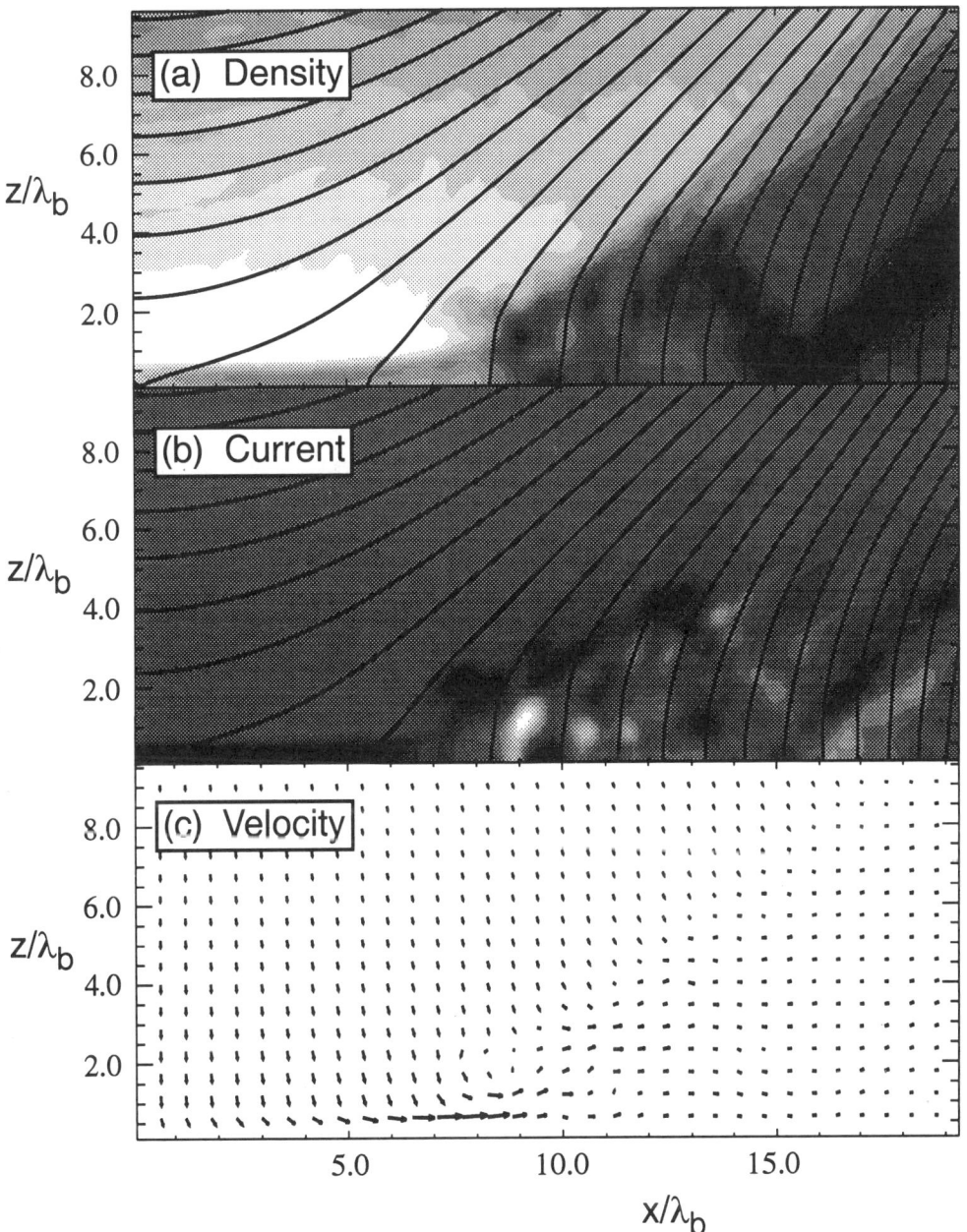

Fig. 1. The density (a), current (b) and bulk velocity (c) in steady-state neutral line configuration, where the neutral line is located at the bottom left corner. Field lines are superimposed on the gray-scale plots of density and current. Black denotes highest intensity and white denotes least intensity. The units of the vertical and horizontal axes is the ion-inertial length at the top left corner.

The results of the analytic theory were compared to steady-state numerical solutions, of which Figure 1 is an example, and general agreement was found.

The reader should bear in mind four important shortcomings of the analytic theory and the calculation depicted in Figure 1. First, the constraint of time independence forced the use of a linear neutral line model for the large-scale fields. In some instances this is acceptable for the immediate neighborhood of the neutral line; however, the external field model is not correct near the peak reconnection rate because the length of the current sheet becomes macroscopic. (This problem is addressed in Section 4.) Second, the current due to the pre-existing pressure profile is not included for the same reason. Unless the lobe fields are undergoing reconnection, the pre-existing magnetotail current can add an additional B_x gradient not included in the model. Third, time independence precludes wave-particle interactions, which may well broaden the current layer; and fourth, electron dynamics and charge separation are not included. The last point is discussed in more detail in the next section.

3. MODIFICATIONS DUE TO ELECTRON DYNAMICS

For the theory discussed in the previous section it was assumed that the ions are the principal current carriers and that the effects of charge separation are unimportant. Within the dissipation region the ions are unmagnetized and can be accelerated by the reconnection electric field. In most of that region, however, the electrons are magnetized; therefore, one would assume that the electrons are tied to the flux surfaces and and that they decelerate as they convect into larger B_z. The two species' different behaviors result in charge separation, Hall currents and field-aligned currents [cf. *Drake and Burkhart*, 1992; *Mandt et al.*, 1994].

Recent hybrid simulations have shown that in the regime where the system size is very small, reconnection can occur too quickly for the ions to respond. This electron reconnection is controlled by whistler waves rather than by the Alfvén waves that usually control reconnection in MHD [*Mandt et al.*, 1994]. It is interesting that although the whistler waves have a much faster phase velocity than Alfvén waves, the maximum reconnection rate given by *Mandt et al.* [1994] is identical to the ion reconnectoin rate given by *Burkhart et al.* [1991].

For a system with a larger scale size, *Drake and Burkhart* [1992] extended the results of *Burkhart et al.* [1991] by including a massless electron fluid with a small resistivity to limit the electron current at the x-line. They found that the different motions of the electrons and ions lead to field-aligned currents and perturbation B_y magnetic fields (i.e. whistler waves) just as in *Mandt et al.* [1994], but they also found that the different motions of the two species could lead to the electrostatic trapping of ions within the dissipation region.

This effect, which is similar to the mechanism that leads to the stabilization of the linear tearing instability with $B_z \neq 0$ [*Lembége and Pellat*, 1982; *Pellat et al.*, 1991], leads to a much smaller peak reconnection rate and a peak electric field than were found by *Burkhart et al.* [1991]. However, it was concluded that a small amount of cross-field diffusion, due perhaps to wave-particle interactions, would remove this ion trapping and allow the faster ion reconnection rates.

Although *Drake and Burkhart* suggested that the details of processes on electron time scales will not change the configuration too much, they probably will completely control the electron energization, and these energized electrons can be observed remotely. A model of the charge separation layer within the dissipation region has been developed. Future work will use this model to investigate these processes.

4. THE OUTFLOW REGIONS

Because the structure of the dissipation region in MHD is determined by the dissipation model, one of the principal focuses for MHD studies of reconnection is the structure of the outflow region. *Petschek* [1964] proposed that the outflow region should be composed of two back-to-back slow shocks. These shocks turn off the x-component of the magnetic field and they heat and compress the plasma. In Petschek's model, both the length and the width of the dissipation region were small compared to the macroscopic system size; the shocks in the outflow region are solely responsible for the conversion of magnetic energy to plasma heat and flow. In the description of the collisionless dissipation region above, however, it is found that the dissipation region is of macroscopic length, particularly near the peak reconnection rate, and the acceleration of charges within the dissipation region accommodates most of the conversion of magnetic energy to plasma heat and flow.

In Petschek's model there is no bulk flow across the outflow region; therefore it is inapplicable to the magnetopause, where plasma flows from the magnetosheath to the magnetosphere during reconnection. To allow for this flow, *Levy et al.* [1964] constructed a model in which a rotational discontinuity and a slow-mode expansion replace the slow-mode shocks.

If the possibility of an anisotropic plasma is allowed, however, a new type of rotational discontinuity is available, one in which the bulk flow across the discontinuity is zero [*Hudson*, 1970]. This possibility occurs because in marginal firehose conditions (i.e. $P_\parallel - P_\perp = B^2/4\pi$), the propagation speed of Alfvén waves is zero. (This type of rotational discontinuity will hereafter be called "MFRD" (marginal-firehose rotational discontinuity).) Unlike rotational discontinuities in isotropic plasma, an MFRD need not have a strong B_y component at the current-sheet center. MFRDs become relevant to the magnetotail if we recall that particles passing through a current sheet with a dawn-to-dusk electric field become accelerated; this leads to a two component distribution outside the current sheet, which consists of unaccelerated particles that have not yet encountered the current sheet and accelerated particles that have encountered the current sheet and are streaming outward along the field lines [*Cowley*, 1978; *Lyons and Speiser*, 1982]. The two distributions form an effective pressure anisotropy, which can allow a rotational discontinuity in marginal firehose conditions. (Another way to describe this configuration is to say that the particles, which are accelerated out along the field lines by the current sheet interaction, carry away momentum and help balance the field-line tension force.) It has been suggested that, rather than two slow mode shocks, the outflow region in magnetic reconnection in collisionless plasma should contain MFRD's [*Coroniti*, 1985]. If this is correct, there are important consequences. In MHD, the field-line tension force of the newly reconnected field lines acts to accelerate the plasma, with the greatest acceleration occurring at the point of maximum curvature. Since the plasma is tied to the field lines, the acceleration of the plasma near the center causes the relaxation of the field-line curvature as the plasma and field lines move into the outflow region. In the collisionless model, on the other hand, particles are accelerated out along the field lines as a result of their interaction with the current sheet [*Cowley* 1978; *Lyons and Speiser*, 1982]. This current sheet acceleration can result in transport of momentum away from the point of maximum field-line curvature, resulting in less bulk acceleration and less of a tendency for field lines to relax.

MFRD configurations and current sheets that form within the dissipation region are fundamentally the same. Studies of MFRD's [*Eastwood*, 1972; *Francfort and Pellat*, 1976; *Burkhart et al.*, 1992a] have found identical relations for the thickness of the current sheet and the current within the current sheet as are obtained near a neutral line. This is because the fast z component of particle motion within the dissipation region and the particle motion within the MFRD are the same provided $\kappa_A \ll 1$. (Here κ_A is the value of $\kappa = (R_c/\rho_z)^{1/2}$ for ions of average energy, where R_c is the field-line radius of curvature and ρ_z is the gyroradius at the current sheet midplane.) Also, they are similar because in both cases the current is due to the acceleration of ions by E_y. Coroniti's suggestion that the dissipation region should match to MFRD's is strengthened by the similarity of the two types of current sheets. Matching MFRD's to the dissipation region current sheet also strengthens the conclusion that the reconnection current sheet is very long, where we now define the reconnection current sheet to be both the dissipation region and MFRD's that are in the outflow region.

As for the region of larger κ, [*Burkhart et al.*, 1992a] suggested that no thin current sheet equilibria exist for κ_A larger than about 0.7. It was suggested that in this regime, the field-line tension force and the gradient of the magnetic pressure cannot be balanced simultaneously by the particle motion. In any regime of κ_A, the momentum carried away by the accelerated particles causes the field lines to stretch into a taillike configuration until the continual removal of momentum is balanced by the field-line tension force. The stretching of the field lines into a taillike configuration also leads to an increase in the northward gradient of the magnetic pressure. In the regime of $\kappa_A \lesssim 0.7$, the gradient of the magnetic pressure can be balanced by the oscillation of ion trajectories across the current sheet. In fact, this consideration gives rise to the half-thickness of the current sheet discussed above. As κ_A is increased, the frequency of fast z-oscillations is reduced until for $\kappa_A \simeq 0.7$ they disappear entirely. At this point the gradient of the magnetic pressure would be unbalanced and the current sheet would turbulently relax to a configuration with no current (i.e. the configuration would "dipolarize").

The lack of a quasi-steady equilibrium in this regime was called the "Current Sheet Catastrophe" by *Burkhart et al.* [1992a]. *Burkhart et al.* [1992b] later suggested that the turbulent relaxation that would accompany the current sheet catastrophe could account for observations of current disruption in region near the apogee of AMPTE/CCE.

Burkhart et al. [1993] investigated the time-dependent consequences of the current sheet catastrophe. They found that many of the characteristics of these simulations agreed with AMPTE/CCE observations of current disruption at around $x = -9R_E$. In particular, the simulations found similar magnetic fluctuations, preferential perpendicular heating of the ions, and similar distribution function characteristics. For example, in Figure 2 we show a compareson between, in (a) the fluctuations of the maximum and intermediate variance components measured by AMPTE/CCE during the event of Day 240, 1986 [*Takahashi et al.*, 1987] and in (b) the fluctuations found in a hybrid simulation of a thin cur-

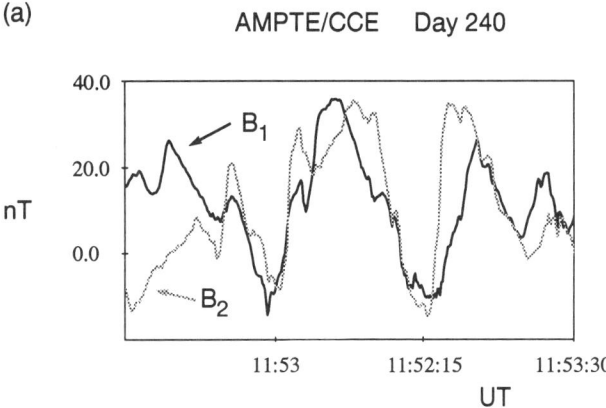

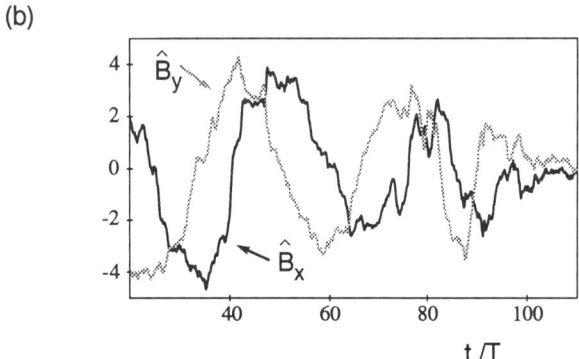

Fig. 2. (a) The maximum and intermediate variance components measured by AMPTE/CCE at the beginning of the Day 240, 1986 current disruption event and (b) the fluctuations found in the magnetic field of simultations of the current sheet catastrophe [*Burkhart et al.*, 1993]. (For (b), the time unit, $T = \lambda_0/v_T$ is less than one second and the normalization of the magnetic field, $\hat{\mathbf{B}} = (e\mathbf{B}/m_p c)T$ means that $\hat{B} = 1$ corresponds to about $B = 10$ nT.)

rent sheet during the current sheet catastropy for close to the same time period. (For Figure 2b, the time unit, $T = \lambda_0/v_T$ is less than one second and the normalization of the magnetic field, $\hat{\mathbf{B}} = (e\mathbf{B}/m_p c)T$ means that $\hat{B} = 1$ corresponds to about $B = 10$ nT.) Further evidence for this model is the observation of *Lui et al.* [1992] that the values of κ_A in the current disruption region prior to current disruption are between 0.5 and 0.9.

This work indicates that incursion of a large dawn-to-dusk electric field into the region where κ_A is near 1 will not result in a thin current sheet, as it would in regions of smaller κ_A, because no local, quasi-steady equilibrium exists in that region. Rather, in that region the electric field will cause the development of fluctuations characteristic of the AMPTE/CCE current-disruption events. In addition the the incursion of this convection electric field, the dipolarization itself implies a large electric field in the same direction [c.f., *Delcourt*, 1991] that must self-consistently also energize the ions in that region.

Thus we are led to a model that exhibits both thin current sheets tailward of the current disruption region and current disruption itself: Reconnection begins significantly tailward of the current disruption region and the thin reconnection current sheet/reconnection electric field expands until the electric field reaches the current disruption region, which has been found to be characterized by values of κ between 0.5 and 0.9 by *Lui et al.* [1992]. Here the reconnection current sheet is defined to be the current sheet within the dissipation region plus the current sheet in the outflow regions. Just prior to the onset of current disruption, the incursion of the dawn-to-dusk electric field into the current-disruption region may cause the local current to increase suddenly, as has been observed by *Ohtani et al.* [1992]. In the current-disruption region, however, the reconnection current sheet cannot remain in force balance because of the current sheet catastrophe, and we would assume that the current sheet turbulently relaxes to a dipolar configuration. Note that the relationship between the thin reconnection current sheet deep in the tail and the current disruption closer in that is presented here is a hypothesis that is based upon the local studies listed above. No global simulations have yet been performed that can theoretically validate the hypothesis, nor have satellite observations confirmed it. Also, it should be stressed that as a cause of current disruption, the reconnection deeper in the tail is only important because it is a mechanism whereby a strong electric field may extend into the near-Earth region. It has the advantage that it would seem to also explain the thin current sheets observed deeper in the tail and the abrupt increase in the current before current disruption, but other mechanisms could be invoked. Furthermore, other promising mechanisms besides the current sheet catastrophe have been invoked to explain current disruption [*Hesse and Birn*, 1991; *Lui et al.*, 1993], however, none have been shown to be capable of reproducing the large amplitude and the spectrum of the fluctuations that is observed, the rapid time scales of the events, or the degree of perpendicular heating of the ions.

The expected ground signatures should be mostly related to the formation of field-aligned currents. Tailward of the current-disruption region a substorm current wedge is not expected, because the current should be largest near midnight. Within the current-disruption region, on the other hand, the dipolarization near midnight would lead to a substorm current wedge and intensification of the electrojet in the ionosphere. Timing studies have found that intensifications of the electrojet closely follow current-disruption events in the near-Earth tail [*Lopez et al.*, 1992].

5. SUMMARY

Several different models that relate to magnetic reconnection in the magnetotail have been presented: a dissipation region model that includes only the ion current, a model for the modifications to the dissipation region due to frozen-in electrons (outside of a small region near the neutral line where the electrons are unmagnetized), and many results concerning the current sheets in the outflow region.

The results show that in collisionless magnetic reconnection there is a tendency for long, thin current sheets to form within the dissipation region and it has been suggested that thin current sheets might form in the outflow region as well [*Coroniti*, 1985]. Their half-thickness is less than the ion-gyroradius, and the length of the current sheet is comparable to global length scales. Such thin current sheets are often observed during the growth phase of substorms [*Fairfield*, 1984; *Mitchell et al.*, 1990; *Pulkkinen et al.*, 1992; *Sergeev et al.*, 1993; *Sanny et al.*, 1994].

Deep in the tail, where κ is small during substorm growth phase, a thin current sheet equilibrium can be formed and maintained on time scales of a few to tens of minutes. Closer in, where κ is near 1, results indicate that no quasi-steady current-sheet configurations exist. In the latter region, a large electric field will instead result in turbulent fluctuations similar to what is seen in AMPTE/CCE current disruptions. Thus I suggest a model that incorporates both thin current sheets deep in the tail and current disruption closer in: Outside of the AMPTE/CCE orbit, magnetic reconnection will be accompanied by current sheets that are macroscopic in length (half-lengths may be several to 10's of R_E) but very thin (half-thicknesses less that the ion-gyroradius). Closer in, the thin current sheet configuration will be unable to form, and the reconnection electric field will instead result in current disruption.

Acknowledgments. It is my pleasure to acknowledge discussions with J. F. Drake. This work was supported the National Research Council.

REFERENCES

Burkhart, G. R., J. F. Drake and J. Chen, Magnetic reconnection in collisionless plasmas: Prescribed fields, *J. Geophys. Res.*, 95, 18,833, 1990.

Burkhart, G. R., J. F. Drake and J. Chen, Structure of the dissipation region during magnetic reconnection in collisionless plasma, *J. Geophys. Res.*, 96, 11,539, 1991.

Burkhart, G. R., J. F. Drake, P. B. Dusenbery, and T. W. Speiser, A particle model of magnetotail neutral sheet equilibria, *J. Geophys. Res.*, 97, 13,799, 1992a.

Burkhart, G. R., R. E. Lopez, P. B. Dusenbery, and T. W. Speiser, Observational support for the current sheet catastrophe model of substorm current disruption, *Geophys. Res. Lett.*, 19, 1635, 1992b.

Burkhart, G. R., P. B. Dusenbery, T. W. Speiser and R. E. Lopez, Hybrid simulations of thin current sheets, *J. Geophys. Res.*, 98, 21,373, 1993.

Coroniti, F. V., Explosive tail reconnection: The growth and expansion phases of magnetospheric substorms, *J. Geophys. Res.*, 90, 7427, 1985.

Cowley, S. W. H., A note on the motion of charged particles in one dimensional magnetic current sheets, *Planet. Space Sci.*, 27, 991, 1978.

Delcourt, D. C., Tracing and acceleration of mid-tail ions during substorms, in *Magnetospheric Substorms*, Geophys. Monogr. Ser., vol. 64, edited by J. R. Kan, T. A. Potemra, S. Kokubun, T. Iijima, p. 225, AGU, Washington, D. C., 1991.

Drake, J. F., and G. R. Burkhart, Magnetic blowout during collisionless reconnection, *Geophys. Res. Lett.*, 19, 1077, 1992.

Eastwood, J. W., Consistency of fields and particle motion in the Speiser model of the current sheet, *Planet. Space Sci.*, 20, 1555, 1972.

Francfort, P., and R. Pellat, Magnetic merging in collisionless plasma, *Geophys. Res. Lett.*, 3, 433, 1976.

Hesse, M. and J. Birn, On dipolarization and its relation to the substorm current wedge, *J. Geophys. Res.*, 96, 19,417, 1991.

Hudson, P. D., Discontinuities in an anisotropic plasma and their identification in the solar wind, *Planet. Space Sci.*, 18, 1611, 1970.

Lembège, B., and R. Pellat, Stability of a thick two-dimensional quasineutral sheet, *Phys. Fluids*, 25, 1995, 1982.

Levy, R. H., H. E. Petschek, and G. L. Siscoe, Aerodynamic aspects of the magnetospheric flow, *AIAA J.*, 2, 2065, 1964.

Lopez, R. E., H. E. J. Koskinen, T. I. Pulkkinen, T. Bösinger, T. A. Potemra, and R. W. McEntire, Simultaneous observation of the poleward expansion of substorm electrojet activity and the tailward expansion of current sheet disruption in the near-Earth magnetotail, *J. Geophys. Res.*, in press, 1992.

Lui, A. T. Y., R. E. Lopez, B. J. Anderson, K. Takahashi, L. J. Zanetti, R. W. McEntire, T. A. Potemra, D. M. Klumpar, E. M. Green, and R. Strangeway, Current disruptions in the near-Earth neutral sheet region, *J. Geophys. Res., 97*, 1461, 1992.

Lui, A. T. Y., P. H. Yoon, C.-L. Chang, Quasi-linear analysis of ion Weible instability in the Earth's neutral sheet, *J. Geophys. Res., 98*, 153, 1993.

Lyons, L. R., and T. W. Speiser, Evidence for current sheet acceleration in the geomagnetic tail, *J. Geophys. Res., 87*, 2276, 1982.

Mandt, M. E., R. E. Denton, J. F. Drake, Transition to whistler mediated magnetic reconnection, *Geophys. Res. Lett., 21*, 73, 1994.

Ohtani, S., K. Takahashi, L. J. Zanetti, T. A. Potemra, R. W. McEntire, and T. Iijima, Initial signatures of magnetic field and energetic particle fluxes at tail reconfiguration *J. Geophys. Res., 97*, 19,311, 1992.

Pellat, R., F. V. Coroniti, and P. L. Pritchett, Does ion tearing exist?, *Geophys. Res. Lett., 18*, 143, 1991.

Petschek, H. E., Magnetic field annihilation, in *The Physics of Solar Flares*, W. N. Hess, ed., NASA SP-50, 425, 1964.

Sonnerup, B. U. Ö., Adiabatic particle orbits in a magnetic null sheet, *J. Geophys. Res., 76*, 8211, 1971.

Takahashi, K., L. J. Zanetti, R. E. Lopez, R. W. McEntire, T. A. Potemra, and K. Yumoto, Disruption of the magnetotail current sheet observed by AMPTE/CCE, *Geophys. Res. Lett., 14*, 1019, 1987.

Dr. G. R. Burkhart, Vexcel Corp., 2477 55th Street, Boulder, CO 80301.